8th International Conference on Environment Science and Engineering (ICESE 2018)

IOP Conference Series: Earth and Environmental Science
Volume 167

Barcelona, Spain
11 - 13 March 2018

Editors:

Kei Eguchi
David Quanrud
Hiroshi Takagi

ISBN: 978-1-5108-6817-5
ISSN: 1755-1307

Printed from e-media with permission by:

Curran Associates, Inc.
57 Morehouse Lane
Red Hook, NY 12571

Some format issues inherent in the e-media version may also appear in this print version.

Printed by Curran Associates, Inc. (2018)

For permission requests, please contact the Institute of Physics
at the address below.

Institute of Physics
Dirac House, Temple Back
Bristol BS1 6BE UK

Phone: 44 1 17 929 7481
Fax: 44 1 17 920 0979

techtracking@iop.org

Additional copies of this publication are available from:

Curran Associates, Inc.
57 Morehouse Lane
Red Hook, NY 12571 USA
Phone: 845-758-0400
Fax: 845-758-2633
Email: curran@proceedings.com
Web: www.proceedings.com

Table of contents

Volume 167

8th International Conference on Environment Science and Engineering (ICESE 2018)

11–13 March 2018, Barcelona, Spain

Accepted papers received: 15 June 2018
Published online: 23 July 2018

Preface

Papers

Hydrology and Water Resources Management

Air Pollution and Climate Change Management

Chemical Engineering and Biotechnology

Urban and Civil Engineering

PREFACE

It is our great pleasure to welcome you to 2018 8th International Conference on Environment Science and Engineering (ICESE 2018) which was held in Barcelona, Spain, March 11-13, 2018. ICESE 2018 is dedicated to issues related to Environment Science and Engineering.

The major goal of the conference is to bring academic scientists, engineers and industry researchers together to exchange their experiences and research results. Prof. Zhijun Peng from University of Bedfordshire, UK is invited to deliver the keynote speech titled "Environmental and Energetic Impacts of Ground Vehicle Electrification" and Prof. Raf Dewilead from KU Leuven - Department of Chemical Engineering, Belgium is invited to deliver keynote speech titled "Current trends in advanced oxidation processes (AOP) for the degradation of pharmaceutical micro-pollutants in wastewater", regarding latest information in their respective expertise areas. It is a golden opportunity for students, researchers and engineers to interact with the experts and specialists to get their advice or consultation on technical matters.

This proceeding presents a selection from papers submitted to the conference from universities, research institutes and industries. All papers were subjected to peer-review by Prof. David Quanrud from University of Arizona, Prof. Asher Brenner from Ben-Gurion University of the Negev, Israel, Prof. Kei EGUCHI from Fukuoka Institute of Technology, Japan and other invited conference committee members and international reviewers. The papers selected depended on their quality and their relevancy to the conference. The volume tends to present to the readers the recent advances in the field of Hydrology and Water Resources Management, Waste Water Treatment, Air Pollution and Climate Change Management, Environmental Monitoring and Management, Chemical Engineering and Biotechnology, Energy and Power Engineering, Urban and Civil Engineering, etc.

We would like to thank all the authors who have contributed to this volume, and also the organizing committee, reviewers, speakers, chairpersons, sponsors and all the conference participants for their support to ICESE 2018.

Prof. Zhijun Peng
University of Bedfordshire, UK
March 30, 2018

IOP Conf. Series: Earth and Environmental Science **167** (2018) 011002 doi:10.1088/1755-1315/167/1/011002

Organizing Committee

Conference Chair

Prof. Zhijun Peng, University of Bedfordshire, UK

Prof. Ignacio Javier Acosta García, School of Architecture, University of Seville, Spain

Program Chair

Prof. Prof. Raf Dewil, KU Leuven - Department of Chemical Engineering, Belgium

Prof. Miguel Ángel Campano, School of Architecture, University of Seville, Spain

Guest Editors

Prof. Kei EGUCHI, Fukuoka Institute of Technology, Japan

Assoc. Prof. David Quanrud, University of Arizona, USA

Assoc. Prof. Hiroshi Takagi, Tokyo Institute of Technology, Japan

International Technical Committee

Prof. Indah Prihartini, University of Muhammadiyah Malang, Indonesia

Prof. Chad Marchand, University of Arizona, USA

Prof. AZIZA Majda Amina, Renewable Energies Development Centre (CDER), Algeria

Prof. Muhammad Yaseen, University of Agriculture, Parkistan

Prof. Asher Brenner, Ben-Gurion University of the Negev, Israel

Prof. Abdeltif Amrane, University of Rennes 1, France

Prof. Leonid Plyatsuk, Sumy State University, Ukraine

Assoc. Prof. Xingxing Zhang, Dalarna University, Sweden

Assoc. Prof. Eivind Brodal, Norges Arktiske Universitet, Norway

Assoc. Prof. ABD HALID BIN ABDULLAH,Universiti Tun Hussein Onn Malaysia, Malaysia

Assoc. Prof. Haliza Abdul Rahman, Universiti Putra Malaysia, Malaysia

Assoc. Prof. Rajat Mazumder, Curtin University Sarawak Malaysia, Malaysia

Assoc. Prof. Nilofar Asim, National University of Malaysia (UKM), Malaysia

Dr. Kakoli Karar Paul, National Institute of Technology, India

Dr. Angel Torriero, Deakin University, Australia

Dr. Chang Choo Khean, Universiti Kuala Lumpur Malaysia France Institute, Malaysia

Dr. Youn-Suk Son, Pukyong National University, South Korea

Peer review statement

All papers published in this volume of *IOP Conference Series: Earth and Environmental Science* have been peer reviewed through processes administered by the proceedings Editors. Reviews were conducted by expert referees to the professional and scientific standards expected of a proceedings journal published by IOP Publishing.

IOP Conf. Series: Earth and Environmental Science **167** (2018) 012001 doi:10.1088/1755-1315/167/1/012001

Mechanisms of Rapid Flow caused by Tidal–Fluvial Flow Interaction in Inland Waterways of the Mekong Delta

H Takagi[1], T T Anh[2], L T Anh[1], N D Thao[3], T Takabatake[4] and R Nakamura[5]

1 Tokyo Institute of Technology, Tokyo, Japan

2 Vietnam Institute of Seas and Islands, Hanoi, Vietnam
3 Ho Chi Minh City University of Technology, Ho Chi Minh, Vietnam
4 Waseda University, Tokyo, Japan
5 Toyohashi University of Technology, Aichi, Japan
E-mail: takagi@ide.titech.ac.jp

Abstract. The present study investigates the river flow intensifications in the urban areas of the Mekong Delta, which are characterized by tributaries, channels, and low-lying lands in addition to the main stream of the Mekong Delta system. A hydraulic model is used to simulate velocities under the influence of both ocean tide and river discharge. A series of numerical simulations estimate that a rapid flow occurs along the tributary in flood season, reaching up to 1.8 m/s during the 2013 historical flood, which can be reproduced by assigning the net downstream river discharge of 9200 m^3/s as well as tidal current. A unique mechanism of flow intensification caused by "tidal blockage", which occurs in the flood tide phase during the flood season, is also identified. Such locally intensified flows could be further strengthened particularly during abnormal high water events such as storm surges and tsunamis, resulting in ship handling difficulties or a risk of overturning small ships.

1. Introduction

The Mekong River flows southward, traversing a distance of approximately 4,800 km from its source to the sea, draining a total catchment area of 795,000 km^2 that falls within six different countries: China, Myanmar, Lao People's Democratic Republic, Thailand, Cambodia, and Vietnam. On the basis of its mean annual flow at the river mouth, the Mekong River ranks 10th amongst the world's great rivers [1]. The Mekong Delta begins at Phnom Penh, Cambodia, where the river divides into its two main distributaries (Figure 1): Tien River and Hau River. The flow of the Tien River is much greater than that of the Hau River, where the annual discharge at Tan Chau accounts for about 83% and the annual discharge at Chau Doc accounts for about 17%. These distributaries trend roughly parallel to each other for most of their journey to the sea (Figure 1). The Tien River then divides into six main channels and the Hau River divides into three channels. These river systems form the Mekong Delta, which is made up of a vast triangular plain with an area of approximately 55,000 km^2 [1]. There is a noticeable difference between the channel network morphologies of the Tien and Hau branches; the former divides into a number of smaller distributaries before discharging into the sea, whereas the latter more or less maintains a single straight course to the sea [2].

The hydrological regime within the Mekong Delta is a product of the interaction among river discharge, tides, the landform, and configuration of the delta [2], [3]. Therefore, it is generally difficult to obtain a reasonable estimate of the discharge distribution over the branches [3]. Nevertheless, the hydrology of the Mekong Delta has been studied by many researchers [4]-[10] and an understanding of the flood inundations or sea-level rise impacts and their mechanisms have been substantially improved over the last couple of decades. Flood, inundation, and increasing sea-levels draw significant attention of authorities as well as researchers mainly because they would have direct and visible impacts on local communities [11], [12].

On the other hand, it appears that the river flow characteristics and patterns in the delta system have been overlooked and have not been sufficiently studied in the previous flood risk studies. Given the importance of oceanic influences on deltaic regions, the authors developed a tidal model and investigated the mechanisms of tidal propagation and its influence on flooding in the urban areas of the Mekong Delta [13]. Their model indicates that the flow velocity largely varies depending on the location, which exceeded 1 m/s in the tributary. However, the model can only be applied to simulations during the dry season because it cannot reproduce an abnormal water level rise induced by a high river discharge. In addition to tidal current, it is important to consider a rapid flow induced by a high river discharge during the flood season because small ships are likely to encounter handling difficulties or risk being overturned owing to the unexpectedly rapid flows.

The present study attempts to extend the tidal model [13] to enable application to both flood and dry seasons and thereby investigates the influence of combined flow characteristics of oceanic and fluvial flows on the development of a rapid flow in the complex waterway. The developed model is expected to be useful for identifying the locations subjected to unusual rapid flows in the Mekong Delta.

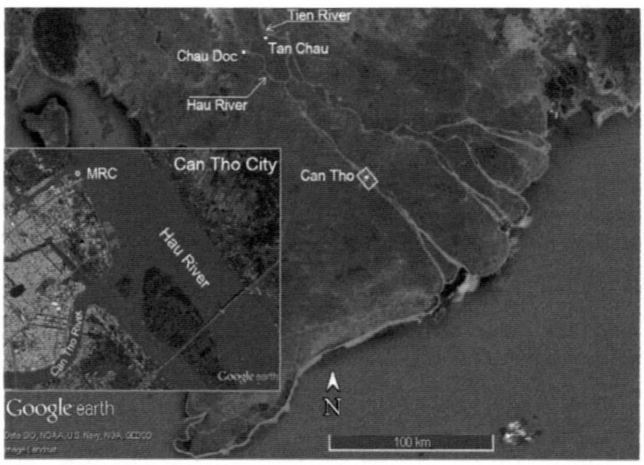

Figure 1. The close-up image of Can Tho, the largest city in the Mekong Delta, which is located in southern Vietnam. The yellow rectangle in the satellite image shows the computational domain dealt with in this study. The study area encompasses the main river (Hau River), the tributary (Can Tho River), and several small canals.

2. Methodology

The present study focuses on Can Tho City, which is the regional capital and has the highest population in the Mekong Delta (around 1.2 million). Can Tho is characterized by the main river (Hau River), tributary (Can Tho River), and many canals, demonstrating a complex river system that appears to create complicated hydraulics in its system (Figure 1). Although Can Tho is located 80 km inland from the river mouth, the ocean tides predominantly determine the water elevation [14], [15].

This study applied Delft3D-FLOW to reproduce ocean tides propagating upstream of the Hau River, which is one of the main streams of the Mekong system. This model is a multi-dimensional (2D or 3D) hydrodynamic simulation program which calculates non-steady flow that result from tidal and/or meteorological forcing on a rectilinear or a curvilinear grid. Although the model is applicable to a much wider region, this study limits the computational domain to encompass only Can Tho and its surrounding area in order to reduce the computational cost and enable a detailed analysis, which considers the local topography and bathymetry characterized by tributaries, channels, and low-lying lands in addition to the main stream. The present study extends the model [13] in order to incorporate river discharge upstream of Hau River as well as the tidal forcing downstream.

IOP Conf. Series: Earth and Environmental Science **167** (2018) 012001 doi:10.1088/1755-1315/167/1/012001

Figure 2 illustrates two different boundary conditions according to tropical seasons: one for the dry season and the other for the flood season. Ocean tides can be generated by assigning the boundary condition with a total of 60 tidal constituents [13]. Amongst all tidal components, the semidiurnal tidal constituent with a period of around 12 h appears to have the strongest influence. In the case of the dry season condition, when the pluvial and fluvial runoff can be negligible, the upstream boundary as well as the other tributaries was set as the Riemann condition, which enables progressive ocean tides to pass through with less reflection (Figure 2). In the case of the flood season condition, the upstream river discharge is determined by a trial method, which finds the discharge that best describes water levels observed by the Mekong River Commission (MRC) and the Southern Regional Hydro-Meteorological Centre, as discussed in the following section.

In March 2015, the author measured the river bathymetry using an echo sounder [13]. The computational grid size is set to be as small as 10 m throughout the domain with 442,704 nodes, while a sufficiently short time step of 0.6 s is assigned to run the simulation as stably as possible. The Manning's n value was set as 0.02 for the water and 0.05 for the land area [16]. The simulation is run for two seasons, which considers the days of 10–11 March 2012 as a typical dry season and 20–21 October 2013 as a typical flood season. The former period is chosen because the author performed the field survey (Figure 3 (left)), while the latter period is selected because a historical flood occurred during this period (Figure 3 (right)). Velocity changes with time are outputted at four locations (points a to d in Figure 2) in order to examine a wide area of the river system including both the main stream (Hau River) and the tributary (Can Tho River). The point referred to as MRC indicates the location of the water-elevation monitoring station.

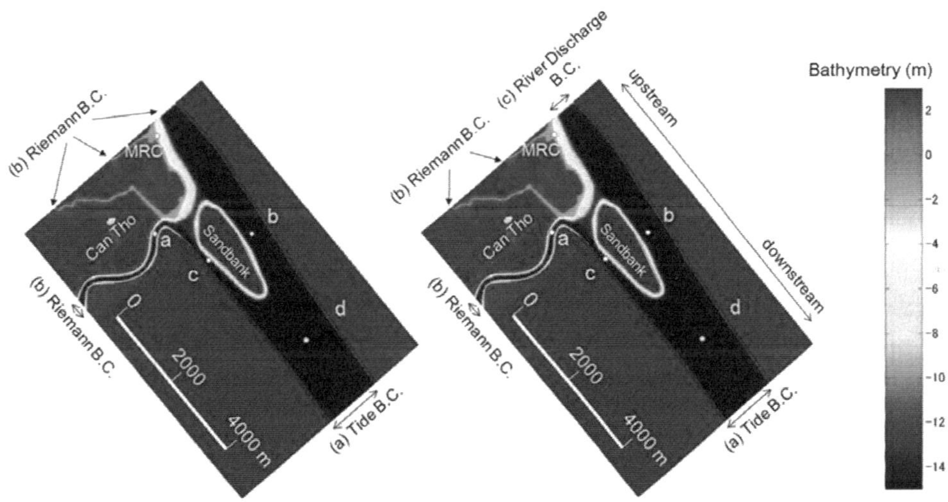

Figure 2. Boundary conditions assigned in the numerical simulations for the dry (left) and flood seasons (right).

10 March 2012 20 October 2013

Figure 3. Photos taken near point a in Figure 2 during dry (left) and flood seasons (right). In October 2013, many local communities across the Mekong Delta suffered from a historical flood [17].

3. Results and Discussion

3.1. Dry season analysis

Given that March lies in the midst of the dry season, the river discharge is typically very small [14]. Therefore, the simulation for the dry season only deals with tides and neglects the river discharge generated by fluvial and pluvial influences. Figure 4(a) shows the tidal fluctuations at one of the output points indicated as MRC in Figure 2, which demonstrates that the simulated and observed water levels are in good agreement in terms of both amplitude and phase. Figure 5(a) depicts the simulated depth-averaged velocities at four different locations in the river basin (Figure 2). The depth averaged velocity significantly varies depending on the location in the river basin. In particular, the flow velocity at point *a*, which reaches a maximum of 1.1 m/s, was almost three times that at point *d* in the main stream. It is obvious that the flow intensification that occurs as tidal current downstream of the river enters into the narrower channels and vice versa, causing a rapid flow. Figure 6a demonstrates that flow intensifications occur during both ebb and flood tides in the dry period. In the ebb tide phase, a strong flow developed in the tributary (Can Tho River) first impinges on the sandbank and then splits into two streams, merging again about 2.5 km downstream (Figure 6a (i)). In the slack tide phase when the tidal current almost ceases, two whirlwinds develop near the edge of the sandbank, indicating that the tide is about to reverse its direction (Figure 6a (ii)). In the flood tide phase, the tide from the sea propagates upstream and a part of the water departs from the main stream and flows into the narrower tributary, causing a rapid flow (Figure 6a (iii)).

3.2. Flood season analysis

It appears that the treatment of the boundaries is more difficult for simulating the hydraulics in the flood season than that in the dry season. Although the Southern Regional Hydro-Meteorological Centre measures river velocity with a current meter in Hau River, it is inevitable that gross discharge rates are substantially influenced by tidal currents. Thus, the net downstream river discharge on the dates studied needs to be estimated by some other method. In this study, the rate is determined by repeating the simulation with a number of different river discharge rates in order to identify the condition which best reproduces the observed water levels at the MRC observatory.

Figure 4(b) demonstrates that the maximum water level up to 2 m on 20 October 2013 cannot be accounted for by only tidal fluctuations, showing about a 60 cm difference between the actual observed levels and tide levels derived by tidal harmonic analysis [14] or the present numerical simulation. The water level abnormally rises above the predicted tidal levels, implying significant pluvial and fluvial influences in addition to tidal influence. The numerical simulation with the boundary condition for the flood season shown in Figure 2 has been repeatedly performed until this gap becomes sufficiently small; the optimal river discharge rate is determined to be 9200 m^3/s. The red continuous line in Figure 4(b) indicates the best fitted water level calculated by inputting a river discharge rate of this volume, demonstrating that the model can reproduce water levels during the flood season with a reasonable accuracy.

Figure 5(b) presents the simulated flow velocities in the flood period, including the points *a* at the tributary, *b* and *c* at the narrower strips in the main stream, located on both sides of the sandbank, and *d* at the mains tream of the Hau River, where the river is the widest in the computational domain. Numerical simulations were repeated for three different assignments of boundary conditions. While the maximum flow velocity is estimated to be only about 0.6 m/s for (iii) the river discharge condition, it reaches up to 1.3 m/s and 1.8 m/s for (ii) the tide condition and (i) the tide and river discharge combined condition, respectively. This observation corroborates that the tidal regime is generally more predominant than the fluvial regime in the lower Mekong Delta, such as in Can Tho [14].

Given that the tidal regime in this delta is typically semi-diurnal, both ebb and flood tides occur approximately once in the course of half a day, producing the flow velocity peak that occurs twice a half day, as shown in Figure 5b (ii). In the flood tide phase, the flow at point *a* in Figure 2 becomes remarkably fast owing to the influence of tides (Figure 5b (ii)), which is further excited by the fluvial inflow (Figure 5b (i)). In contrast, in the ebb tide phase, it appears that the flow at point *a* reduces to the lowest owing to an opposite flow direction between the tide and river discharge.

The stagnation of flow in the tributary can also be observed in the contour and vector maps (Figure 6b (i)), which are snapshots taken when the tide is at the lowest level. The flow in the tributary is not stagnated even in the slack tide phase (Figure 6b (ii)) because of a constant inflow of river discharge. A unique mechanism that could be found in the flood tide phase (Figure 6b (iii)) is that the tidal current interferes with the fluvial downflow at the confluence of the main stream and tributary. This phenomenon, which could be referred to as *tidal blockage*, forces a combined water flow of tidal and fluvial waters to rush into the narrower tributary. In this way, the diverted river discharge from the main stream of the Hau River exclusively flows into the tributary, Can Tho River, owing to the tidal–fluvial flow interaction. On the other hand, in the ebb and slack phases, it splits into two flows: one towards the tributary and the other towards downstream.

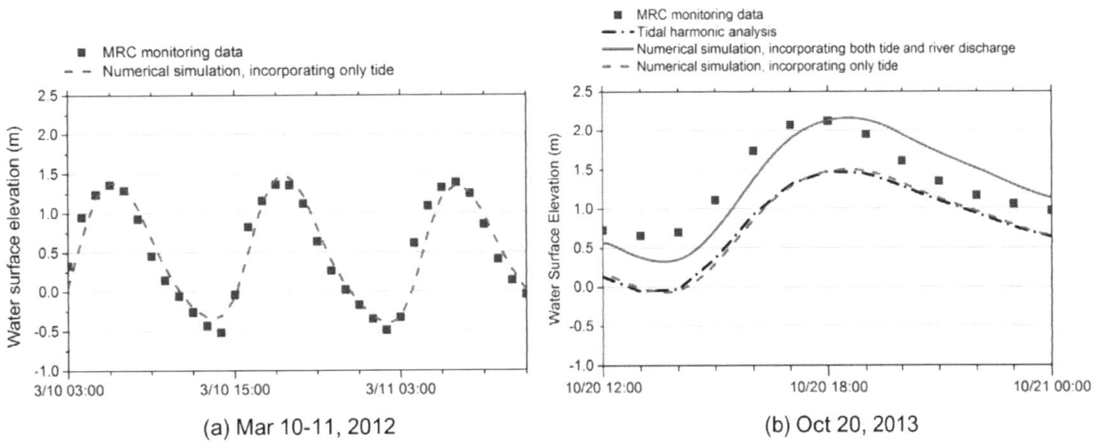

Figure 4. Comparison between the numerically simulated and measured water levels at the observatory, indicated by MRC in Figure 2, covering a typical dry period (a) and flood period (b).

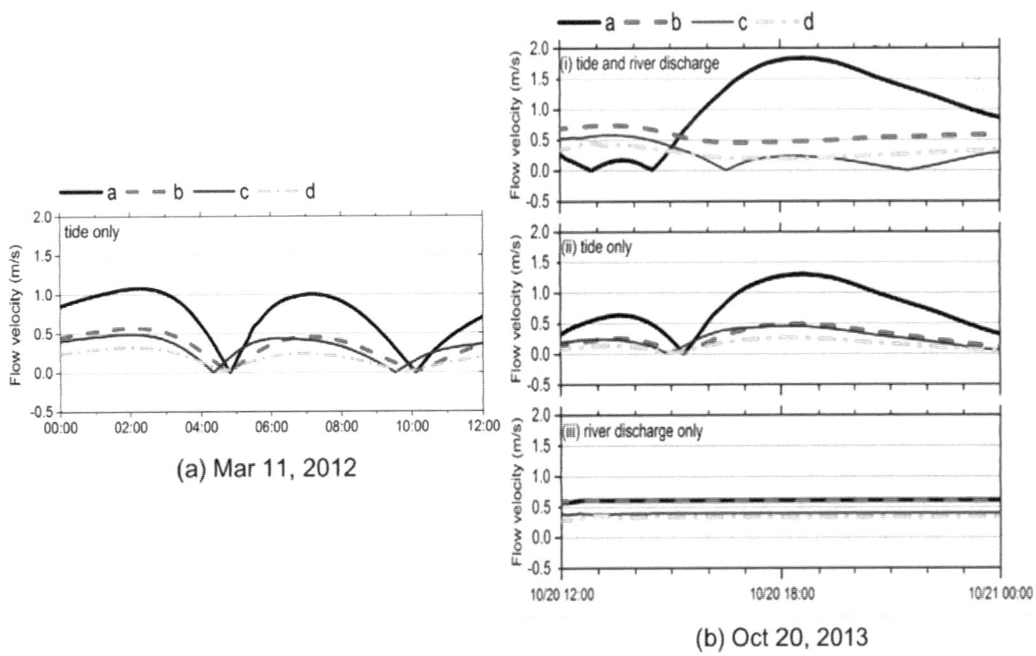

Figure 5. Simulated depth averaged velocity during a semi-diurnal cycle in a typical dry period (a) and a flood period (b). The outputs correspond to the four locations indicated in Figure 2.

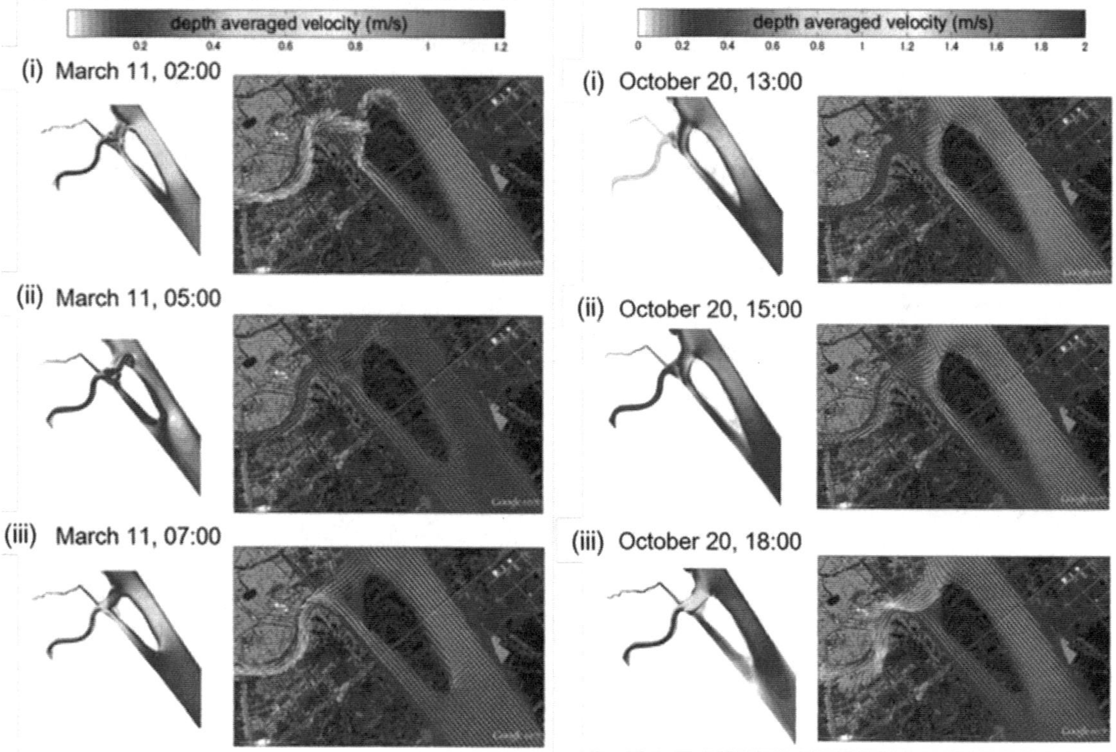

(a) Dry season (b) Flood season

Figure 6. Tidal flow intensifications observed in the numerical simulation. (i) ebb tide phase, (ii) slack tide phase, and (iii) flood tide phase.

3.3. Flow classifications

A rule of thumb suggests that the maximum tidal velocity of the spring tide may not exceed more than 1 m/s in estuaries [18]. Indeed, this rule is also applicable to the main stream of the Mekong Delta system, as found in Figures 5. However, this study corroborates that this is the case for the main stream (point *a*), but not the case for narrower tributaries or channels which connect to a wider main stream (points *b*, *c*, *d*). This kind of complex waterway may create flow intensifications due to a contraction of the waterway, causing unusual rapid flows which sometimes exceed 1 m/s.

To sum up, Figure 7 illustrates how flow intensifications can be developed according to different seasons and different tidal phases. Flow intensification occurs along the tributary under all these conditions. As already discussed, the most intensified flow can be induced when the tidal and fluvial flows meet at the confluence of the main stream and tributary, induced by tidal blockage.

Although the phenomena found in this study may not be simply generalized, it is considered that such flow intensifications could occur at many other waterways in deltas or estuaries. Urban development increases flood risk in cities owing to local changes in the hydrological and geological conditions that increase flood hazard as well as urban concentrations that increase vulnerability e.g. [8], [19].

Furthermore, the rapid economic developments appear to have complicated the river system as a result of construction of man-made channels. It is feared that while comprehensive flood control measures will reduce flooding, the developed embankments could cause more severe riverbank erosion due to increased flow speeds in the rivers. Although the proposed model focuses on a specific area in the Mekong Delta, this methodology can also be applied to other stretches of the delta by assigning appropriate river conditions such as bathymetry, river discharge, and tidal regime.

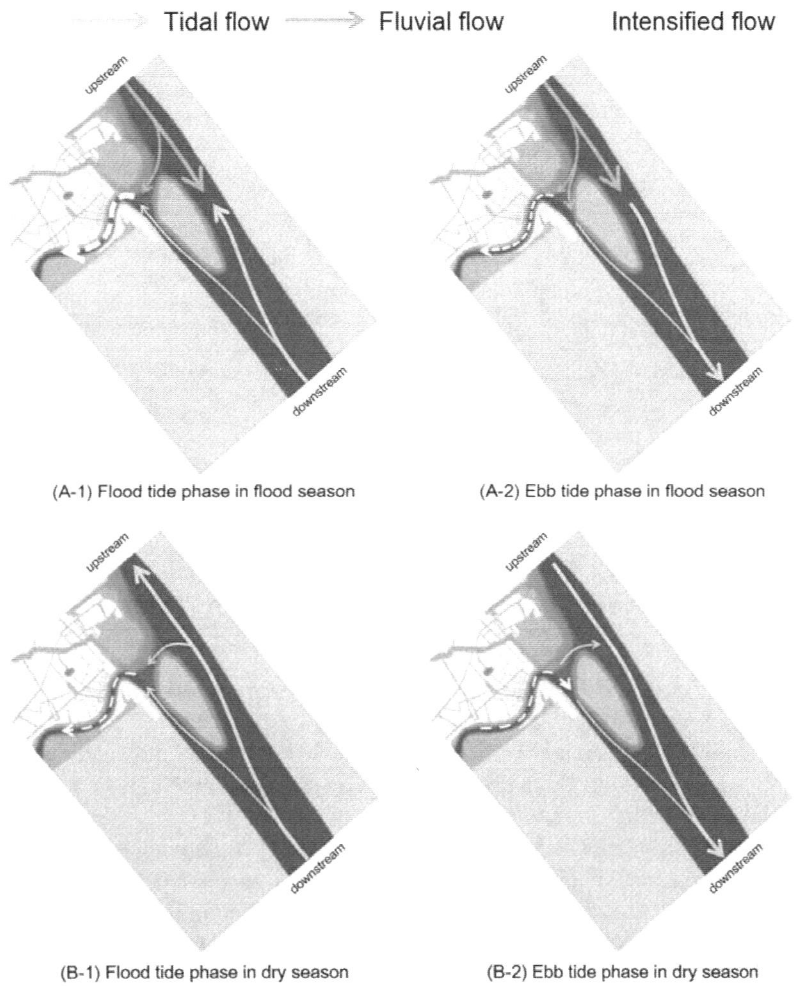

Figure 7. Classification of flow intensification induced in accordance with two seasons and two tidal phases.

4. Conclusions

The hydraulic model was used to simulate velocities in the most populated city of the Mekong Delta, Can Tho, under the combined influence of ocean tides and river discharge. The computational boundaries were differently modelled for the flood and dry seasons to identify unique flow characteristics in the two distinctive seasons. The results from the numerical simulations suggest that a very rapid flow likely occurs along the tributary in both flood and dry seasons. In particular, a unique mechanism of flow intensification, *tidal blockage*, was observed during the flood tide phase in the flood season. Tidal blockage is a phenomenon wherein the tidal current interferes with the fluvial flow at the confluence of the main stream and tributary and causes a rapid flow up to 1.8 m/s in the study river. This local intensification of the flow velocity could be an important characteristic when considering the flood risks in the urban areas of the Mekong Delta or other deltas, which could be further amplified during unusually high tides or typhoon-induced storm surges. Because a large number of local fishermen, tourist guides, and floating market merchants use small wooden ships, there is a considerable risk that they may encounter ship handling difficulties or risk being overturned owing to these unexpected unusually rapid flows. Although the present study only deals with a small area of the Mekong Delta, the proposed model is expected to be applicable to other stretches of the delta in order to identify inland waterways subjected to rapid flows caused by the tidal–fluvial flow interaction.

5. References

[1] Mekong River Commission (MRC) 2005 Overview of the hydrology of the Mekong Basin.

[2] Hashimoto T K 2001 Environmental issues and recent infrastructure development in the Mekong Delta: review, analysis and recommendations with particular reference to large scale water control projects and the development of coastal areas, Working Paper No. 4, p 70.

[3] Nguyen A D, Savenijie H H G, Pham D N and Tang D T 2008 Using salt intrusion measurements to determine the freshwater discharge distribution over the branches of a multi-channel estuary: The Mekong Delta case, Estuarine, *Coastal and Shelf Science*, 77.

[4] Wassmann R, Hien N X, Hoanh C T and Tuong T P 2004 Sea level rise affecting the Vietnamese Mekong delta: water elevation in the flood season and implications for rice production. *Climatic Change*, 66.

[5] Hoa L T V, Nhan N H, Wolanski E, Cong T T and Shigeko H 2007 The combined impact on the flooding in Vietnam's Mekong River delta of local man-made structures, sea level rise, and dams upstream in the river catchment, *Estuarine, Coastal and Shelf Science*, 71.

[6] Dasgupta S, Laplante B, Meisner C, Wheeler D and Yan J 2007 The impact of sea level rise on developing countries: a comparative analysis. *World Bank Policy Research Working Paper*, 4136.

[7] Carew-Reid J 2008 Rapid assessment of the extent and impact of sea level rise in Viet Nam, *Climate Change Discussion Paper*. International Centre for Environmental Management, p 74.

[8] Huong H T L and Pathirana A 2013 Urbanization and climate change impacts on future urban flooding in Can Tho city. Vietnam, *Hydrology and Earth System Sciences*, 17.

[9] Thao N D, Takagi H and Esteban M 2014 Coastal disasters and climate change in Vietnam - engineering and planning perspectives. Elsevier, New York, p 424. ISBN: 978-0128000076

[10] Takagi H, Thao N D and Anh L T 2016 Sea-Level Rise and Land Subsidence: Impacts on Flood Projections for the Mekong Delta's Largest City, *Sustainability*, 8 (9).

[11] Knaepen H L 2014 Mainstreaming Climate Change Adaptation into Vietnamese Development as a New Policy Arrangement, IN: Nguyen D. T., Takagi H, Esteban M (eds.) *Coastal disasters and climate change in Vietnam*. Elsevier, 355–377.

[12] Yamamoto L 2014 Environmental displacement in Vietnam, IN: Nguyen D. T., Takagi H, Esteban M (eds.) *Coastal disasters and climate change in Vietnam*, Elsevier, 379–393.

[13] Takagi H, Tsurudome C, Nguyen D T, Le T A, Tran V T and Van P D T 2016 Ocean tidal modelling for urban flood risk assessment in the Mekong Delta, *Hydrological Research Letters*, 10 (1).

[14] Takagi H, Tran V T, Nguyen D T and Esteban M 2015 Ocean tides and the influence of sea-level rise on floods in urban areas of the Mekong Delta. *Journal of Flood Risk Management*, 8 (4).

[15] Takagi H, Nguyen D T, Esteban M, Mikami T, Cong LV and Ca VT 2015 Coastal disasters in Vietnam. In: Esteban M, Takagi H, Shibayama T (eds.) *Handbook of coastal disaster mitigation for engineers and planners*, Elsevier, 235–255.

[16] Takagi H, Li S, Leon M, Esteban M, Mikami T, Matsumaru R, Shibayama T and Nakamura R 2016 Storm surge and evacuation in urban areas during the peak of a storm, *Coastal Engineering*, 108.

[17] Viet Nam News 2013 Mekong tributaries flood Can Tho City, *Viet Nam News*, http://vietnamnews.vn/, Last access 26 October 2017.

[18] Savenije HHGA 2001 Simple analytical expression to describe tidal damping or amplification, *Journal of Hydrology*, 243.

[19] Takagi H, Fujii D, Esteban M and Yi X 2017 Effectiveness and Limitation of Coastal Dykes in Jakarta: The Need for Prioritizing Actions against Land Subsidence. *Sustainability*, 9, 619.

Acknowledgements

This research was funded by the JSPS KAKENHI Grant Number 26702009 and 16KK0121 (Japan Society for the Promotion of Science).

Streamflow response under different climate change scenarios in data-scarce White Volta basin of West-Africa using a semi-distributed hydrologic model

S Abubakari[1, 2*], X Dong[1, 2*], J Liu[1, 2], Y Li[1, 2], T Peng[1, 2] and H Ma[1, 2]

1 College of Hydraulic and Environmental Engineering, China Three Gorges University, Yichang, Hubei, China
2 Collaborative Innovation Center for Territorial Sovereignty and Maritime Rights, Wuhan, 430072, China
E-mail: abubakarisulemanaaz206@yahoo.com, xhdong@ctgu.edu.cn

Abstract. In this study, streamflow response in White Volta basin of West Africa was evaluated using ECHAM5 (MPEH5) GCM for two IPCC climate change scenarios (A1B and B1) and semi-distributed hydrologic model (SWAT) for three future time periods; 2020s, 2050s and 2080s using 1987-1996 as baseline. Both climate change scenarios project decrease in rainfall in 2020s but increases in 2050s and 2080s. Decrease in rainfall in 2020s was higher in B1 scenario compared to A1B; whereas for 2050s and 2080s, increases in rainfall were higher in A1B scenario than B1. Projected average changes indicate that overall the climate will be warmer and wetter in future. Projected annual streamflow volumes show disproportionate change (~220%) in response to slight changes in mean annual rainfall. Streamflow volume shows sharp changes for A1B scenario than B1 in 2020s whilst in 2050s and 2080s, the changes are moderate. Projected average annual streamflow changes are -3.1% in 2020s, 5.1% in 2050s and 13.3% in 2080s. Overall projected streamflow changes for both A1B and B1 scenarios were not so dramatic (less than ±20%). These call for adaptation strategies to reduce vulnerability and ensure water security within the basin.

1. Introduction

Although there are many discussions still ongoing, it is widely recognized that future impacts of climate change (CC) will be difficult to avoid. CC and its variability will significantly affect hydrologic cycles and this is likely to affect availability (both quantity and quality) and distribution of water resources in several regions around the world. According to Fifth Assessment Report (AR5) of Intergovernmental Panel on Climate Change (IPCC)[1], Africa's climate is already changing and the impacts are already being felt with further CC inevitable in the coming decades. The African continent will particularly be vulnerable due to the considerably limited adaptive capacity.

Water plays important economic and social roles in sub-Saharan Africa. It is used for hydropower, agriculture, fishing, industry, livestock watering, transportation, urban development, emergency uses such as fire-fighting and recreation. These activities employ majority of people generating revenues which are fundamental to sustenance of economies of these countries. According to previous studies by Opoku-Ankomah[2]; van de Giesen et al.[3]; Jung and Kunstmann [4] and Sood et al.[5], there have been numerous changes in both spatial and temporal distribution of rainfall in Volta basin of West Africa. This has resulted in some sub-catchments having abundant freshwater leaving others with very limited water to satisfy increasing demands. Uncertainties arising out of CC and its

variability will additionally exacerbate future water resources management, resulting in non-attainment of Millennium Development Goals in areas such as health, eradicating poverty and hunger. This is going to pose challenges to growth and development of inhabitants in the basin.
There have been numerous studies utilizing climate models in quantifying effects of CC on hydrology and water resources management in White Volta River basin of West Africa. Obuobie [6] used climate series for 1991–2000 and 2030–2039 generated by stochastic downscaling model, Long Ashton Research Station Weather Generator (LARS-WG) and reflecting MM5-simulated monthly changes, future CC scenario IS92a and Soil and Water Assessment Tool (SWAT) to evaluate effects of CC on recharge of groundwater in White Volta River basin. Results of this study found increase in mean annual rainfall of about 6% in 40 years, resulting in increase of about 29% in groundwater recharge. Kankam-Yeboah et al. [7] used downscaled CC projections from ensemble of two Global Climate Models (ECHAM4 and CSIRO), A1F1 future CC scenario and SWAT model to estimate impacts of CC on streamflow in White Volta basin. Results of this study found that mean annual streamflow showed a decrease of 22% and 50% for 2006-2035 (2020s) and 2036-2075 (2050s) respectively relative to the baseline (1961-1990). Awotwi et al. [8] assessed impacts of CC on various water balance components of White Volta basin using downscaled CC projections from Regional Climate Model (REMO), A1B future CC scenario and SWAT model. Results of this study revealed that an increase of 8% and 1.7% of mean annual precipitation and temperature respectively in 2030-2043 leads to an increase of 26% in mean annual surface runoff relative to the baseline (1995-2008).
All these studies have focused on the response of streamflow to CC in White Volta River basin under a single future emission scenario. In this study, we attempt to investigate the projected changes in streamflow in White Volta River basin of West Africa under different CC emission scenarios. Results of this study will provide local policy makers with quantitative data for incorporation into CC adaptation strategies to reduce vulnerability and ensure water security within the basin.

2. Materials and Methods

2.1. Study Area

The White Volta River basin (Figure 1) is a major sub-basin of the Volta River basin of West Africa. It is located between latitude 8°N and 15°N, and longitude 1°E and 4°W. It is a transboundary river basin with a drainage area of about 106,300 km^2 and shared mainly by Burkina Faso and Ghana. The topography is predominantly flat with more than half of the basin lying in the elevation range of 200-300m with mean elevation of about 270m. The basin has two major climate zones: semi-arid climate in the north and humid climate in the south. Mean annual rainfall in the semi-arid north is around 600 mm whiles that in the humid south is around 1200 mm [9]. In the semi-arid north, mean monthly temperature ranges from 36°C in March to 27°C in August whiles in the humid south, it ranges from 30°C in March to 24°C in August. The basin wide mean annual temperature and Evapotranspiration (ET) are 26°C and 1650 mm respectively [9]. The soils in the basin are Luvisols, regosols, lithosols, vertisols, planosols, cambisols, and gleysols. The basin is dominated by luvisols, regosols and lithosols which cover about 70% of the basin area. The Land use/cover is predominantly savannah, grassland, and agriculture with small patches of forest.

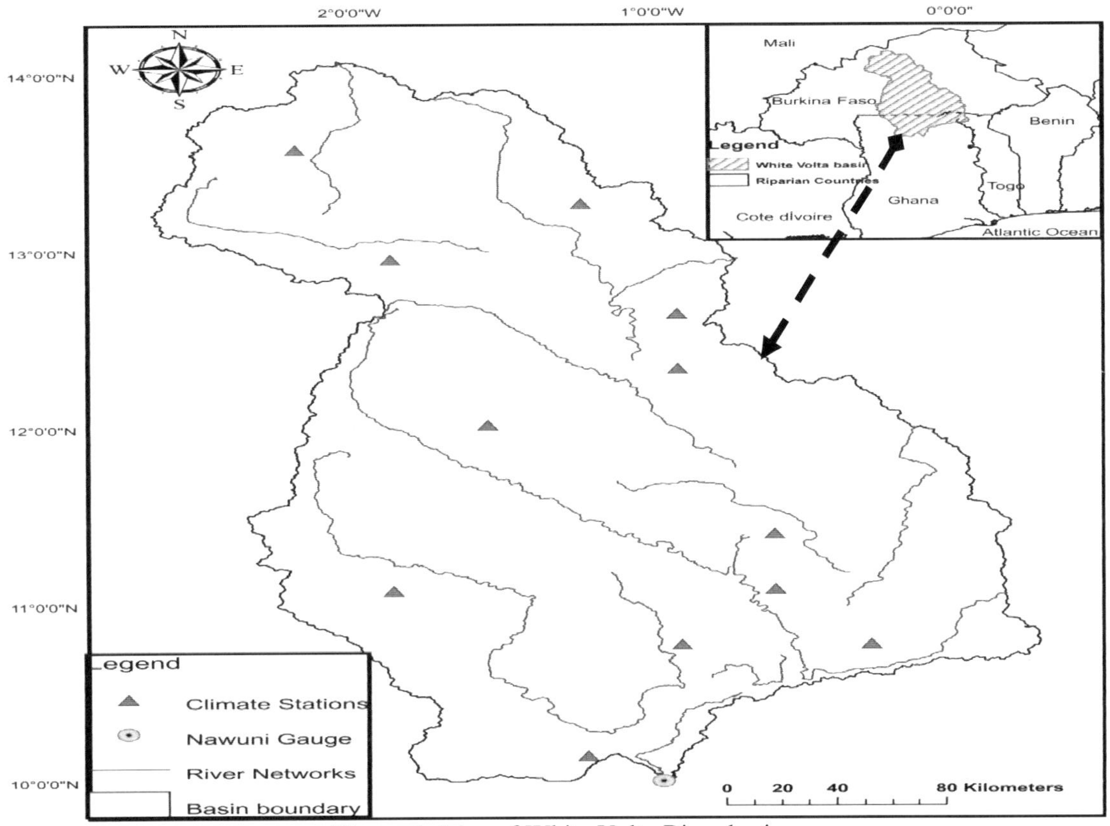

Figure 1. Map of White Volta River basin

2.2. Soil and Water Assessment Tool (SWAT) Model Description

SWAT is a process-based, continuous-time, semi-distributed hydrological model. It was developed by United States Department of Agriculture (USDA) Agricultural research Service (ARS) to predict impact of land use practices on water, sediment and agricultural chemical yields in large and complex watersheds with diverse weather, varying soils, land use and management and topographic conditions over long periods [10]. It uses GIS interface and operates on daily time step; however, daily outputs can be aggregated into monthly and annually outputs. It is a very robust model and has been used for CC studies all around the world.

2.3. Model Input Data

Digital Elevation Model (DEM). The DEM used for this study has spatial resolution of 90 m and was obtained from Shuttle Radar Topographical Mission (SRTM) (http://srtm.csi.cgiar.org/).

Land use/Land cover data. The land-use map is a modified Food and Agricultural Organization (FAO) map with spatial resolution of 400 m

Soil Data. It is a Food and Agricultural Organization (FAO) digitized soil map of the world and derived soil properties. It has spatial resolution of 10 km with almost 5000 distinguishable soil types and was downloaded from WaterBase (http://waterbase.org/download_data.html).

Climate Data. The climate data is Climate Forecast System Reanalysis (CFSR) and was obtained from National Centers for Environmental Prediction (NCEP) (http://cfs.ncep.noaa.gov/cfsr).

River Discharge Data. Monthly discharge data was obtained from Global Runoff Data Centre (GRDC), Koblenz, Germany.

2.4. SWAT Model Set Up

The SWAT model was set-up for White Volta River basin through Arc SWAT 2009 interface following step by step procedure outlined in SWAT user guide. First each basin was divided into sub-basins based on DEM and stream networks of the basin. The number of sub-basins obtained was determined by threshold input value. Sub-basin delineation was followed by automatic parameterization of streams and subdivision of sub-basins into HRUs. Since topography of the basin is predominantly flat, with about 96% of the land having slopes less than 5%; single slope class option was used. This study employed multiple HRUs option where land use class percentage over sub-basin area was set to 20%, soil class percentage over land use area was 10% and slope class percentage over soil area was 20%. Threshold drainage area was set to 5, 50029.2639 km2 and this created 14 sub-basins and 67 HRUs. The model was set up and run using climate data from 1985–1996.

2.5. Sensitivity Analysis

Twenty-one SWAT input parameters (Table 1) relevant to the study area were subjected to sensitivity analysis. The main aim of this exercise was to determine input parameters that had greatest influence on model output.

Table 1. SWAT Input Parameters used for Sensitivity Analysis

No.	Parameter	Range	Description
1	ALPHA_BF	0-1	Baseflow Alpha Factor
2	BLAI	0-1	Maximum Potential Leaf Area Index
3	CANMX	0-10	Maximum Canopy Storage
4	CH_EROD	0-1	Channel Erodibility Factor
5	CH_K2	0-150	Channel Effective Hydraulic Conductivity
6	CH_N2	0-1	Manning's Value for Main Channel
7	CN2	±25	Initial Curve Number II
8	EPCO	0-1	Plant Uptake Compensation Factor
9	ESCO	0-1	Soil Evaporative Compensation Factor
10	GW_DELAY	±10	Groundwater Delay
11	GW_REVAP	±0.036	Groundwater Revap Coefficient
12	GWQMN	±1000	Threshold Water Depth in the Shallow Aquifer for Return Flow
13	RCHRG_DP	0-1	Deep Aquifer Percolation Factor
14	REVAPMN	±100	Threshold Depth of Water in the Shallow Aquifer for "revap" to occur
15	SLOPE	±25	Average Slope Steepness
16	SLSUBBSN	±25	Average Slope Length
17	SOL_ALB	±25	Moist Soil Albedo
18	SOL_AWC	±25	Soil Available Water Capacity
19	SOL_K	±25	Saturated Hydraulic Conductivity
20	SOL_Z	±25	Soil Depth
21	SURLAG	0-10	Surface Runoff Lag Time

2.6. Calibration and Validation

Due to lack of adequate daily discharge data, calibration of SWAT model was carried out using monthly discharge data at Nawuni flow station (see Figure 1) which represents outlet of White Volta basin. The time period selected for calibration was January 1985 to December 1991, using the first two years (1985-1986) as warm-up period. Thus January, 1987-December, 1991 was used for calibration and January 1992 to December 1996 for validation.

Performance indicators such as Nash-Sutcliffe model efficiency coefficient (NSE), coefficient of determination (R2) and PBIAS were used for evaluation of model performance. Model performance was considered acceptable when R2 > 0.60, NSE > 0.50 and PBIAS was within ±25%. For detailed explanation and formulas of NSE, R2 and PBIAS, reference can be made from [11].

2.7. Climate Change (CC) Scenarios and Projections

The IPCC A1B (medium emission) and B1 (low emission) scenarios were used for CC projections. The rainfall and temperature projections were from outputs of ECHAM5 (MPEH5) developed by Max Planck Institute for Meteorology (Germany).

The GCM output was downscaled through the use of the stochastic downscaling model, Long Ashton Research Station Weather Generator (LARS-WG). The downscaling process by LARSWG involved calibration and validation of the model using 30 years (1979-2008) of observed historical daily data at each of the 12 climate stations in the basin through the Site Analysis and Q test functions (step 1). The baseline data (1987-1996) for each station was generated using the Generator function (Step 2). The baseline data (step 2) was then driven by monthly CC scenario information from outputs of the GCM data to produce daily climate scenarios which are representative of the 2020s (2011-2030) and the 2050s (2046-2065) and 2080s (2080-2099) using the Generator tool incorporated in the model (Step 3). The downscaled daily climate outputs (step 3) were then used as SWAT input for CC impact assessment at the basin scale. For detail description of downscaling by LARS-WG, references can be made from [12].

3. Results and Discussion

3.1. Sensitivity Analysis

The results of sensitivity analysis carried on twenty-one (21) SWAT input parameters are presented in Figure 2. The fourteen most sensitive parameters were selected for calibration process.

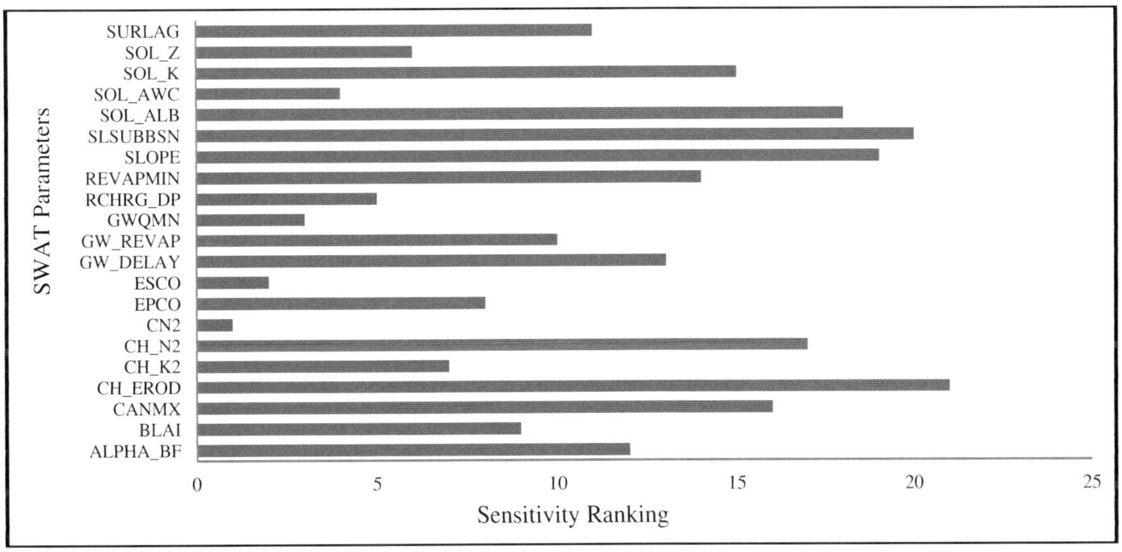

Figure 2. Ranking of twenty-one (21) SWAT Parameters during Sensitivity Analysis

3.2. Calibration and Validation

The simulated streamflow (Table 2) showed 9.10% and 8.60% reduction for both calibration and validation periods respectively, compared to average observed monthly streamflow.

Table 2. Model Performance during calibration and validation periods

	Average obs.flow (m^3/s)		standard deviation(m^3/s)		model performance		
Period	Obs.	Sim.	Obs.	Sim.	NSE	R^2	PBIAS
calibration(1987-1991)	314.6	286.0	506.8	392.7	0.74	0.76	11.0%
validation(1992-1996)	269.9	247.0	445.6	380.1	0.73	0.74	8.6%

A comparison of hydrographs (Figure 3) and model performance (Table 2) shows that objective functions R^2 and NSE were greater than 0.70 for both calibration and validation periods. PBIAS was within ±25% i.e. 11.0% and 8.6% for calibration and validation periods respectively. These indicate good correlation between monthly observed and simulated streamflow; which is a clear indication that physical processes in the basin are well reproduced by the model.

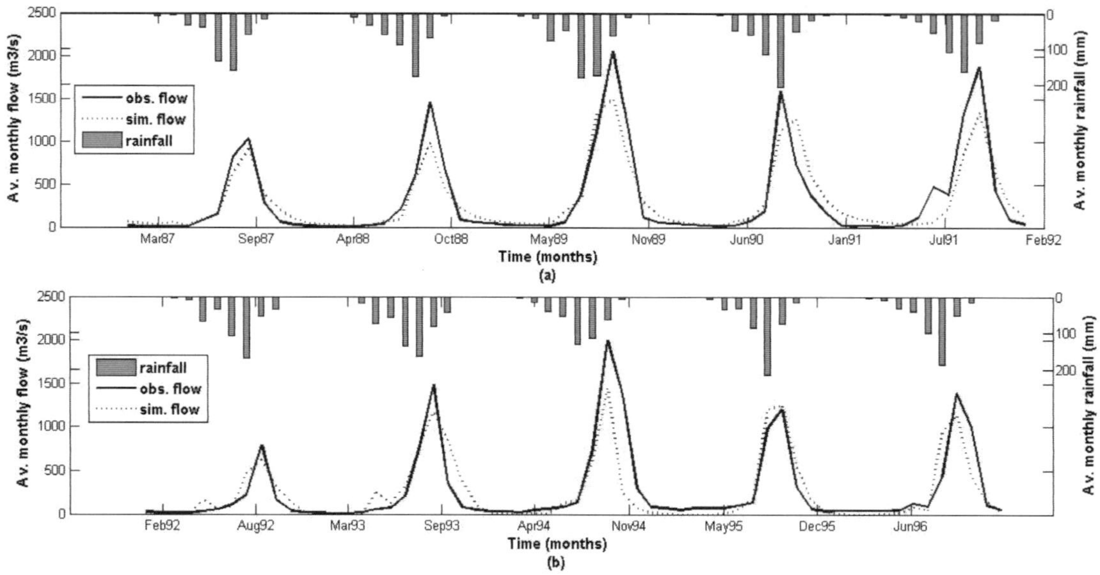

Figure 3. Discharge during a) calibration and b) validation for Nawuni (outlet)

3.3. Calibration and Validation

The performance of LARS-WG in simulating the baseline climate (rainfall, minimum and maximum temperatures) (Figure 4) was assessed using the p value for statistical significance testing provided by the model. The average p values for simulating daily rainfall, minimum and maximum temperature were 0.989, 1.000 and 0.998, respectively. This shows satisfactory reproduction of baseline values indicating minimum model biases and errors.

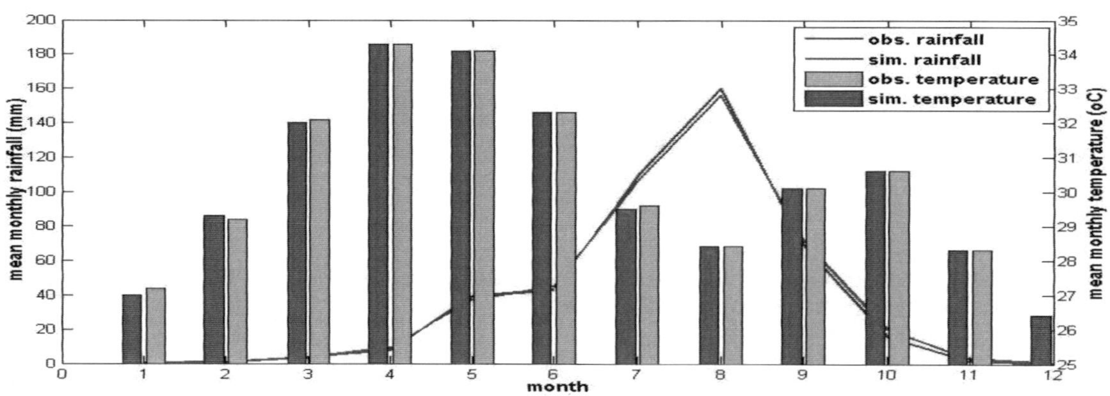

Figure 4. Comparison of observed and simulated mean monthly rainfall and temperature

3.4. Future Climate Change Projections

Projection of future changes in climate were determined from outputs of ECHAM5 GCM for two climate change scenarios, A1B and B1; for three future time periods, 2011-2030 representing the 2020s, 2046-2065 representing the 2050s and 2080-2099 representing the 2080s. The future changes in climate (Figure 5) represent the change in precipitation and temperature respectively in future compared to baseline period (1987-1996).

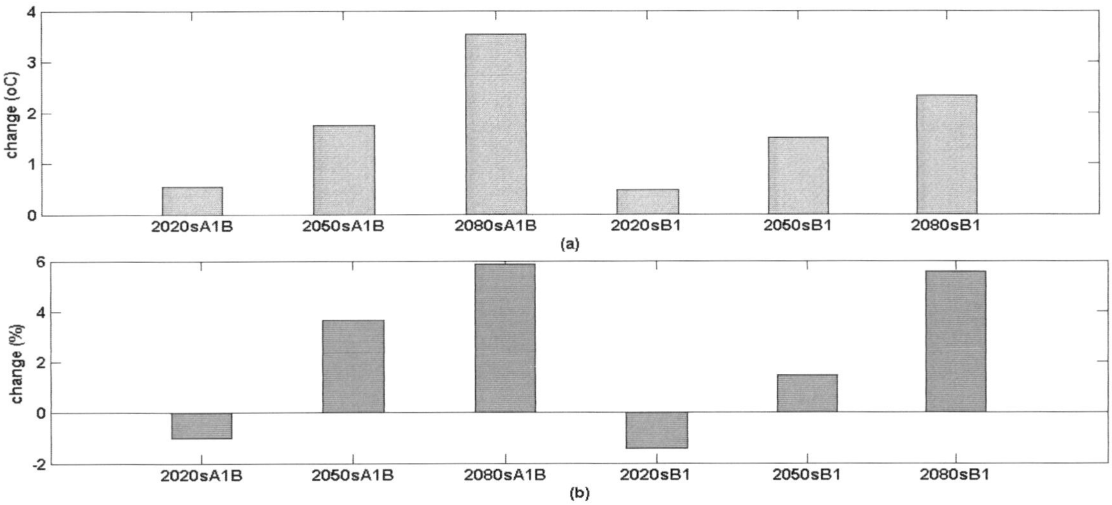

Figure 5. Annual a) temperature and b) rainfall change in 2020s, 2050s and 2080s.

Annual average temperature increase is 0.52°C in 2020s, 1.62°C in 2050s and 2.94°C in 2080s. Further analysis revealed that A1B scenario projected higher future temperatures than B1. The projected changes indicate that overall, temperature will become warmer in future with higher increases anticipated at the end of the century (2080s).

For rainfall, both climate change scenarios projected decrease in 2020s but increases in both 2050s and 2080s. The decrease in rainfall in 2020s was slightly higher in B1 scenario compared to A1B; whereas for 2050s and 2080s, the projected increase in rainfall was higher in A1B scenario than B1. Both scenarios projected more precipitation in 2080s than in 2020s and 2050s. The projected average changes are -1.21% in 2020s, 2.56% in 2050s and 5.73% in 2080s. Unlike temperature changes, precipitation changes show much uncertainty exhibiting both negative and positive trends.

3.5. Streamflow Response to Climate Change

Figure 6 shows the effect of possible future climate change on annual streamflow.

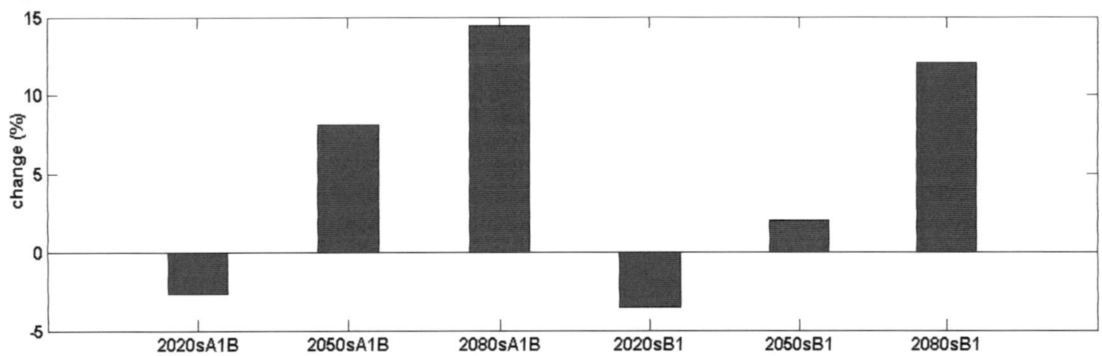

Figure 6. Average annual streamflow change in 2020s, 2050s and 2080s

Projected annual streamflow volume trends are synonymous to that of annual precipitation changes. The projected streamflow changes are expected to range between -2.0% and -4.0% in 2020s, +2.0% and +9.0% in 2050s and +12.0% and +15.0% in 2080s.The projected average changes are -3.1% in 2020s, 5.1% in 2050s and 13.3% in 2080s. Streamflow volume shows sharp changes for A1B scenario than B1 in 2020s whilst in 2050s and 2080s, the change is moderate.

The results also reveals disproportionate change in annual streamflow volumes (~220%) in respond to changes in mean annual rainfall. This finding support Sood[5], Obuobie[6], Awotwi [8] assertions that

IOP Conf. Series: Earth and Environmental Science **167** (2018) 012002 doi:10.1088/1755-1315/167/1/012002

there is a much larger non-linear response of streamflow to smaller changes in rainfall. Overall the projected streamflow changes in future for both A1B and B1scenarios were not so dramatic (less than±20%).

4. Conclusion

For this study, streamflow response under different climate change (A1B and B1) scenarios in data scarce White Volta basin of West Africa was investigated using the semi-distributed hydrologic model SWAT. For rainfall, both climate change scenarios projected decrease in 2020s but increases in both 2050s and 2080s. The decrease in rainfall in 2020s was slightly higher in B1 scenario compared to A1B; whereas for 2050s and 2080s, the projected increase in rainfall was higher in A1B scenario than B1. Both scenarios projected more precipitation in 2080s than in 2020s and 2050s. Projected average changes indicate that overall, the climate will become warmer and wetter in future.

Projected annual streamflow volume trends were synonymous to that of annual precipitation changes with decrease in 2020s and increases in both 2050s and 2080s. There is disproportionate change in annual streamflow volumes (~220%) in respond to changes in mean annual rainfall. Streamflow volume shows sharp changes for A1B scenario than B1 in 2020s whilst in 2050s and 2080s, the change is moderate. Projected average annual streamflow changes are -3.1% in 2020s, 5.1% in 2050s and 13.3% in 2080s. Overall projected streamflow changes in future for both A1B and B1scenarios were not so dramatic (less than ±20%).

Though there are uncertainties concerning CC scenarios and global climate model predictions, these results nonetheless will provide local policy makers with quantitative data for incorporation into CC adaptation strategies so as to foster resilience reduce vulnerability and ensure water security within the basin.

5. Acknowledgments

This research has been supported by Non-profit Industry Financial Program of Ministry of Water Resources of China (No. 201301066), the National Natural Science Foundation of China (40701024, 41101511, 51409152), and Hubei Provincial Collaborative Innovation Center for Water Security.

6. References

[1] IPCC 2013 Climate change: The physical science basis. Contribution of Working Group I to the Fifth Assessment Report of the Intergovernmental Panel on Climate Change

[2] Opoku-Ankomah Y 2000 *Impacts of Potential Climate Change on River Discharge in Climate Change Vulnerability and Adaptation Assessment on Water Resources of Ghana*, Water Research Institute (CSIR), Accra, Ghana

[3] van de Giesen N, Liebe J, and Jung G 2010 *Curr Sci* **98** 1033-1037

[4] Jung G and Kunstmann H 2007 *IAHS Publication* **313** 150

[5] Sood A, Muthuwatta L and McCartney M 2013 *Water Int* **38** 297-311

[6] Obuobie E 2008 *Estimation of groundwater recharge in the context of future climate change in the White Volta River Basin, West Africa*, Doctoral Thesis University of Bonn

[7] Kankam-Yeboah K, Obuobie E, Amisigo B and Opoku-Ankomah Y 2013 *Hydrolog Sci J* **58** 773-788

[8] Awotwi A, Kumi M, Jansson P E, Yeboah F and Nti I K 2015 Predicting Hydrological Response to Climate Change in the White Volta Catchment, West Africa *J Earth Sci Clim Change* **6** 1-7

[9] Oguntunde P G, Friesen J, van de Giesen N and Savenije H H 2006 *Phys Chem Earth* **31** 1180-1188

[10] Neitsch S L, Arnold J G, Williams J R, Kiniry J R and Srinivasan R 2009 SWAT input/output file documentation

[11] Moriasi D N, Arnold J G, Van Liew M W, Bingner R L, Harmel R D and Veith T L 2007 *T ASABE* **50** 885-900

[12] Semenov M A 2007 *Agr Forest Meteorol* **144** 127-138

Experimental investigation of nonlinear flow for single rock fractures

S J Zhang[1], X Z Dong[1], B H Wu[1] and C L Wang[2, 3]

1 Zhejiang Institute of Hydrogeology and Engineering Geology, Zhengjiang Ningbo, China
2 School of Civil Engineering, Chongqing University, Chongqing, China
3 Key Laboratory of Geotechnical Engineering Stability Control and Health Monitoring of Hunan Province, Hunan University of Science and Technology, Xiangtan, Hunan, China
E-mail: 370567882@qq.com

Abstract. The hydraulic behavior of rocks is significantly controlled by fracture geometries, such as aperture, contact areas, roughness, interconnections, and so on. However, these characteristics are strongly influenced by the confining pressure conditions. This paper experimentally investigated the nonlinear flow behaviors through single rock fractures subjected to a wide range confining pressures. A series of transient pulse tests were conducted on three fractured limestone samples by MTS815 Rock Mechanics Test System. The experimental results show that the pulse decay curves diverge from the classical exponential model due to nonlinearity, thus an empirical relationship between differential pressure and time is developed with consideration of nonlinearity. Subsequently, the nonlinear flow coefficient and permeability were calculated based on the Forchheimer equation. The calculated results show that nonlinear flow coefficient increases with confining pressure, and rougher fracture surface helps to stronger nonlinearity. As the confining pressure increases, the permeability first experiences a dramatic decrease and then behaviors a much slow-down drop. The critical confining pressure for climb-rush shifts increases with the fracture rouhghness.

1. Introduction

Fluid flow in fractured rock masses exists in numerous industrial and scientific fields, such as excavation of rock caverns[1], oil or natural gas exploration[2], [3], contaminant pollution control, hazardous wastes isolation[4], [5] and grouting activities [6]-[8]. Microcracking induced by repository excavation and hydraulic pressure will significantly enhance the permeability in the disturbed zone of the surrounding rocks [5], raise the risk of radionuclide migration across the natural barrier system, and hence cause environment pollution. Therefore, the knowledge of fluid flow behavior through fractured rocks is crucial to pollution prevention and proper management of natural resources.

Fluid flow through single rock fractures obeys the basic law of fluid dynamics [9]. It can be well described by Navier-Stokes (NS) equations. However, how to solve the NS equations in rough-walled fracture is a challenging task due to the inertia term of the N-S equations and irregular geometry of rock fracture [10]. To circumvent the problem, early studies assume that laminar flow occurs between two parallel plates and yields the well-known cubic law. However, the real fracture surfaces are rough which makes the evolution of fluid flow deviate from the cubic law. More importantly, deviation from linear Darcy's law (i.e. nonlinear flow) may occur in rock fractures [11], [12]. The mechanisms that trigger nonlinearity in rough-walled fracture have been extensively studied, including fracture

geometries [13], [14], solid-water interaction [15], turbulence, and localized eddy formation [16]. Previous studies suggest that flow through rock fractures generally occurs in the linear and transitional regimes in real hydrogeological environment [11], [13], [17]. Therefore, this paper mainly focuses on the nonlinear flow as a result of heterogeneity of fracture geometries at low Reynolds number.
Steady-flow method and transient-pressure method are the two major techniques for fluid flow tests. The steady-flow method usually consists of imposing a constant pressure gradient to a sample and measuring the flow rate. This method can easily obtain the relation of pressure gradient and flow rate, so it has been widely applied to investigate the nonlinear flow behavior [9], [11], [18]. However, for low permeability geo-materials, steady-flow method requires a relatively long test time to enhance the measurement accuracy. The transient-pressure method is based on the analysis of the differential pressure between upstream and downstream circuits within a sample, which is appropriate for low permeability geo-materials, and has accurate results better than 5% of results from steady-flow method [19], [20]. Although pulse transient technique has been applied for permeability measurement in rocks fractures [18], [15], few attempts have been made at measuring the nonlinear flow behavior due to the non-steady flow state of the method. In this paper, we develop a new methodology to analyse nonlinear flow for pulse transient tests by calculating instantaneous velocity.
In this paper, the water flow behavior through rough-walled fractures is investigated experimentally using transient pulse method. The observed differential pressure data is analysed to estimate the nonlinearity. Then, the nonlinear flow coefficient and permeability are calculated using Forchheimer law. The relationship between non-Darcy coefficient and confining pressure is explored, and the regression analysis of experimental data is conducted.

2. Experimental methodology

Three limestone samples (M01, M02 and M03) for flow tests were cored from the Mei Tan-ba coal mine in Hunan Province (China). Samples were cut into cylinders by stone processing machine with dimensions of 50 mm in diameter and 100 mm in length. The mechanical properties of samples are shown in Table 1. An artificial fracture was created along the core axis in the uniaxial compressive apparatus using Brazilian technique as illustrated in Figure 1. Before being pre-fractured, the sample was circumferentially applied with glue and covered with thermo-shrinking plastic pipe to avoid sample shattering. Note that the artificial fractures were generated at a loading rate of 50N/s for sample M01, 100N/s for M02 and 200N/s for M03 respectively to achieve different roughness. Visual inspection showed that initial profile of sample M01 is smoother than for samples M02 and M03. And the fracture roughness of samples is measured by a dimensional parameter *JRC* using the method from [21], which gives *JRC* values of 8.96, 13.3 and 14.6 of sample M01-M03, respectively.

$$JRC = 32.2 + 32.471 \lg Z_2 \tag{1}$$

$$Z_2 = \sqrt{\frac{1}{L}\sum_{i=1}^{n}\frac{\left(y_{i+1}-y_i\right)^2}{\left(x_{i+1}-x_i\right)}} \tag{2}$$

where *JRC* is rock roughness, L is the length of sample, x_i is the ith measurement step, y_i is the ith discrete point of y-axis, n is the number of cross-section of discrete points.

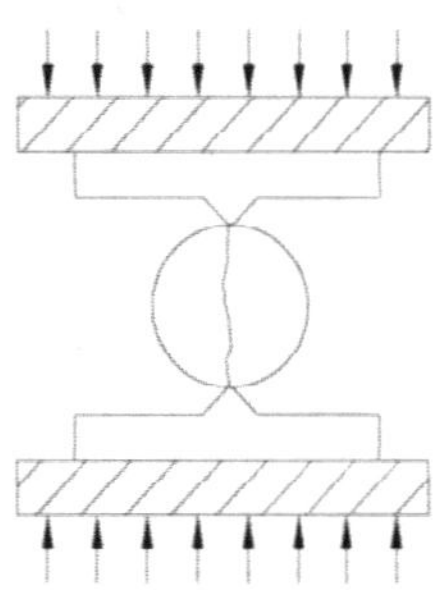

Figure 1 Splitting test

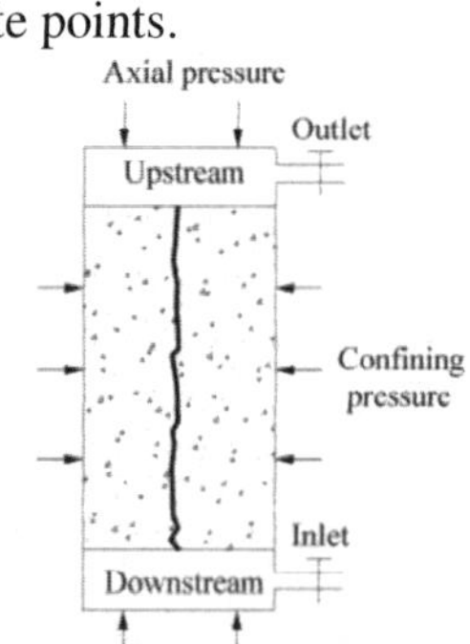

Figure 2 Sketch of transient pulse test

Table 1. Mechanical property of samples

Density (g/cm3)	Porosity	Poisson's ratio	Elastic modulus (GPa)	Uniaxial compressive strength (MPa)
2.61	8.5%	0.23	20.1	44.8

2.1. Experimental setup and procedure

The permeability test was carried out by MTS815 Rock Mechanics Test System using transient pulse method. The system is mainly composed of four units including triaxial cell, loading unit, pressure monitoring unit, water supply unit. The experimental setup has the maximum loading capacity of 4600 kN, and the maximum confining pressure and pore water pressure both are 140 MPa.

Figure 2 shows the sketch of transient pulse test, and the test procedures are summarized as follows:

(1) Produce fractured samples as introduced earlier, saturate the prepared samples with water for 24h.

(2) Install the sample into the triaxial cell, apply the axial and confining pressure to predetermined value at a rate of 0.5MPa/min.

(3) Apply the initial water pressure of 0.2 MPa to the upstream and downstream reservoirs.

(4) Increase the pressure of upstream reservoirs by 2 MPa to form a differential pressure, monitor the change of differential pressure with time until equilibrium is attained.

(5) Increase confining pressure to the next predetermined value and repeat the process (4) to conduct next flow test. For these flow tests, confining pressures were increased from 8 MPa to 20 MPa in gradual levels of 2 MPa.

3. Experimental methodology

3.1. Analysis of pulse decay curves

The evolution of all measured differential pressure is analysed for three samples successively, with the main results being exemplified by sample M01. Simplified analytical methods approach the differential pressure with an exponential formulation [19],

$$\Delta P(t) = \Delta P_0 \exp(-ct) \tag{3}$$

where c is a constant depending on the experimental arrangement. Theoretically, parameter c is a function of reservoir volumes, fluid compressibility, sample accumulation coefficient and external factors such as stress conditions [19]. Figure 3 shows the pulse decay curves of sample M01 at varying confining pressures. The figure demonstrates that the pulse decay curves diverge from the classical exponential model (Equation (3)), especially under high confining pressure. Similar experimental results have also been reported in laboratory tests [15]. The most reasonable explanation is that a nonlinear flow behavior occurs in real rock fractures for their rough walls and variable surface [9]. Previous literatures manifest that the presence of asperities and obstructions or sharp changes in fracture profile will change flow velocity or direction along the flow path, causing inertial losses which make flow regimes deviate from linearity. Meanwhile, the fracture closure and contact areas increase with the confining pressure, which lead to more inertial losses and hence cause nonlinearity.

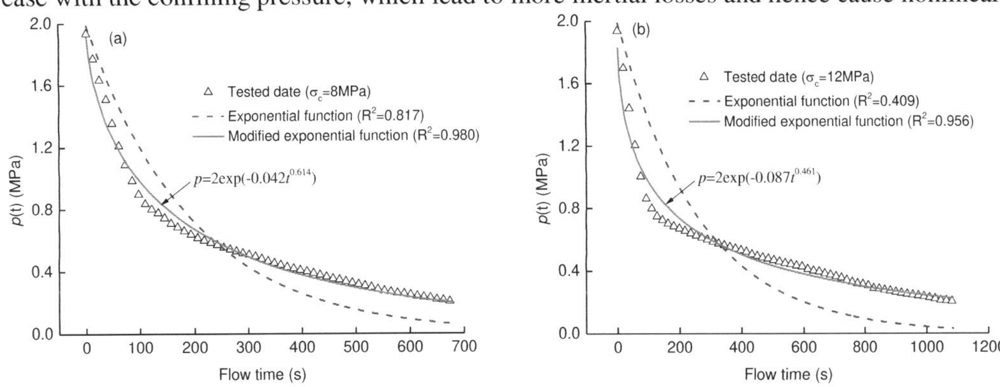

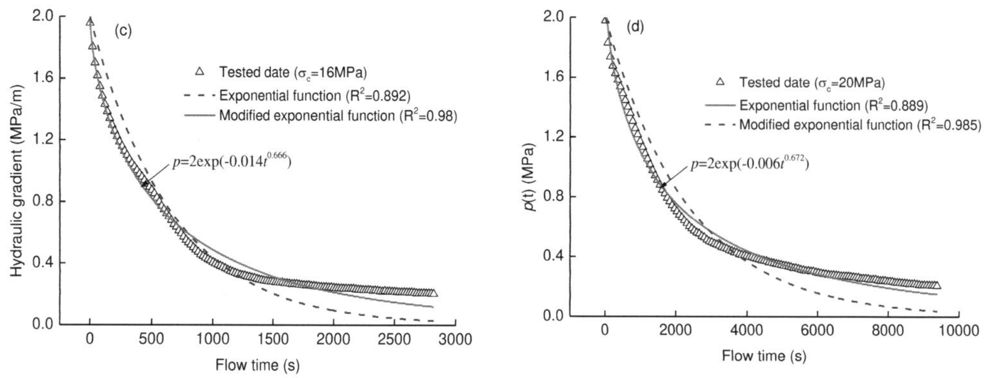

Figure 3 Differential pressure-flow time data and fitting curves for sample M01 at a confining pressure of (a) 8 MPa; (b) 12 MPa; (c) 16 MPa and (d) 20 MPa

Ignoring the volume change of the reservoirs due to the applied water pressure, the variation of the water mass (m_w) in the upstream reservoir can be expressed as [15], [22]

$$\frac{dm_w}{dt} = V_u \frac{d\rho}{dt} = -\rho A v \tag{4}$$

where A is the sectional area of sample; V_u is the volume of reservoir; ρ is the fluid density; and v is the flow velocity through the fracture. The compressibility of water C_f can be expressed as [22]

$$\frac{1}{C_f} = \rho \frac{dp_1(t)}{d\rho} \tag{5}$$

Combination of Equations (4) and (5) gives

$$\frac{dp_1(t)}{dt} = -\frac{Av}{C_f V_u} \tag{6}$$

Similarly, we obtain the following equation for the downstream reservoir:

$$\frac{dp_2(t)}{dt} = \frac{Av}{C_f V_d} \tag{7}$$

Combination of Equations (6) and (7) yields

$$\frac{dp(t)}{dt} = \frac{dp_1(t) - dp_2(t)}{dt} = -\frac{Av}{C_f}\left(\frac{1}{V_u} + \frac{1}{V_d}\right) \tag{8}$$

Or

$$v = -\frac{C_f}{A\left(\frac{1}{V_u} + \frac{1}{V_d}\right)} \frac{dp(t)}{dt} \tag{9}$$

The hydraulic gradient j at time t can be simply calculated using the experimental pulse decay data, i.e.

$$j = \frac{p(t)}{L} = \frac{p_1(t) - p_2(t)}{L} \tag{10}$$

It can be seen from Equations (9) and (10) that the experimental pulse decay data can be used as a criterion of linear or nonlinear flow. If the pulse decay data follows the classical exponential formulation (Equation (3)), the relationship between the velocity and pressure gradient is linear, which means linear flow occurs. Otherwise, nonlinear flow occurs. Obviously, in this paper, the relationship between flow velocity and hydraulic gradient doesn't meet the Darcy's law, indicating that a nonlinear flow behavior occurs. The nonlinear flow characteristics will be further discussed in detail later.

To have a best-fit regression analyses of the experimental pulse decay data, $p(t)$, a modified exponential equation is proposed, i.e.

$$p(t) = p_0 \exp\left(-ct^n\right) \tag{11}$$

where n is a fitting parameter depending on the nonlinearity. By selecting an appropriate value of n, a high correlation coefficient can be obtained for the fitting equation. Note that the general exponential formulation (Equation (11)) can reduce to traditional formulation (Equation (3)) for value of n=1 when the flow occurs in the linear regime. The experimental data are fitted using this modified and more general exponential formulation and the fitting curves are shown in the Figure 3. It can be clearly seen that the generalized model provides a better fit than the exponential model given by Equation (3).

3.2. Analysis of non-Darcy flow

Previous studies [11], [18] indicate that Forchheimer law has proved to fit the flow in fractured rocks well. For a one-dimensional non-Darcy flow, the relationship between pressure and flow velocity can be expressed as [23]

$$\rho c_a \frac{dv}{dt} = -j - \frac{\mu}{k} v + \rho\beta v^2 \tag{12}$$

where c_a is acceleration factor, k is permeability; μ is kinematic viscosity of the fluid. The flow velocity and $\eta = \frac{dv}{dt}$ can be obtained with Equation (9). The differential pressure $(p_1 - p_2)_{t=i} (i = 1,2\cdots n)$ were measured every 1.0 s. Therefore, the following equitation is satisfied

$$\rho c_a \eta_i = -j_i - \frac{\mu}{k} v_i + \rho\beta v_i^2 \ (i = 1,2\cdots n-2) \tag{13}$$

Establishing a fonctionelle $\prod$

$$\prod = \sum_{i=2}^{N-2} \left(\rho\beta V_i^2 - \frac{\mu}{k} V_i - \rho c_a \eta_i - \xi_i \right)^2 \tag{14}$$

By solving the fonctionelle $\prod$ on basis of extremum conditions, we can obtain β and k.

$$\frac{\partial \prod}{\partial(\rho\beta)} = 0; \left(\sum_{i=2}^{N-2} V_i^4\right)\rho\beta - \left(\sum_{i=2}^{N-2} V_i^3\right)\frac{\mu}{k} - \left(\sum_{i=2}^{N-2} \eta_i V_i^2\right)\rho c_a - \left(\sum_{i=2}^{N-2} \xi_i V_i^2\right) = 0 \tag{15}$$

$$\frac{\partial \prod}{\partial\left(\frac{\mu}{k}\right)} = 0; \left(\sum_{i=2}^{N-2} V_i^3\right)\rho\beta - \left(\sum_{i=2}^{N-2} V_i^2\right)\frac{\mu}{k} - \left(\sum_{i=2}^{N-2} \eta_i V_i\right)\rho c_a - \left(\sum_{i=2}^{N-2} \xi_i V_i\right) = 0 \tag{16}$$

$$\frac{\partial \prod}{\partial(\rho c_a)} = 0; \left(\sum_{i=2}^{N-2} V_i^2 \eta_i\right)\rho\beta - \left(\sum_{i=2}^{N-2} V_i \eta_i\right)\frac{\mu}{k} - \left(\sum_{i=2}^{N-2} \eta_i^2\right)\rho c_a - \left(\sum_{i=2}^{N-2} \xi_i \eta_i\right) = 0 \tag{17}$$

3.2.1 Relation of k and confining pressure.

Figure 4a-c shows the relation of permeability and confining pressure for tested samples. As the confining pressure increases, permeability of three samples first experiences a dramatic decrease and then behaviors a much slow-down drop. This can be integreted that the fracture aperture decreases sharply due to the favourable fracture compressibility, and then decreases at a much slower speed due to fracture closure when the confining pressure is higer than the fracture closure stress. Comparing with Figure 4(a), Figure 4(b) and Figure 4(c), it is found that the sample M03 with the largest *JRC* value shows the minimum permeability, M02 takes the second place, while M01 shows the maximum permeability. This further emphasis the common knowledge that rocks with smoother fracture surfaces have a larger permeability. Note that the drop shift of $k - \delta_c$ curve for three samples are different. In general, the critical confining pressure for drop shift increases with the fracture rouhghness. Specificly, the threshold value of Sample M01<Sample M02<Sample M03. This phenomenon would possibly result from that a rougher fracture needs a larger stress to make the fracture surfaces joint closely. To characterize the relation between permeability and confining pressure for fractured rock, Ma et al. [18] proposed a power function to fit the obserted data.

$$k = a_1 \sigma_c^{b_1} \tag{18}$$

where k is the permeability, σ_c is the confining pressure, the coefficient a_1 and b_1 are scaling parameters. As shown in Figure 4a-c, the power function given by Equation (18) fits the experimental data well, supporting the usefulness of the power function.

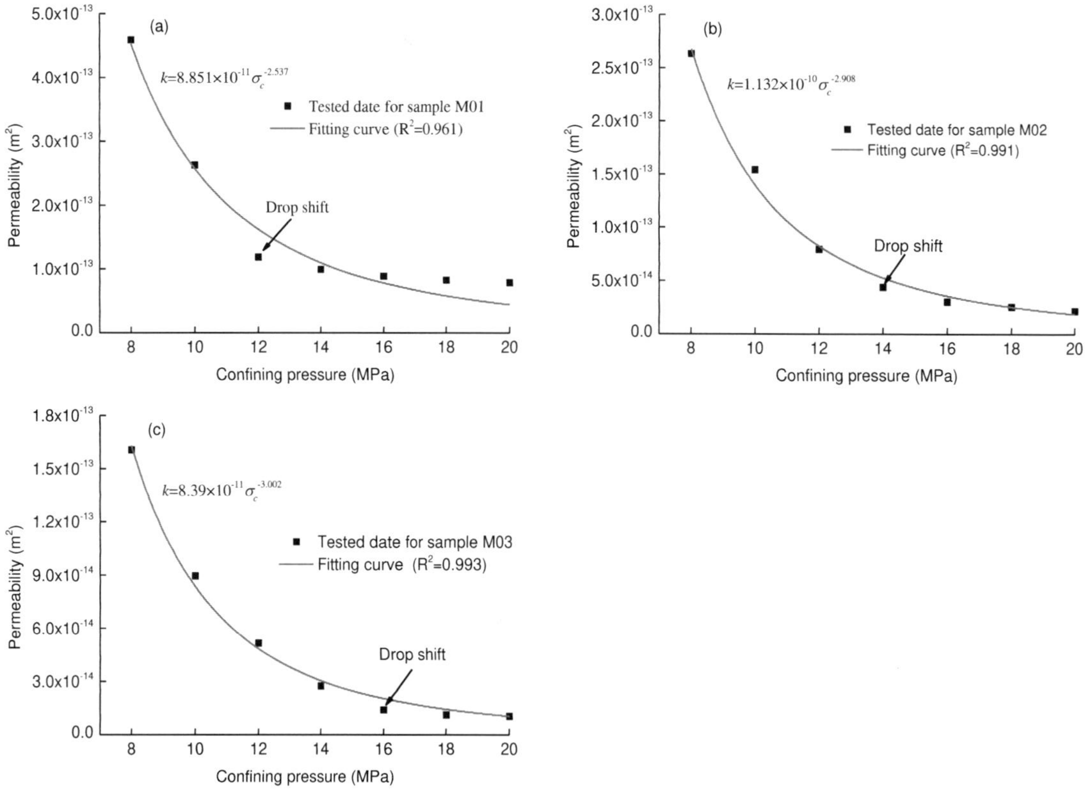

Figure 4 Permeability k-confining pressure curves of sample (a) M01; (b) M02 and (c) M03

3.2.2 Relation of β and confining pressure.

Figure 5 demonstrates that the non-Darcy flow coefficient β increases with confining pressure at a decreasing change rate, indicating that higher confining pressure helps to stronger nonlinearity. The values of non-Darcy flow coefficient β , in decreasing order, are those of test samples M01 (minimum), M02 and M03 (maximum) for a given confining pressure. This can be integrated as a rougher fracture shows a stronger nonlinear flow behavior at the same condition of confining pressure. Such trend is in qualitative agreement with the experimental results from Zhou et al. [11], in which he suggests that a rougher fracture surface usullay induces a more dramatic nonlinear flow behavoir due to solid-water interaction.

Ma et al. [18] introduced a logarithmic function and a polynomial function to analyze the relationship between the non-Darcy coefficient β and confining pressure. Figure 5(a)-(c) shows that the logarithmic function didn't match tested data well. The average value of R^2 of the polynomial function is less than 0.9.

$$\beta = a_2 \ln(\sigma_c) + b_2 \tag{19}$$

where σ_c is confining pressure, a_2 and b_2 are fitting parameters. To have a best-fit analysis of tested data, an empirical formula is proposed in this paper.

$$\beta = a_3 \sigma_c \exp(\lambda \sigma_c) + b_3 \tag{20}$$

where a_3 is the scaling parameters, λ gives the slope, and parameter b_3 gives the intercept. It can be seen that the new function provides a better fit than the logarithmic function. For the new function, the average value of R^2 is close to unit (over 0.99), while for logarithmic function, the average is less than 0.9, indicating that the performance of new function is better than the logarithmic function.

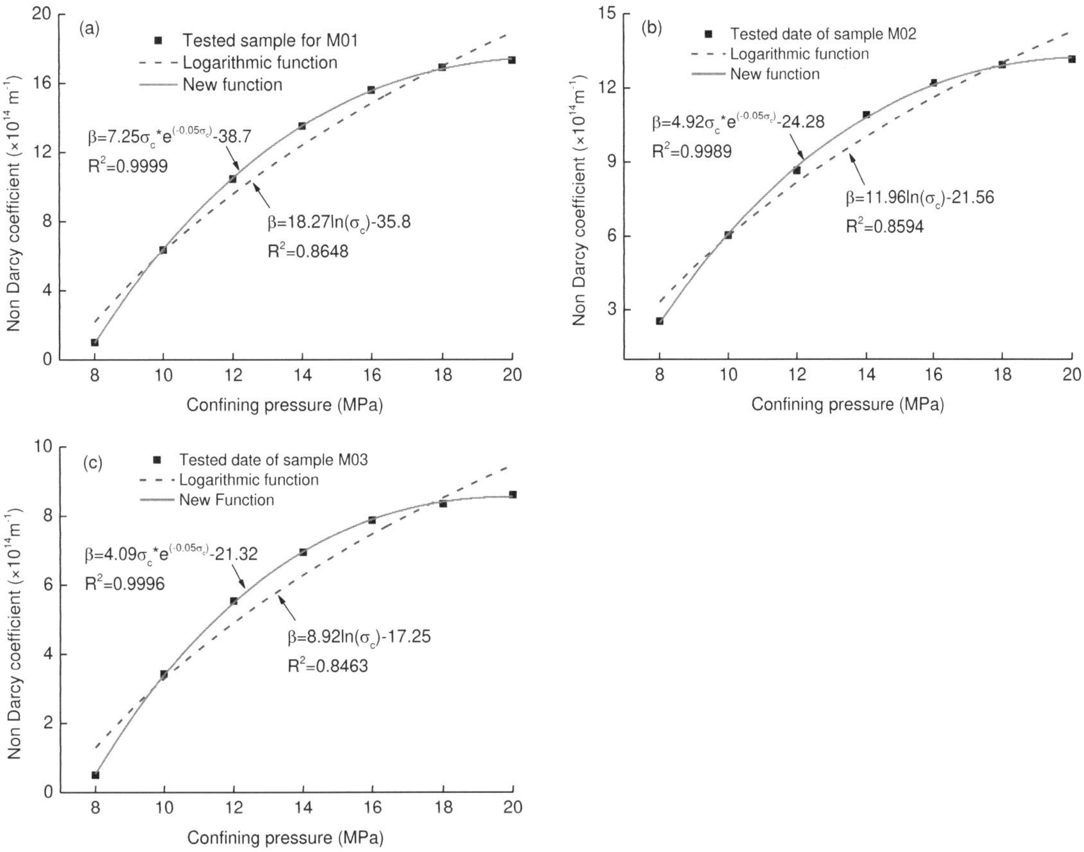

Figure 5 Non-Darcy coefficient -confining pressure curves of sample (a) M01; (b) M02 and (c) M03

4. Conclusions

In this paper, the effects of variable confining pressure on nonlinear flow behaviors of rough-walled fractures are investigated through transient pulse tests, the main conclusions are as follows:

(1) The observed pulse decay data versus time diverges from the classical exponential model due to nonlinear flow behavior, and then a modified exponential relationship is developed with consideration of nonlinearity. Theoretical derivation gives that the experimental pulse decay data can be used as a criterion of nonlinear flow.

(2) The nonlinear flow coefficient β and permeability k were calculated based on the Forchheimer equation. The calculated results show that β increases with confining pressure, and rougher fracture surface helps to stronger nonlinearity.

(3) As the confining pressure increases, the permeability first experiences a dramatic decrease and then behaviors a much slow-down drop. The critical confining pressure for climb-rush shifts increases with the fracture rouhghness.

5. References

[1] Ren F, Ma G, Fu G, et al. Investigation of the permeability anisotropy of 2D fractured rock masses. *Eng Geol*, 2015, 196, 171-182.

[2] Karpyn Z T, Alajmi A, Radaelli F, et al. X-ray CT and hydraulic evidence for a relationship between fracture conductivity and adjacent matrix porosity. *Eng Geol*, 2009, 103(3):139-145.

[3] Wu Y S, Lai B, Miskimins J L, et al. Analysis of multiphase non-Darcy flow in porous media. *Transport Porous Med*, 2011, 88(2): 205-223
[4] Quinn P M, Parker B L and Cherry J A. Using constant head step tests to determine hydraulic apertures in fractured rock. *J Contam Hydrol*, 2011;126(1)85-99.
[5] Chen Y, Hu S, Wei K, Zhou C and Jing L. Experimental characterization and micromechanical modeling of damage-induced permeability variation in Beishan granite. *Int J Rock Mech Min Sci*, 2014, 71:64-76.
[6] Fernandez G and Moon J. Excavation-induced hydraulic conductivity reduction around a tunnel-Part 1: Guideline for estimate of ground water inflow rate. *Tunn Undergr Space Technol*, 2010, 25(5):560-566.
[7] Yang W, Lin B Q, Qu Y A, et al. Stress evolution with time and space during mining of a coal seam. *Int J Rock Mech Min Sci*, 2011, 48(7): 1145-1152.
[8] Shen C M, Lin B Q, Sun C, et al. Analysis of the stress-permeability coupling property in water jet slotting coal and its impact on methane drainage. *J Pet Sci Eng*, 2015, 126:231-241.
[9] Rong G, Yang J, Cheng L, et al. Laboratory investigation of nonlinear flow characteristics in rough fractures during shear process. *J Hydrol*, 2016, 541, 1385-1394.
[10] Brush D J and Thomson N R. Fluid flow in synthetic rough-walled fractures: Navier-Stokes, Stokes, and local cubic law simulations. *Water Resour Res*, 2003, 39 (4), 1-15.
[11] Zhou J Q, Hu S H, Fang S, et al. Nonlinear flow behavior at low Reynolds numbers through rough-walled fractures subjected to normal compressive loading. *Int J Rock Mech Min Sci*, 2015, 80:202-218.
[12] Zou L, Jing L, Cvetkovic V. Roughness decomposition and nonlinear fluid flow in a single rock fracture. *Int J Rock Mech Min Sci*. 2015;75:102-118.
[13] Zimmerman R W, Bodvarsson G S. Hydraulic conductivity of rock fractures. *Transport Porous Med*, 1996, 23(1):1-30.
[14] Tenchine S and Gouze P. Density contrast effects on tracer dispersion in variable aperture fractures. *Adv Water Resour*, 2005 28(3): 273-289
[15] Zhao YL, Zhang LY, Wang W J, et al. Transient pulse test and morphological analysis of single rock fractures. *Int J Rock Mech Min Sci*, 2017, 91:139-154.
[16] Zhang Z, Nemcik J and Ma S. Micro- and macro-behavior of fluid flow through rock fractures: an experimental study. *Hydrogeo J*, 2013, 21(8):1717-1729.
[17] Schrauf T W, Evans D D. Laboratory studies of gas flow through a single natural fracture. *Water Resour Res*, 1986; 22(7)1038-1050.
[18] Ma D, Miao X X, Chen Z Q, et al. Experimental investigation of seepage properties of fractured rocks under different confining pressures. *Rock Mech Rock Eng*, 2013, 46(5):1135-1144.
[19] Davy C A, Skoczylas F, Barnichon J D, et al. Permeability of macro-cracked argillite under confinement: Gas and water testing. *Phys Chem Earth Parts A/B/C*, 2017, 32:667-680.
[20] Wang H L, Xu W Y and Shao J F. Experimental researches on hydro-mechanical properties of altered rock under confining pressures. *Rock Mech Rock Eng*, 2014, 47:485-493.
[21] Tse R and Cruden D M. Estimating joint roughness coefficients. *Int J Rock Mech Min Sci Geomech Abstr*, 1979, 16:303-307.
[22] Dana E and Skoczylas F. Gas relative permeability and pore structure of sandstones. *Int J Rock Mech Min Sci*, 1999, 36(5):613-625.
[23] Ma D, Bai H, Chen Z, et al. Effect of Particle Mixture on Seepage Properties of Crushed Mudstones. Transport Porous Med, 2015, 108(2):257-277.

Acknowledgments
This work was supported by the National Natural Science Foundation of China [grant numbers No.51374257 and No. 50804060] and the Geotechnical Engineering Stability Control and Health Monitoring of Hunan Province Key Laboratory, Hunan University of Science and Technology [grant numbers No. E21619].

Trends in Agro-Meteorological Parameters as Groundwater Exploitation Indicators

A A Pathak[1], S Nizar[1] and B M Dodamani[2]

1 Research Scholar, National Institute of Technology Karnataka, Surathkal, India
2 Associate Professor, National Institute of Technology Karnataka, Surathkal, India
E-mail: abhipathak2013@gmail.com

Abstract. Rainfall being a major component of the hydrologic cycle, influences the agricultural practices in an area. Thus, trends in rainfall as well as rainy days are of major concern to farmers. Present study focusses on understanding the rainfall trends and its spatial distribution along with the trends in vegetation. An approach where Normalized Difference Vegetation Index (NDVI) procured from MODIS NDVI as an indicator for vegetation was used in this study. Mann Kendall trend test was performed on a 0.25-degree gridded data and the trends were then compared with the distribution of groundwater stress map of the study area. The study tries to examine the coupled use of NDVI and rainfall trends to decrypt the groundwater exploitation in the region. Further Ghataprabha river basin being susceptible to drought by hosting most of the significantly decreasing trend was investigated further. The propagation of severe drought return periods within the basin resembles the agro-meteorological trends. Even within the limitations of the present study, the methodology with further modifications promises to portray strong indication of groundwater exploitation.

1. Introduction

Economy of the Indian subcontinent is mainly controlled by agricultural sector where 65% of the total cultivable land is rain fed. Rainfall being the major source of water for agriculture is confined to the monsoon season from June to September. A small variation in its temporal or spatial distribution may lead to drastic changes in the agricultural productivity. Extensive studies have been carried out in India by different researches to analyse rainfall and rainy-day trend at monthly, seasonal and annual time scale [1]-[5]. Specifically, Goswami and Ramesh (2007) observed decreasing trends in both number of rainy days and rainfall over India [6]. Joshi and Pandey (2011) studied rainfall trend and its spectral analysis over India for a longer period (1901-2000) and observed negative trends in rainfall over the Himalayan and other regions of India [7]. Decreasing trend in annual rainfall and an increasing trend in rainy days over Mahanadi and Krishna river basins were noted by Kumar and Jain (2010) [8].
Recent developments in the remote sensing and its allied applications enables to monitor vegetation dynamics from space with different indices. Among them, Normalized Difference Vegetation Index (NDVI) is widely used to capture variations of vegetation. Temporal and spatial trends in NDVI indicates its changes over an area and helps to explore regional interactions between vegetation and different climatic variables [9]. Many studies were reported on inter relationship between rainfall and NDVI at different regions of the world [10]-[13]. In the semi-arid and arid climatic condition most of the vegetation depends on rainfall with an increasing vegetation trend with increasing rainfall. Hoscilo et al (2014) reported increasing trend in vegetation in the parts of Africa even though the rain fall reduces significantly which indicates increase in the irrigation activity with other source of water in that region [14]. Similar conditions can also be observed in the study conducted by Herrmann et al.,

(2005) [15] and Huber et al., (2011) [16]. Application of NDVI is not only limited to its relationship with precipitation but is also extended to analyse crop yield, soil moisture, temperature [17] and drought assessment [18], [19]. The complex association between groundwater levels and vegetation at regional scales has also been inspected with the aid of NDVI [20] Effect of groundwater level fluctuation on different vegetation were carried out by Froend and Sommer, 2010 [21]; Chen et al., 2014 [22].

The present study aims to understand the spatiotemporal variation of rainfall, rainy days and vegetation trends over the study area by employing the non-parametric Mann Kendall trend test. The study further attempts to identify severe groundwater exploitation regions with the aid of rainfall and NDVI trends. The area experiencing significant negative trends in rainfall and rainy days was further explored for its susceptibility towards drought. The study is conducted in the south Indian state of Karnataka with an average annual rainfall that varies from 3456 mm at Coastal Karnataka to 731 mm over North Interior Karnataka. In terms of area prone to drought, Karnataka ranks second in India after Rajasthan. Recently, the Govt. of Karnataka has declared 157of the 177 taluks as affected by the drought (Karnataka state natural disaster monitoring centre 2012).

2. Study area

Karnataka which is one of the prominent state in India is situated in the Deccan plateau bounded between 11^0 30' N and 18^0 30' N latitudes and 74^0 E and 78^0 30' E longitudes. Figure 1 shows the location of Karnataka with the green shaded region masking the Western Ghats mountain ranges running parallel to the west coast. Accounting for about 5.05% of the total population of the country, Karnataka is the 9th most populated state in India (Census 2011). Covering a total land area of 1,91,791 km^2, Karnataka alone concentrates five of the six major climate types of the entire Indian Union [23]. The coastal strip, the Western Ghats and the Deccan plateau are the three principal physical zones in the state. The southwest monsoon from June to September (JJAS) accounts for almost 80% of the annual rainfall. The spatial distribution of rainfall reveals the distinct interaction of Western Ghats mountain ranges with the south west monsoon. The 1500 km long Western Ghats also known as the Great Indian Escarpment runs parallel to the western coast through Kerala, Karnataka, Tamil Nadu and Maharashtra. These Western Ghats mountain belts provide a unique orographic framework inducing excess turbulence which steers the atmospheric circulation triggering precipitation [23]

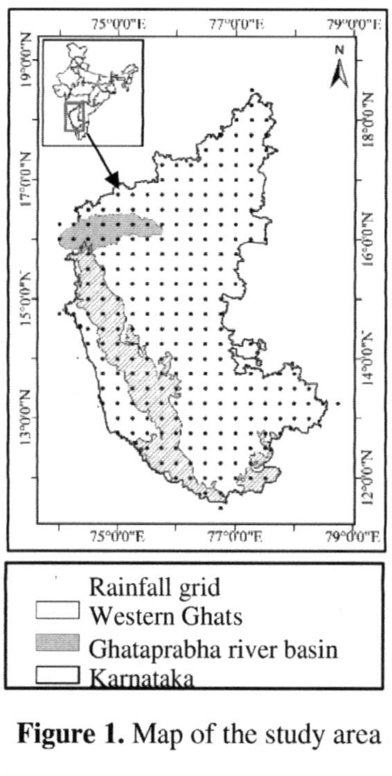

Figure 1. Map of the study area

3. Data and methodology

To assess the possible relationship between rainfall, rainy day and NDVI trends at a finer resolution, the Indian Meteorological Department (IMD) 0.25^0 Gridded Rainfall Dataset is adopted. The dataset comprises of over 6955 rain gauge stations, covering a longer period of 110 years over the Indian main land. The detailed description of the dataset has been discussed by D. S. Pai et al. (2014) [24]. They have also compared the Gridded dataset with other existing datasets showing comparable results. In addition, their results show that the low rainfall in the lee ward side of the Western Ghats and heavy rainfall areas in the orographic region of the west coast are realistically represented in the dataset. MODIS NDVI data products with a resolution of 250 m from 2000 to 2013 was downloaded and was considered for the present study. These data were further processed with ArcGIS-9.3 software.

3.1. Mann Kendall trend test

The Mann-Kendall (MK) trend test [25], [26] is used in this study. This test is commonly employed to detect monotonic trends in environmental, climate or hydrological data series. Being a non-prismatic approach, it requires no assumption about the distribution of the data. The null hypothesis for the test that there is no trend in the series is checked against the alternate hypothesis that the data follows a monotonic trend. The computations assume that the observations are independent. MK test was performed for rainfall, rainy day and for NDVI for the period of post monsoon, monsoon and post monsoon seasons. The magnitudes of these trends were then estimated by Sen's-Slop method. Before comparing rainfall trends and NDVI trends graphically, the MODIS NDVI was resampled to the resolution of rainfall grid.

3.2. Reconnaissance Drought Index (RDI)

Reconnaissance Drought Index (RDI) was established [27] and was then categorized as meteorological index. RDI describes climatic water deficiency by considering the ratio of aggregated Precipitation (P) to the total Potential Evapotranspiration (PET) for a certain time period. RDI considers both rainfall and temperature in the form of PET to frame drought characteristics and thus overcomes one of the major drawback of the most popularly used Standardize Drought Index (SPI). Calculation of RDI involves, fitting appropriate distribution (usually 2 parameter Gamma distribution) to the P / PET ratio and CDF of the distribution is converted in to standardized normal distribution with mean 0 and standard deviation 1 to arrive RDI value. Negative values of RDI indicates dry condition whereas positive values indicate wet period. RDI classifies extreme severe, moderate and mild condition when its value will be less than 2, between -1.5 to -2, between -1 to -1.499 and between 0 to -1 respectively [28]. Temperature based Penman-Monteith reference evapotranspiration [29] was calculated with the aid of NASA's NEX-GDDP temperature data [30].

4. Results and discussions

Mann Kendall trend test have been performed at grid level of 0.25^0 to identify trend in rainfall time series. Sen's slope analysis was then performed at each grid to determine the magnitude of change. Since the rainfall varies seasonally, pre-monsoon, monsoon and post-monsoonal trends were found separately. Figure 2 and 3 shows the results of rainfall and rainy-day trends over the study area. During the pre-monsoon season North Interior Karnataka and Coastal Karnataka exhibits a decreasing trend whereas South Interior Karnataka shows an increasing trend. Sen's slope estimate shows that out of 252 grid points 131 points have a decreasing magnitude whereas 121 have increasing magnitude. Out of the 131 decreasing points 17 are significantly (95% significant) decreasing and are concentrated around Ghataprabha river basin.

Decreasing trend No trend Increasing trend + 95% Significant

Figure 2. Mann Kendall trend and Sen's slope estimate of Rainfall from 1960-2013. (a) Pre-Monsoon (b) Monsoon (c) Post-Monsoon.

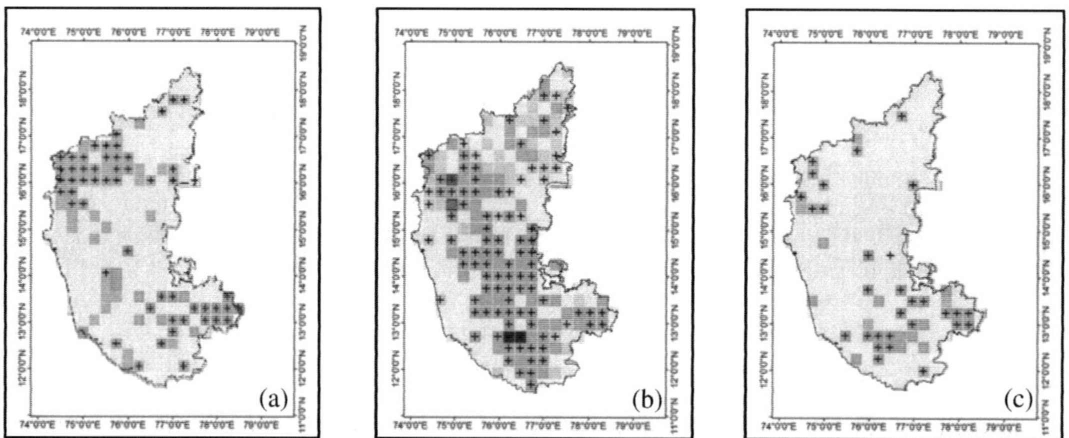

Heavily decreasing Decreasing trend No trend Increasing trend
Heavily Increasing + 95% Significant

Figure 3. Mann Kendall trend and Sen's slope estimate of Rainy days from 1960-2013. (a) Pre-Monsoon (b) Monsoon (c) Post-Monsoon.

The Ghataprabha river basin is one of the important southern sub basins of Krishna River in its upper reach. It's a semi-arid region in which rainfall is confined to monsoon season from June to October where the annual average rainfall varies from 5000 mm (at Western Ghats) to 600 mm and average annual rainy day varies from 120 days to 44 days. In the analysis of rainy day trends during the monsoon season four grids stood out with almost 2.5 times more increasing/decreasing trend. These locations are marked as heavily decreasing and heavily increasing trend in figure 3b. Not only the amount of precipitation but also its frequency is decreasing in the basin as evident from the rainy-day trends during the pre-monsoon season (Figure 3a). Being an agriculture dominated basin this could forecast serious consequences in its yield and may even be an indicator of draught. But as a contradiction to the above decreasing trends, the NDVI shows increasing trends (Figure 4) suggesting an alternate source of water for the crops in the basin. According to Shah (2009) [31], India experiences a steep increase in agricultural groundwater use since 1960. Thus the increase in NDVI despite a decrease in precipitation could be accounted to the use of groundwater. This is also evident from the groundwater exploitation map (Figure 5) published by the Central Groundwater Board (2011).

The map portrays most of the Ghataprabha basin to be either over exploited or in a semi-critical state. Thus in an agrarian country like India a decrease in rainfall trend coupled with an increase in NDVI could indicate an over exploitation of groundwater to compensate the agricultural water demand. A similar situation occurs during the monsoon season over the north interior Karnataka indicating ground water stress.

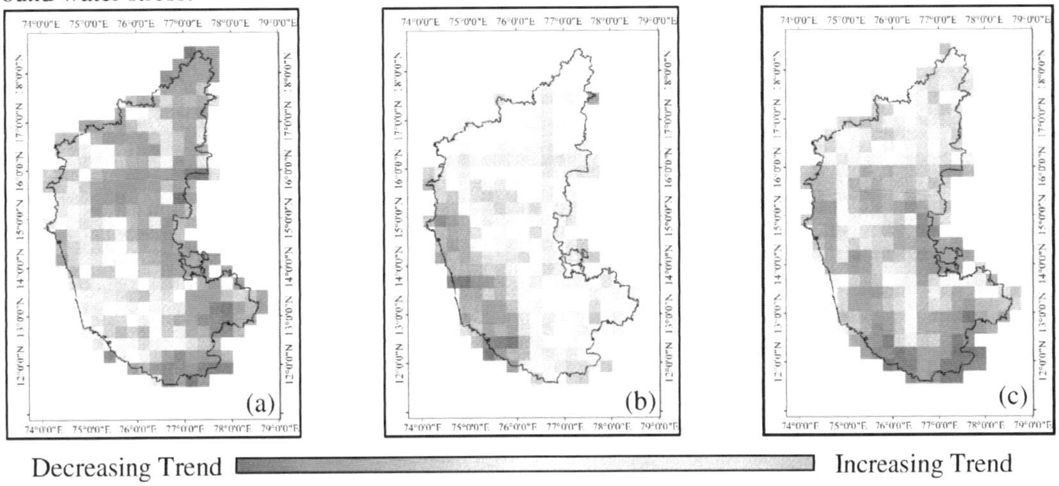

Figure 4. Mann Kendall trend and Sen's slope estimate of NDVI from 2000-2013. (a) Pre-Monsoon (b) Monsoon (c) Post-Monsoon.

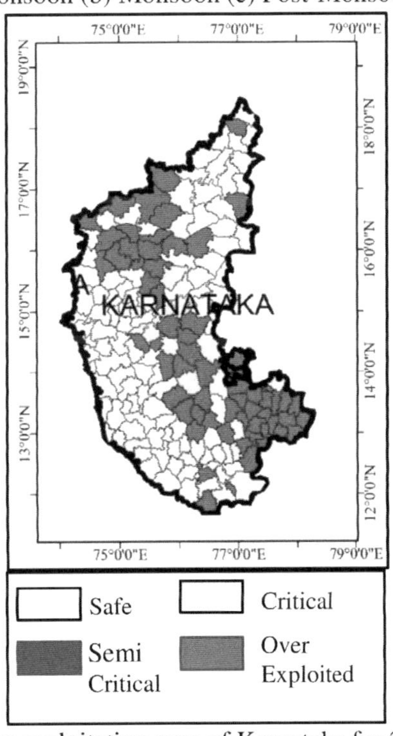

Figure 5. Groundwater exploitation map of Karnataka for 2011(Source: CGWB).

During the monsoon season there is a line of significantly increasing trend parallel to the Western Ghats mountain ranges indicating a significant increase in orographic precipitation along the mountains. These Western Ghats mountain belts provide a unique orographic framework inducing excess turbulence which steers the atmospheric circulation triggering precipitation [23]. North interior Karnataka shows decreasing Sen's slope magnitudes with few significantly decreasing trends. Significant decreasing trends are visible over coastal as well as north interior Karnataka. Even during the monsoon season, certain areas around Ghataprabha river basin persists with significantly decreasing trends though the NDVI is increasing (Figure 4). This indicates the continued dependence on groundwater even during the monsoon season. During the post-monsoon season there are few significant trends, most of which are increasing. Coastal Karnataka shows an increasing trend with significant trends concentrated along the Western Ghats. North and South interior Karnataka is devoid of significant trends. Thus, being a hotspot of decreasing trends, the Ghataprabha basin is further investigated for its vulnerability to drought.

Assessment of drought and its return periods has been conducted using RDI over a Ghataprabha river basin. Impact of drought on different sectors water resources can be ascertain with the different time scales of drought. 6 month drought, being an intermediate time scale, which could reflect its impact on Irrigation as well as on other water sources sectors. To quantify drought characteristics, RDI of 6 month time scale has been calculated for all the grid points of the basin and average severity of the basin has been depicted in the Figure 6.

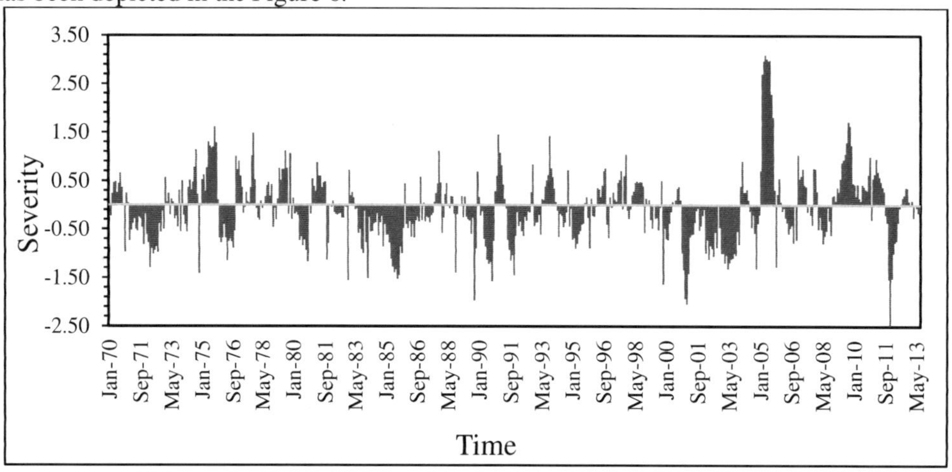

Figure 6. RDI series for Ghataprabha river basin at 6 month time Scale.

The basin experience severe meteorological drought with high intensity of -2.45 during 2011 and this was the main cause for sever groundwater exploitation during 2011 in the region. Along with that major drought episodes were observed during years 1970-1972, 1982-1985 and in 2000-2003. Among them 2000-2003 drought was incensed with the duration of 37 months and the cumulative severity of -31.28.

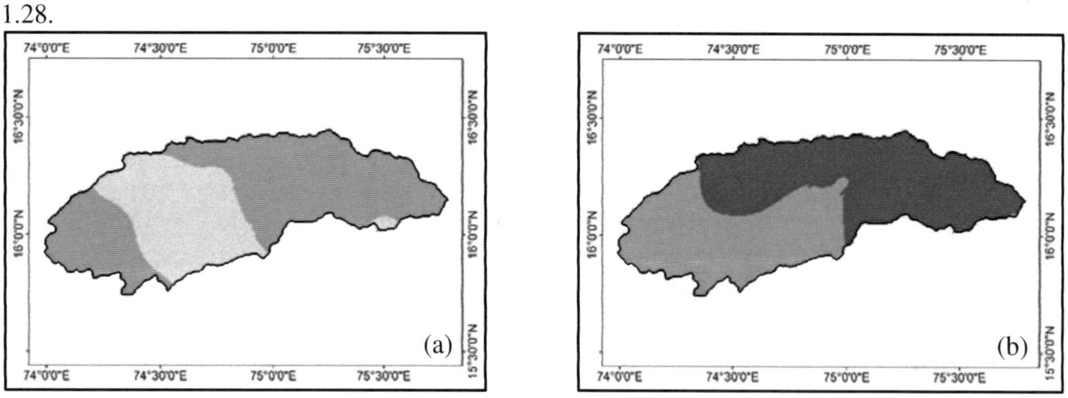

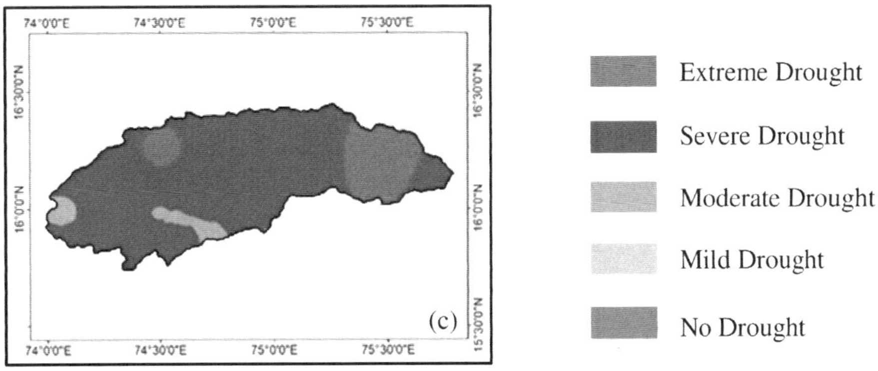

Figure 7. Severity of RDI return period at (a) 2 year (b) 5 year (c) 10 year.

The severity of the drought time scale (Figure 6) revels that the frequency of severe drought were increasing with time and is an alarm for upcoming drought events. Spatial variation of drought severity with the return period of 2, 5 and 10 years were presented in the Figure 7 where the propagation of drought severity from eastern part of the basin to west can be observed clearly. Major portion of the basin is prone moderate drought, at least once in a 2 year whereas the eastern part of the basin is prone to severe drought with recurrence period of 5 years. The entire basin undergoes severe drought at least once in a 10 years and trace of the extreme drought can also be observed in certain pockets of the basin.

5. Conclusion

In this study rainfall and rainy-day trends were analysed along with vegetation trends for the pre monsoon, monsoon and post monsoon seasons in Karnataka. Trends of rainfall and rainy days during monsoon periods were increasing significantly in southern parts of Karnataka highlighting the importance of Western Ghats towards increase in orographic precipitation. Whereas significant decreasing trends were observed in certain regions of costal and north interior Karnataka. During pre-monsoon and post monsoon, both rainfall and rainy days were decreasing in the north western part of the area, which covers most of the area in Ghataprabha river basin. Even though with decreasing rainfall trends, the NDVI trends were increasing contradictorily during pre and post monsoon, which pertains severe groundwater exploitation in those seasons. Significant decreasing trends of rainfall and rainy day over Ghataprabha basin, which forces the region to suffer more recurrent meteorological drought episodes leading to severe groundwater exploitations. Return period of severe drought once in a decade emphasises the need of profound drought preparedness strategies and conjunctive use of water resources for the agricultural practise without effecting groundwater levels during deficit rainfall conditions.

6. References

[1] Sarker R P and Thapliyal V 1988 Climate change and variability. *Mausam.* 39 127-138.
[2] Lal M 2001 Climate change – implications for India's water resources. *J Ind Water Resour Soc.* 21(3) 101–109.
[3] Dash S K, Jenamani R K & Shekhar M S 2004 On the decreasing frequency of monsoon depressions over the Indian region. *Current Sci.* 86 1404–1411.
[4] Jain S K, & Kumar V 2012 Trend analysis of rainfall and temperature data for India. *Current Science.* 102(1) 37–49.
[5] Raju B C K, & Nandagiri L 2017 Analysis of historical trends in hydrometeorological variables in the upper Cauvery Basin , Karnataka , India. *Current science.* 112(3) 577-587.
[6] Goswami P, and K V Ramesh 2006 A comparison of interpolated NCEP (I-NCEP) rainfall with high-resolution satellite observations, *Geophys. Res. Lett.* 33 L19821.

[7] Joshi M K, Pandey A C 2011 Trend and spectral analysis of precipitation over India during 1901–2000.*Journal of Geophysical Research.*116.
[8] Kumar V, Jain S K and Singh Y 2010 Analysis of long-term rain- fall trends in India. *Hydrol. Sci*. J. 55 484–496
[9] Mabuchi K, Sato Y and Kida H 2005 Climatic impact of vegetation change in the Asian tropical region. Part I: Case of the Northern hemisphere summer, *J. Climate*. 18 410–428.
[10] Davenport M L and Nicholson S E 1993 On the relation between rainfall and the Normalized Difference Vegetation Index for diverse vegetation types in East Africa, *Int. J. Remote Sens.* 14 2369–2376.
[11] Santos P and Negri A J 1997. A comparison of the normalized difference vegetation index and precipitation for the Amazon and Northeastern Brazil. *Journal of Applied Meteorology*. 36(7) 958-965.
[12] Mao J F, Shi X Y, Thornton P E, Piao S and Wang X 2012 Causes of spring vegetation growth trends in the northern mid-high latitudes from 1982 to 2004. *Environmental Research Letter*. 7 014010.
[13] West T O, Page Y L, Huang M, Wolf J and Thomson A M 2014 Downscaling global land cover projections from an integrated assessment model for use in regional analyses: results and evaluation for the US from 2005 to 2095, *Environmental Research Letter*. 9 064004.
[14] Hoscilo A, Balzter H, Bartholomé E, Boschetti M, Brivio P A, Brink A Clerici M and Pekel J F 2015 A conceptual model for assessing rainfall and vegetation trends in sub-Saharan Africa from satellite data, *Int. J. Climatol*. 35 3582–3592.
[15] Herrmann S M, Anyamba A, Tucker C J 2005 Recent trends in vegetation dynamics in the African Sahel and their relationship to climate. *Glob. Environ. Change*. 15 394–404.
[16] Huber, S, Fensholt, S, and Rasmussen, K, 2011 Water availability as the driver of vegetation dynamics in the African Sahel from 1982 to 2007 *Global and Planetary Change*. 76 186–195.
[17] Lakshmi Kumar T V, Koteswara Rao K, Humberto Barbosa and Emily Prabha Jothi 2013 Studies on spatial pattern of NDVI over India and its relationship with rainfall, air temperature, soil moisture adequacy and ENSO, *GEOFIZIKA* 30
[18] Kogan, F N 1995 Application of vegetation index and brightness temperature for drought detection. *Adv. Space Res.* 15 91–100.
[19] Karnieli A, Agam N, Pinker R T, Anderson M., Imhoff M L, Gutman G, Panov N, and Goldberg A 2010 Use of NDVI and land surface temperature for drought assessment: Merits and limitations. *Journal of Climate*. 23(3) 618–633.
[20] Doble R, Simmons C, Jolly I and Walker G 2006 Spatial relationships between vegetation cover and irrigation-induced groundwater discharge on a semiarid floodplain, Australia. *Journal of Hydrology*. 329 (1) 75-97.
[21] Froend R and Sommer B 2010 Phreatophytic vegetation response to climatic and abstraction-induced groundwater drawdown: Examples of long-term spatial and temporal variability in community response. *Ecological Engineering*.
[22] Chen Y, Li W, Xu C, Ye Z, Chen Y 2014 Desert riparian vegetation and groundwater in the lower reaches of the Tarim River basin, *Environ Earth Sci*.
[23] Gunnell Y 1997 Relief and Climate in South Asia : the Influence of the Western Ghats on the Current Climate Pattern of Peninsular India, *International Journal of Climatology*. 17 1169–1182.
[24] Pai D S, Sridhar L, Rajeevan M, Sreejith O P, Satbhai N S, & Mukhopadyay B 2014 Development of a new high spatial resolution (0 . 25 ° × 0 . 25 °) Long Period (1901-2010) daily gridded rainfall data set over India and its comparison with existing data sets over the region. *Mausam*. 65(1) 1–18.
[25] Kendall M G 1955 Rank Correlation Methods. *Griffin,London*.
[26] Mann H B 1945 Nonparametric tests against trend, *Econometrica*. 13 245-259
[27] Tsakiris G, Vangelis H 2005 Establishing a drought index incorporating evapotranspiration. *Eur Water*. 9 10 1–9

[28] Zarch M, Malekinezhad H, Mobin M H, Dastorani M T, and Kousari M R 2011 Drought Monitoring by Reconnaissance Drought Index (RDI) in Iran. *Water Resources Management.* 25(13) 3485–3504

[29] Allen R G, Pereira L S, Raes, D, and Smith M. 1998 Crop Evapotranspiration, (Guidelines for Computing Crop Water Requirements, *Irrigation and Drainage Paper 56, FAO*, Rome).

[30] Thrasher B, Xiong J, Wang W, Melton F, Michaelis A, Nemani R, 2013 Downscaled climate projections suitable for resource management. *Eos Trans. Am. Geophys. Union.* 94 321–323.

[31] Shah T 2009. Climate change and groundwater: India's opportunities for mitigation and adaption. *Environ. Res. Lett.* 4 035005.

Evaluation of water quality in mangrove ecosystem for the Beilun Gulf and Zhenzhu Bay: A Complex Approach

Minghui Ou[1,2], Shan Liang*[1,2], Qingyu Xiong[1,3] and Ru Zhang[1,2]

1 Key Laboratory of Dependable Service Computing in Cyber Physical Society, Ministry of Education, China
2 College of Automation, Chongqing University, Chongqing, China
3 College of Software Engineering, Chongqing University, Chongqing China
E-mail: omhcumt_1992@163.com, lightsun@cqu.edu.cn, cquxqy@163.com

Abstract. Water quality evaluation is vital for water monitoring and management in mangrove ecosystem of Beilun gulf and Zhenzhu bay in Guangxi of china. Over the past few years, with the expansion of urbanization and economic development, the coastal water quality in mangrove ecosystem is becoming worse. This literature applied the Fuzzy C-Means (FCM) and principal component analysis (PCA) techniques to evaluate the water quality in the study area and to evaluate the overall rating of stations with respect to the 14 water quality parameters. The PCA analysis is based on the precise clustering of FCM analysis. More reasonable evaluation of water quality by PCA was obtained. Our numerical analysis has verified the result of the current paper is correct. In addition, those results can provide some selectable methods to improve station number for interrelated management department with the limit funds and resources.

1. Introduction

Mangroves have global distribution within coastal tropical and subtropical regions, which are particularly sensitive to changes in the surrounding sea environment and provide many ecosystem services. It is well known that the mangrove ecosystem possesses many ecological functions [1], [2]. What's more, mangroves and other coastal vegetation play a vital role in protecting the safety of human inhabitation and preventing wealth from destroying by tsunami and typhoon. The knowledge that mangroves may protect coastal communities from coastal hazards (coastal erosion, typhoon, tidal bores and salt spray, tsunami, etc.) is well known in tropical and subtropical coastal ecology and increasingly by coastal managers [3]. Accordingly, much attention have been paid to discuss the ecological effects of mangrove ecosystem from researchers in recent years, such as attenuate wave energy [4], [5], provide optimal breeding, feeding and nursery grounds for many ecologically and the valuable sources of fuel wood, fodder, timber, tannin and other natural products for local people [6], [7].

However, with the socioeconomic development and expansion of urbanization, mangrove ecosystem is facing a serious threat. Mangrove forest is extremely sensitive to the health of the growing environment, such as the soil, water and microbial environment. Indeed, the water quality has a great effect on the growth of mangrove forests and is prone to changed by the external environment. Many literatures have reported on the water quality parameter and monitoring station by some statistical methods. For example, the principal component analysis (PCA) and principal factor analysis (PFA) techniques have been used to evaluate the effectiveness of the surface water quality-monitoring network in a river and to analysis the water quality for the Nakdong river watershed [8], [9]. PCA seeks to establish combinations of variables capable of describing the principal tendencies observed while studying a given matrix [10].

According to the long coastline for Beilun Gulf and Zhenzhu Bay in Guangxi of China, the monitoring devices have been distributed along with the long coastline. It is not hard to know that the effect factors of water quality between monitors which have long distance are different. To find a more precise method reflects the water quality is vital importance obviously.

In this paper, the monitoring station's location information is known. And we can't partition those stations into small clustering by eyes or experience exactly. The location information (Longitude and Latitude) were clustered into smaller groups using Fuzzy C-Means (FCM) algorithm. And then the monitoring data of stations have strong correlation in each group. PCA is a multivariable statistical technique which is used to identify important components that explain most of the variances of a complex system. This technique was used to process the data. Analysis the water quality of each clustering group, and the results will helpful to management the mangrove environment.

The contributions of this work are four-fold: 1) Get more precise clustering of location information by FCM algorithm. 2) Clustering process to enhance the correlation of monitoring data. 3) More reasonable evaluation of water quality for each partition group by PCA. 4) Provide some selectable method to reduce station number for interrelated management department with the limit funds and resources.

2. Method

In this section two methods of data process will briefly be introduced. The water quality of mangrove forest has many influencing factors (human activities, environment pollution and other uncertainties) along with the long coastline. Finding strong correlation monitoring data is a vital issue. This is very important, especially, to get the mainly principal component of water parameter in mangrove forest by PCA. Indeed, the FCM is a partitioning-based clustering algorithm and the clustering group will be high-efficiency for PCA analysis. It has the property to group a set of objects into classes or clusters so that the objects within a cluster have high similarity. And the FCM has clear advantage over crisp and probabilistic clustering methods [11].

2.1. Clustering

FCM is a fuzzy clustering algorithm which assigns memberships to each data corresponding to all clusters based on the distances of the data and the cluster centres [12]. Due to the characteristics of location information (Longitude and Latitude), it is not accurate to group location information into exclusive clusters. Thus, FCM is very suitable to solve the location information clustering problem as it use a fuzzy membership which assigns a number of membership for every cluster. The accurate description of FCM theory can be found in [12], [13], [14]. The basic FCM algorithm aims to minimize the objective function (1) iteratively.

$$J_o(\mathrm{U},\mathrm{G}) = \sum_{i=1}^{N}\sum_{j=1}^{C} u_{ij}^{m} \left\| x_i - g_j \right\| \tag{1}$$

where $X = (x_1, x_2, \cdots, x_N)$ denotes the data samples set and $G = (g_1, g_2, \cdots, g_C)$ denotes the clustering groups, and $U = (u_{ij})_{N\times C}$ is a fuzzy partition matrix composed of the membership of each feature vector x_i in each cluster j and u_{ij} should satisfy $\sum_{j=1}^{C} u_{ij} = 1$.The $m = 1, 2, \cdots, \infty$ is an index that controls the fuzziness of the resulting partition. $\|*\|$ denotes the distance between data x_i and the cluster centre g_j. If C is an integer, $2 \le C < N$, a conventional C-partition of X is a C-tuple $(X_1, X_2, \cdots, X_C)$ of subsets of X that satisfies three conditions:

$$X_i \neq \Phi \quad 1 \le i \le C; \tag{2}$$

$$X_i \cap X_j = \Phi \quad i \neq j; \tag{3}$$

IOP Conf. Series: Earth and Environmental Science **167** (2018) 012005 doi:10.1088/1755-1315/167/1/012005

$$\bigcup_{j=1}^{C} X_j = X \tag{4}$$

In these equations, Φ stands for the empty set, and $(\bigcap, \bigcup)$ are intersection and union, respectively.
The objective function J_o is minimized when the sampled data set are assigned to closer to the cluster centres, and vice versa. The memberships and the cluster centres are computed by:

$$u_{ij} = \frac{1}{\sum_{k=1}^{C} \left(\frac{\|x_i - g_j\|}{\|x_i - g_k\|} \right)^{\frac{2}{m-1}}} \tag{5}$$

$$g_j = \frac{\sum_{i=1}^{N} u_{ij}^m x_i}{\sum_{i=1}^{N} u_{ij}^m} \tag{6}$$

If m is set to 1, the memberships u_{ij} converge to 0and 1, which transforms to hard clustering method.

2.2. Principle component analysis
PCA was first performed combined with the FCM to identify the potential for reducing the number of monitoring stations by analysing the sampled data in the monitoring area.
Mathematically, PCA is a multivariable statistical techniques used to identify components that explain most of the variances of a complex system by linear transportation. The procedures of PCA are shown as follows.

(i) Standardization of the measurements to ensure that they all have equal weights in the analysis,
(ii) Calculate the covariance matrix C,
(iii) Find the eigenvalues $\lambda_1, \lambda_2, \cdots, \lambda_n$ and the corresponding eigenvectors $\alpha_1, \alpha_2, \cdots, \alpha_n$ and extract the principal component,
(iv) Find the principal component comprehensive evaluation function.

The sampled data have strong correlation in every cluster. We should select high standard for the factor correlation coefficient. And the detailed procedure of the experiment is illustrated in Figure 1.

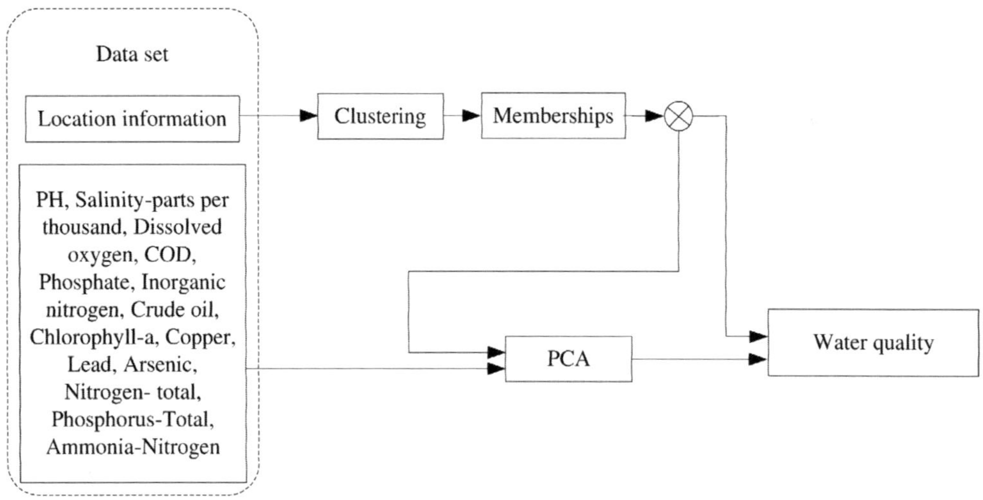

Figure 1. Block diagram of the experiment procedure.

3. Result and discussion

3.1. Location information clustering

In practice, the monitoring devices were installed along with the coastline and the estuaries to monitor water pollution. We got the location information of each monitoring station as shown in Table 1.

Table 1. Detail information (longitude and latitude) of monitoring station.

Stations	Longitude	Latitude	Stations	Longitude	Latitude
1	108.02194	21.54889	12	108.23889	21.58917
2	108.03500	21.54833	13	108.16139	21.52417
3	108.04333	21.54500	14	108.16111	21.52528
4	108.05528	21.53278	15	108.16056	21.52583
5	108.05972	21.52889	16	108.18361	21.61472
6	108.06639	21.52500	17	108.20917	21.60194
7	108.08611	21.53000	18	108.20454	21.59634
8	108.08833	21.53778	19	108.20500	21.58570
9	108.09056	21.54528	20	108.23000	21.61389
10	108.24806	21.59944	21	108.22944	21.61194
11	108.24583	21.59389	22	108.23000	21.60667

Those stations were clustered into 7 groups by using FCM algorithm. And the centre information also was obtained as shown in Table 2.

Table 2. The clustering result and centre coordinate

Groups	I		II		III		IV		V		VI		VII	
Stations	1,2,3		4,5,6		7,8,9		13,14		15,16,17, 18,19		10,11,12		20,21,22	
	108.03 Lon	21.55 Lat	108.06 Lon	21.53 Lat	108.09 Lon	21.54 Lat	108.16 Lon	21.53 Lat	108.20 Lon	21.60 Lat	108.24 Lon	21.59 Lat	108.23 Lon	21.61 Lat

Where Lon and Lat are represent longitude and latitude, respectively. In order to get better observation for the clustering result. We have painted the seven groups on the map of study area by circle as shown in Figure 2.

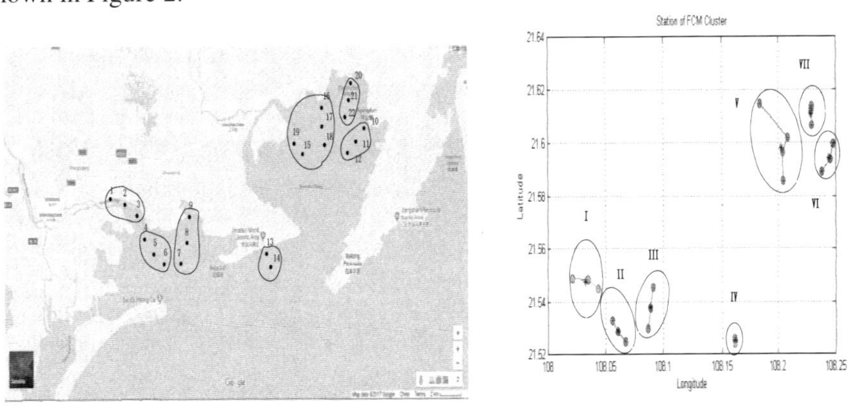

Figure 2. Map information of study area and clustering result.

Figure 2 schematically illustrates the study area which is covered by mangrove forest and each cluster group was partitioned. The PCA analysis will be explored based on the partition.

3.2. PCA analysis for each cluster

Software IBM SPSS Statistic 19 and MATLAB 2014a are used to perform the analysis. However, the dissolved oxygen (DO) is a positive index. We need use the equation (7) to pre-processing the data of each DO.

$$\overline{x}_{DO} = \max x_{DO} - x_i \tag{7}$$

One of our goals is to reduce the number of monitoring station of each cluster. The value Kaiser-Meyer-Olkin(KMO) is 0.743, which indicate to analysis the correlation and other properties are reasonable. The variance distribution in PCA for each cluster group is shown in Table 3. And the cumulative variance value is greater than 0.85(or 85%) was considered significant.

Table 3. The variance of the components.

Components	Total	Initial eigenvalue variance %	Cumulative value %
1	7.2525	51.802	51.802
2	2.828	20.203	72.005
3	1.492	10.656	82.661
4	0.756	5.403	88.064

From Table 3, we selected four components and their variance together accounted for about 88.064% of the total variance. Therefore, our discussions should focus only on the first four components.
The loading matrix of four components is shown in Table 4. Compared the loading absolute value of four components that we can know the fourteen water quality parameters information is reflected by the four components properly.

Table 4. The loading matrix of components about parameters

Water quality parameters	1	2	3	4
PH	-0.289	-0.349	0.013	-0.053
Salinity	-0.359	-0.089	-0.003	0.154
DO	-0.118	0.492	0.005	0.112
COD	0.188	-0.403	0.172	-0.455
Phosphate	0.345	0.167	0.043	-0.009
Inorganic-nitrogen	0.342	0.159	0.039	0.187
Crude oil	0.231	-0.318	0.222	-0.223
Chlorophyll-a	0.127	-0.434	-0.173	0.535
Cu	-0.177	0.034	0.633	0.018
Pb	-0.185	0.164	0.564	0.157
As	-0.309	-0.222	-0.054	0.311
Nitrogen-total	0.361	0.103	0.034	-0.002
Phosphorus-total	0.202	-0.180	0.401	0.253
Ammonia-Nitrogen	0.316	-0.048	0.065	0.444

From the loading matrix in the analysis, the four components score, P_1, P_2, P_3, P_4, can be given as follows, respectively.

$$P_1 = -0.289x_1 - 0.359x_2 - 0.118x_3 + 0.188x_4 + 0.354x_5 + 0.342x_6 + 0.231x_7 + 0.127x_8 - 0.177x_9 - 0.185x_{10} - 0.309x_{11} + 0.361x_{12} + 0.202x_{13} + 0.316x_{14} \quad (8)$$

$$P_2 = -0.349x_1 - 0.089x_2 + 0.492x_3 - 0.403x_4 + 0.167x_5 + 0.159x_6 - 0.318x_7 - 0.434x_8 + 0.034x_9 + 0.164x_{10} - 0.222x_{11} + 0.103x_{12} - 0.180x_{13} - 0.048x_{14} \quad (9)$$

$$P_3 = 0.013x_1 - 0.003x_2 + 0.005x_3 + 0.172x_4 + 0.043x_5 + 0.039x_6 + 0.222x_7 - 0.173x_8 + 0.633x_9 + 0.564x_{10} - 0.054x_{11} + 0.034x_{12} + 0.401x_{13} + 0.065x_{14} \quad (10)$$

$$P_4 = -0.053x_1 + 0.154x_2 + 0.112x_3 - 0.455x_4 - 0.009x_5 + 0.187x_6 - 0.223x_7 + 0.535x_8 + 0.018x_9 + 0.157x_{10} + 0.311x_{11} - 0.002x_{12} + 0.253x_{13} + 0.444x_{14} \quad (11)$$

And then, in order to understand the comprehensive level of the water quality of the monitoring stations in the study area. We use the four components corresponding variance value as the coefficient to construct a comprehensive evaluation function(12), where $P_i,(i=1,2,3,4)$ represents the four components score.

$$S = 0.51802P_1 + 0.20203P_2 + 0.10656P_3 + 0.05403P_4 \quad (12)$$

According to the comprehensive evaluation function, the water quality of 22stations in study area in 2014 was ranked by the comprehensive score. And the relevant results for each partition also were shown in Table 5.

Table 5. The rank of water quality and the relevant partitions

Partitions	Monitoring stations	Overall rating
IV	14	-1.317
V	15	-1.211
VI	12	-1.179
VI	11	-0.863
II	6	-0.819
III	7	-0.727
IV	13	-0.725
V	18	-0.700
VII	22	-0.684
V	17	-0.653
III	8	-0.580
VI	10	-0.485
VII	21	-0.332
VII	20	-0.326
II	5	-0.251
V	19	0.014
V	16	0.153
II	4	0.304
III	9	0.383
I	3	2.567
I	1	3.542
I	2	3.887

Based on the score of the main components and the comprehensive score, the water quality of mangrove forest can be quantitatively described. Since the water quality index is a low-quality index, the higher the score, the more serious water pollution. On the other hand, we can know each monitoring station belong to a specific partition. And the water quality of each station for every cluster can be obtained easily. For example, the second station is the worst and the third station is the best in the first cluster. As the same way, we can get which is best or worst for other cluster.

According to the overall rating in Table 5, some guidance can be obtained from these results to improve the coastal water quality of the mangrove forest when the funds and resources are limited. For example, it is a prior task for us to monitor the water pollution of the station that have bad ranking in each cluster if the management with a limited budget.

4. Conclusion

This paper has demonstrate the application of the FCM and PCA techniques to evaluate the water quality monitoring stations located in Beilun Gluf and Zhenzhu Bay of Guangxi and to evaluate the overall rating of stations with respect to the 14 water quality parameters is reasonable. And analysing the strategy of mangrove forest water resource management based on clustering was considered significant. Using the FCM and PCA techniques to analysis water pollution in our study area have profound significance. Not only can save the government's economic costs, but also can save the manpower resources. For another, the methods will be useful for other similarity research.

5. References

[1] Nagelkerken I, et al., The habitat function of mangroves for terrestrial and marine fauna: A review. Aquatic Botany, 2008. 89(2): p. 155-185.

[2] Zhang K, et al., The role of mangroves in attenuating storm surges. Estuarine, Coastal and Shelf Science, 2012. 102-103: p. 11-23.

[3] Wells S A R C, In the front line: shoreline protection and other ecosystem services from mangroves and coral reefs. 2006.

[4] Massel S R, K Furukawa and R M Brinkman, Surface wave propagation in mangrove forests. Fluid Dynamics Research, 1999. 24(4): p. 219-249.

[5] Blankespoor B, Dasgupta S and Lange G M, Mangroves as a protection from storm surges in a changing climate. Ambio, 2017. 46(4): p. 478-491.

[6] Rasolofo M V, Use of mangroves by traditional fishermen in Madagascar. Mangroves and Salt Marshes, 1997. 4(1): p. 243-253.

[7] Spaninks F A V B, Economic valuation of mangrove ecosystems: potential and limitations. 1997.

[8] OUYANG Y, Evaluation of river water quality monitoring stations by principal component analysis. Water Research, 2005. 39(12): p. 2621-2635.

[9] Jung K Y, et al., Evaluation of water quality for the Nakdong River watershed using multivariate analysis. Environmental Technology & Innovation, 2016. 5: p. 67-82.

[10] Bengraı¨ne K and T F Marhaba, Using principal component analysis to monitor spatial and temporal changes in water quality. Journal of Hazardous Materials, 2003. 100(1-3): p. 179-195.

[11] Sun H, S Wang and Q Jiang, FCM-Based Model Selection Algorithms for Determining the Number of Clusters. Pattern Recognition, 2004. 37(10): p. 2027-2037.

[12] Bezdek J C, R Ehrlich and W Full, FCM: The fuzzy c-means clustering algorithm. Computers & Geosciences, 1984. 10(2-3): p. 191-203.

[13] Liu L, et al., A modified Fuzzy C-Means (FCM) Clustering algorithm and its application on carbonate fluid identification. Journal of Applied Geophysics, 2016. 129: p. 28-35.

[14] Jamshidi A, et al., Dynamic risk assessment of complex systems using FCM. International Journal of Production Research, 2017: p. 1-19.

Acknowledgements

The authors would like to thank experts for their helpful comments and suggestions, and this work is supported by the Science and Technology Major Special Project of Guangxi (GKAA17129002).

Assessment of total mercury content in water of the Balkyldak Lake-reservoir, Pavlodar, Kazakhstan

A V Korobeinyk[1,2*], V R Yapyiev[1], S Azat[1] and V J Inglezakis[1]

1 Environmental Science & Technology Group (ESTg), Chemical Engineering Department, School of Engineering, Nazarbayev University, Astana, Kazakhstan
2 O. O. Chuiko Institute of Surface Chemistry of NAS of Ukraine, Kyiv, Ukraine
E-mail: alina.v.korobeinyk@gmail.com

Abstract. This study aimed at determining the levels of total mercury in water samples collected in the vicinity of the Lake-reservoir Balkyldak in Pavlodar, Kazakhstan. Legacy of the chloralkali process in Pavlodar which involved the mercury cell method while producing a chlorine-free sodium hydroxide leads to serious environmental problems. At the time of the unit continuous operation in between 1971-1993 approximately 1000 metric tons of metallic mercury has been released into the wastewater holding pond at the chemical plant. For many years, mercury collected in the wastewater pond was a subject of the Kazakhstan sharply continental climate fluctuation and overflow to the Balkyldak Lake-reservoir, creating a significant mercury contamination issue for the whole region. Mercury emitted from industrial eluents in environment carries predominantly anthropogenic pollutants. Due to its natural cycle through atmosphere, water and soil in different forms prompts the widespread global mercury pollution. In present investigation, samples collected in a close proximity of the Balkyldak Lake were analyzed in respect to the geological location of collection.

1. Introduction

The USSR heritage in terms of industrial activity, environmental pollution with waste of chloralkali plant inputs into the Pavlodar region has disturbed the equilibrium of its ecosystem. The chloralkili plant was on operation for 18 years and estimated mercury losses are in order of 1000 metric tons of Hg [1]. According to authors [2] more than 700 metric tons of metallic Hg is speed through the soil below the plant and significant amount of mercury entered the Lake-reservoir Balkyldak [3]. Mercury as a global and highly mobile pollutant affects ecosystem[4] and human health[5]. The first study dedicated to health effects awareness of Hg was published in the early 1950s based on the Minatama (Japan) case study where Hg contaminated fish were consumed as a main food source [6]. Mercury exposure [7] even at low concentrations (>2 mg/Kg body weight per day) is responsible for many fatal diseases. Considering the amount of mercury released in to the environment and the fact that the last field research was conducted ten years ago the primary objective of this study was to monitor the Hg concentration in aqueous samples collected in close proximity from ex-chloralkali plant.

2. Materials and methods

2.1. Apparatus

The apparatus used was a Lumex RA–915+ atomic absorption (AA) spectrometer coupled with a PYRO-915 pyrolysis attachment (Lumex Instruments) [8]. The setup consists of a first atomizer section, where at a temperature near 750 °C, the mercury compounds in samples are decomposed into volatile Hg species in molecular or elemental forms. A sample is placed into a quartz boat, which is inserted into the evaporator, where temperature is varied from ambient to 850° C. The atomized mercury and is transferred through the catalytic converter to analytical AA detector.

2.2. Sampling area and sample collection

The lake-reservoir Balkyldak is located North of Pavlodar city (20 km North), so-called Northern industrial area of Pavlodar, in Pavlodar's region of Kazakhstan. A "Khumprom" chemical plant is a major and largest source of mercury pollution in North-Kazakhstan region, which operated before 1989 for 14 year employed mercury as a cathode in production of chlorine and caustic soda. The Balkuldak lake-reservoir was created to settle the mercury bearing sediments. Figure 1 shows the study area and sampling points.

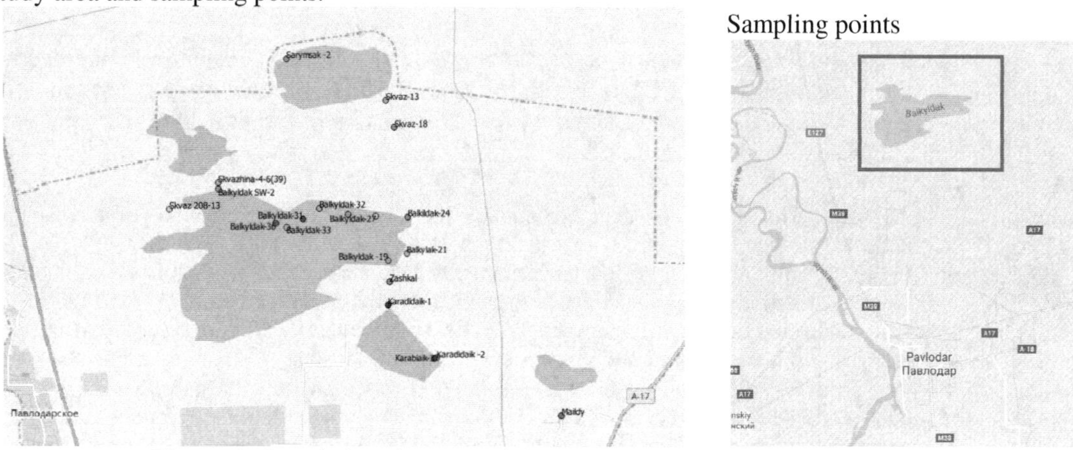

Figure 1. Sample location map of the study area (street view map. 2017)

Twenty five (17 Balkyldak lake-reservoir, 5 revision well and 3 wetland) samples were collected from legacy chloralkali plant tailings dams and their surrounding areas in summer (August) period in 2017. Water samples were collected by dipping thoroughly acid washed glass scintillation bottle 5 cm below the water surface. The precautions as prescribed by the USEPA Method 7471B and 245.1 for sampling procedures were carefully employed throughout the sampling processes [9]. The collected samples were transported to the laboratory and refrigerated until analysis commenced.

2.3. Data analysis

All the samples were prepared in triplicate and from the triplicate measurements the average and standard deviations were calculated. Statistical analysis to test for variations and relationships of concentrations were calculated using Microsoft office excel 2010.

3. Results and discussion

Basing of the commenced analysis, the average Hg concentration in water samples of Pavlodar region is evaluated as ≈30 ppb, which is almost fifteen times higher than MCLG [10]. The maximum contaminant level goal (MCLG) is a non-enforceable level that is based solely on possible health risks and exposure. In the National Primary Drinking Water Regulations formulated by US Environmental Protection Agency (EPA), MCLG for mercury in drinking water is set at 2 ppb [10]. The total mercury content in the studied samples exhibits high variability and changes in a range of 10 to 100 ppb (Figure 2). Figure 2 represent left to right: total mercury content in all aqueous samples collected from 5 revision wells; in all 17 lake-reservoir Balkyldak samples; and wetlands 3 samples. All samples were measured in triplicates and the averages are presented. The average (10 ppb) of total Hg concentration in all 17 analyzed samples collected from Balkyldak is lower as compared to that in all 5 wells water samples having average 65 ppb. It should be noted that the elevated mercury content in revision wells samples is governed by samples having the highest Hg concentration and geographically situated right next to ex-chloralkali plant which confirms continuous leakage of mercury from the building [2].

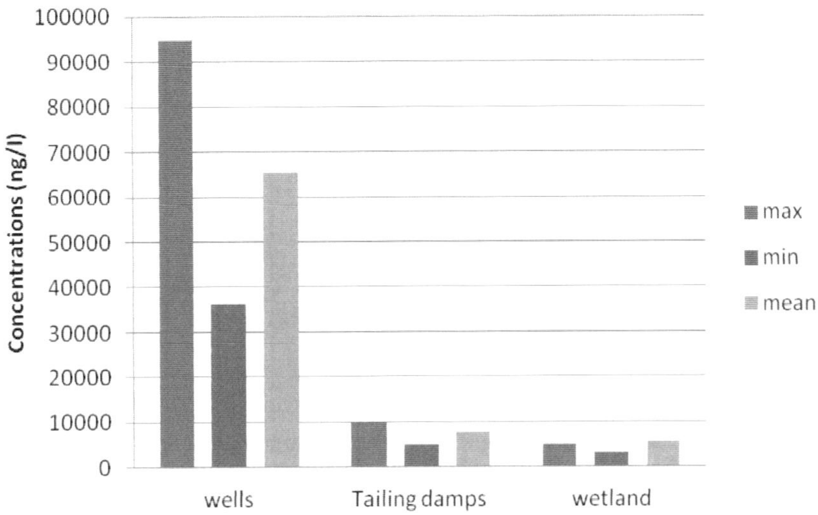

Figure 2. Concentrations of THg in aqueous samples from revision wells, tailings dams, and wetlands (1000 ng/l= 1 ppb).

As can be seen in Figure 2, THg of lake-reservoir Balkyldak samples were lower than those collected from nearby wells. This can be explained by static condition of mercury in the lake and a movement of the underground water in the revision wells. This may be attributed to increased deposition of runoff eroded Hg on sediments in the lake. The study conducted so far have confirmed the presence of Hg in legacy chloralkali plant and these may pose health threat to the communities living within the vicinity of the ex-chloralkali plant.

4. Conclusion

This study has shown the presence of Hg in the lake-reservoir Balkyldak, revision wells and water samples collected from wetlands of the ex-chloralkali plant surrounding area where metallic Hg was used as cathode in producing of a sodium hydroxide. The highest concentration was observed in the revision wells in close proximity from the ex-plant building compared to the other environmental samples. On the basis of collected data, it can be inferred that the legacy chloralkali plant is primary source of Hg contamination in the study area. The majority of concentrations of THg determined were found to be above the recommended limits, and, therefore, it is of concern to surrounding communities. Several adsorbents can be used in the extraction of this pollutant from the solution such as sulfurized activated carbon, mineral zeolites, Hg-specific resins, polymer membranes and etc. Nevertheless, utilization of those filter materials in a field where concentration of Hg is high is limited due to high cost, bio/mechanical filter clogging or low filter capacity.

5. References

[1] Lushin E N, Krakhaleva T E and Krakhalev A F, *Results of the determination of the level of mercury contamination at the site of chlorine and caustic soda production of the Pavlodar Chemical Plant.* 1990, Pavlodar Hydrogeological Expedition and Scientific-Research Centre "Tekhnolog": Pavlodar. p. 194.

[2] Ullrich S M, et al., *Mercury contamination in the vicinity of a derelict chlor-alkali plant. Part I: Sediment and water contamination of Lake Balkyldak and the River Irtysh.* Science of The Total Environment, 2007. **381**(1): p. 1-16.

[3] Ullrich S M, et al., *Mercury contamination in the vicinity of a derelict chlor-alkali plant: Part II: Contamination of the aquatic and terrestrial food chain and potential risks to the local population.* Science of The Total Environment, 2007. **381**(1): p. 290-306.

[4] Unep, *Global Mercury Assessment 2013: Sources, Emissions, Releases, and Environmental Transport*. UNEP, 2013: p. 42.
[5] Ballester F, et al., *Prenatal exposure to mercury and longitudinally assessed fetal growth: Relation and effect modifiers*. Environmental Research, 2018. **160**: p. 97-106.
[6] Kurland L T, Faro S N and Siedler H, *Minamata disease. The outbreak of a neurologic disorder in Minamata, Japan, and its relationship to the ingestion of seafood contaminated by mercuric compounds*. World neurology, 1960. **1**: p. 370-395.
[7] Cadet J L and Bolla K I, *CHAPTER 111 - ENVIRONMENTAL TOXINS AND DISORDERS OF THE NERVOUS SYSTEM A2 - Schapira, Anthony H.V*, in *Neurology and Clinical Neuroscience*, A. Editors, et al., Editors. 2007, Mosby: Philadelphia. p. 1477-1488.
[8] Kolker A, et al., *Mercury and trace element contents of Donbas coals and associated mine water in the vicinity of Donetsk, Ukraine*. International Journal of Coal Geology, 2009. **79**(3): p. 83-91.
[9] Agency, U.S.E.P., *Method 7471B: mercury in solid or semisolid waste (manual cold-vapor technique)*. 2007. p. 1-11.
[10] Wakuda D, Kim K S, and Suganuma K, *Room temperature sintering of Ag nanoparticles by drying solvent*. Scripta Materialia, 2008. **59**(6): p. 649-652.

Acknowledgements
The financial assistance of Nazarbayev University (project ORAU "HyperActiv"), Kazakhstan, European Commission, Horizon 2020 MSCA-RISE-2016 project 'NanoMed', and the Royal Academy of Engineering (UK), projects IAPP/1516/46 and IAPP/1516/3 research grants towards this research is hereby acknowledged.

Quantitative analysis the role of reactive oxygen species (ROS) in a Fe/Cu/O_3 process for *p*-nitrophenol treatment

Z K Xiong[1], B Lai[1] and P Yang[1]

1 Department of Environmental Science and Engineering, School of Architecture and Environment, Sichuan University, Chengdu 610065, China

Abstract. In this study, a highly efficient catalytic ozonation with Fe/Cu bimetallic catalyst was developed for *p*-nitrophenol (PNP) aqueous solution treatment. The result shows that the high COD removal (94.1%) was obtained by the Fe/Cu/O_3 process, owing to the strong synergetic effect between Fe/Cu and ozone. In addition, the COD removal (94.1%) obtained by the Fe/Cu/O_3 process after 30 min treatment is twice more than the sum (45.2%) of COD removal efficiencies obtained by sole ozone and sole Fe/Cu. These results confirm that the synergetic effect between Fe/Cu and ozone played a vital role in the degradation of PNP in aqueous solution. Additionally, the formed corrosion products in exhibited high catalytic ability for ozone decomposition, which could generate more •OH to degrade PNP in aqueous solution. Finally, the role of various reactive oxygen species (ROS) was quantitative analyzed by a series of trapping experiments.

1. Introduction

Zero valent iron (ZVI or Fe^0) is a cost-effective technology for the degradation of toxic and refractory pollutants [1], [2]. Due to the high standard reduction potential difference between Fe^0 and Cu^0 (0.777V), Cu^0 plating on the Fe^0 surface could noticeably improve the reactivity of Fe^0 and its reduction efficiency for the pollutants [3], [4]. The reactivity of Fe/Cu bimetallic particles under the anoxic or aerobic condition has been investigated thoroughly [5], [6]. Likewise, ozone as a stronger oxidizing gas than oxygen, it can directly attack the chromophoric groups or indirectly form smaller molecules, and obtain high color removal in wastewater treatment [7]. However, the sole ozonation does not cause the complete oxidation of organic pollutants since some oxidative reactions with ozone are relatively selective and slow [8].

In recent years, there is an increasing interest in the study of catalytic ozonation which utilizes catalysts to accelerate the decomposition of ozone, and its oxidative capacity is much higher than the sole ozonation [7]. Several Fe-based materials were investigated as the catalysts in catalytic ozonation process, which shown satisfactory performance for organic pollutants degradation [9], [10]. In our previous work, we found that the synergistic effects between Fe^0 and ozone could improve the degradation efficiency for the PNP treatment [11]. Furthermore, ozonation in the present of Fe/Cu bimetallic particles (i.e., Fe/Cu/O_3) process was also developed to degrade the pollutants in automobile coating wastewater, and the satisfactory results indicated that Fe/Cu/O_3 process is a cost-effective technology for the treatment of toxic and refractory automobile coating wastewater [12]. The reaction mechanism of the Fe/Cu/O_3 process consist of four parts: (a) homogeneous catalytic ozonation of Fe^{2+}/Fe^{3+}, (b) heterogeneous catalytic ozonation of Fe_3O_4, Fe_2O_3 or FeOOH, (c) Fenton-like reaction, and (d) peroxone reaction [11]. The generation of reactive oxygen species (ROS) such as superoxide anion radical ($O_2^{\bullet-}$), hydrogen peroxide (H_2O_2), and hydroxyl radical (•OH) would be involved in these reactions. However, the relative contribution of these ROS toward pollutants removal has not been investigated yet to the best of our knowledge. Therefore, it is of great importance to quantitative analysis the role of the ROS in $Fe/Cu/O_3$ system.

In summary, *p*-nitrophenol (PNP) as a probe contaminant due to its high toxicity, carcinogenicity. Thus, the original objectives of this study were to (i) confirm the superiority of Fe/Cu/O3 process toward PNP removal, (ii) characterize the structure and composition of the corrosion products during PNP removal by $Fe/Cu/O_3$ system, (iii) quantitative analysis the contribution of the ROS in $Fe/Cu/O_3$ system.

2. Methods

2.1. Materials

Micron-sized zero valent iron powders, $CuSO_4{\cdot}5H_2O$ (analytical reagent), Na_2SO_4 (analytical reagent), isopropyl alcohol (IPA) and PNP (99%) procured from Chengdu Kelong chemical reagent factory were used in this experiment. Superoxide dismutase (SOD) and catalase (CAT) were purchased from Shanghai Kayon biological technology Co. The zero valent iron powders have mean particle size of approximately 120 μm, and their iron content reaches approximately 99%. Other chemicals used in the experiment were of analytical grade. Deionized water was used throughout whole experiment process.

2.2. Preparation of mFe/Cu bimetallic particles

The mFe/Cu bimetallic particles were prepared through a simple Fe-Cu displacement reaction via adding micron-sized zero valent iron powders into a $CuSO_4$ aqueous solution. The main prepared process was the same with that in our previous work [3].In this study, the main prepared conditions including theoretical Cu mass loading (TML_{Cu}), temperature, mixing speed and Cu^{2+} concentration ($CuSO_4$) of planting solution were 0.41 g Cu/g Fe, 40 °C, 250 rpm and 3 g/L, respectively.

2.3. Batch experiments

In this study, 300 mL certain concentration of PNP aqueous solution was placed into a 500 mL beaker. Meanwhile the desired dosage of prepared Fe/Cu^{0} particles was added, and ozone was continuously dispersed into the reacting solution through a gas diffuser. Ozone was generated onsite from pure compressed dry oxygen (99.9%, v/v) by a laboratory model ozone generator (the maximum generation capacity is 3 g/h, Chengdu Yifeng, China), and the inlet concentration of ozone ($[O_3]_0$) was 5.42 mg/L. In addition, to ensure the complete fluidization of Fe/Cu particles in reaction solution, the slurry was stirred by a mechanical stirrer (300 rpm). The whole experiment process was performed at a running temperature of 25±1 °C by heating in thermostatic bath for 30 min.

In order to affirm the superiority of $Fe/Cu/O_3$ process, three control experiments including (a) O_3, (b) Fe/Cu, and (c) Fe/Cu/air were set up. Meanwhile, the operating conditions of these control experiments were in accordance with the optimal conditions of the $Fe/Cu/O_3$ process. During the reaction process, the samples were withdrawn at pre-determined time intervals, and diluted by deionized water and filtered through a PTFE syringe filter disc (0.45 μm). And then, the COD concentrations of PNP in the samples were measured by COD rapid analyzer.

2.4. Analytical methods

COD of the effluents were measured by COD rapid analyzer (Lianhua, China). Ozone concentrations were measured by the iodometric and indigo method. The compound composition of particles before and after $Fe/Cu/O_3$ system treatment were analyzed by XRD (Xpert Pro, Phillips, Netherlands). The concentrations of dissolved Fe^{2+}, Fe^{3+} and total iron were measured by using the 1,10-phenanthroline colorimetric method at 510 nm with a UV–vis spectrophotometer (UV-2550, Shimadzu, Japan).

3. Results and discussion

3.1. Superiority of the $Fe/Cu/O_3$ process

In order to evaluate the superiority of the $Fe/Cu/O_3$ process and synergetic effect between Fe/Cu and ozone, three control experiments including (a) O_3, (b) Fe/Cu, (c) Fe/Cu/air were setup to treat the PNP aqueous solution under the same conditions (i.e., initial PNP concentration of 500 mg/L, initial pH of 5.4, gas flow rate of 200 mL/min, Fe/Cu dosage of 20 g/L, and treatment time of 30 min). In particular, the two control experiments of O_3 and Fe/Cu were used to confirm the synergetic effect between ozone and Fe/Cu in Fe/Cu/O_3 process. In addition, dissolved oxygen (DO) could cause Fenton-like reaction in Fe/Cu bimetallic system, which could enhance the yield of •OH and improve its treatment efficiency [5]. Therefore, it is necessary to compare the treatment efficiencies of Fe/Cu/air and Fe/Cu/O_3 process.

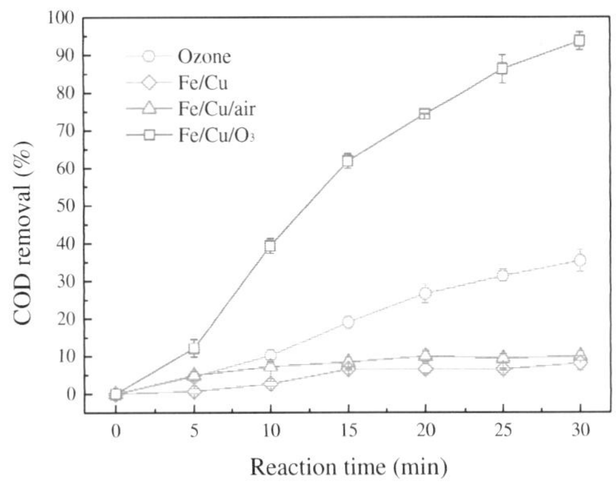

Figure 1. Comparison of performance on COD removal efficiencies by different treatment processes ($[PNP]_0$ = 500 mg/L, $[Fe/Cu]_0$ = 20 g/L, TML_{Cu} = 0.41 g Cu/g Fe, $[Na_2SO_4]$ = 50 mM, initial pH = 5.4, gas flow rate = 200 mL/min, stirring speed = 300 rpm, reaction time = 30 min).

Figure 1 shows the COD removal obtained by the Fe/Cu/O_3 process was higher than those of three control experiments. In addition, the COD removal (94.1%) obtained by the Fe/Cu/O_3 process after 30 min treatment is twice more than the sum (45.2%) of COD removal efficiencies obtained by sole ozone and sole Fe/Cu. These results confirm that the synergetic effect between Fe/Cu and ozone played a vital role in the degradation of PNP in aqueous solution. Furthermore, the COD removal obtained by the Fe/Cu/O_3 process (94.1%) was 9 times higher than that of the Fe/Cu/air process (9.7%). The present of DO could facilitate the direct reduction on the catalytic activity site of the iron surface and formed the Fenton-like reaction in Fe/Cu/air system [5], [6]. However, the corrosion products generated in Fe/Cu/O_3 process could be used as catalyst for the decomposition of ozone, which could generate more •OH to degrade PNP in aqueous solution [9], [13]. Besides, the generated organic acids in the decomposition of catalytic ozonation could promote the corrosion rate of Fe/Cu, which could enhance the catalytic rate of ozone decomposition in return [7]. Therefore, there is strong synergetic effect between Fe/Cu and ozone, and the Fe/Cu/O_3 process is much superior to the existing Fe/Cu/air process.

3.2. Characterization of mFe/Cu particles

In order to confirm the corrosion products, the Fe/Cu bimetallic particles were further analyzed by XRD. The XRD pattern in Figure 2(a) indicates that the chemical composition of the fresh Fe/Cu bimetallic particles were mainly iron and copper. The same chemical composition was obtained in our previous work even if their preparation conditions were different [3]. Figure 2(b) indicated Fe^0 was transformed to lepidocrocite (γ-FeOOH) and goethite (α-FeOOH) after 30 min of Fe/Cu/O_3 treatment. It can be observed that γ-FeOOH was the predominant corrosion product in Fe/Cu/O_3 system, which could generate by the continuous supply of oxygen (equation (1)) [14]:

$$4Fe^0 + 3O_2 + 2H_2O \rightarrow 4\gamma\text{-FeOOH} \quad (1)$$

Additionally, the transformation of γ-FeOOH into α-FeOOH is important in the process of iron corrosion, because the initial corrosion product of γ-FeOOH is gradually transformed to the more stable α-FeOOH [15]. Thus, a slight quantity of α-FeOOH could be formed in Fe/Cu/O_3 system. The formed α-FeOOH and γ-FeOOH exhibit high catalytic ability for ozone decomposition [16], [17], which could generate more •OH to degrade PNP in aqueous solution.

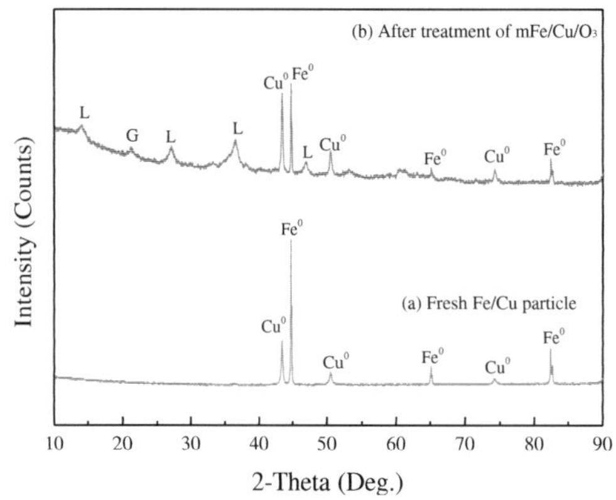

Figure 2. XRD patterns of (a) fresh Fe/Cu bimetallic particles, (b) Fe/Cu bimetallic particles after mFe/Cu/O_3 treatment (G: goethite (α-FeOOH), L: lepidocrocite (γ-FeOOH)).

In order to evaluate the corrosion products released into the solution, dissolved iron ions and total iron ions (i.e., Fe^{2+} and Fe^{3+}) concentrations were detected during 30 min of Fe/Cu/O_3 treatment. As shown in Figure 3, it can be observed that the total iron ions concentration continuously increased and eventually achieved to about 1277 mg/L after 30 min of Fe/Cu/O_3 treatment, which was much higher than the concentration of dissolved iron ions. Furthermore, the generated Fe^{2+}/Fe^{3+} also could catalyze the decomposition of ozone and generate the •OH, which result in the formation of homogeneous catalytic ozonation in solution.

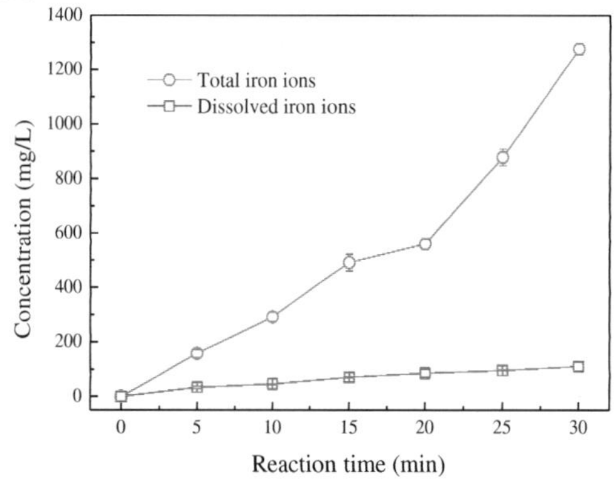

Figure 3. The total iron ions and dissolved iron ions concentration during 30 min of Fe/Cu/O_3 treatment.

3.3. The role of ROS on PNP degradation

In order to clarify the roles of ROS on PNP degradation in Fe/Cu/O_3 system, we therefore carried out a series of trapping experiments to compare the contribution of ROS in Fe/Cu/O_3 system by adding excess isopropyl alcohol (IPA) (•OH scavenger), catalase (CAT, H_2O_2 scavenger), and superoxide dismutase (SOD, $O_2^{\bullet-}$ scavenger) as scavengers. Herein, CAT and SOD were selected as scavengers for H_2O_2 and $O_2^{\bullet-}$ rather than *p*-benzoquinone and *p*-nitro-blue tetrazolium chloride due to their better stability and selectivity [18]. As shown in Figure 4(a), The PNP degradation was largely depressed by the addition of IPA, revealing the leading role of •OH on the degradation of PNP. Interestingly, the degradation rates of PNP also decreased after adding CAT and SOD, indicating that H_2O_2 and $O_2^{\bullet-}$ were also involved in Fe/Cu/O_3 system. It can be observed from Figure 4(b) that k_{obs} for PNP degradation was 0.0917 min^{-1} in the absence of scavengers, while the addition of IPA, CAT, and SOD decreased k_{obs} to 0.0078, 0.0463, and 0.0682 min^{-1}, respectively. To preferably manifest the contribution of different ROS for PNP degradation, we thus estimated the inhibitory efficiency (η) of different scavengers through equation (2).

$$\eta = ((k_0 - k_s) / k_0) \times 100\% \quad (2)$$

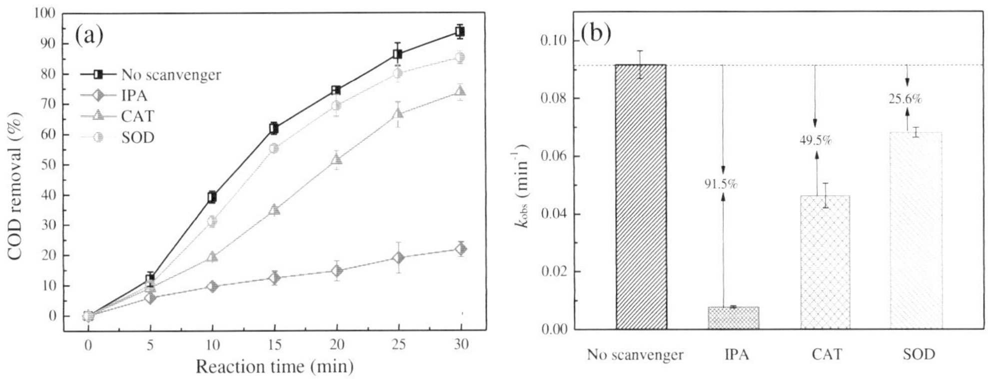

Figure 4. (a) Degradation of PNP with the addition of different scavengers and (b) k_{obs} for PNP treatment with the addition of different scavengers in ECP system (IPA = 30 mM, CAT ≥ 500 U/mL, SOD ≥ 500 U/mL).

Where k_0 and k_s were the degradation rate constant in the absence or presence of scavengers, respectively. The inhibitory efficiencies of IPA, CAT, and SOD were found to be 91.5%, 49.5%, and 25.6%, respectively, as shown in Figure 4(b). Besides, the residual degradation rate in the presence of IPA was calculated about 8.5%, which mainly was attribute to direct molecular ozone reaction. Consequently, the contribution of various ROS could be analyzed by the inhibitory efficiencies of different scavengers. It can be calculated that the role of •OH, H_2O_2, $O_2^{\bullet-}$ and sole ozone were 91.5%, 49.5%, 25.6%, and 8.5%.

3.4. Reuse of Fe/Cu in the reaction system

To evaluate the reusability of the Fe/Cu in Fe/Cu/O_3 process, the reuse of Fe/Cu particles were investigated by running five successive catalytic ozonation experiments under the same optimal conditions (i.e., $[Fe/Cu]_0$ = 20 g/L, TML_{Cu} = 0.41 g Cu/g Fe, $[Na_2SO_4]$ = 50 mM, initial pH = 5.4, gas flow rate = 200 mL/min, stirring speed = 300 rpm, reaction time = 30 min). At the end of each cycle, Fe/Cu particles were separated from the reaction solution and collected, and then were gently rinsed three times with deionized water and dried at 25 °C in the vacuum drying oven. As shown in Figure 5, the COD removal efficiencies were all above 87% in each cycle, and no evident decline was observed after five successive Fe/Cu/O_3 experiments, which indicate the approving recyclability of the Fe/Cu particles.

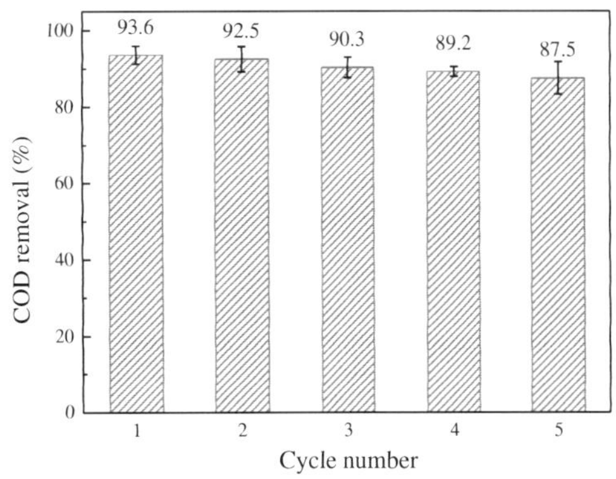

Figure 5. The COD removal for five successive Fe/Cu/O_3 experiments.

4. Conclusion

The high efficient catalytic ozonation of Fe/Cu/O_3 process was developed to degrade PNP in aqueous solution. The result shows that the COD removal obtained by the Fe/Cu/O_3 process was much higher than those of sole ozone, sole Fe/Cu and Fe/Cu/air processes, owing to the strong synergetic effect between Fe/Cu and ozone. Additionally, Fe^0 was transformed to γ-FeOOH and α-FeOOH after 30 min of Fe/Cu/O_3 treatment, and corrosion products (i.e., Fe^{2+} and Fe^{3+}) could release into the solution.

These corrosion products exhibited high catalytic ability for ozone decomposition, which could generate more •OH to degrade PNP in aqueous solution. Finally, the contribution of •OH, H_2O_2, $O_2^{\bullet-}$ and sole ozone were analyzed to about 91.5%, 49.5%, 25.6%, and 8.5% in $Fe/Cu/O_3$ system for PNP treatment.

5. References

[1] Kim D H, Kim J and Choi W 2011 *J. Hazard. Mater.* **192** 928-31.
[2] Xu C, Zhang B, Wang Y, Shao Q, Zhou W, Fan D, Bandstra J Z, Shi Z and Tratnyek P G 2016 *Environ. Sci. Technol.* **50** 11879-87.
[3] Lai B, Zhang Y, Chen Z, Yang P, Zhou Y X and Wang J 2014 *Appl. Catal. B Environ.* **144** 816-30.
[4] Liu X, Fan J H, Hao Y and Ma L M 2014 *Chem. Eng. J.* **250** 354-65.
[5] Xiong Z, Lai B, Yang P, Zhou Y, Wang J and Fang S 2015 *J. Hazard. Mater.* **297** 261-68.
[6] Sun L, Song H, Li Q and Li A 2016 *Chem. Eng. J.* **283** 366-74.
[7] Nawrocki J and Kasprzyk H B 2010 *Appl. Catal. B Environ.* **99** 27-42.
[8] Kasprzyk H B, Ziółek M and Nawrocki J 2003 *Appl. Catal. B Environ.* **46** 639-69.
[9] Ziylan A and Ince N H 2015 *Catal Today* **240** 2-8.
[10] Quiroz A C, Barrera C D, Morales G R, Hernández P B, Romero R and Natividad R 2011 *Ind. Eng. Chem. Res.* **50** 2488-94.
[11] Xiong Z, Lai B, Yuan Y, Cao J, Yang P and Zhou Y 2016 *Chem. Eng. J.* **302** 137–45.
[12] Xiong Z, Cao J, Yang D, Lai B and Yang P 2016 *Chemosphere* **166** 343-51.
[13] Zhang X, Dong W and Yang W 2013 *Chem. Eng. J.* **233** 14-23.
[14] Liu A, Liu J, Pan B and Zhang W 2014 *RSC Adv.* **4** 57377-82.
[15] Ishikawa T, Takeuchi K, Kandori K and Nakayama T 2005 *Colloid Surface* A **266** 155-59.
[16] Yuan L, Shen J, Chen Z and Guan X 2016 *Appl. Catal. B Environ.* **180** 707-14.
[17] Sui M, Sheng L, Lu K and Tian F 2010 *Appl. Catal. B Environ.* **96** 94-100.
[18] Ding X, Wang S, Shen W, Mu Y, Wang L, Chen H and Zhang L 2017 *Water Res.* **112** 9-18.

Acknowledgments

The authors would like to acknowledge the financial support from National Natural Science Foundation of China (No. 21207094), and Fundamental Research Funds for the Central Universities (No. 2015SCU04A09).

Chlorine dioxide catalytic oxidation by zero valent iron for the degradation of PNP in aqueous solution

J Li[1] and B Lai[1]

1Department of Environmental Science and Engineering, School of Architecture and Environment, Sichuan University, Chengdu 610065, China
E-mail address: dreamslijun@163.com (Jun Li)

Abstract. In order to evaluate the degradation of *p*-nitrophenol (PNP) in aqueous solution by chlorine dioxide oxidation process catalyzed by Fe^0 micro-particles, the effects of Fe^0 dosage (0-5.0 g/L), ClO_2 dosage (0-2.0 g/L), initial pH (3.0-13.0) and air flow rate (0-5.0 mL/min) on chemical oxygen demand (COD) removal of PNP in aqueous solution were investigated comprehensively. In particular, the maximum COD removal was obtained under the optimal conditions (i.e., Fe^0 dosage of 4.0 g/L, ClO_2 dosage of 1.2 g/L, initial pH of 5.0, stirring speed of 300 rpm and reaction time of 30 min). Meanwhile, the two control experiments (i.e., Fe^0 alone and ClO_2 alone systems) were carried out to confirm the performance of chlorine dioxide oxidation process catalyzed by Fe^0 micro-particles. In a word, the study provided an effective and promising catalytic oxidation technology for toxic and refractory pollutants.

1. Introduction

Phenolic compounds are broadly produced in the chemical and processing industries [1]. In particular, *p*-nitrophenol (PNP), as one of the most widely used nitrophenolic compounds, is an important organic chemical raw material and intermediate used in manufacturing pesticides, dyestuffs, explosives, wood preservatives and rubber chemicals [2]. Although a widespread use has appeared in the production of the chemical industry, PNP is highly toxic both for the environment and humans, such as injury to blood cells caused by repeated exposure, damage to the central nervous system and mutagenic effects. Meanwhile, because of its high solubility, low biodegradability and persistency in wastewater, it is commonly detected in surface water, and both agricultural and industrial wastewater. Thus, the US and European Environmental Protection Agencies (EPA and EEA) have included it in their lists of priority pollutants to be monitored in industrial effluents [3]. In this study, PNP has been chosen as the model pollutant.

Chlorine dioxide is a powerful one-electron oxidant (E^0 = 0.936V), which has comparable biocidal efficacy, but less pH dependence and DBP formation potential, compared with free chlorine [4]. In the 1940s, ClO_2 was firstly used in the United States for water treatment. Nowadays, ClO_2 in widely applied in drinking water treatment include disinfection, nitrification and bromate control [5] and broadly used to bleach flour and wood pulp [6]. According to the literatures, the application of ClO_2 in the wastewater treatment has been drew great attention [7], [8]. In addition, zero value iron (Fe^0) has been has been extensively applied for the remediation/treatment of groundwater and wastewater because of its effectiveness, low cost and benign environmental impact[9]. In this study, the zero value iron micro-particles were used as catalyst in the ClO_2 catalytic oxidation process for the degradation of PNP in aqueous solution. Herein, the main objective of our work is investigating the effects of key parameters on chemical oxygen demand (COD) removal of PNP in aqueous solution and determining the optimal conditions of experimental parameters for the practical application of toxic organic wastewater by ClO_2 catalytic oxidation process.

2. Materials and methods

2.1. Reagent

p-Nitrophenol (PNP, 99.0%), zero valent iron (ZVI, Fe^0) powders from Chengdu Kelong chemical reagent factory were used in the experiments. Fe^0 particles have a mean particle size of approximately 120 μm, and their Fe content is above 99%. Other chemicals used in the experiment were of analytical grade. Distilled water was used throughout the whole experiment process. All solutions in the study were prepared by the distilled water. The preparation of different concentration of chlorine dioxide solution was through mixing the different concentration of reagent A and B reagent from Chengdu Kelong chemical reagent factory.

2.2. Experimental procedures

In our experiment, PNP stock solution (500 mg/L) was prepared by simple dissolution in distilled water. 300 mL solution of PNP was transferred into 500 mL beaker and the pH value of PNP aqueous solution was adjusted with 0.1 M sulfuric acid or sodium hydroxide solution. The reactions were initiated after adding the desired amount of Fe^0 micro-particles and ClO_2 aqueous solution into the reactor synchronously. The experiment was performed at 25 ± 2 °C by water batch heating and mixed by a mechanical stirrer (300 rpm). After the reaction process, aqueous samples were withdrawn with syringe and immediately filtered through a PTFE syringe filter disc (0.45 μm). All experiments were performed in tripartite, and all data in the figures were averaged values of 3 replicate samples and the error bars represented standard deviation of the means.

2.3. Analytical methods

The chemical oxygen demand (COD) of treatment effluents of the experiments were analyzed by COD analyzer (Lianhua, China). The solution pH was measured by a PHS-25 meter (Rex, China)

3. Results and discussion

3.1. Parameters optimization

3.1.1. Effect of Fe^0 dosage on COD removal

According to the literature [10], [11], as the catalyst, Fe^0 plays an important role in the catalytic oxidation process. In our study, effects of Fe^0 dosage (0, 1.0, 2.0, 3.0, 4.0, 5.0 and 5.5 g/L) on the COD removal of PNP in aqueous solution were evaluated comprehensively (experiment conditions: ClO_2 dosage of 1.2 g/L, initial pH of 5.3, reaction time of 30 min and stirring speed of 300 rpm). Figure 1 shows that COD removal dramatically increased from 49.2% to 62.2% with Fe^0 dosage enhancing from 0 to 2.0 g/L. The result shows that, in the presence of Fe^0 catalyst, the COD removal was much higher than without catalyst, which demonstrates that Fe^0 could improve the degradation of PNP in aqueous solution in ClO_2 oxidation system. In addition, the COD removal continually rose to maximum (64.3%) when Fe^0 dosage increased 4.0 g/L. The result could be explained that the amount of active sites increased with the increased Fe^0 dosage. What's more, COD removal leveled off with the Fe^0 dosage further increasing to 5.5 g/L. Considering from the view of economic cost, the Fe^0 dosage of 4.0 g/L was selected in the following experiments.

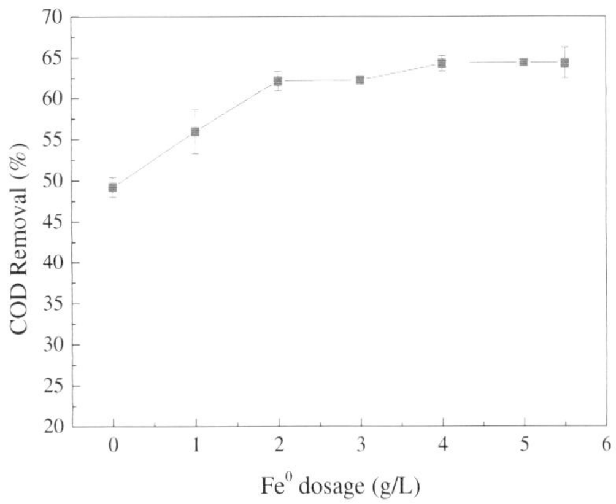

Figure 1. Effects of Fe^0 dosage on the COD removal of PNP in aqueous solution.

3.1.2. Effect of ClO_2 dosage on COD removal

Effects of ClO_2 dosage (0, 0.2, 0.5, 0.8, 1.0, 1.2, 1.5 and 2.0 g/L) on the COD removal of PNP in aqueous solution were investigated (experiment conditions: Fe^0 dosage of 4.0 g/L, initial pH of 5.3, reaction time of 30 min and stirring speed of 300 rpm). As shown as in Figure 2, in the absence of ClO_2, only 0.1% COD removal was obtained after 30 min treatment process, which proved that the adsorption of Fe^0 micro-nanoparticles were weak and could be ignored in this system. Furthermore, when ClO_2 dosage added from 0 to 1.2 g/L, the COD removal dramatically increased after 30 min reaction time. In particular, the COD removal achieved the maximum value (74.9%) when ClO_2 dosage was 1.2 g/L. However, COD removal decreased to 51.4% with ClO_2 dosage further increasing to 2.0 g/L which might be due to the self-consumption between active species. Therefore, the optimal ClO_2 dosage was selected as 1.2 g/L in the subsequent experiments.

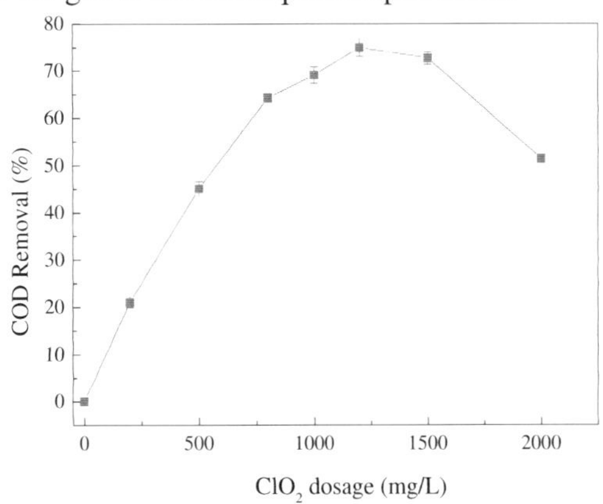

Figure 2. Effects of ClO_2 dosage on the COD removal of PNP in aqueous solution.

3.1.3. Effect of initial pH on COD removal

According to literature, the initial pH is an important parameter influencing the performance of ClO_2 oxidation process [5]. In order to investigate the influence of pH values on the oxidative process, effects of initial pH (3.0, 5.0, 7.0, 9.0, 11.0, 12.0 and 13.0) on the COD removal of PNP in aqueous solution were studied (experiment conditions: Fe^0 dosage of 4.0 g/L, ClO_2 dosage of 1.2 g/L, reaction time of 30 min and stirring speed of 300 rpm). The results are presented in Figure 3, it shows that the COD removal depends strongly on the pH. The COD removal quickly increased from 69.5% to 74.9% when initial pH increased from 3.0 to 5.0. According to literatures [12], [13], based on the redox potential of ClO_2, chlorine dioxide is typically more reactive under acidic conditions than in neutral and alkaline media. Meanwhile, the COD removal only changed slightly with the initial ranged from 5.0 to 11.0. Beyond pH value of 11.0, the COD removal begun to decrease, which was due to the fact that at alkaline conditions, ClO_2 reacts with OH^- to form chlorite (ClO_2^-) and chlorate (ClO_3^-) species [14], leaving less ClO_2 available for pollutants removal. The above result indicates that the optimum pH value in catalytic system ranges from 3.0 to 11.0. Therefore, the optimal initial pH was selected at initial solution pH (5.3) without changing the pH.

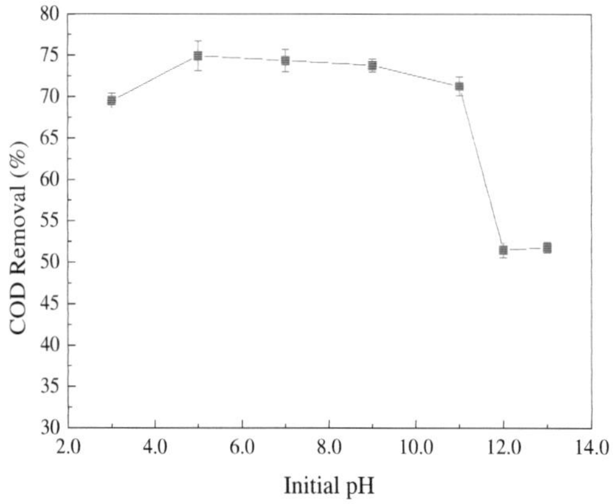

Figure 3. Effects of initial pH value on the COD removal of PNP in aqueous solution.

3.1.4. Effect of air flow rate on COD removal

According to the literature [15], the influence of dissolved oxygen on the chlorine dioxide catalytic oxidation process for the degradation of pollutants was significant. Thus, effects of air flow rate (0, 1.0, 2.0, 3.0, 4.0 and 5.0 L/min) on the COD removal of PNP in aqueous solution were explored thoroughly (experiment conditions: Fe^0 dosage of 4.0 g/L, ClO_2 dosage of 1.2 g/L, initial pH of 5.3, reaction time of 30 min and stirring speed of 300 rpm).

As can be seen from Figure 4, the COD removal remained between 74.6% and 76.2% under different aeration conditions (i.e., air or without aeration). The results suggest that effect of dissolved oxygen (DO) on the COD removal by chlorine dioxide oxidation system catalyzed by Fe^0 micro-particles could be neglected completely. Therefore, the chlorine dioxide catalytic oxidation process was performed without aeration.

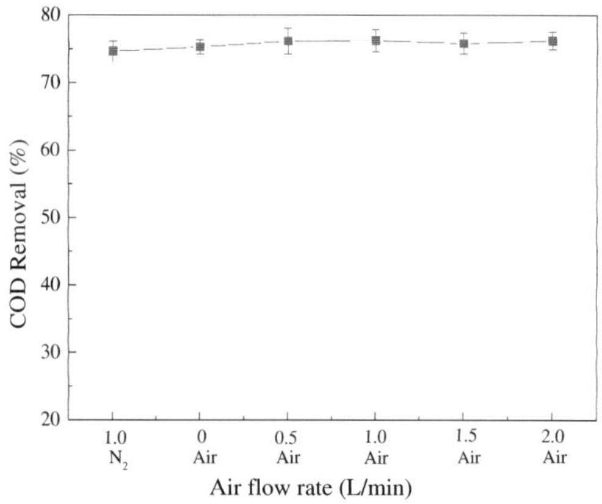

Figure 4. Effects of air flow rate on the COD removal of PNP in aqueous solution.

3.2. Control experiments

In our study, two control experiments (i.e., Fe^0 alone and ClO_2 alone systems) were performed to evaluate the performance of COD removal by Fe^0/ClO_2 catalytic oxidation system thoroughly under the optimal conditions (i.e., Fe^0 dosage of 4.0 g/L, ClO_2 dosage of 1.2 g/L, initial pH of 5.3, reaction time of 30 min and stirring speed of 300 rpm).

Figure 5 shows the COD removal of PNP in aqueous solution in Fe^0 alone, ClO_2 alone and Fe^0/ClO_2 systems. In particular, the combination of Fe^0 and ClO_2 enhanced the COD removal significantly, achieving a COD removal of 74.9% after 30 min treatment, whereas the only addition of Fe^0 led to the COD removal of 0.2% and ClO_2 alone system achieved a COD removal of 55.0%. Considering the above result, the Fe^0 micro-particles could be used as catalyst in the chlorine dioxide oxidation and improve the performance of degradation of pollutants.

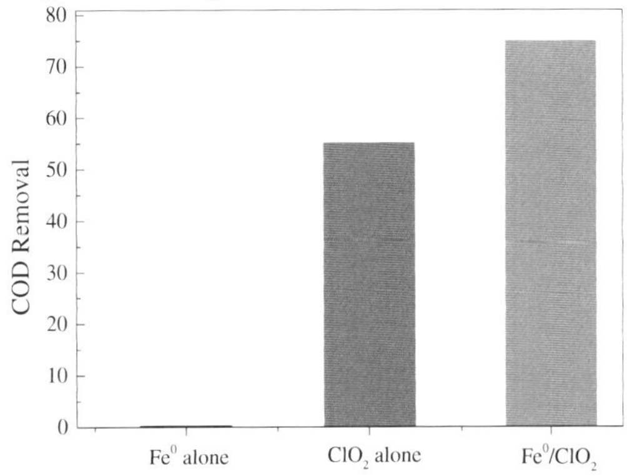

Figure 5. Comparative study of different systems.

4. Conclusions

In this study, we investigated the performance of degradation of PNP in aqueous solution by ClO_2 catalytic oxidation process. Meanwhile, Fe^0 micro-particles as the catalyst exhibited the excellent catalytic activity in the ClO_2 catalytic oxidation process (i.e., Fe^0/ClO_2 system). Firstly, key parameters (i.e., Fe^0 dosage, ClO_2 dosage, initial pH and air flow rate) were investigated to obtain the optimal conditions. It shows that a high COD removal (74.9%) was obtained after 30 min treatment under the optimal experiment conditions (i.e., Fe^0 dosage of 4.0 g/L, ClO_2 dosage of 1.2 g/L, initial pH of 5.3, stirring speed of 300 rpm). Compared with Fe^0 alone and ClO_2 alone systems, it was noted that the COD removal (74.9%) achieved in ClO_2 oxidation process catalyzed by Fe^0 micro-particles was much higher than others (i.e., Fe^0 alone (0.2%) and ClO_2 alone (55.0%) systems), which indicated that the Fe^0 could efficiently improve the performance of pollutant removal in ClO_2 oxidation process. In conclusion, the results manifested that the ClO_2 catalytic oxidation process with Fe^0 micro-particles was an efficient and cleaning technology for the toxic and refractory industrial wastewater.

5. Reference

[1] Kulkarni P M, Effect of shock and mixed loading on the performance of SND based sequencing batch reactors (SBR) degrading nitrophenols, *Water. Res.* **46** (2012) 2405-2414.

[2] Chen X, Murugananthan M and Zhang Y, Degradation of p-Nitrophenol by thermally activated persulfate in soil system, *Chem. Eng. J.* **283** (2016) 1357-1365.

[3] Petrova B, Tsyntsarski B, Budinova T, Petrov N, Velasco L F and Ania C O, Activated carbon from coal tar pitch and furfural for the removal of p-nitrophenol and m-aminophenol, *Chem. Eng. J.* **172** (2011) 102-108.

[4] Wang Y, Liu H, Liu G and Xie Y, Oxidation of diclofenac by aqueous chlorine dioxide: Identification of major disinfection byproducts and toxicity evaluation, *Sci. Total. Environ.* **473–474** (2014) 437-445.

[5] Wang H, Dong J and Jiang W, Study on the treatment of 2-sec-butyl-4,6-dinitrophenol (DNBP) wastewater by ClO_2 in the presence of aluminum oxide as catalyst, *J. Hazard. Mater.* **183** (2010) 347-352.

[6] Li B, Deng Q, Lv J and Zhang Z, Homogeneous chemiluminescence system in neutral and near neutral aqueous solution with ClO_2 as oxidant and its analytical application, *Anal. Chim. Acta.* **560** (2006) 128-133.
[7] Wang C T, Qi-Huan H E and Peng H E, Catalytic Oxidation Treatment of Reactive Brilliant Red Dye Wastewater by Chlorine Dioxide, *Environ. Sci. Technol.* **29** (2006) 65-67.
[8] Wang Y, Liu H, Xie Y, Ni T and Liu G, Oxidative removal of diclofenac by chlorine dioxide: Reaction kinetics and mechanism, *Chem. Eng. J.* **279** (2015) 409-415.
[9] Li J, Ji Q, Lai B and Yuan D, Degradation of p-nitrophenol by Fe^0/H_2O_2/persulfate system: Optimization, performance and mechanisms, *J. Taiwan. Inst. Chem. E.* **80** (2017) 686-694.
[10] Li J, Liu Q, Ji Q and Lai B, Degradation of p-nitrophenol (PNP) in aqueous solution by Fe^0-PM-PS system through response surface methodology (RSM), *Appl. Catal. B.* **200** (2017) 633-646.
[11] Wei X, Gao N, Li C, Deng Y, Zhou S and Li L, Zero-valent iron (ZVI) activation of persulfate (PS) for oxidation of bentazon in water, *Chem. Eng. J.* **285** (2016) 660-670.
[12] Cseko G and Horvath A. K, Kinetics and mechanism of the chlorine dioxide-trithionate reaction, *J. Phys. Chem. A.* **116** (2012) 2911-2919.
[13] Horváth A K, Nagypál I and Epstein I R, Kinetics and Mechanism of the Chlorine Dioxide-Tetrathionate Reaction, *J. Phys. Chem. A.* **107** (2003) 10063-10069.
[14] Kull T P J, Backlund P H, Karlsson K M and Meriluoto J A O, Oxidation of the Cyanobacterial Hepatotoxin Microcystin-LR by Chlorine Dioxide: Reaction Kinetics, Characterization, and Toxicity of Reaction Products, *Environ. Sci. Technol.* **38** (2004) 6025-6031.
[15] Wang Y, Liu H, Liu G, Xie Y and Liu X, Kinetics for diclofenac degradation by chlorine dioxide in aqueous media: Influences of natural organic matter additives, *J. Taiwan. Inst. Chem. E.* **56** (2015) 131-137.

Acknowledgment
The authors would like to acknowledge the financial support from National Natural Science Foundation of China (No. 21207094), and Fundamental Research Funds for the Central Universities (No. 2015SCU04A09).

IOP Conf. Series: Earth and Environmental Science **167** (2018) 012009 doi:10.1088/1755-1315/167/1/012009

Evaluation of different structures of moving bed biofilm reactors (MBBR) for synthetic wastewater treatment

L Yang[1], H Q Li[1] and P Yang[1]

1 Department of environment science and Engineering, School of architecture and Environment, Sichuan University, 610065, China

Abstract. Performance of three different structures of moving bed biofilm reactors (MBBR) treating synthetic wastewater was evaluated. The three systems were operated under the same conditions of dissolved oxygen (DO) 3 mg/L at 26℃. The effluent of chemical oxygen demand (COD) decreased from 111.6 mg/L in reactor 1 (R1), 128.8 mg/L in reactor 2 (R2), and 154.5 mg/L in reactor 3 (R3), to less than 100 mg/L, respectively, with concentration of influent NH_4^+-N rising from 30 mg/L to 50 mg/L. The concentration of effluent NH_4^+-N for R1, R2 and R3 were close to 2 mg/L, 4 mg/L and 5 mg/L, respectively. The highest TN removal rate in R1 achieved average 84.4 %, which was 6.0 % and 9.7 % higher than those of R2 and R3, respectively. R1 with two baffles in it exhibited well fluidization and abundant microbes growing on carrier.

1. Introduction

In recent years, water eutrophication has aroused widespread concern. Nitrogen and phosphorus are the main factors that cause water eutrophication. Therefore, the emission standards of nitrogen and phosphorus are becoming more and more stringent at home and abroad. Ammonia removal is often achieved using nitrification/denitrification system. In the processes, nitrifying bacteria oxidize ammonia to nitrate under oxic conditions, and then denitrification occurred under anoxic conditions [1]. However, the operational control of aerobic and anaerobic conditions needed for nitrification and denitrification is hard to implement. To resolve these problems, various kinds of bioreactors have been studied for promoting the efficiency of nitrogen removal [2]. A biofilm reactor can accumulate high concentrations of biomass by way of microorganism immobilization on carriers compared to activated sludge reactors, which are comprised of bacterial suspensions [3]. Biological processes based on biofilms have been proved to be a better pathway for the removal of organic components and nitrogen from wastewater, avoiding some of the problems regarding activated sludge process such as large reactor size, require for settling tank, and biomass recycling [4]-[6]. Microbes with slow specific growth rates can attach to the carriers and circumvent being easily washed out [3], [7].

Moving Bed Biofilm Reactors (MBBR) was developed and applied in the 1990s and is an efficient wastewater treatment method that combines the advantages of both conventional fluidized bed and biological contact oxidation processes. MBBRs rely on biofilms that are grown on small (1-4 cm diameter) plastic carriers which are suspended and mixed in a reactor [8]. Microorganisms attached to the carriers are in fluidized state by aeration (aerobic) or agitation (anoxic or anaerobic), and the aerobic/anoxic conditions of the biofilm enable nitrification and denitrification to take place simultaneously in MBBR [9]. MBBRs which possess high resistance of load fluctuation, high nitrogen removal efficiency, and high effluent quality have been widely used to treat both urban and industrial wastewaters for its. This technology allows biochemical oxygen demand (BOD_5) and N removal rates similar to those of activated sludge-based processes, with the advantage of a smaller tank volume [4], [10].

The effect of different structures of MBBR on the removal efficiency of pollutants is rarely reported. In the present study, MBBRs with different structures were developed, the systems included two parts: a major reactor and a settle tank. We evaluated the effect of structure of MBBR on synthetic wastewater treatment and to nitrogen removal under different NH_4^+-N concentrations.

2. Materials and methods

2.1. Reactor and operational strategy

This experiment was carried out in three plastic Plexiglas MBBR (Figure 1), each one had a working volume of 8.1 L and a settle tank with a volume of 3 L. The first reactor (R1) had two baffles in the reactor, the second one (R2) had one baffle, and the last one had no baffle in the reactor as the control reactor. Plastic carrier was used as the biomass carrier and the packing rate was 30% (V/V). The carrier was 10 mm tall with diameter of 10 mm. The temperature was fixed at 26℃ during the whole experiment by means of a resistance wire heating. Mixing inside the reactor was performed using an aerator at the bottom of the reactor. The MBBR was running with dissolved oxygen (DO) concentration controlled at about 3.0 mg/L by adjusting the air flow rate. After the starting up of the MBBRs, MBBRs were operated for about 70 days. In phase I (1-39 d), the reactors operated at NH_4^+-N concentration of 30 mg/L, and the residual days (II) at 50 mg/L, with the same HRT of 14 h, respectively.

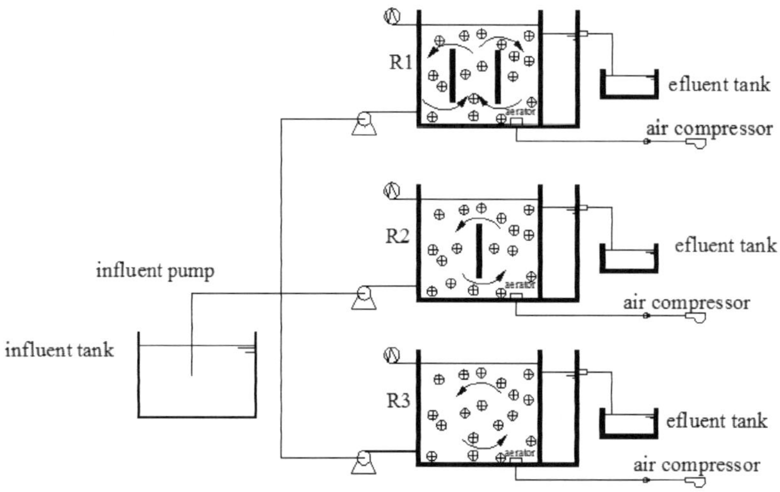

Figure 1. Schematic diagram of experiment equipment

2.2. Inoculation

The reactor was inoculated with activated sludge from a domestic wastewater treatment plant in Chengdu, China. The initial MLVSS (mixed liquor volatile suspended solid) concentration of the sludge was approximately 20,500 mg/L.

2.3. Feed composition

The MBBRs were fed with synthetic wastewater containing glucose as substrate. The composition of the synthetic wastewater was as follows: glucose, 1000 mg/L; NH_4Cl, 125-191 mg/L; K_2HPO_4, 44 mg/L; and 1 mL/L of a trace element solution. One liter of trace element solution contained 1.5 mg/L $CaCl_2$, 10 mg/L EDTA,1.1 mg/L $CuCl_2 \cdot 2H_2O$, 0.003 mg/L H_3BO_3, 0.003 mg/L Na_2SeO_3, 1.2 mg/L $MgSO_4 \cdot 7H_2O$, 0.28 mg/L $FeSO_4 \cdot 7H_2O$, 0.2 mg/L $ZnSO_4 \cdot 7H_2O$, 0.11 mg/L $MnSO_4 \cdot H_2O$, 0.06 mg/L $CoSO_4 \cdot 7H_2O$. $NaHCO_3$ with 100 mg/L was added to the influent to maintain the pH of the MBBRs suspension between 6.5 and 7.5.

2.4. Analytical method
The influent and effluent samples were collected on daily basis and were immediately analyzed. The determination of ammonium, nitrite, nitrate, COD, TN, TSS and VSS were measured were analysed according to Standard Methods [11].

3. Results and Discussion

3.1. Organic removal
The COD removal efficiency of the MBBRs at the same conditions was summarized in Figure 2. In the first 39 days (I), the average COD removal efficiency were about 89%, 87.1%, and 84.7%, and effluent COD concentration were about 111.6 mg/L, 128.8 mg/L and 154.5 mg/L, respectively. In phase II, the NH_4^+-N rose to 50 mg/L, the COD of effluent quickly decreased to less than 100 mg / L of the three MBBRs accordingly, and the optimal effluent quality was obtained in R1 with about 50 mg/L. The efficiency trend of reactors was R1>R2>R3. On one hand, MLVSS concentration was increased from 847 mg/L to 1447 mg/L in R1, from 565 mg/L to 1416 mg/L in R2, and from 443 to 1039 g/L in R3, respectively. This phenomenon indicated that the enrichment of microbes on the carriers, which caused the further decreasing of COD. On the other hand, denitrification enhances the COD removal rate. Therefore, it was concluded that the baffle in the reactor could promote the removal efficiency of organic because of the highest COD removal efficiency and the substantial microbes that had been obtained in R1.

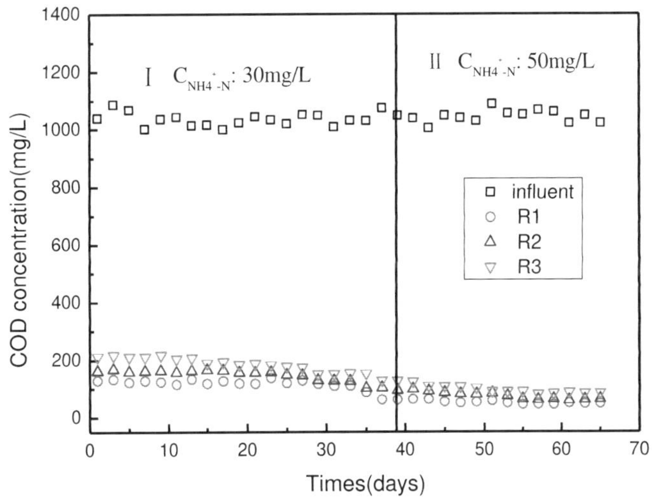

Figure 2. COD concentrations in influent and effluent.

3.2. Nitrogen removal
Figure 3 illustrates the NH_4^+-N (A) and total nitrogen (TN) removal efficiency (B) of different MBBRs during the whole operation. In phase I, DO concentration was about 3.0 mg/L, the effluent ammonia concentration was below 1.0 mg/L, and the average removal rate of NH_4^+-N was more than 96%. When concentration of influent NH_4^+-N was switched to 50 mg/L in phase II, the removal rate of NH_4^+-N decreased and then increased until the biomass gradually adapted to the imposed conditions. The concentration of effluent NH_4^+-N for R1, R2 and R3 were close to 2 mg/L, 4 mg/L and 5 mg/L, respectively. The DO content is a limiting variable in biological nitrification-denitrification treatment processes. R1 with two baffles promoted sufficient carrier movement and obtained high oxygen transfer co-efficient, which favored the oxidation of ammonia.

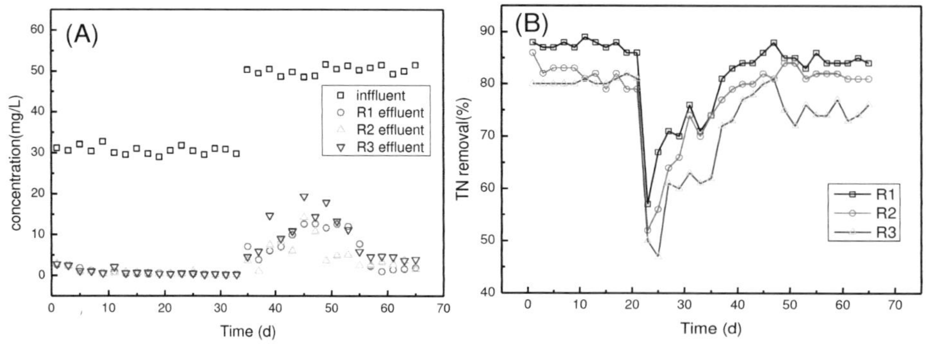

Figure 3. Typical concentration profiles of NH_4^+-N (A) and TN removal efficient of the MBBRs (B).

In the conventional nitration–denitrification nitrogen removal process, ammonium is oxidized to nitrate sequentially by aerobic ammonium oxidizing bacteria (AOB) and aerobic nitrite-oxidizing bacteria (NOB). And then denitrifying bacteria utilize organic substrate as electron donor to reduce nitrate to nitrogen gas [12]. The C/N ratio had great influence on R2 and R3, and the TN removal rate decreased 3.0% in R1, 3.3% in R2 and 7.0 % in R3, respectively, when the ratio decreased from 13.8 in phase I to 8.3 in phase II. The highest TN removal rate in R1 achieved average 84.4 %, which was 6.0 % and 9.7 % higher than those of R2 and R3, respectively. This result indicated that R1 with two baffles can resist load fluctuation. The NH_4^+-N and TN removal results showed that the effect of reactor structure on the nitrogen removal cannot be negligible.

4. Conclusion

Comparison among different structures of MBBRs, higher effluent quality was obtained for R1 with two baffles. R1 exhibited well fluidization and abundant microbes on carriers. The result further indicated that MBBR structure had a significant effect on the pollution removal; structural optimization by adding baffles could enhance the mass transfer efficiency leading to MBBR performance improvement. Finally, we can further reduce energy consumption by adding baffles in practice.

5. References

[1] Olav Sliekersa N D A, Campos Gomezb J L, Strousa M, Kuenena J G and Jetten M S M 2002 *Water Res.* **36** 2475–82.
[2] Khin T and Annachhatre A P 2004 *Biotechnol. Adv.* **22** 519-32.
[3] Shi Y, Huang C, Gamal El-Din M and Liu Y 2017 *Sci. Total. Environ.* **598** 690-96.
[4] Calderon K, Martin-Pascual J, Poyatos J M, Rodelas B, A Gonzalez-Martinez and J Gonzalez-Lopez 2012 *Bioresour. Technol.* **121** 119-26.
[5] Luostarinen S, Luste S, Valentín L and Rintala J 2006 *Water Res.* **40** 1607-15.
[6] McQuarrie J P and Boltz J P 2011 *Water Environ. Res.* **83** 560-75.
[7] Gu Q, Sun T, Wu G, Li M and Qiu W 2014 *Bioresour. Technol.* **166** 72-78.
[8] Casas M E, Chhetri R K, Ooi G, Hansen K M, Litty K, Christensson M, Kragelund C, Andersen H R and Bester K 2015 *Water Res.* **83** 293-302.
[9] Yuan Q, H Wang Q, Hang Q, Deng Y, Liu K, Li C M, and Zheng S Z 2015 *Environ. Sci. Pollute. Res.* **22** 1-10.
[10] Andreottola R F G, Ragazzi M and Tatàno F 2000 *Water Sci. Technol.* **41** 375–82.
[11] APHA 1998 *Standard Methods for the Examination of Water and Wastewater, 20th ed* (USA: United Book Press).
[12] Zhang D, Cai Q and Cong L 2009 *J. Chem. Technol. Biot.*

Acknowledgement
The authors wish to thank all my teachers who have helped me to develop the fundamental and essential academic competence.

Color removal by ozonation process in biological wastewater treatment from the breweries

S Sripiboon[1] and K Suwannahong[2]

1 Department of Environmental Engineering, Faculty of Engineering, Rangsit University, Thailand
2 Department of Occupational Health, Faculty of Public Health , Western University, Thailand
E-mail: siriporn.sr@rsu.ac.th

Abstract. Colors from the final stage of processing of brewery biological wastewater treatment by Upflow Anaerobic Sludge Blanket (UASB) reactors have caused extremity problems. The aim of this research was to investigate the ozone dosage required to remove the color from brewery industrial wastewater effluent. The dosage of ozone is varied between 50 to 300 mg/L. for 18 minutes, flow rate 45 L/hr., pH 8.4. As the result, the increasing depolarization efficiency is obtained with the increase in an ozone dose. Maximum depolarization of 90% is achieved at an ozone measurement of 300 mg/L. Moreover, the kinetic of depolarization is fitted by the first order equation. The constant of the reaction (k) was 0.120 1/minute. Electrical energy required to move color was 6 kW-h/m^3.

1. Introduction

Organic compounds are one of the most significant components from wastewater products of brewery operations [1] which are a result of the fermentation process of malt (usually barley malt), hops, fermented proteins and waste yeas lead to the water discharged being found with highly organic waste, especially lignin. Biological wastewater treatments in the brewing industry have two (2) continuous systems. 1. Upflow anaerobic sludge blanket (UASB) reactor systems have the efficiency to reduce organic pollutant concentration up to 75-85% [1], [2] and 2. Activated Sludge (AS) system gave a maximum removal efficiency of Chemical Oxygen Demand (COD) range from 50 to100 mg/l being under value according to Thai Industrial Effluent Standards (COD not exceeding 120 mg/l) [3]. However, although the wastewater effluents were passed both UASB and AS systems had COD less than 120 mg/l, color due to a last stage process of brewery biological wastewater treatment was still more than 500 ADMI color index (The ADMI color index is American Dye Manufacturer's Institute) , making it over the limit according to Thai Industrial Effluent Standards (Color not exceeding 300 ADMI color index) [3]. The purpose of this research is to reduce the colour of the effluent to obtain a standard and unobjectionable effluent from the surrounding communities by using the ozonation method as an oxidation process. Ozone is one of the most powerful and commercially available oxidants and is proven to be one of the most efficient methods of treatment of coloured aqueous effluent [4]. Furthermore, the ozone affects removal of true color and destroys the structure of organic compounds in the same time.

2. Materials and Methods

2.1. The materials
The wastewater samples used in this study were collected from the wastewater effluent after passing both the Upflow Anaerobic Sludge Blanket (UASB) and Activated Sludge (AS) systems from industrial breweries in the Pathum Thani province, one of the central provinces of Thailand. The wastewater samples characteristics were pH 8.4, COD and ranged from 56.96 to 68.35 mg/l where the true color was 500 ADMI.

2.2. The methods
The experimental setup was used the ozonation process on color reduction and studied factors influencing ozone dosages of 50, 60, 75, 100, 150, and 300 mg/l where the true color was 500 ADMI. The production of ozone was created by using an ozone reactor as seen in Figure 1. By pumping water into the system. Then opened an Ozone generator. The reactor would drain the water through water inlet and vacuum the air through the air duct. Both water and air were pumped into an Ozone generator for producing ozone. After that the water was sprayed out a sprinkler at the tip of the water tube. The system would circulate the water back into contact with the ozone again, repeating the cycle until it reaches the equilibrium condition

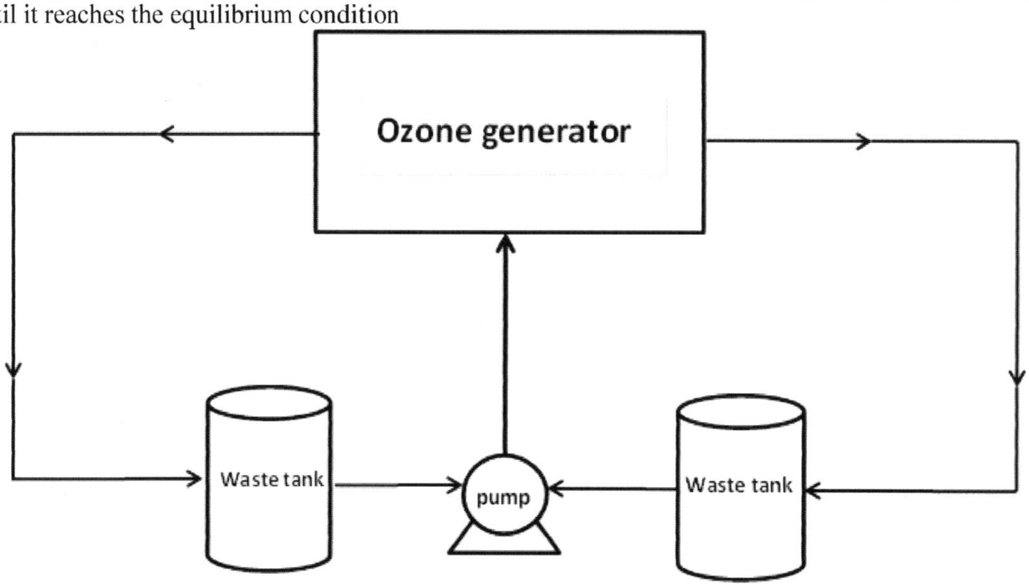

Figure 1. Experimental Setup for Ozonation of waste water treatment.

2.3. kinetics of degradation of color with ozonation process experiment
The study was conducted in the three models as follows 1.Zero-order reaction 2.First-order reaction and 3.Second-order reaction. The wastewater samples used in this study were controlled the pH at 8.4 ± 0.1. and 5 ml. were collected at 2, 4, 6, 8, 10, 15, 20, 25, 30, 35, 40, 45, 50, 55, and 60 minutes, respectively. The obtained data was analyzed for ozone degradation kinetics based on color (ADMI) and COD, respectively. The reaction tanks used in this study were 30 cm in diameter, 45 cm in height, 38 liters in total, and 5-30 liters in capacity as depicted in Figure 1. EE /O (that is the electrical energy required to remove a pollutant by one order of magnitude in one m^3 of Waste water) used in the degradation of colour by using ozone has been calculated by applying the following empirical formula equation (1) [5].

$$\mathrm{EE}\ /O\ (\frac{kWh}{m^3}) = \frac{P\times t\times 1000}{V\times 60\times \log{(C_i/C_e)}} \quad (1)$$

When P (kW) is the power input, t is the oxidation time (in min), V is the volume of the effluent sample (in litter) and Ci and Ce are the initial and final concentrations of the contaminant.

3. Results and Discussion

3.1 Efficiency of Color and COD removals

The results in Figure 2 and Figure 3 show that Color and COD removals efficiency were increased with the increase of the ozone dosages [6]. The effect of ozonation on color removal were grew from 50 to 300 mg/l during the 18th minute of ozone contact time and on Color and COD removals were increased from 57% to 80% and 28% to 50% (Figure 4) respectively.

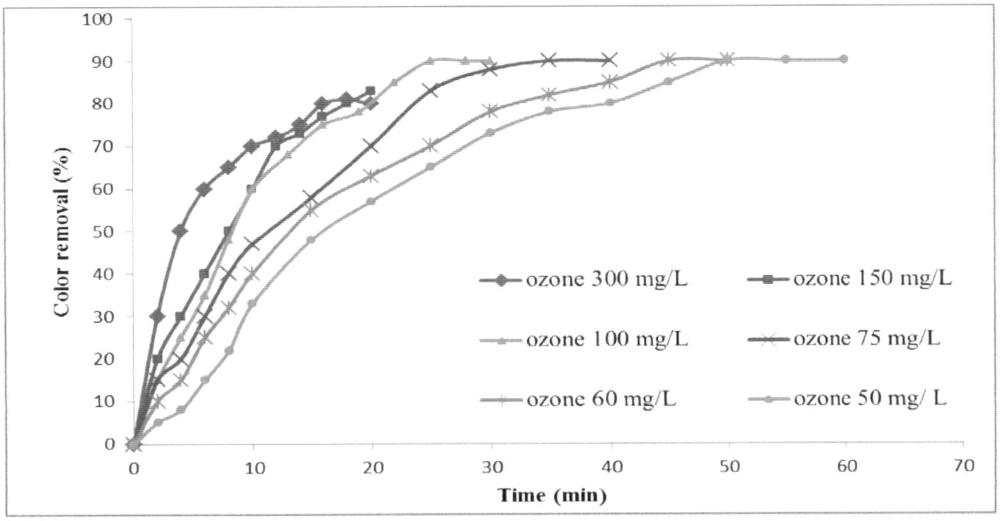

Figure 2. Effect of ozone dosage on the efficiency of ozonation for color removal of wastewater.

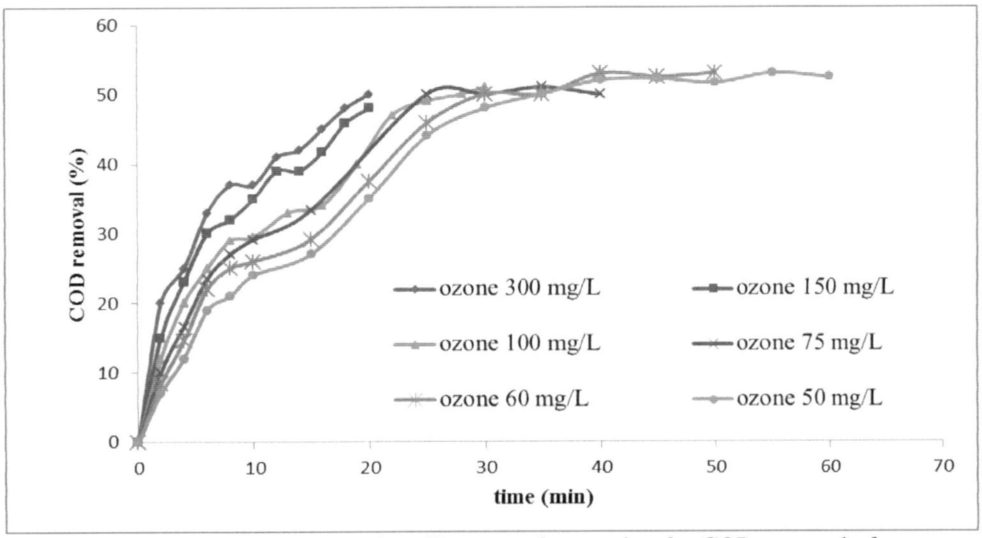

Figure 3. Effect of ozone dosage on the efficiency of ozonation for COD removal of wastewater.

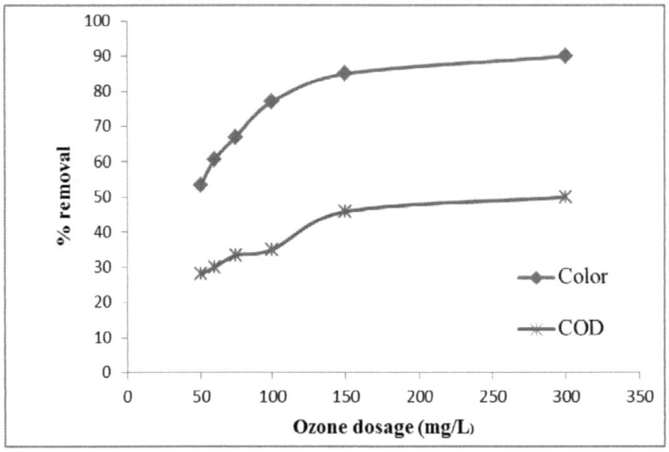

Figure 4. Effect of ozone on the efficiency for Color and COD removal of wastewater.

3.2 The kinetics of degradation of color with ozonation process

Degradation kinetics describe the rate of color degradation in the ozonation process in three models as follows 1.Zero-order reaction 2.First-order reaction and 3.Second-order reaction are shown in Equations (2) - (4), respectively.

$$-\frac{dc}{dt} = k \rightarrow [C] = [C]_0 - k_0 t \quad (2)$$

$$-\frac{dc}{dt} = kC \rightarrow ln[C] = ln[C]_0 - k_1 \quad (3)$$

$$-\frac{dc}{dt} = kC^2 \rightarrow \frac{1}{[C]} = \frac{1}{[C]_0} + k_2 t \quad (4)$$

When $\frac{-dC}{dt}$ = rate of degradation (ADMI/min), k_0, k_1 and k_2 = constants of reactions of a zero order, first order and second order respectively, [C] and $[C]_0$ = concentration of Color in effluent and influent (ADMI) and t = time (minutes).

The regression analysis Multiple (R^2) of the plotted data by Equations (2) - (4) shown in Figure 5 and Table 1, which found that wastewater being oxidized by OH • with the ozonation process was fitting the first reaction in all ozone concentrations and the initial wastewater pH was 8.4.

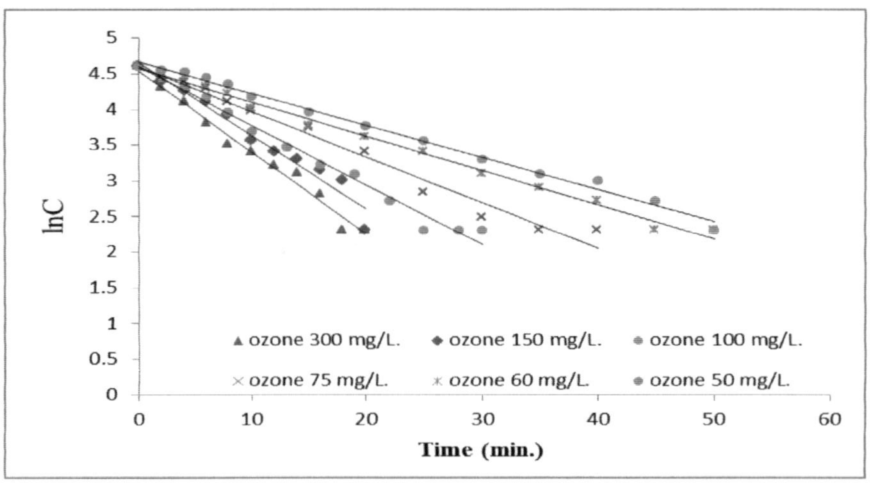

Figure 5. 1st order reaction

Table 1. The values of constant K, and R^2 of the kinetics of ozonation process.

Ozone concentration (mg/L)	0th Order Reaction K_0 (mg/L.min) R^2	1st Order Reaction K_1 (1/min) R^2	2nd Order Reaction K_2 (L/mg.min) R^2
300	-3.7136	-0.12	-0.0042
	0.837	0.9841	0.7971
150	-4.1682	-0.1023	-0.0034
	0.9439	0.9674	0.751
100	-2.8999	-0.0832	-0.0034
	0.8921	0.9864	0.92
75	-2.6663	-0.0639	-0.0025
	0.9133	0.9809	0.9349
60	-1.7752	-0.0479	-0.0018
	0.8984	0.9933	0.9208
50	-1.5837	-0.0424	-0.0016
	0.9025	0.9894	0.9176

In this study, it seems that k rate was the first order reaction and was consistent with [7] research. k rate was increased with an increase in the graph of ozone concentration. In the other hand, the k rate in color removal was higher than the k rate in COD removal almost 10 times due to the wastewater having a very high initial color while the COD is very low because the wastewater had the characteristics of effluent discharging similar to Industrial Effluent Standards already [8]. The EE /O was used in the degradation of color was 3-6 kW-h/m^3. As the amount of water increased, the EE /O was reduced. The amount of ozone was lower and the cause of the efficiency of color and COD removal were decreased. The use of electrical energy is not very high compared to the performance that can be treated which was consistent with Yasar's research [5] that used ozone removal of color and COD showing energy requirements for using pre ozonation at 20.8 and 20.4 kWh/m^3, respectively and energy requirements for using post ozonation were 1.3 and 7 kWh/m^3, respectively.

4. Conclusion

The ozonation process has been used for the successful removal of color from brewery wastewater. The concentration of ozone that can affect the performance impact of wastewater treatment was 300 mg/l. which can increase the efficiency of the treatment up to 90%. In addition, mathematical equations can be calculated for evaluating the effectiveness of the ozonation process which had the multiple regression coefficient (R^2) being 0.9841.

5. References

[1] Tchobanoglous G, Burton F and Stensel H 1991 *Wastewater Engineering Treatment Disposal and Reuse* (Singapore: McGraw-Hill) pp 428

[2] Berta R, Monica C, Mari G 2014 Evaluation of Fenton method and ozone-bases processes fpr colour and organic matter removal from biologically pre-treated swine manure. *Chemosphere* **117** p 193-9

[3] Savanayana U 2017 *Notification of Ministry of Industry Industrial Effluent Standards* (Bangkok: JETRO Bangkok) pp 11-3

[4] Mustapha A 2015 Colour Removal Technology Using Ozone in Textile Industrial Wastewater Effluent: and overriew *IJISETR* **3** p 45-51

[5] Yasar A, Ahmad N, Chaudhry M, Rehman M and Khan A 2007 Ozone for Color and COD Removal of Raw and Anaerobically Biotreated Combined Industrial Wastewater *Polish J. of Environ. Stud* **16** p 289-94

[6] Preethi V, Kalyani K, lyappan K, Srinivasakannan C, Balasubramaniam N, Vedaraman N 2009 Ozonation of tannery effluent for removal of COD and color J *Hazard. Mater.* **166** p150-4

[7] Selcuk H 2005 Decolorization and Detixification of Textile Wastewater by Ozonatio and Coagulation Processes *Dyes and Pigments* **64** p 217-22

[8] Zou L, Zho B 2008 The synergistic effect of ozonation and photocatalysis on color removal from reused water *J. Photochem.***196** p 24-32

Acknowledgments
The authors would like to express gratitude to Rangsit University, Thailand for the financials support.

IOP Conf. Series: Earth and Environmental Science **167** (2018) 012011 doi:10.1088/1755-1315/167/1/012011

Characterization of coke oven wastewater

L Mishra[1], K K Paul[2] and S Jena[3]

1Research Scholar, Civil Engineering Department, NIT Rourkela, India, E-mail: lipsa.mis@gmail.com
2Assistant Professor, Civil Engineering Department, NIT Rourkela, India, E-mail: kkpaul@nitrkl.ac.in
3Assistant Professor, Civil Engineering Department, NIT Rourkela, India, E-mail: sjena@nitrkl.ac.in

Abstract. During manufacture and processing of iron in iron and steel industry various wastewaters are generated, out of which the coke oven wastewater is contemplated to be the most toxic byproduct. Untreated discharge of coke oven wastewater to environment will increase the contamination and may threaten the flora and fauna. Physico-chemical characterization is required to assess the pollution level and to identify the treatment method. In the present research work, coke oven wastewater was collected from iron and steel plant and various physico-chemical characteristics were studied. The study found that concentrations of BOD_5 (73.13 mg/l), COD (540.25 mg/l) and cyanide (27.9 mg/l) exceeded the tolerance limit as per IS: 2490 effluent water standard (inland surface water). Biological treatment may supersede the conventional technologies for safe disposal to inland surface water.

1. Introduction

The Steel and Iron industry is one of the most energy consuming sector of the Indian economy and is scrutinized both locally and globally. In the last decade the production of steel in India has surged by a growth rate of 8% compounded annually [1]. India is 3rd largest producer of steel and largest producer of sponge iron in the world. The world produces 1630 million tons of steel annually of which India contributes 91.46 million tons [2]. In India relatively large quantity of water is consumed to produce steel as compared to developed countries. The reason is for high usage of water is coke; a major component for producing steel consumes around 4000m^3 of freshwater for producing 1000 ton of coke. The whole process at the end produces over 1000m^3 of extremely hazardous wastewater [3].

The rising demand of steel due to speedy urbanization occurring in the developing world has established a huge demand for metallurgical coke. Carbonization of Coal generates gas and other by-products such as anthracene, toluene, naphthalene, benzene and coal-tar products which comprises the unprocessed materials for the production of various drugs, synthetic dyes and explosives [4]. The effluent produced from the quenching of coke contains large quantities of organic and inorganic pollutants, such as ammonium, sulphate, cyanide, thiocyanate, phenol and polycyclic aromatic hydrocarbons (PAHs) [5]. Due to the complexity in removing the organic matter, treated wastewater discharged from the steel plant cannot achieve the required effluent standard; hence the released effluent becomes an objectionable matter [6].

The coke plants produce polluted effluent generated due to washing of the ammonia stills, which accumulates from the gas coolers as a result of condensation/sedimentation. The effluents contain cyanide, phenol, ammonia and sulphide which are very toxic to the marine life. The indissoluble pollutants mostly from tar; form a layer in the surface which obstructs the contact of oxygen with air.

Fine suspended particles choke the respiratory organs of fish making it unsuitable to survive. Heavy suspended particles get settled on the bottom of stream and source for further pollution. Inadequate dilution, pyridine, free lime and thiocyanates can also have far-reaching problems to the marine life. Most of the dissolved oxygen (DO) is used up by sulphur compounds, mainly sulphides which also have direct hazardous impacts on aquatic life. Oil and grease form oil slimy which limits the dispersal of oxygen from air into the water [7].

Various probes have indicated that biological treatment can be used to remove organic pollutants in colossal quantities from the wastewater at comparatively very low cost [8]-[12]. However, the biodegradation is very slow and time consuming. Pre-treatment studies [13], [14] like steam chemical coagulation, sludge concentration, stripping, aeration, settling and sometimes dilution under non-identical conditions (anaerobic, anoxic and aerobic) have been used to reduce unfavourable impact on environmental conditions. Treatment of wastewater comprising different combinations of thiocyanate (SCN^-), ammonia-nitrogen (NH_4^+–N), cyanide (CN^-) and phenol compounds is very tough since the contaminants resist the effective removal of each other.

Wastewater can either be reused or recycled through proper wastewater management. Due to rapid increase of population, industrialization and urbanization the accessibility of fresh water had decreased drastically. Hence an honest wastewater management approach is the need of the hour. The new perspective should be focused on enhancing sustainable treatment technologies. Detailed physio-chemical and bacteriological characterization is required for identifying the best and optimized treatment for coke oven liquid waste and its proper disposal.

Present study was performed to know the concentration of different organic and toxic parameters of coke oven wastewater. As biological treatment is the least cost method for the treatment of large volume of coke wastewater, we proposed for an attached growth bioreactor i.e. rotating biological contactor system (RBCs). RBC systems are comparatively easy to operate, using discs on a rotating shaft that provide a partially effluent-submerged surface that supports biofilm growth responsible for the degradation of the organic compounds in the wastewater[15].

2. Monitoring protocol

2.1. Sampling site

The effluent collected from coke oven plant is located in Rourkela (22^0 26’ 04° N, 84^0 85’ 36° E), India. The plant uses water from river the Brahmani and discharges its effluents to the same. Coking temperature ranges from 1050–1100^{0}C for coking hours: 18-20 h.

2.2. Sample collection

Five litres of liquid effluent was collected in morning hour (10-11am) of every week and brought to the laboratory in a plastic jar. Total five numbers of samples were collected. Temperature and pH were performed immediately. The remaining wastewater was properly marked and preserved in the refrigerator at 4^{0}C for further physico-chemical characterization summarized in Table.1.The wastewater sample was acidified by adding H_2SO_4 and alkaline by adding 1N NaOH for testing of phenol and cyanide concentration where the pH targets 3.9-4. The samples were kept for COD determination by 2 ml/l of H_2SO_4 and were examined following the methods [16].

2.3. Analytical methods

Potassium buffer was used for the chemical analysis in order to maintain the pH constant at the required value and phenol standard (Merck). Deionized water was used during experimental steps. All chemicals and reagents used for various physico-chemical analyses were purchased from Merck and Himedia. The physico-chemical tests were conducted using the untreated coke oven effluent samples. Colorimetric method was used to estimate the phenol content from the effluent sample using 4-aminoantipyrine method. Phenol reacts with 4-aminoantipyrine in the presence of potassium ferricyanide to form a coloured antipyrine dye. Chloroform extracts the dye from aqueous solution. The absorbance of the extract was measured at 460 nm using Perkin Elmer make UV VIS Spectrophotometer. Pyridine barbituric acid colorimetric method was used to verify the amount of cyanide in the effluent sample. The absorbance was read at 578 nm for the complex developed with pyridine-barbituric acid reagent and CNCl. All the procedures followed for the analysis were as per the accordance to the standard [16]. Metallic species of coke oven effluent were determined by using atomic absorption spectrophotometer (Perkin Elmer, AAnalyst 200).

2.4. Data analysis

In this research work, physico-chemical characterized data were used to study spearman rank correlation analysis investigates the relationship among them with the help of software IBM SPSS 24 shown in Table 2. The spearman rank correlation coefficients were determined to measure the relationship between two variables.

3. Results and Discussion

Total 5 numbers of samples were tested for physico-chemical analysis. The procedures followed for analyses were as per the accordance to the standard [16]. The pH of the effluent was found to be moderately alkaline in nature and ranges from 8.2 to 8.5 with an average value of 8.35. The COD concentration of the effluent also exceeded permissible limit of 250 mg/l as per IS code [17]. High value of BOD demonstrated the existence of biologically degradable substances. A high value of dissolved solid content was found as compared to raw water indicating contamination by inorganic salts. Dissolved solid in the effluent was found as 1035 mg/l. The average concentration for phenol and cyanide exceeds the tolerance limit which is 1mg/l and 0.2mg/l, respectively as summarized in Table 1. Mercury, arsenic and iron dominate the metallic species.

Table 1. Characterization of coke plant effluent

Parameters	Min. value	Max. value	Average	Tolerance limit, IS: 2490 Effluent
pH	8.2	8.5	8.35	5.5-9
Temperature	30 ^{0}C	33 ^{0}C	31.5 ^{0}C	40 ^{0}C
BOD_5	63.92 mg/l	82.34 mg/l	73.13 mg/l	30 mg/l
COD	436.7 mg/l	643.8 mg/l	540.25 mg/l	250 mg/l
Dissolved Solids	943.2 mg/l	1128.4 mg/l	1035.8 mg/l	2100 mg/l
Phenol	170.27 mg/l	192.34 mg/l	181.30 mg/l	--
Cyanide (CN-)	25.6 mg/l	30.2 mg/l	27.9 mg/l	0.2 mg/l
Total Hardness	428.7 mg/l	440.2 mg/l	434.45 mg/l	--
Mercury(Hg)	28.9 µg/l	31.08 µg/l	30.35 µg/l	10 µg/l
Lead(Pb)	0.031 µg/l	0.065 µg/l	0.048 µg/l	100 µg/l
Arsenic(As)	10.65 µg/l	15.74 µg/l	13.19 µg/l	200 µg/l
Iron(Fe)	0.743 µg/l	1.338 µg/l	1.04 µg/l	--
Zinc(Zn)	0.013 µg/l	0.028 µg/l	0.020 µg/l	5000 µg/l
Nickel(Ni)	0.048 µg/l	0.053 µg/l	0.050 µg/l	3000 µg/l

Coke plant effluent consists of huge amount of suspended solids. After quenching the coke, huge quantity of effluents are generated that mainly constitute suspended coke breeze [18]. At the discharge point, river water was found to be turning black due to deposition of coke breezes. So, along with surface water pollution and siltation on the riverbed, a large quantity of valuable coke breeze is being lost every day [19]. Various environmental circumstances will increase the toxicity of phenol. The toxicity level of phenol increases due to lower concentration of dissolved oxygen, increased salinity and increased temperature [20]. High BOD, COD, phenol contents of effluent is prompting acute surface water pollution. The test results unveil that proper treatment of coke oven effluent and strict adherence of the norms is required before discharging to surface water bodies.

After identifying the different parameters present in coke oven wastewater, a low cost biological treatment method for the better removal of the organic pollutants can be proposed.

Spearman rank correlation (Table 2) shows significantly good correlation between phenol and COD. Cyanide shows mainly inverse correlation with other physico-chemical parameters.

Table 2. Spearman rank correlation matrix between different parameters present in coke oven wastewater

	pH	Temperature	BOD	COD	Phenol	Cyanide
pH	1	0.162	0.359	0.289	0.162	-0.379
Temperature		1	-0.527	-0.73	-0.667	-0.111
BOD			1	0.205	0	0.527
COD				1	.892*	-0.46
Phenol					1	-0.667
Cyanide						1

* Correlation is significant at the 0.05 level (2-tailed)

Spearman rank correlation has performed between different parameters such as pH, temperature, BOD, COD, Phenol and cyanide to identify the relationship and common sources between them. In this study the effluent is collected from coke-oven plant. This correlation analysis will help to determine dependency on existence for parametance. It is one of the vital step to select its appropriate treatment method as removal of one parameter shows the possible removal of related parametance.

4. Conclusion

The physico-chemical characterization was analysed using real coke oven wastewater collected from steel plant. Most of the physico-chemical parameters such as COD, BOD, Phenol and Cyanide concentration exceeded the limit prescribed for disposal on inland surface as per IS code [16]. There is a need to identify the low cost eco-friendly technology to treat the coke oven wastewater for its further disposal or reuse.

5. References

[1] Sharma N K and Philip L 2016 Combined biological and photo-catalytic treatment of real coke oven wastewater *Chemical Engineering Journal* **295** 20-28

[2] Ahmed S, Chandra U and Rathi R K 2010 Waste water treatment technologies commonly practiced in Major Steel Industries of India. In *16th Annual International Sustainable Development Research Conference, The University of Hong Kong* **1(16)**

[3] Pal P and Kumar R 2014 Treatment of coke wastewater: a critical review for developing sustainable management strategies, *Sep. Purif. Rev.* **43** 89–123

[4] Bone W A 1919 *Coal and its scientific uses*. Longmans, Green and Company.

[5] Kim Y M, Park D and Lee D S 2008 *j. Hazard Mater* **152(3)** 915-921

[6] Chang E E, Hsing H J, Chiang P C, Chen M Y and Shyng J Y 2008 The chemical and biological characteristics of coke-oven wastewater by ozonation *Journal of Hazardous Materials* **156(1)** 560-567

[7] Ghose M K 2002 Complete physico-chemical treatment for coke plant effluents *Water Research* **36(5)** 1127-1134

[8] Gonzalez G, Herrera M G, Garcia M T and Pena M M 2001 Biodegradation of phenol in a continuous process: comparative study of stirred tank and fluidized bed bioreactors *Bio-resource Technol*ogy **76** 245–251

[9] Ballinger S J, Head I M, Curtis T P and Godley A R 1998 Molecular microbial ecology of nitrification in an activated sludge process treating refinery wastewater *Water Sci. Technol.* **37(4–5)** 105–108

[10] Fang H Y, Chou M S and Huang C W 1993 Nitrification of ammonia–nitrogen in refinery wastewater. *Water Res.*, **27** 1761–1765

[11] Gupta S K and Sharma R 1996 Biological oxidation of high strength nitrogenous wastewater. *Water Res.* **30** 593–600.
[12] Hirata A, Nakamura Y and Tsuneda S 2001 Biological nitrogen removal from industrial wastewater discharged from metal recovery processes *Water Sci. Technol.* **44(2–3)** 171–179
[13] Chakraborty S and Veeramani H 2002 Anaerobic–anoxic–oxic sequential degradation of synthetic wastewaters *Appl. Biochem. Biotechnol.***102–103** 443–451
[14] Chakraborty S and Veeramani H 2006 Effect of HRT and recycle ratio on removal of cyanide, phenol, thiocyanate and ammonia in an anaerobic–anoxic–aerobic continuous system. *Proc. Biochem.*, **41** 96–105
[15] Tawfik A, Klapwijk A, El-Gohary F and Lettinga G 2005 Potentials of using an RBC for post treatment of anaerobically pre-treated domestic wastewater *Biochemical engineering journal* **25(1)** 89-98
[16] APHA. Standard methods for the examination of water and wastewater, 16th ed. Washington, DC: American Public Health Association 1992
[17] IS 2490, Part VIII. Tolerance limits for industrial effluent discharged into inland surface water. Indian Standard, 1986
[18] Ghose M K and Kumar A 1993 Impact on surface water quality due to the discharge of coal washery effluents and dispersion profile of pollutant in Damodar River *Asian Environ.* **15(1)** 32–40
[19] Ghose M K and Roy S 1996 Status of water pollution due to coke oven effluent, a case study *J. Ind. Pub. Health Eng* **3** 1–9
[20] Prasad B and Singh G 1989 Characterisation of coke plant effluents with subsequent treatment—an appraisal *Indian Journal of Environmental Protection* **9** 525-529.

IOP Conf. Series: Earth and Environmental Science **167** (2018) 012012 doi:10.1088/1755-1315/167/1/012012

Modular Eco -Percolation Tank R & D and Engineering Application

X T Cui[1], Z Q Liu[2], Q X Huang[3] and Q Miao[4]

1 Master student, School of Environment and Municipal Engineering, Qing'dao University of Technology,11 Fushun Road,Qing'dao, Shandong 266000,P.R.China.
2 Full professor, Corresponding author, School of Environment and Municipal Engineering, Qing'dao University of Technology,11 Fushun Road,Qing'dao, Shandong 266000,P.R.China.
3 Master student, School of Environment and Municipal Engineering, Qing'dao University of Technology,11 Fushun Road,Qing'dao, Shandong 266000,P.R.China.
4 Full professor, School of Environment and Municipal Engineering, Qing'dao University of Technology,11 Fushun Road,Qing'dao, Shandong 266000,P.R.China.
E-mail:lab506@163.com

Abstract. The dispersed sewage in some areas which are far from the city's municipal pipelines has not been treated properly. That the effluents are directly discharged without treatment can cause adverse effects on the surrounding environments. In order to treat the dispersed sewage, which has characteristic of high concentration of BOD_5 or NH_3-N, the designers have developed an ecological treatment technology for sewage, which is called Modular Eco-Percolation Tank Process and applied to a highway service area in Shandong Province. The statistics of treated sewage has shown that the Modular Eco-Percolation Tank Process has good applicability for the sewage treatment in the service area. The treated effluents have a stable quality which has reached the water quality standard and can be directly discharged into the environment.

1. Introduction

The scattered drainage area is far away from the urban areas thus its drainage system is usually not in the city municipal sewage pipe network coverage If not handled directly, the discharge of sewage could cause pollution to the surrounding environment [1]. Although the laying of sewage pipe network could be transported to the city sewage treatment plant, the high cost is not realistic. Therefore, decentralized treatment of decentralized sewage is very necessary. What's more, with the improvement on people's living standards, the demand for domestic miscellaneous water (such as urban greenbelt water use and flushing, etc.) is also growing [2]. It will result in a waste of water resources if we all use tap water, which could lead to a fact that the current shortage of water supply situation becomes more serious. Accordingly, under the current shortage of water resources, the research and development for decentralized sewage treatment and reuse technology are even more meaningful.

Dispersed sewage treatment and reuse have the dual meaning of protecting the environment and conserving water resources. It can not only provide new "water source" for resource-strained areas but also to prevent and control pollution even in water-rich areas [3]. Meanwhile, it reduces the burden of urban sewage pipe network and sewage treatment plant and truly realizes the protection of the environment and save the unity of resources, which is in line with China's current advocacy of the "energy-saving emission reduction" and building "resource-saving and environment-friendly" society [4].

Designers have analyzed processing facilities of the decentralized domestic sewage (such as expressway service area, remote neighborhoods), the selection of technology, equipment operation, treatment effects and developed a running low energy consumption, which has characteristics of resisting the anti-load change, low construction and operation costs of sewage treatment and reusing decentralized sewage [5].

The application of Modular Eco-Percolation Tank Process provides technical support for the disposal of dispersal wastewater in China, which has important practical significance.

1.1. Research at home and abroad

Decentralized sewage treatment technology is developed relatively early in foreign countries. They generally use decentralized treatment and some small sewage treatment system for processing with particular emphasis on water reuse, the most representative of which is the United States and Japan. The application of a wide range of processing technologies includes membrane bioreactor and artificial wetland treatment process [6]. The water is mainly used for green irrigation, flushing toilets and so on.

China's research on decentralized wastewater treatment and reuse is not systematic enough comparing with foreign countries. The current decentralized sewage treatment process for urban residents mainly uses buried biochemical treatment equipment, SBR, membrane bioreactor, ecological infiltration deep treatment, artificial wetlands, soil infiltration and so on. However, the corresponding research is less in decentralized sewage due to the characteristics of it [7].The water quality, operation cost, management control and so on are not satisfactory which have affected the use of the enthusiasm of the operator and the sewage treatment effect.

2. Interpretation

2.1. Research and development

The principle of Modular Eco-Percolation Tank Process comes from soil infiltration system. The soil infiltration system belongs to the land treatment system and is a practical and inexpensive engineering measure for the decentralized sewage treatment. It makes full use of the microbes, animals, plants, and soil physical and chemical properties of the soil, and can effectively remove the pollution from the sewage substance. Soil infiltration system has many advantages in treating sewage, but the researchers have been researching and developing the corresponding improvement process because of the shortcomings of the system [8]. For example, the system is easy to plug; the treatment load is low; the denitrification effects are not good and the backwashing is not easy.

Among domestic researchers, the majority is to study the removal of pollutants in the sewage effect of the best filler. To accept more sewage treatment with less areas and have the good effect are indeed very few in the research [9]. In view of this defect, researchers developed a set of the treatment process of the recycled water, which is easy to construct and simple to manage---Modular Eco-Percolation Tank Process. The Modular Eco-Percolation Tank Process is based on the soil infiltration system. Modular ecological soil infiltration system based on percolation tank process, treatment structures filled with porous media, using ecological engineering principle by means of engineering [10]. Through arranging the distributing and collecting pipes in the filler, the sewage doses to the filler under the control and being collected effectively. Using the physical and chemical characteristics of Filler-Plant-Microbe composite system to recycle the fertilizer resources of the sewage, and purify the water. Modular Eco-Percolation Tank Process can integrate different number of processing unit modules according to actual sewage quantity and sewage load. This process shortens the time of ecosystem establishment and stabilization, and has better effect on pollutant removal in sewage.

2.2. Technological advantages

Compared with the complex construction process of the traditional soil infiltration system, modular modules of the various modules have been assembled in the factory; only need to locate the

combination and the construction is simple and fast. Modular equipment processing is more durable and has a high degree of standardization and strength compared to the traditional construction of the soil impermeable membrane and will not create a problem due to the traditional way which could format the cracks in the foundation of the leakage and result in groundwater pollution.
The Modular Eco-Percolation Tank Process has the advantages of low investment, low running energy consumption and high efficiency of pollutant removal, and it is beneficial to be used in small and medium-sized sewage treatment. Modular Eco-Percolation Tank Process can really be energy-saving emission reduction, flexible design and can be integrated with the surrounding landscape environment. While doing the sewage treatment, it is also regarded as the role of landscape and ecological protection; truly respects and protects nature.

2.3. Structural diagram

The structural diagram of the process are as shown in the Figure 1, 2 and 3.

Figure 1. Front view of modular ecological flotation tank.

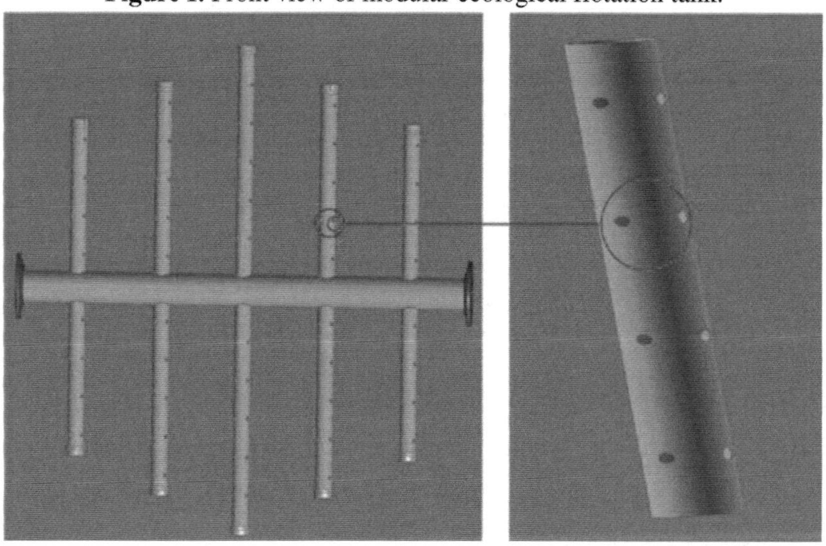

Figure 2. Internal pipe network diagram.

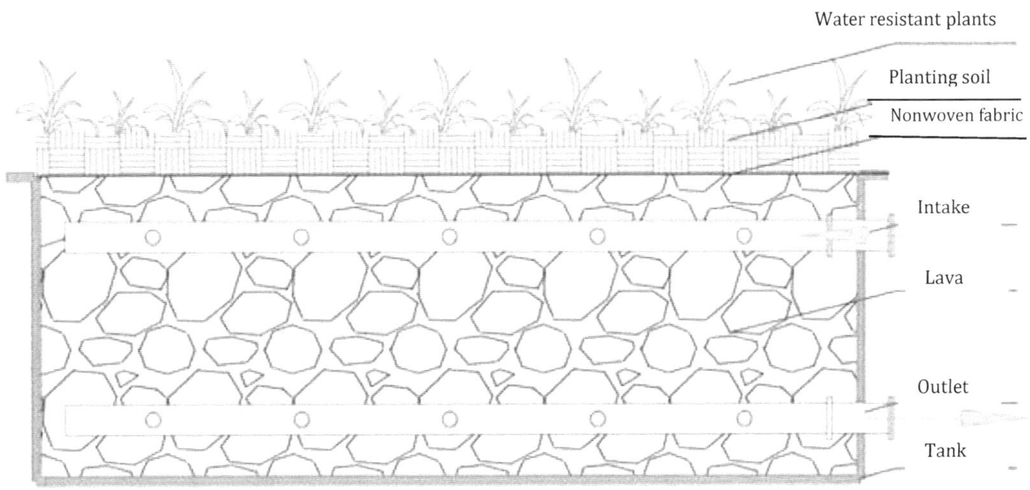

Figure 3. Section of the modular ecological flotation tank system.

2.4. The principle of modular ecological flotation tank

Ecological infiltration tank uses an impermeable material and the filler is filled with the whole ecological infiltration tank. Ecological infiltration tank is buried in the set, covered with non-woven fabrics, and then backfilled planted soil; the surface is planted moisture-resistant plants. The ecological leaching tank relies on the combined action of microorganisms, aquatic plants and fillers to remove the pollutants such as suspended solids, organic matter, nitrogen, and phosphorus. The main removal principles of pollutants are as followed:

(1) From the bottom of the ecological infiltration tank to the top, the filler in the microbial is in the anaerobic, anoxic and aerobic state step by step. The researchers use the metabolic activity of microorganisms to oxidize organic compounds in sewage into inorganic substances so as to purify the water.

(2) Water-tolerant plants absorb some nutrients by themselves. Leaves and branches can transport oxygen from the atmosphere to the root in order to provide dissolved oxygen for the microbial growth. The roots of the water-resistant plants extend vertically downward and have very strong penetrability. Thus the ability to transmit oxygen is strong. The uptake of the surface plant and the decomposition of the underground microbial are an artificial strengthening on natural ecological processes.

(3) After allocating the sewage, the filler can remove nitrogen and phosphorus and other pollutants through physical and chemical pathways such as absorption, filtration, ion exchange, complex reaction. The selected filler has characteristics such as inert, corrosion resistance, strong surface hydrophilic, rough and porous. The attached biofilm's growth speed is fast and there's no inhibitory effect for the immobilized microbial elements therefore it does not affect the activity of micro-organisms. Consequently, it is easy to format a stable micro-ecological balance system in the tank. Even if the packing is caused by improper use, the filler can be used for backfilling after cleaning.

3. Application example

3.1. Project overview

A highway service area in Shandong Province covers an area of 100 acres. Housing construction area is more than 5500 square meters. It is a comprehensive service area with parking, refueling, car maintenance, restaurants, shopping malls, toilets, guest rooms, leisure and so on. The peak daily discharge of sewage in the north and south part of service area reaches 200 tons. The treatment process used in original sewage treatment plant was "contact oxidation + precipitation + filtration". Not only it required a higher management, but also the water was difficult to reach a stable A Standard. Therefore, it resulted in that sewage station had been abandoned because of long-term idle. After abandoning the sewage treatment station, the sewage in the service area is treated by suction truck.

The project covers an area of about 216m^2 ; regional service population is about 8,000; the maximum daily sewage volume is 200t/d; the maximum sewage flow is 8.5t/h. After about 15 days of construction, the project has been successfully completed on March 20, 2017, and began to commission the operation of the system in early April. After a month of debugging operation, the water quality can be achieved "urban sewage treatment plant pollutant discharge standards" (GB18918-2002) In the A-level emission standards. The water quality of the project is shown in the following Table 1.

Table 1. Engineering water quality.

Project	pH	COD_{Cr}(mg/L)	BOD_5(mg/L)	NH_3-N(mg/L)	SS(mg/L)	Colority
Water in	6~9	≤350	220	≤30	≤150	≤70
Water out	6~9	≤50	≤10	≤5	≤10	≤30

3.2. Treatment process

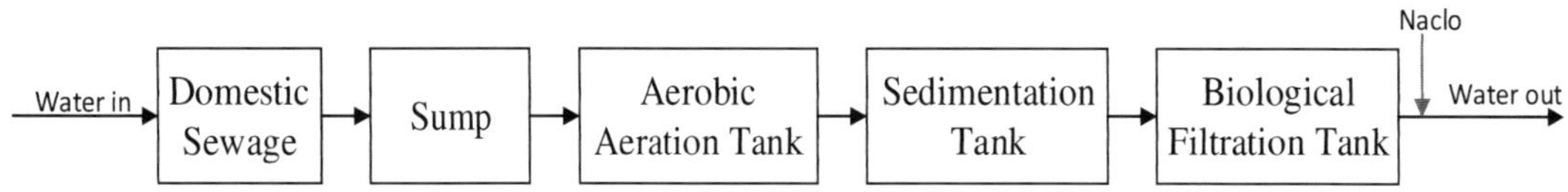

Figure 4. Process flow chart.

The sewage in north and south service areas of which is collected by the service area drainage network flows into the sump by gravity. The process of the treatment process is shown in the Figure 4. Sewage lifting pump is installed in sump and sewage pumps through the water pipe lift the sewage into the 4 pretreatment tanks (3 aerobic aeration tanks, 1 vertical flow sedimentation tank). Pre-treated sewage is distributed by gravity into 20 ecological infiltration tanks and disposed in the ecological infiltration tank through the complex biochemical decomposition and diafiltration process. The way of water disinfection is to release chlorine regularly. The chematic diagram is shown in the Figure 5.

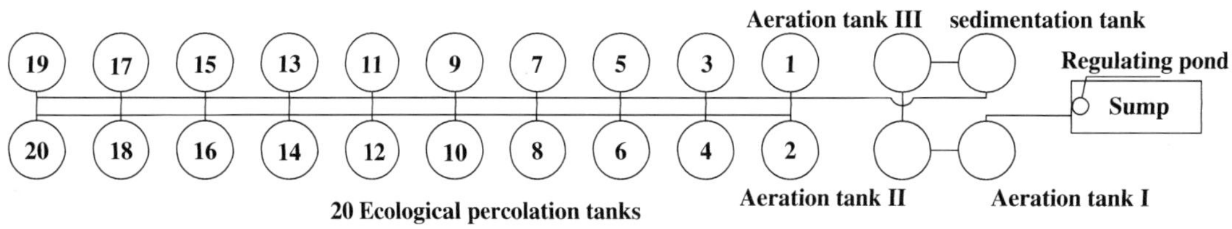

Figure 5. Schematic diagram of the plane layout.

3.3. Composition and parameters

Sump: taking anti-seepage measures and using the service area northeast side of the reservoir to dredge and for its masonry. The size of the sump is 30 m × 6 m × 1.5 m, and the sump collects the sewage in the service area and adjusts the water quality and water quantity effectively, and plays the role of regulating the pool to ensure the stable operation of the follow-up process. The regulation pool with steel structure is set in the sump; the effective volume is 150m^3; the sludge flows back to the regulator pool by gravity then lifted back to the aeration tank through the pump.

Aerobic aeration tank: the main materials use anti-corrosion steel plate, part of it buried in the ground; the diameter is 2.8m; the height is 5.6m; total of 3 are in series. Aerobic aeration tank pretreats the sewage with hypoxia/ aerobi; ation tank aeration uses the underwater aerator. Researchers use steel ring to fix three aerobic aeration tank and sedimentation tank together to prevent the subsidence of the pool body.

Sedimentation tank: the materials and specifications are same with aeration tank; sedimentation tank uses the method of vertical precipitation.

Ecological infiltration tank: The main body of the flotation tank is made of glass fiber reinforced plastic; the diameter is 2.8m; the height is 1.4m; the perforated pipe is used to distribute and drain water. The filler uses 3~5 cm massive volcanic rocks.
Water resistant plants: the project obtains materials locally, choosing iris or reed for moisture-resistant plants.

3.4. Structure of ecological infiltration system

The ecological infiltration tank uses FRP as its material, filled with volcanic rocks. Ecological infiltration tank is buried in the set, covered with non-woven fabrics then backfilled planted soil with the thickness of 100 ~ 150 mm, planting water resistant plant on the surface such as yellow iris or reed. Firstly, sewage is pretreated by flowing through the pool, then aeration tank and the sedimentation tank. The sewage flows to the ecological drainage tank water pipe buried in the underground, and then infiltrate the surrounding volcanic rock from top to bottom, finally to the next ecological infiltration tank through the bottom of the porous collection tube. The comparion diagrams of the front and rear effect are as shown in the Figure 6.

Figure 6. Before and after the works of soil cover.

3.5. System construction cost and benefit analysis

The costs statistics of the treatment construction in the Geran high-speed service area are shown in the table below. 1 ton of water investment is about 3,500 yuan. The construction costs are shown in the Table 2.

Table 2. Construction costs.

Serial number	project name	Total price (million Yuan)
1	Civil part	0.14
2	Plant part	0.06
3	Supporting facilities	0.5
Total		0.7

Direct operating costs:
Electricity: The total installed capacity of the project is 3kW; the starting time is 24 h, the power factor is 0.7; the electric charge is 1 yuan/kW·h then:
Electricity charges: $3 \times 24 \times 0.7 \times 1 = 50.4$ yuan/day
Labor cost: The function of ecological diafiltration is a maintenance-free process, no need for management, only with a small amount of mechanical and electrical equipment in the pretreatment stage. The equipment is all automatic. The property management personnel can do the patrol inspection job thus it will not incur additional costs.
Per ton of water direct operating costs:

(Electricity + labor costs)/daily treatment of water = 50.4/200 = 0.252 yuan/ per ton

4. Benefit analysis

1) Sewage treatment costs

Under the premise that the project does not take the cost of labor into account, the direct operation of water costs 0.252 yuan per ton. Compared to the costs of the sewage to the municipal drainage, the fee of municipal sewage treatment is 1.0 yuan per ton. The sewage treatment capacity considered in accordance is about 200 yuan per ton for a day, and the daily processing fee is 150 yuan for a day. Therefore it can save sewage commission processing costs more than 50,000 yuan for a year.

2) Cost of self-contained wells

Under normal operating conditions, the process can be used for service areas of greening, spraying and other use. The cost of self-provided wells is about 2 to 3 yuan per ton if water costs, sewage charges, management fees and equipment maintenance costs are considered in it. If used daily 200 tons of recycled water to replace these high-quality water, it could save 70,000 tons of groundwater resources per year and reduce water costs more than 100,000 yuan.

3) Environmental benefits

It can fully improve the service area and the surrounding environmental health through the implementation of this project. Annual COD emissions are reduced by 35 tons; BOD emissions are reduced by 20 tons, and total nitrogen emissions are reduced by 2 tons. The project adopts the way of ecological treatment to reduce pollution and integrate into nature.

4) Social benefits

The regional environment has been improved through the implementation of the project. The social problems caused by sewage pollution have been eliminated. The image has been established for the development of service area, and the conditions for sustainable development have also been established.

5. Innovative features

(1) The theme of the wastewater treatment design and application is "Energy saving and emission reduction". In the practical application, the ecological management can be used to reduce the investment cost and the pollutant discharge, also save energy. The water quality is better, and it is more respectful for nature and protects the ecological environment.

(2) The pretreatment equipment and the ecological percolating tank can be combined flexibly and conveniently according to the water treatment capacity.

(3) Compared with the rectangular box structure, the round tank structure has reasonable force structure, uniform hydraulic conditions, and avoids the common distribution dead angle of the structure to ensure the stability of the treatment effect.

(4) Commercial production is more favorable.

6. Conclusion

The research and development are mainly used for the highway service area away from the municipal facilities, remote villages, towns and other scattered domestic sewage treatment; it has the characteristics of low investment cost, convenient operation and management and low maintenance cost in the later stage. The research and development of this project are of great significance to speed up the construction of "resource-saving and environment-friendly" society and to improve the efficiency of water resources utilization and respond to the "13th Five-Year Plan" of energy conservation and emission reduction issued by the State Council. At the same time, the reuse of sewage treatment technology can effectively alleviate the current situation of water resources and provide strong support for the construction of ecological civilization.

7. References

[1] Zhang Z Y, Wen K, Wang D B, Wang Y J, Yu G J and Yang B 2010 Application of Ecological-Artificial Diafiltration Technology in Domestic Sewage Treatment of Toll Station. *Chinese Journal of Environmental Engineering*. **(02)** 360-364.

[2] Zhang Z Y, Kang J, Wang X H and Xiong H 2009 Application of Ecological Diafiltration Technology in Domestic Sewage Treatment of Hengyang Expressway Service Area and Management Office. *Hunan Transportation Science and Technology.* **(04)** 158-161.

[3] Hu K, Xu H, Zhang Y L, Li X Y, Chen W, Wu H Y, Lu X and Pe C Y 2017 Discussion on Operation Mode of Dispersed Rural Domestic Sewage Treatment Facilities. *Water Resources Protection.* **(02)** 63-66.

[4] Song F C, Zhao Q Q and Si Q 2005 Highway Ancillary Facilities Sewage Treatment Process Selection *East China Highway.* **(02)** 54-56.

[5] Sun Q B, Zhen X Y, Fang R, Wang X Y, Dou Z R, Tao B and Chuan J 2007 Research and Application of Wastewater Ecological Treatment Combination in Service Area along Anchu Highway in Yunnan Province. *Highway Traffic Technology (application technology version)***.** **(12)** 34-36.

[6] Huang J 2004 Brief Description of the Sewage Treatment Facilities of Expressway *Communications Standardization.* **(01)** 57-59.

[7] Wang Z 2016 Soil Infiltration - Artificial Wetland - Ecological Floating Bed Combination Process to Treat Domestic Sewage *Academic Dissertation of Hefei University of Technology.*

[8] Liu D H 2011 Study on Sewage Ecological Treatment Process in Highway Service Area *Academic Dissertation of Shandong University.*

[9] Dong B C 2010 Study on the Design and Operation Control of Sewage Treatment in Highway Service Area *Academic Dissertation of Hebei University of Science and Technology.*

[10] Ma Y, Wang H J and Wang Y Y 2007 The Whole Process Management of Sewage Treatment Facilities in Highway Service Area. *Journal of wuhan university of technology (traffic science and engineering version).* **(06)** 1053-1056.

IOP Conf. Series: Earth and Environmental Science **167** (2018) 012013 doi:10.1088/1755-1315/167/1/012013

Removal of methyl violet dye via adsorption using activated carbon prepared from *Randu* sawdust (*Ceiba pentandra*)

Achmad Chafidz[1,*], Widi Astuti[2], Venitalitya Augustia[1], Dinda Tri Novira[2] and Nur Rofiah[2]

1 Department of Chemical Engineering, Universitas Islam Indonesia, Yogyakarta 55584, Indonesia
2 Department of Chemical Engineering, Universitas Negeri Semarang, Semarang 50229, Indonesia
*E-mail: achmad.chafidz@uii.ac.id

Abstract. In this study, randu sawdust carbon used as raw material for synthesis of activated carbon via KOH chemical activation assisted to adsorb methyl violet dye in the water. Activation by microwave radiation was conducted to accelerate the activation time. Randu sawdust char/carbon and randu sawdust activated carbon were characterized by FTIR, SEM, and BET. The results showed that the randu sawdust carbon and the activated carbon have a functional group which can adsorb methyl violet. Randu sawdust activated carbon has a larger pore and surface area than the randu sawdust carbon/char. The optimum adsorption occured at pH 9 and 90 minute of contact time. The maximum adsorption capacity of methyl violet dye by the randu sawdust activated carbon was approximately 531.16 mg g^{-1}.

1. Introduction

Dye production in the world is estimated at 7 x 10^5 to 10^6 tons per years, and approximately 15% of the dye wastes have been discharged to the environment [1]. One type of dyes that is most commonly found is methyl violet, which generally comes from industrial inks, paints and textiles [2]. The existence of this dye is dangerous for aquatic life because it is considered as mutagenic and mitotic poison, and thus carcinogenic. Various methods have been developed to overcome the pollution of this dye in the waters, including photocatalytic [3], membrane [4], and adsorption [5]. Adsorption is a promising alternative method that relatively simple, inexpensive, and can be applied at low concentrations. Several types of adsorbents have been used to reduce dye concentration in the waters. Activated carbon-MnOx hydrothermally from coconut shell is able to adsorb azo dye at 99.1% [6]. However, this process requires a relatively high energy, and therefore the active carbon is expensive and difficult to be applied in the industry. Therefore, the synthesis of activated carbon which is simple and inexpensive is needed. One of the example is the synthesis of activated carbon assisted by microwave radiation as heating system and with KOH as the activator.

Microwave is an electromagnetic wave that gives a heat effect when the wave is absorbed on the body object. Microwave radiation offers several advantages over conventional heating methods, including rapid temperature rise, a uniform temperature distribution, and reduction of processing time, as well as the energy consumption [7]. In this study, the synthesis of activated carbon using *randu* sawdust as a raw material has been investigated. *Randu* is widely available on the Java Island, but it has limited use and it has a low of economic value because it is generally used for casting of concrete for the building, but then go into the waste. Therefore, the potentials of this *randu* to be used as a raw material for the synthesis of an activated carbon have been investigated. *Randu* as an adsorbent has an ability to give the adsorption percentage over 90% [8], [9]. In the current study, synthesis of activated carbon prepared from *randu* sawdust was investigated. The prepared char/carbon and activated carbon were characterized for surface area and pore size distribution, Fourier Transform Infra-Red (FTIR) spectroscopy, and Scanning Electron Microscopy (SEM). The adsorption capacity of activated carbon was evaluated using methyl violet dye as the adsorbate. The effect of operating parameters, such as pH, contact time, initial methyl violet concentration were investigated.

2. Materials and methods

2.1. Adsorbate

Methyl violet (chemical grade reagent) was used as the adsorbate without further purification. The sample solution is prepared by dissolving methyl violet powder into distilled water at different initial concentrations (50, 100, 150, 200, 250, 300, 350, 400, 450, and 500 mg/L). The chemical structure of methyl violet shown in Figure 1.

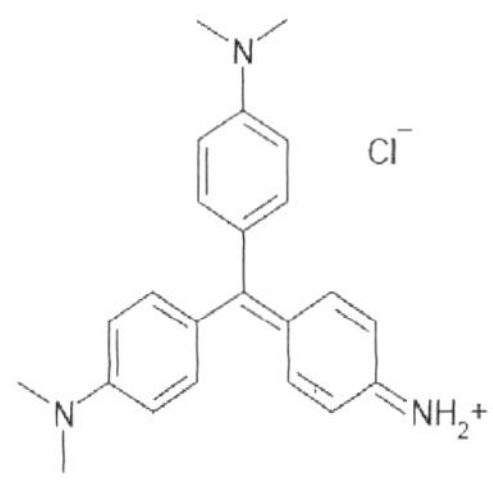

Figure 1. Chemical structure of methyl violet

2.2. Carbonization of Randu wood dust

Randu sawdust used as a raw material in this study was obtained from the local sawmill. To prepare the activated carbon, the *randu* sawdust was washed and then dried in the oven at 105°C to remove the moisture. The dried *randu* sawdust was then crushed and sieved using a 150 mesh sieve. Afterward, carbonization of the *randu* sawdust was achieved by heating it in a furnace at temperature of 700°C for 2 hours.

2.3. Microwave-induced Activation Process

The prepared char/carbon was mixed with potassium hydroxide (KOH) with an impregnation ratio of 1:1 (wt/wt%). The resulted char slurry was then placed in a glass reactor installed in the microwave chamber. The microwave was operated with an input power of 180 W and frequency of 2.45 GHz. The heating time was set at 10 min. Pure nitrogen flow was maintained at a flow rate of 1 cm^3 min^{-1} during the activation and cooling processes. The resulted activated carbon was washed with 0.1 M HCl and then rinsed with distilled water until pH of 6-7 was attained and finally dried in an oven at 120°C. The dried activated carbon is ready to use for the adsorption tests.

2.4. Characterization of char and activated carbon

The surface area of the char and activated carbon was determined using the Brunauer-Emmet-Teller (BET) method. Whereas, the average width of the pores and pore size distribution was determined using the Barrett-Joyner-Halenda (BJH). The surface functional groups were characterized by Fourier Transform Infra-Red (FTIR) spectroscopy in the scanning range of 4000-400 cm^{-1}. The surface morphology of char and activated carbon was characterized using a Scanning Electron Microscopy (SEM).

2.5. Adsorption experiments

Several batch adsorption experiments were conducted in 250 mL Erlenmeyer flasks containing 0.06 g of activated carbon and 100 mL of methyl violet solution and placed in a shaker at 120 rpm. The effect of pH in the range of 3, 5, 7, and 9 was studied. The pH was adjusted using 0.1 M HCl and 0.1 M

NaOH. The effect of contact times (5, 10, 15, 20, 30, 60, 90, 120, 150, and 180 min) was also carried out to study the equilibrium adsorption time. Methyl violet with different initial concentrations (50, 100, 150, 200, 250, 300, 350, 400, 450, and 500 mg/L) were also studied. The concentrations of methyl violet before and after the adsorption were analyzed via absorbance measurement using a UV-Vis Spectrofotometer at 581 nm. The methyl violet adsorbed at equilibrium q_e (mg/g), was calculated using Eq. (1):

$$q_e = \frac{(C_o - C_e)}{M} V \tag{1}$$

where C_o (mg/l) and C_e (mg/l) are the initial and equilibrium concentrations of methyl violet, respectively, V (L) is the volume of solution and M (g) is the dry mass of *randu* sawdust activated carbon [10].

3. Results and discussion

3.1. Characterization results of char and activated carbon

The FTIR analysis was conducted to confirm the presence of functional groups in the char and activated carbon of *randu* sawdust are capable to adsorb the methyl violet. The FTIR spectrum are shown in Figure 2.

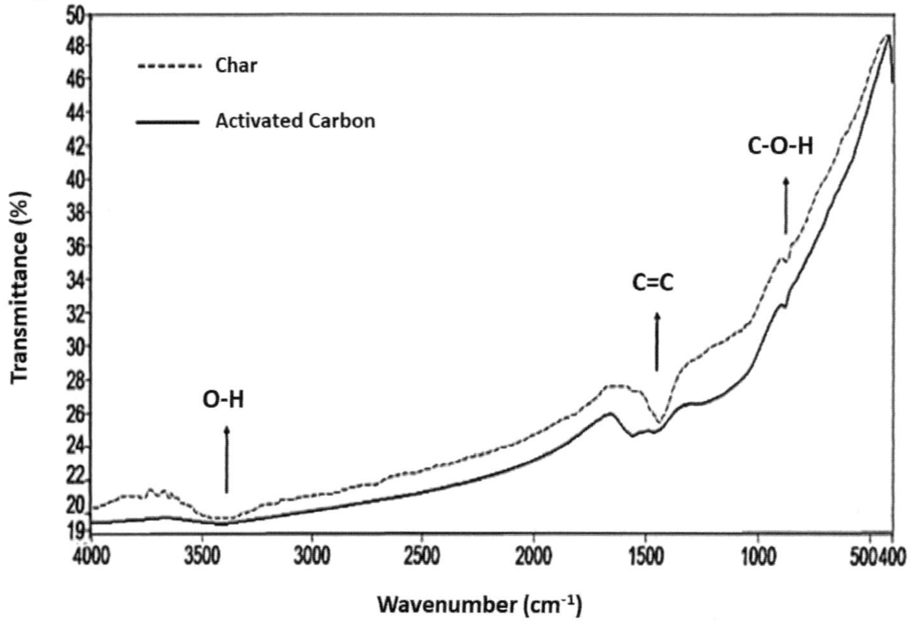

Figure 2. FTIR spectrum of char and activated carbon

The results of the randu carbon FTIR spectra showed broad absorption peak at 3431.32 cm^{-1} indicating the presence of O-H group, which were derived from the alcohol compounds. Absorption peaks at 1435.12 cm^{-1} indicating the presence of aromatic C=C groups. Whereas, the presence of group C-O-H was indicated by absorption peak at 877.15 cm^{-1}. There was also a decrease in absorption peaks at wavenumber of 3431.32 cm^{-1} and 1435.12 cm^{-1}. The reduced intensity of absorption peaks at wave number of 1435.12 cm^{-1} that belongs to C=C group was probably caused by a reaction between some Carbon with the KOH. Whereas, the reduced OH group at wave number of 3431.32 cm^{-1} was likely caused by the partial OH group, which reacted with KOH. While, the widening of the peak at about wave number of 1500 cm^{-1} was probably caused by the group C-O formation of aldehyde numbers waves almost coincide with the C=C. Additionally, Scanning Electron Microscopy images of the char and activated carbon at 5000 X magnification are shown in Figure 3.

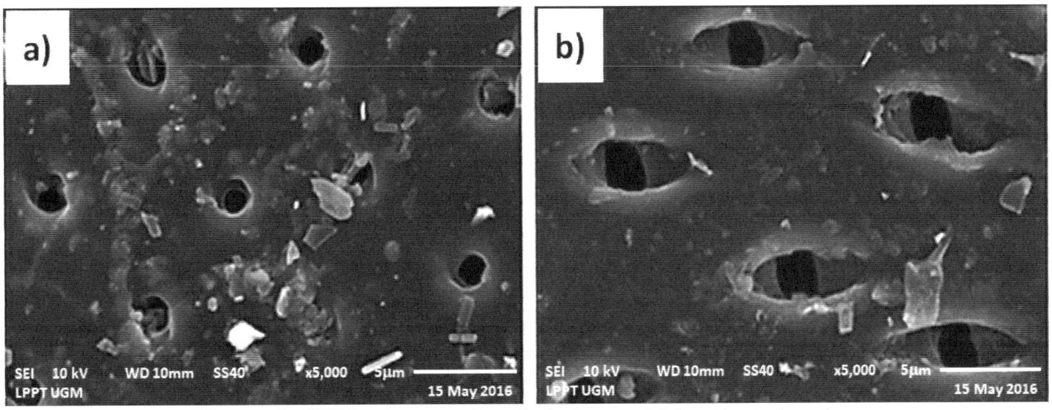

Figure 3. SEM micrographs of **a)** char and **b)** activated carbon

As seen in Figure 3a, the char seemed to have a lot of impurities product of carbonization. While the activated carbon was more pure, which is likely due to the N_2 flowing during the activation process (see Figure 3b). Additionally, the BET analysis shows that the *randu* sawdust activated carbon has a larger of specific surface area, i.e. 69.486 m^2/g. The pore size distribution of the activated was analyzed using the method of Barret Joyner Halenda (BJH), and was found to be approx. 17.230 nm. Whereas, the *randu* char/carbon has a specific surface area of 5.587 m^2/g with a pore size distribution approx. 17.185 nm.

3.2. Adsorption of methyl violet

Effect of pH on the adsorption of methyl violet studied at various pH 3, 5, 7, and 9 is shown in Figure 4. The figure shows that at pH 9 dye adsorbed amount reached the highest rate (99.76%). *Randu* sawdust activated carbon inclined protonated at low pH so that the active sites to be positively contents. This causes the repulsion between the active sites with methyl violet that was also positively contents, and hence the adsorption process is difficult to occur. It can be concluded that the optimum pH of methyl violet adsorption by *randu* sawdust activated carbon is at pH = 9.

The effect of contact time on the adsorbed of methyl violet by *randu* sawdust activated carbon at initial concentration of 100 mg/L is shown in Figure 5. Both the curves showed that the adsorption increased rapidly in the first 10 min, and thereafter gradually slowed down until it reached the equilibrium time. This phenomena was likely due to a large number of vacant surface sites were available for adsorption during the initial stage, and hence the tendency of methyl violet adsorbed by activated carbon was still high. Once the equilibrium was attained, the amount of methyl violet adsorbed did not change with the further increase of time.

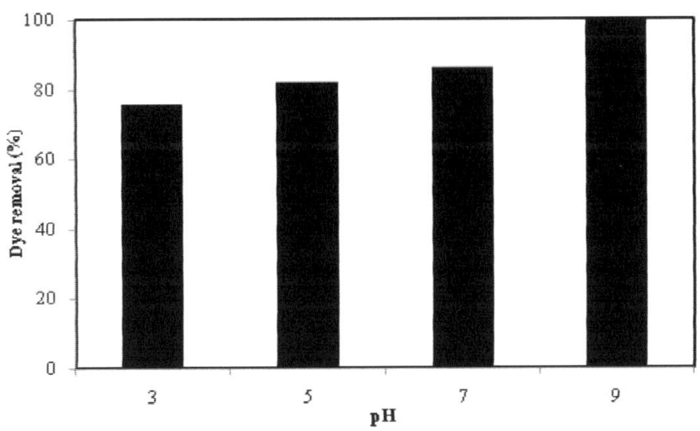

Figure 4. Effect of pH on the adsorbed of methyl violet (condition: initial concentration = 100 mg L^{-1}, contact time = 120 min, activated carbon dose = 0.6 g L^{-1})

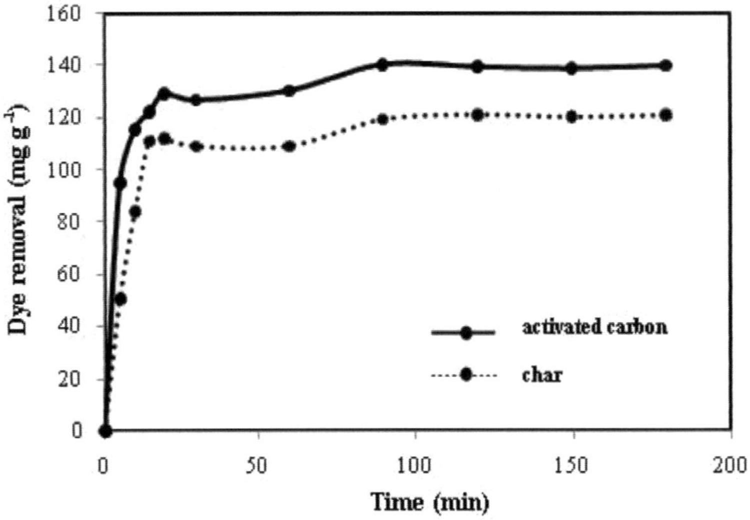

Figure 5. Effect of contact time on the adsorbed of methyl violet (condition: initial concentration = 100 mg L^{-1}, pH = 9, activated carbon dose = 0.6 g L^{-1})

Figure 6 shows that the amount of methyl violet adsorbed increased with the increase of methyl violet initial concentration and remained constant after reached the equilibrium. This was likely due to the fact that increasing concentration will also increase the driving force, so the rate of mass transfer of adsorbate onto the activated carbon will also be faster [11].

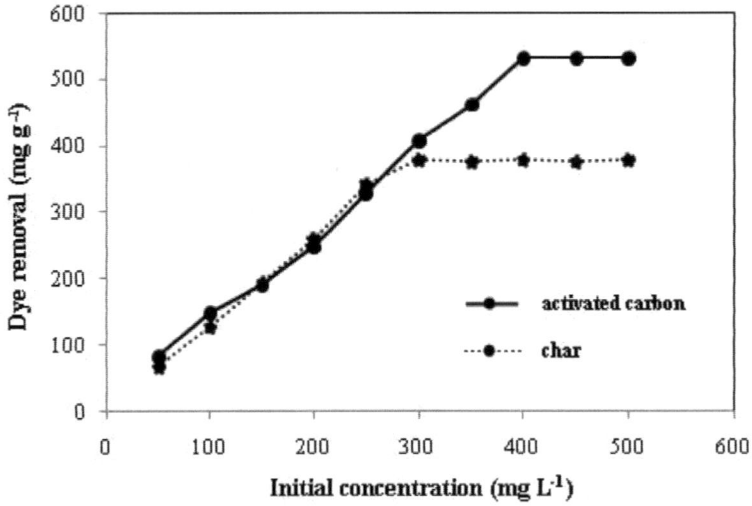

Figure 6. Effect of initial concentration on the adsorbed of methyl violet (condition: pH = 9, contact time = 90 min, activated carbon dose = 0.6 g L^{-1})

4. Conclusion

The current study has demonstrated the suitability of *randu* sawdust for the preparation of activated carbon by microwave heating-induced KOH activation. The high surface area activated carbon was produced in a short activation time by the microwave heating system. The optimum adsorption of the methyl violet was occured at pH 9 and the equilibrium time of 90 minute. The maximum adsorption capacity of methyl violet dye by *randu* sawdust activated carbon was approximately 531.16 mg g^{-1}.

5. References

[1] Gupta V K and Suhas 2009 Aplication of Low-cost Adsorbent for Dye Removal – A Review *Journal Enviromental Management* 90: 2313-2342.
[2] Christina M, S Mu'nisanatun, Saptaji R, Marjanto D 2007 Studi Pendahuluan Mengenai Degradasi Zat Warna Azo (Metil Orange) dalam Pelarut Air Menggunakan Mesin Berkas Elektron 350 kev/10 ma *Jurnal Forum Nuklir* Vol. 1 No. 1.
[3] Zhang Y, Sun X, Yang G, Zhu Y, Si H, Zhang J, and Li Y 2016 Preparation and Characterization of Bifunctional $BiOCl_xI_y$ Solid Solutions with Excellent Adsorption and Photocatalytic Abilities for Removal of Organic Dyes *Materials Science in Semiconductor Processing* 41: 193-199.
[4] Shi J, Wu T, Teng K, Wang W, Shan M, Xu Z, Lv H, and Deng H 2016 Simultaneous Electrospinning and Spraying Toward Brach-Like Nanofibrous Membrans Functionalised with Carboxylated MWCNTs for Dye Removal *Materials Letters* 166: 26-29.
[5] Fontana K B, Chaves E S, Sanchez J D S, Watanabe E R L R, Pietrobelli J M T A, and Lenzi G G 2016 Textile Dye Removal from Aqueous Solution by Malt Bagasse: Isotherm, Kinetic and Thermodynamic Studies *Ecotoxicology and Evironmental Safety* 124: 329-336.
[6] Xu L, Li X, Ma J, Wen Y, and Liu W 2014 Nano-MnO_x on Activated Carbon Prepared by Hydrothermal Process from Fast and Highly Efficient Degradation of Azo Dyes *Applied Catalysis A: General* 485: 91-98.
[7] Yuen F K and Hameed B H 2009 Recent Developments in the Preparation and Regeneration of Activated Carbon by Microwaves *Advances Colloid Interface Science* 149: 19-27.
[8] Astuti W, Sulistyaningsih T, and Maksiola M 2016 Chemical Modified Kapok Sawdust as Adsorbent of Methyl Violet Dye from Aqueous Solution *Jurnal Teknologi* 78(9): 35-42.
[9] Maziyyah D F 2015 Modifikasi Adsorben Berbasis Kayu Randu dengan Metode Pemanasan dan Aplikasinya sebagai Penjerap Zat Warna Methyl Violet pada Limbah Industri Batik *Tugas Akhir Jurusan Teknik Kimia Universitas Negeri Semarang*.
[10] Njoku V O, Foo K Y, Asif M, and Hameed B H 2014 Preparation of Activated Carbon from Rambutan (Nephelium lappaceum) peel by Microwave-induced KOH Activation for Acid Yellow 17 Dye Adsorption *Chemical Engineering Journal* 250: 198-204.
[11] Zou W, Bai H, Gao S, Li K 2013 Characterization of modified sawdust, kinetic and equilibrium studyabout methylene blue adsorption in batch mode. *Korean Journal of Chemical Engineering* 30 (1): 111-122.

IOP Conf. Series: Earth and Environmental Science **167** (2018) 012014 doi:10.1088/1755-1315/167/1/012014

Application of copper-zinc metal as a catalytic converter in the motorcycle muffler to reduce the exhaust emissions

Achmad Chafidz[1,*], Megawati[2], Catur Rini Widyastuti[3], Venitalitya Augustia[1], Khairatun Nisa[2], Ratnaningrum[2]

1 Department of Chemical Engineering, Universitas Islam Indonesia, Yogyakarta 55584, Indonesia
2 Department of Chemical Engineering, Universitas Negeri Semarang, Semarang 50229, Indonesia
3 Technische Universität Darmstadt, Heidelberger Landstraße 246, 64297 Darmstadt, Germany
*E-mail: achmad.chafidz@uii.ac.id

Abstract. One of the main sources of air pollution is exhaust emission from vehicles. In the developing countries such as Indonesia, the most highest number of vehicles is motorcycle. Some of gas components from the exhaust emission that are commonly measured are carbon monoxide (CO) and hydrocarbon (HC). One of technology that can reduce the emission of CO and HC is the use of catalytic converter in the motorcycle's exhaust system/muffler to assist the oxidation of carbon monoxide (CO) and hydrocarbon (HC). In the present study, Copper-Zinc (Cu-Zn) metal was selected as the catalyst. The selection of this combination of non-noble metals is due to their low cost materials, abundant materials, low cost production, and low-temperature oxidation). The objective of this study are to prepare the catalytic converter based on non-noble metal (i.e. Cu-Zn) and to study the performance of the catalytic converter installed in the muffler in converting exhaust gases of CO and HC into less harmful gases. The effect of two different motor engine rotation speeds (i.e. 2000 and 2500) on the conversion of CO and HC was also studied. The analysis of the emission from the muffler was carried out using gas analyzer. Additionally, the kinetic of oxidation reaction of CO and HC was also studied using pseudo homogeneous approach. The emission analysis results show that the use of Cu-Zn catalyst was effective to decrease the emission level of CO and HC. The CO and HC concentration level decreased up to 47.71 % at 2000 rpm and 55.34 % at 2500 rpm, respectively.

1. Introduction

One of the main sources of air pollution is exhaust emission from vehicles. In the developing countries such as Indonesia, the most highest number of vehicles is motorcycle. The data from Central Bureau of Statistics (*Badan Pusat Statistik*) of Indonesia about the number of vehicles (by type) in the range of year 1990-2015 is presented in Figure 1. As noticed in the figure, the number of motorcycle was about 100,000,000 and it was four times of the total of the other types in 2015. Whereas, the total number of all the types of vehicle was about 120,000,000. This number itself is a major concern, since the main source of air pollutant in Indonesia is the exhaust emission from vehicles. Data from the Ministry of Environment of Indonesia shows that the major components of air pollution from motor vehicles, especially *Premium*^{TM} fuel (spark ignition engines) are carbon monoxide (CO) and hydrocarbon (HC).

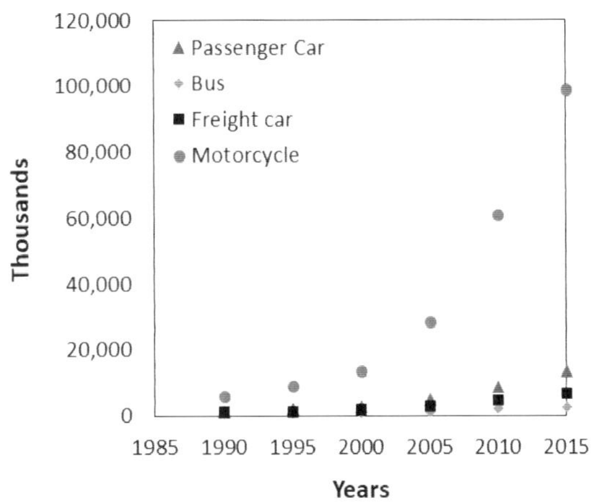

Figure 1. Number of Motor Vehicles in Indonesia by types (1990-2015) [1]

Hydrocarbon (HC) is a class of partially burned fuel. HC is considered to be toxic, since it can cause asthma, lung and liver diseases, and cancer if exposed for long time. Whereas, the carbon monoxide (CO) is a product of incomplete combustion of fuel. The CO is also toxic and harmful to human. Because, if it is inhaled, it will reduce the blood's ability to carry oxygen, hence overdosage of CO in the body may cause death. Therefore, permitted level of HC and CO in the exhaust emission of vehicles are often regulated by the government.

Several alternative technologies have been considered to reduce the emission levels of the vehicles, including engine design improvement, fuel pre-treatment, exhaust treatment, etc [2]. Among those technologies, the one that considered to be the best method that can effectively reduce the concentration level of CO and HC of motor vehicle (especially gasoline-fueled engine) is by installing a catalytic converter in the vehicle's exhaust system/muffler [3], [4]. The use of catalytic converter was first introduced in automobiles in the United States market in 1975 [4]. A catalytic converter is a device that installed in the vehicle's exhaust system/muffler to help reducing the toxicity level of emission produced by the internal combustion engine [2], [5]. There are two types of catalysts that work in the catalytic converter, a reduction and an oxidation catalysts. The reduction catalyst helps to reduce the NOx emissions (e.g. NO or NO2), whereas the oxidation catalyst helps to reduce the unburned HC and CO [2]. In this study, the type of catalyst used in the catalytic converter was oxidation catalyst.

The purpose of installing a catalytic converter in the motorcycle's exhaust system/muffler is to assist the oxidation of CO and HC emission resulted from the combustion chamber of the engine to decrease their level before released to the atmosphere [3]. The objective of oxidation reaction is to convert CO and HC emission into less harmful gases such as carbon dioxide (CO2) and water vapor (H2O) by reacting them with the remaining oxygen (O2) in the exhaust gas. The oxidation reactions of CO and HC are shown in Eq. (1) and Eq. (2), respectively [3], [4].

$$2\,CO + O_2 \rightarrow 2\,CO_2 \qquad (1)$$

$$HC + O_2 \rightarrow CO_2 + H_2O \qquad (2)$$

The common materials used as catalysts for the catalytic converter are noble metals such as Palladium (Pd), Platinum (Pt), and Rhodium (Rh), Iridium (Ir), etc. The drawback of the use of such metals is of course the price of the metals, which is relatively expensive. They are also not abundantly available, and operated at high temperature [5], [6]. Therefore, it is necessary to investigate the prospective of non-noble materials to replace the noble ones. Transition metals (non-noble metals) with high availability and relatively inexpensive can become one of the alternatives for the catalyst [2].

Copper-Zinc (Cu-Zn) material has been selected as catalyst because it has high activity in the exhaust system, low cost production, the availability is abundant, and operated at a lower temperature than the noble material. Research studies about the influence of transition metal catalyst to reduce the emission level of CO and HC have been carried out. Nevertheless, the study about the reaction kinetic was still limited. Several literatures have reported the study of oxidation of CO and HC by using other types of catalyst including silver (Ag) [7], Nickel (Ni) [2], Platinum (Pt) [8], Gold/Silver/Zirconia (Au5AgZr14) [9], Iridium-Gold (Ir-Au) [10], and Ir/FeOx [11].

To the best of our knowledge, research studies on the kinetic of catalytic converter made of Cu-Zn metal alloy to reduce emission of CO and HC was very limited and close to none. Therefore, the objective of this study are to prepare the catalytic converter based on non-noble metal (i.e. Cu-Zn) and to study the performance of the catalytic converter installed in the muffler in converting exhaust gases of CO and HC into less harmful gases. Additionally, the kinetic of oxidation reaction of CO and HC was also studied using pseudo homogeneous approach.

2. Experimental

2.1. Preparation of catalyst

Design innovation of Copper-Zinc (Cu-Zn) metal as a catalyst was a cylindrical shape with 6,6 cm of length; 2,4 cm of diameter and 10 tubes in the inner of the cylinder (see Figure 2).

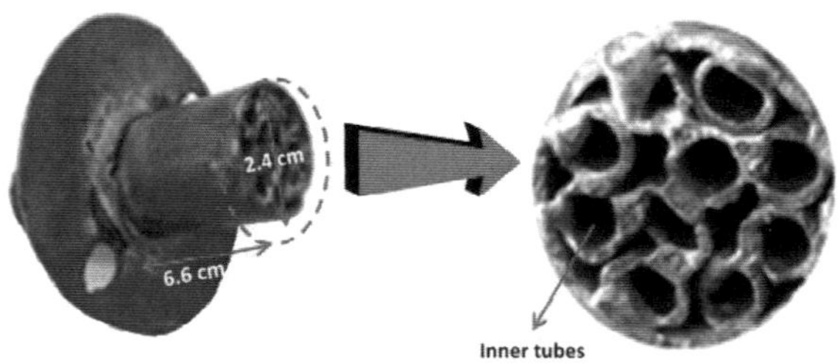

Figure 2. Design of the Copper-Zinc (Cu-Zn) catalyst

2.2. Preparation of motorcycle muffler

In this study, motorcycle's muffler of Honda Supra Fit 2004 was used for experiment. The muffler has been modified by installing copper-zinc (Cu-Zn) catalytic converter (see Figure 2).

2.3. Exhaust emissions analysis

Emission test was conducted to determine the concentration level (CO and HC) of exhaust emission from the motorcycle muffler. Measurements of exhaust emission were carried out at motor engine rotation speeds of 2000 and 2500 rpm. The schematic experiment procedure is shown in Figure 3.

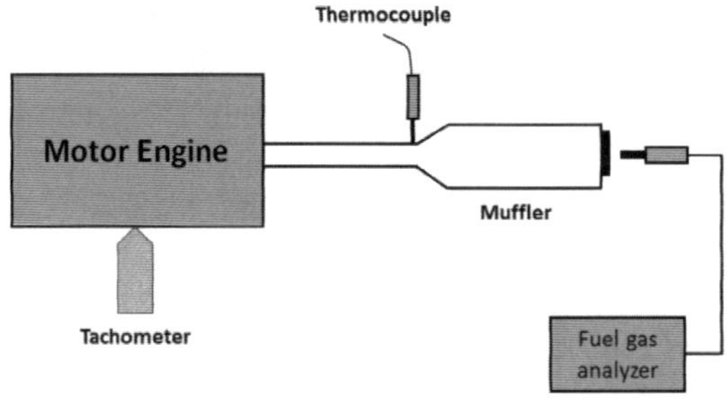

Figure 3. Schematic procedure of the emission test

2.4. Reaction kinetic study

Oxidation kinetics model of CO and HC was approximated by pseudo-homogeneous model. Model of pseudo homogeneous reaction was used with the assumption that the mass transfer of reactants in the solid (catalyst) was very fast, hence it could be ignored on the speed of the reaction [12]. Another basic assumption of modelling pseudo-homogeneous in the reactor was only one phase, but the real condition were two phases [12].

3. Results and discussion

3.1. Analysis results of exhaust emission

Figure 4a shows the analysis results of CO concentration in the exhaust emission from the motorcycle muffler. The figure shows the effect of using Cu-Zn catalyst on the reduction of CO concentration at two different motor engine rotation speeds (i.e. 2000 and 2500 rpm). The motor engine rotation speed depends on the fuel combustion process. As indicated in Figure 4a, the more fuel used for the combustion, the higher the engine rotation speed, and the higher the CO concentration in the exhaust emission from the muffler. Additionally, as seen in the figure, there is a significant decrease of CO concentration in the exhaust emission from the muffler that used the Cu-Zn catalyst, for both engine speeds (i.e. 2000 and 2500 rpm). The decreasing percentage was about 47.71% and 38.86% for the engine speed of 2000 rpm and 2500 rpm, respectively. Therefore, by applying the Cu-Zn catalyst as a catalytic converter in the motorcycle muffler we can reduce the emission of CO from the motorcycle, by converting the CO into CO2 (according to Eq. (1)). The regulation from the Ministry of Environment about exhaust emission standard of old motor vehicle stated that the maximum level of CO allowed is 5.5 (% vol) at motor engine rotation speed of 2400 rpm at idle condition. As expected, as seen in Figure 4a, the CO level in exhaust gas from the muffler that used Cu-Zn catalyst was much lower than the CO level allowed by the Ministry of Environment.

Additionally, the analysis results of HC concentration in the exhaust gas from the motorcycle muffler is shown in Figure 4b. The figure shows the effect of using Cu-Zn catalyst on the reduction of HC concentration (in ppm vol.) at two different engine rotation speeds (i.e. 2000 and 2500 rpm). As seen in the figure, the concentration of HC decreased with the increase of motor engine rotation speed. It happened because the HC is the main component of the fuel, the higher the engine rotation speed, the more amount of fuel is needed, and thus the HC concentration decreased. Furthermore, as seen in Figure 4b, there were significant decreases of HC concentration in the exhaust gas from the muffler that used the Cu-Zn catalyst. The HC concentration level decreased from 1564.11 ppm to 925.78 ppm (i.e. 40.81 %) and from 1241.78 ppm to 554.56 ppm (i.e. 55.34 %), for the engine rotation speed of 2000 rpm and 2500 rpm, respectively. Therefore, by applying the Cu-Zn catalyst as a catalytic converter in the motorcycle muffler we can reduce the concentration level of HC from the motorcycle exhaust gas, by converting the HC into CO_2 and H_2O. As expected, the HC concentration level in exhaust gas from the muffler that used Cu-Zn catalyst was much lower than the HC level allowed by the Ministry of Environment. Additionally, as noticed, the decreasing percentage of the HC concentration for the engine rotation speed of 2500 rpm was higher than that of 2000 rpm. It is because, the higher the engine rotation speed (i.e. 2500 rpm), the higher temperature of the exhaust system, and thus the rate of oxidation reaction of hydrocarbon in the Cu-Zn catalytic converter also become faster.

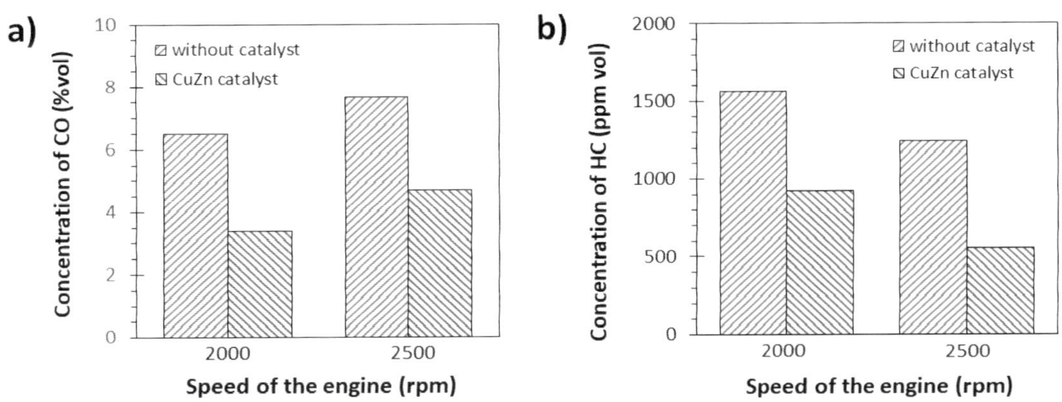

Figure 4. The effect of Cu-Zn catalyst toward **a)** CO concentration and **b)** HC concentration

Additionally, the analysis results of CO_2 concentration in the exhaust gas from the motorcycle muffler is shown in Figure 5. The figure shows the effect of using Cu-Zn catalyst on the reduction of CO2 concentration (in %vol.) at two different engine rotation speeds (i.e. 2000 and 2500 rpm). As seen in the figure, the concentration of CO_2 for both engine rotation speeds were similar. In other hand, there were significant increases of CO_2 concentration in the exhaust gas from the muffler that used the Cu-Zn catalyst. The CO2 concentration level increased from 4.96 %vol. to 6.22 %vol (i.e. 25.40 %) and from 4.42 %vol to 6.33 %vol (i.e. 43.21 %), for the engine rotation speed of 2000 rpm and 2500 rpm, respectively. In the motor engine, during the combustion reaction, a complete reaction between the fuel and oxygen will produce CO2 along with water vapor. While, an incomplete combustion will produce CO. This CO is converted into CO_2 according to Eq. (1) with the help of the Cu-Zn catalytic converter. Therefore, by applying the Cu-Zn catalyst as a catalytic converter in the motorcycle muffler will decrease the CO concentration level (as discussed in the previous paragraph), and in the same time will increase the CO_2 concentration level (as a result of CO conversion).

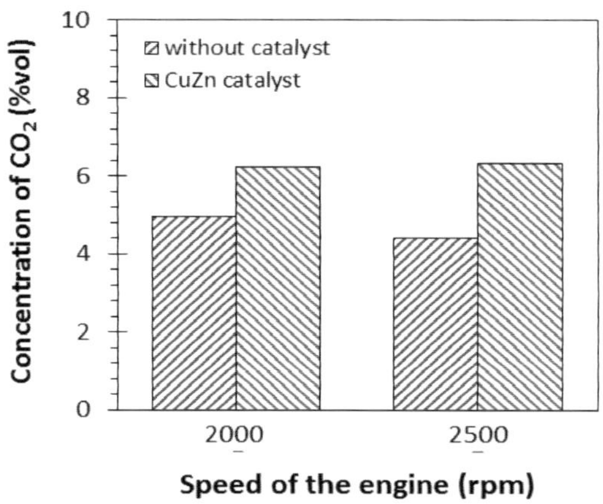

Figure 5. The effect of Cu-Zn catalyst toward CO_2 concentration

3.2. Kinetics study

In this work, the kinetics of the oxidation reaction of CO and HC into CO_2 and H_2O (i.e. Eq. (1) and Eq. (2)) with the help of the Cu-Zn catalyst was studied. The oxidation kinetics model was approximated by pseudo-homogeneous model.

3.2.1. Kinetic of oxidation reaction of CO

The oxidation reaction was assumed to be first order reaction, in which the reaction rate is directly proportional to the concentration of reactant. Figures 6a and 6b show the data fitting of order reaction for CO oxidation reaction at engine rotation speeds of 2000 and 2500 rpm, respectively. As seen in Figure 6, the data fitting of CO oxidation reaction by assuming the first order reaction gives the values of correlation coefficient (R^2) of 0.9669 and 0.9177 at engine rotation speeds of 2000 rpm and 2500 rpm, respectively. Another assumption that can be made is the use of second order reaction. The difference between first and second order reaction is the change of reactant concentration. In second order, the reaction is faster than the first order. Another factor affecting the order of reaction is temperature, because the kinetic energy of the molecules increase with increasing temperature. Hence it will increase the number of collisions between molecules per unit time, which leads to the faster reaction. According to research of Westley 1980 [13], oxidation reaction of CO at temperature of 1500 – 3000 K used second order reaction by Eq. (3), as follows:

$$CO + O_2 \rightarrow CO_2 + O \tag{3}$$

Whereas, the temperature of the oxidation reaction measured in the muffler was approximately 336 – 338 K, which was much lower than the Wesley (i.e. 1500 – 3000 K). Therefore, it can concluded that the CO oxidation reaction occurred in the muffler with the Cu-Zn catalytic converter has followed the first order reaction.

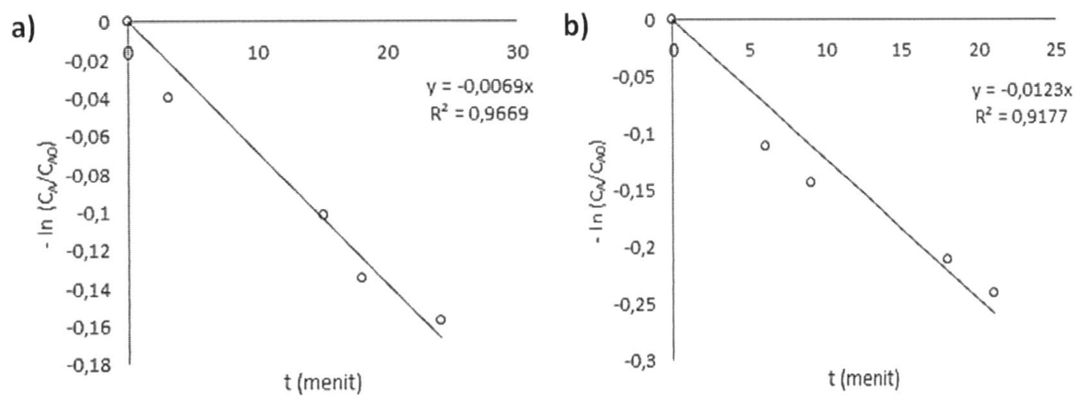

Figure 6. Data fitting of first order reaction of CO oxidation: **a)** 2000 rpm; and **b)** 2500 rpm

3.2.2. Kinetic of oxidation reaction of HC.
According to research of Westley 1980 [13], oxidation reaction of HC at temperature of 298 – 650 K is following second order reaction. Whereas, the temperature of the oxidation reaction measured in the muffler was approximately 336 – 338 K, which was in the range of the Wesley (i.e. 298 – 650 K). Therefore, it can be assumed that the HC oxidation reaction occurred in the muffler with the Cu-Zn catalytic converter has followed the second order reaction. To validate this assumption, data fitting of HC oxidation for both engine rotation speeds (i.e. 2000 rpm and 2500 rpm) using second order reaction was carried out, and the data fitting results are presented in Figures 7a and 7b, respectively.

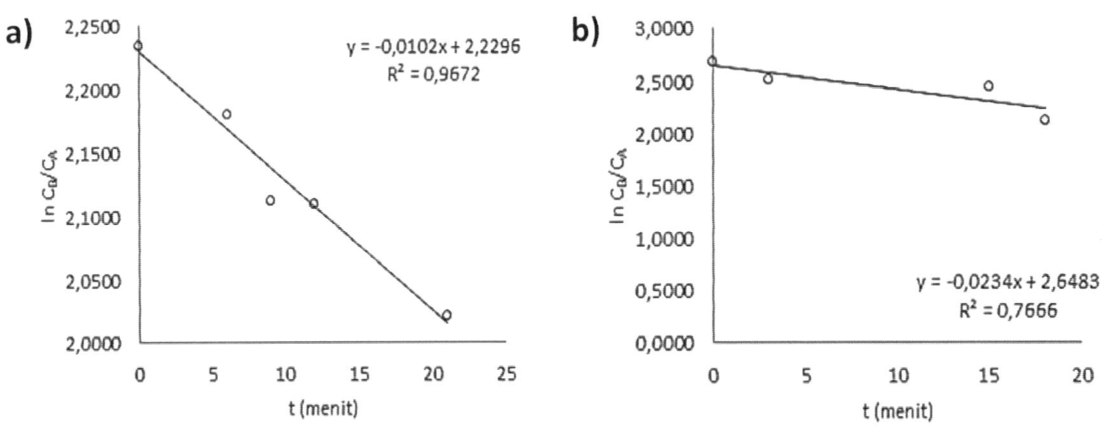

Figure 7. Data fitting of second order reaction of HC oxidation: **a)** 2000 rpm; and **b)** 2500 rpm

3.2.3. Activation energy of CO oxidation.
Generally, the temperature change can affect the value of reaction rate constant (k). When the temperature increases, the value of reaction rate constant (k) also increases. Factors influencing of activation energy are temperature (T), pre-exponential factor (A), and a catalyst. The value of activation energy (E) in this study was 132.54 kJ/mol. According to research of Delgado 2014, the value of activation energy (E) to CO oxidation reaction using a nickel (Ni) catalyst at 373 – 1273 K was 123.60 kJ/mol. The value of E obtained in this study was higher than Delgado's research. It was likely because the activation energy is getting smaller as the temperature rises. Figure 8 shows the effect of (ln k) and 1/T on the first order reaction of CO oxidation.

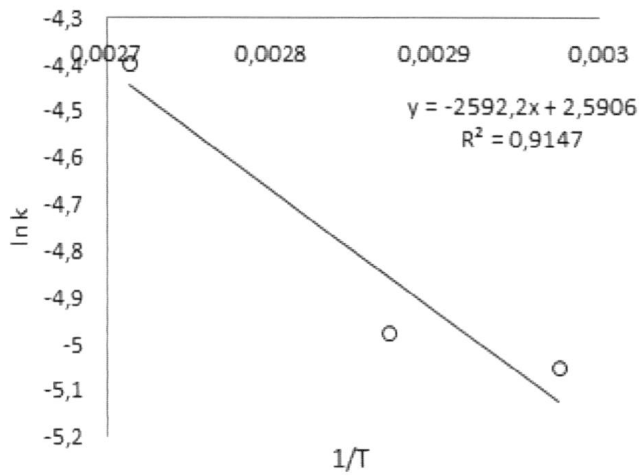

Figure 8. (ln k) versus $1/T$ on the first order reaction of CO oxidation.

3.2.4. Activation energy of HC oxidation.
The value of activation energy (E) in this study was 37.55 kJ/mol. According to research of Westley 1980 [13]. The value of activation energy (E) for HC oxidation reaction at 298 – 650 K was 24.27 kJ/mol. The value of E obtained in this study was higher than Westley's research because the range of temperature from Westley's research was higher than in this study. The activation energy decreases as the temperature increases. Figure 9 shows the effect of (ln k) and $1/T$ on the second order reaction of HC oxidation

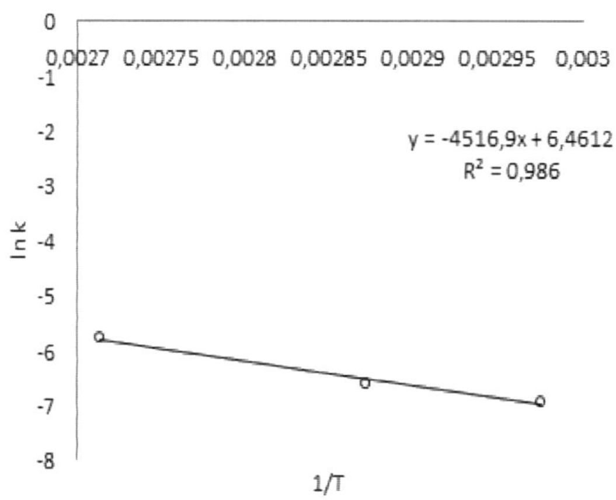

Figure 9. (ln k) versus $1/T$ on the second order reaction of HC oxidation

4. Conclusion

In the present work, the application of catalytic converter based on copper-zinc (Cu-Zn) to reduce the exhaust emission, especially the carbon monoxide (CO) and hydrocarbon (HC) was investigated. The emission test results clearly exhibited the effectiveness of Cu-Zn as a catalyst for the catalytic converter to reduce CO and HC emission level. The CO and HC emission level decreased up to 47.71 % at 2000 rpm and 55.34 % at 2500 rpm, respectively. Additionally, in the kinetic study, the activation energy of oxidation CO and HC were found to be approximately 132.54 kJ/mol and 37.55 kJ/mol, respectively. Based on these results, Cu-Zn based on non-noble material can been considered as a prospective catalyst used in the catalytic converter to reduce the emission level of CO and HC.

5. References

[1] https://www.bps.go.id/linkTableDinamis/view/id/1133 (accessed on January 3, 2018)
[2] Makwana N R, Amin C M and Dabhi S K 2013 *International Journal of Advanced Engineering Technology* 4, 10.

[3] Sharma S, Goyal P and Tyagi R 2016 *International Journal of Ambient Energy* 37, 507.
[4] Bera P and Hegde M 2010 *J. Indian Inst. Sci.* 90, 299.
[5] Patel N B, Patel S B, Patel M S and Patel D 2013 *International Journal for Scientific Research & Development.*
[6] Kalam M, Masjuki H, Redzuan M, Mahlia T, Fuad M, Mohibah M, Halim K, Ishak A, Khair M, Shahrir A 2009 *Sadhana* 34(4), 467-481.
[7] Keulks G W and Chang C C 1970 *The Journal of Physical Chemistry* 74, 2590.
[8] Aghalayam P, Park Y and Vlachos D 2000 *Proceedings of the Combustion Institute* 28, 1331.
[9] Baiker, A.; Maciejewski, M.; Tagliaferri, S.; Hug, P., J. Catal. 151, 407 1995.
[10] Bokhimi X, Zanella R, Angeles-Chavez C 2010 *Journal of Physical Chemistry C* 114, 14101-14109.
[11] Zhang Y, Wu Y, Wang H, Guo Y, Wang L, Zhan W, Guo Y, Lu G Catal 2015 *Commun.* 61, 83.
[12] Elnashaie S S, Elshishini S S 1993 *Modelling, Simulation, and Optimization of Industrial Fixed Bed Catalytic Reactor* (Gordon and Breach Science Publisher).
[13] Westley, Francis 1980 *Table of Recommended Rate Constant for Chemical Reactions Occuring in Combustion* (Washington D C).

Towards Smart Cities Development: A Study of Public Transport System and Traffic-related Air Pollutants in Malaysia

S N Brohi[1*], T R Pillai[1], D Asirvatham[1], D Ludlow[2], J Bushell[2]

1 School of Computing & IT, Taylor's University, Selangor, Malaysia
2 Department of Architecture and the Built Environment, University of the West of England, Bristol, UK
E-mail: SarfrazNawaz.Brohi@taylors.edu.my

Abstract. Increasing number of privately owned vehicles are depicting Malaysians preferred mode of mobility and lack of interest in the public transport system. In most developing countries such as Malaysia, motorized vehicles are the major contributors to air pollution in urban zones. Air pollution is a silent killer as it infiltrates the vital organs, leading to serious diseases and death. This research critically analyses the emissions of air pollutants such as CO, NO_2, SO_2, hydrocarbon, and PM from various sources in Malaysia with emphasis mainly on the emission of pollutants from motor vehicles. This research also discusses the public transport initiatives undertaken by the government of Malaysia such as enhancing the bus and rail system, transforming Malaysia's taxi system, managing travel demand and enhancing the integration of urban public transport system. Furthermore, considering the smart cities initiatives, this research identified that weather, safety, security and inappropriate infrastructure are major barriers in Malaysia's move towards the implementation of smart and eco-friendly mobility practices such as cycling, carpooling and car sharing.

1. Introduction

Malaysian citizens are less interested in non-motorized modes of mobility such as walking or cycling and not fully aware of the harmful effects of motorized transportation [1]. There is no doubt that transport sector is fulfilling the mobility requirements and contributing to strengthening the economy but pollution from transport is also an emerging concern that needs to be addressed [2], [3]. Public Transport System (PTS) in Malaysia is not advanced yet as in well-developed European countries. Therefore, there is massive use of privately owned vehicles. According to Malaysian Institute of Road Safety Research (MIROS), the registration of vehicles in Malaysia is significantly growing every year, as of 2016, over 27 million vehicles have been registered. The increase in the number of vehicles registration leads to accidents and crashes causing deaths, serious and slight injuries as shown in table 1 [4]. Apart from accident and crashes, air pollution is one of the major problems that is caused by growing number of vehicles on the road. Traffic congestion pollutes the environment, health, and lifestyle of the citizens. Most of the Carbon Monoxide (CO) emissions in Malaysian atmosphere are generated from motor vehicles. Malaysia is second biggest per capital Greenhouse Gas (GHG) emitter among the ASEAN nations [5]. In spite of the fact that, Malaysia shares just 0.3% of worldwide GHG emission [6]. The real concern lies in the regularly expanding pattern of GHG emission. At the point when many developed nations have effectively reduced the GHG discharge, in Malaysia emission level keeps expanding [7].

Table 1. MIROS: Vehicle Statistics and Accidents in Malaysia (1997 – 2016).

Year	Registered Vehicles	Population	Road Crashes	Road Deaths	Serious Injury	Slight Injury
1997	8,550,469.00	21,665,600.00	215,632.00	6,302.00	14,105.00	36,167.00
1998	9,141,357.00	22,179,500.00	211,037.00	5,740.00	12,068.00	37,896.00
1999	9,929,951.00	22,711,900.00	223,166.00	5,794.00	10,366.00	36,777.00
2000	10,598,804.00	23,263,600.00	250,429.00	6,035.00	9,790.00	34,375.00
2001	11,302,545.00	23,795,300.00	265,175.00	5,849.00	8,680.00	35,944.00
2002	12,068,144.00	24,526,500.00	279,711.00	5,891.00	8,425.00	35,236.00
2003	12,819,248.00	25,048,300.00	298,653.00	6,286.00	9,040.00	37,415.00
2004	13,828,889.00	25,580,000.00	326,815.00	6,228.00	9,218.00	38,645.00
2005	15,026,660.00	26,130,000.00	328,264.00	6,200.00	9,395.00	31,417.00
2006	15,790,732.00	26,640,000.00	341,252.00	6,287.00	9,253.00	19,885.00
2007	16,813,943.00	27,170,000.00	363,319.00	6,282.00	9,273.00	18,444.00
2008	17,971,907.00	27,730,000.00	373,071.00	6,527.00	8,868.00	16,879.00
2009	19,016,782.00	28,310,000.00	397,330.00	6,745.00	8,849.00	15,823.00
2010	20,188,565.00	28,910,000.00	414,421.00	6,872.00	7,781.00	13,616.00
2011	21,401,269.00	29,000,000.00	449,040.00	6,877.00	6,328.00	12,365.00
2012	22,702,221.00	29,300,000.00	462,423.00	6,917.00	5,868.00	11,654.00
2013	23,819,256.00	29,947,600.00	477,204.00	6,915.00	4,597.00	8,388.00
2014	25,101,192.00	30,300,000.00	476,196.00	6,674.00	4,432.00	8,598.00
2015	26,301,952	31,190,000	489,606	6,706	4,120	7,432
2016	27,613,120	31,660,000[e]	521466[a]	7152[a]	NA	NA

[e] Estimated value from Department of Statistics Malaysia.
[a] Media statement.
[NA] Not available (The official figures are not available yet).

The high demands of mobility and ineffective PTS resulted in an increase of cars in contrast to growth In 1995, approximately three individuals had one vehicle that reduced to 1.4 people having one vehicle in 2010 [8]. With the increasing number of vehicles, the consumption of diesel and petrol has rapidly increased. Consequently, 36% of the national energy is consumed by the transport sector. Therefore the sector is the major emitter of Carbon Dioxide (CO_2) and other air pollutants [9], [10]. The rest of this research paper is organized as follows: the emission of air pollutants from traffic are discussed in Section 2. The reasons behind Malaysian citizens not using public transport are explained in Section 3. Section 4 contains discussion on the initiatives of the Malaysian government to improve PTS by 2020. The era of smart mobility and its implementation challenges in Malaysia are discussed in Section 5, and the research is concluded in Section 6 with the discussion of its future direction.

2. Emissions of Traffic-related Air Pollutants

It was estimated that in 2015, the combined air pollutant emission load accumulated to 2,001,195 metric tonnes of CO, 835,927 metric tonnes of Nitrogen Oxides (NO_2), 209,156 metric tonnes of Sulphur Dioxide (SO_2) and 23,904 metric tonnes of Particulate Matter (PM). Emission load for CO had increased in 2015 compared to 2014 due to high consumptions of fuel oil and coke which were used as fuels in industries and, power and heat generation plants. The declination of 0.10% in CO emission load, 5.56% in SO_2 emission load and 6.89% in PM emission load compared to 2014 could be due to lower fuel consumption in agricultural activities. Power plants contributed to the highest SO_2 emission load (54%), followed by others (32%), industries (7%) and motor vehicles (7%). The highest contributor of PM was industries (32%) followed by power plants (31%), others (20%) and motor vehicles (17%). The highest contributor of NO_2 was power plants (65%) followed by motor vehicles (27%), others (6%) and industries (2%). However, motor vehicles remain the highest contributor of CO (95%) as shown in Figure 1 [11]. Others in Figure 1 represents air pollutant sources from residential, commercial, non-energy use and agriculture.

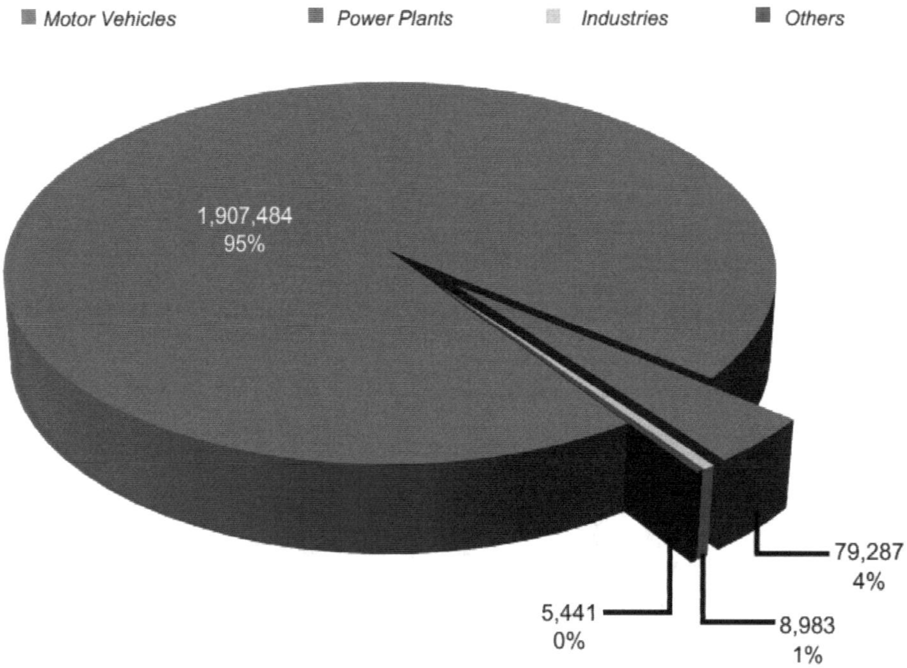

Figure 1. CO Emission from Various Sources (Metric Tonnes) – 2015.

In 2015, like the past years, emission from motor vehicles remained the major source of air pollution, especially in urban areas. There was an overall increase in the number of motor vehicles registered. The number of registered passenger cars, motorcycles, goods vehicles, taxis and buses increased by 4.40%, 3.10%, 2.47%, 1.68% and 2.39% respectively as compared to 2014. The number of in-use vehicles has also shown an ascending trend in which the number of passenger vehicles, motorcycles, goods vehicles, and taxis increased by 4.90%, 0.86%, 0.54% and 0.57% respectively as compared to 2014. Meanwhile, for the buses, it is still a descending trend of 7.13% compared to 2014. The overall statistics of vehicles in-use in Malaysia as reported by the Road Transport Department for 2014 and 2015 is shown in Figure 2 [11].

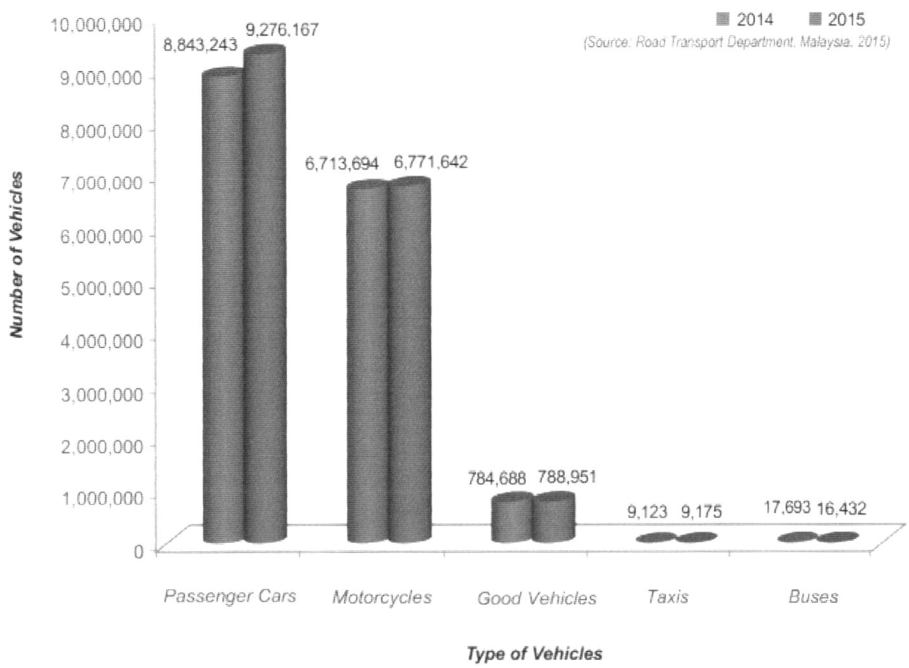

Figure 2. Number of In-use Vehicles in Malaysia: 2014 – 2015.

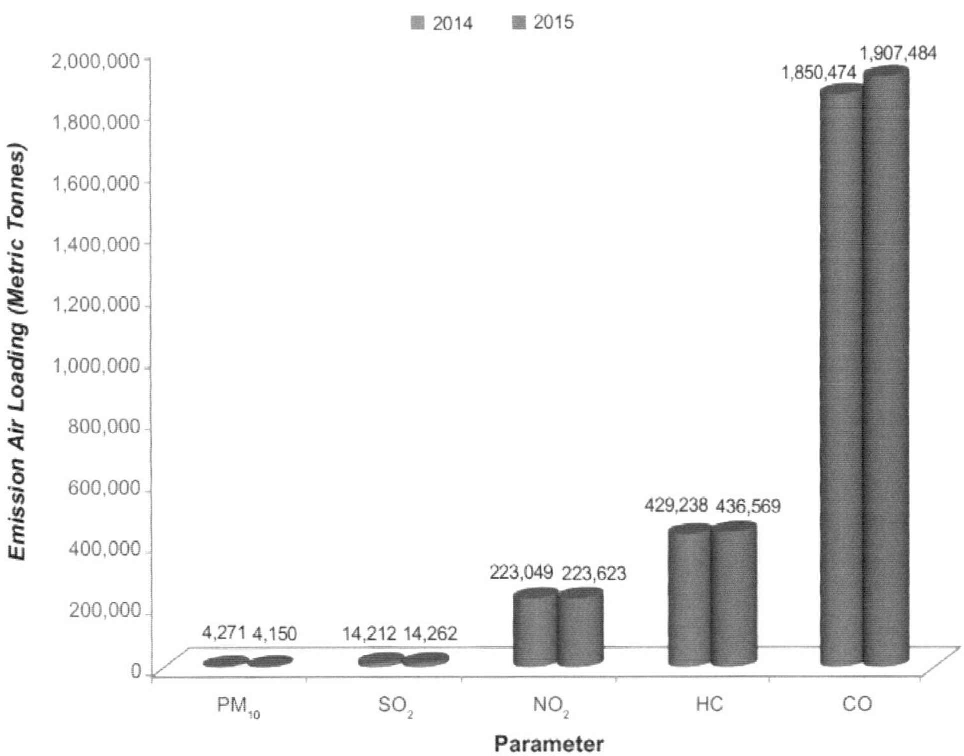

Figure 3. Air Pollutant Emission Load from Motor Vehicles in Malaysia: 2014 – 2015.

The estimated annual air pollutant emission load of Hydrocarbon (HC), CO, PM, NO_2, and SO_2 from motor vehicles for 2014 and 2015 is shown in Figure 3 [11]. In 2015, the emission load of HC and CO was estimated to be 436,569 metric tonnes and 1,907,484 metric tonnes respectively. There was an increase of 3.08% of CO emission load compared to 2014. A similar increasing trend was also observed for the emission load of SO_2, NO_2 and HC from motor vehicles in 2014. SO_2 was 14,262 metric tonnes in 2015 as compared to 14,212 metric tonnes in 2014 (0.35% increased), and NO_2 was 223,623 metric tonnes as compared to 223,049 metric tonnes in 2014 (0.26% increased). However, for the emission load of HC, 436,569 metric tonnes (increased by 1.70%). Emission load of PM was 4,150 metric tonnes in 2015 as compared to 4,271 metric tonnes in 2014 (decreased by 2.83%). CO emission is dangerous as it decreases the flow of oxygen into the bloodstream and it is severely threatening for people having heart disease. The NO_2 and Nitric Oxide (NO) increase exponentially with highest flame temperatures if endured long enough with the concurrent availability of oxygen. In the existence of sunlight and nitrogen oxides, HC reacts to form ground-level ozone that is a main element of pollution. Hydrocarbons are toxic and have the probability to cause cancer whereas ozone damages the lungs, aggravate respiratory problems and irritates the eyes [12].

3. Malaysians Lack of Interest in Public Transport

It has become imperative research activity for the responsible authorities in Malaysia to control the level of pollutants produced by the transport sector mainly in urban zones. One of the key steps in reducing pollution is to reduce the number of vehicles by improving the PTS that could be shared by many commuters as compared to private vehicles. However, in the perception of community, the PTS in Malaysia has been declining over the years. It is a key challenge for the government to make public transport a preferred choice for the citizens of Malaysia by undertaking significant transformations. Several improvements are required such as increasing capacity, infrastructure, integration, cost-effectiveness, technology, etc. By reviewing the existing research, this section critically discusses the limitations of the PTS which are core reasons for the public to use privately owned vehicles as their preferred mode of mobility.

3.1. Infrastructure

The government of Malaysia has spent tremendous amounts of money especially in the development of rails such as Light Rail Transit (LRT), monorail and Keretapi Tanah Melayu (KTM) in an attempt to make the railway service the preferred mode of transportation for citizens and to overcome the problems of traffic congestion in urban areas such as Kuala Lumpur. However, the majority of citizens interested in using rail service as their preferred mode of mobility are still unsatisfied with the service [13]. Lack of integration represents the limitations of the infrastructure. The buses from different companies are not coordinating with each other, which makes the transfers very difficult. Moreover, schedules of buses are not fixed, as they have to complete trip cycles, some travelers find it difficult to manage transit from trains to buses. Connectivity is also a major concern as some train stations are not directly connected to bus stations via facilities such as air-conditioned bridges. People find it difficult to walk across the road to continue their journey due to factors such as time-consumption, hot and rainy weather, safety and security.

3.2. Quality of Service

Usually, three attractive words cheap, convenient and comfort are used to describe the PTS in Malaysia. Commuters are expecting comfortable and convenient services at affordable cost. Although the 3Cs sound attractive to represent the public transport, however quality and efficiency of the service have not been up to the expectations of commuters [14]. [13] conducted a study to analyze the people's satisfaction and perception regarding service quality attributes of public transport. The study was based on a set of questionnaires using statistical models by including variables such as comfort, quality of service and facilities to examine the overall satisfaction of the commuters and the factors that influence their satisfaction. Various analytics methods such as descriptive, factor analysis, correlation and multiple regression were used for analysis. The results depict that PTS mainly buses need to improve their quality of service to achieve customer satisfaction.

3.3. Technology

To improve commuters' mobility and to address the major issues of public transport in Kuala Lumpur, [15] studied the level of Urban Public Transport (UPT) institutional framework. The research identified that mobility in Kuala Lumpur could be easier if traffic flow runs smoothly. However, it requires smart planning with major transformations and improvements to reduce congestion on Kuala Lumpur roads. Currently, people avoid public transport due to heavy congestion on roads and lack of accurate as well as up-to-date information on arrivals and departures. There is a need for implementing journey planning and vehicle tracking applications. It will greatly motivate people to use public transport if they could plan their journey, record their mobility patterns, undertake real-time mobility decisions, and determine effective and efficient modes as well as routes of mobility as per their preference. Technology plays an integral role in making mobility via public transport easier, and it could be used as an approach to encourage people to use public transport. The government needs to consider facilitating citizens with the smart applications such as journey planners using modern technology.

3.4. Cost and Comfort

Cost is one of the most critical factors in selecting mobility methods. Many people want to travel via public transport such as bus and train due to low cost and efficiency of service [16]. Higher cost and less comfort in public transport enforce people to own private vehicles such as cars or motorcycles. Public Bus Transport (PBT) in Malaysia is underutilized, and there is an opportunity to improve the service provided. [17] conducted a study to explore the aspects that impact the usage of PBT in Malaysia. The results showed that ticket price was a great contributing factor in commuters' decision to use public transport. Efficiency and accuracy of schedule also play an integral role in attracting commuters. Therefore, stakeholders must consider these factors to increase the use of PTS. For instance, for the commuters who are using public transport for long-journey, they should have comfortable journeys with basic facilities such as wider seats, user-friendly interchanges, air condition and toilets for disabled, etc.

3.5. Safety and Security

[18] and [19] found in their research that safety and security is a serious issue in convincing citizens to choose public transport as their method of mobility. In Malaysia, safety has been a major concern for people to use private vehicles for mobility due to the alarming rate of street crime. A study conducted by [20] discovered that in Klang Valley, safety is a major concern for the public to avoid PTS. Furthermore, [21] conducted a study that revealed that in Petaling Jaya, only 2.7% of the respondents agreed that public transport in their region is safe. Factors such as robbery, sexual harassment of female passengers and snatch-thieves were cited as the serious concerns in mobility via PTS. Apart from that, drivers of the public transport drive recklessly in a bid to meet schedule. It can be seen from the studies that government authorities need to increase the level of security standards to eradicate crime from streets of Malaysia. Insufficient security will remain a barrier to using the PTS. Therefore, responsible authorities must undertake measures to overcome the safety and security issues.

4. Government Initiatives

With the growth of population, the number of private vehicles is increasing. By 2020, the population of greater Kuala Lumpur (KL) is projected to be 10 million [22]. Government is undertaking precautions to improve the PTS in a bid to reduce private vehicle ownership that would eventually lead to reduced traffic congestion and air-pollution. The government of Malaysia has recognized the current problems related to the PTS and increased emission of traffic-related pollutions that are affecting the health and lifestyle of Malaysian citizens. Therefore, the government is addressing these concerns, and planning improvements as well as enhancements in Government Transformation Programme (GTP) 2.0. From transport sector, GTP 2.0 is mainly based on the goals discussed in the following sub-section.

4.1. Enhancement of Bus System

Bus service is still an important source of mobility for the citizens of Malaysia. However, their inefficiency has been proved as an obstacle to their adoption. In greater KL/Klang Valley, UPT-National Key Resul Area (NKRA) has implemented new initiatives to improve bus service. At present, several bus companies are operating via the same routes that create congestion in Central Business District (CBD) due to overlapping services and unhealthy competition. The first initiative is to implement a systematic bus network by downsizing the number of operators per route and by requiring the buses to stop at dedicated Inter-Urban Transport Terminals (IUTT) located at the periphery of the CBD. In an objective to provide express service in Klang Valley, the government also implemented Bus Rapid Transit (BRT). It is more convenient and covers many regions [22].

4.2. Enhancement of Rail System

The improvement of train service will keep on being a state of accentuation in the GTP 2.0 as it is the most highly used method of public transport in the nation. Three noteworthy activities will be propelled under this work stream influencing each of the systems serving the greater KL/Klang valley territory, in particular, the inter-city KTM trains, RapidKL Light Rail Transit and KL Monorail. The GTP 2.0 lab has recognized a few areas of an upgrade to enhance the service of KTM. These are track power upgrades of KTMB's networks, rehabilitation programmes and updating of the communication and electrification system. The objective here is to enhance the efficiency, effectiveness, and reliability of KTM. Moreover, GTP 2.0 plans to expand the Kelana Jaya and Ampang LRT lines. This activity will address the linkages of the high-volume Kelana Jaya and Ampang LRT lines to many areas. The multi-billion ringgit line expansion project will interface the current Sri Petaling LRT station to the Ampang line, and also the Kelana Jaya station on the Kelana Jaya line to Putra Heights. The capacity of monorail has been expanded to diminish bottlenecking in specific stations, for example, Hang Tuah and KL Sentral [22]. Furthermore, the government built Mass Rapid Transit (MRT) system in the Klang Valley that is capable of increasing the use of public transport and help to reduce traffic congestion in the area [23].

4.3. Transformation of Taxi System

The inclusion of taxi work stream into the Urban Public Transport NKRA of the GTP 2.0 is targeted to enhance the quality of service offered by taxis in Malaysia, particularly in urban zones. Specifically, the main objective of NKRA is to enable the taxis in Malaysia to provide services of similar quality as in other ASEAN nations. The initiatives of this program incorporate the implementation of a centralized taxi system and presenting a new business model for taxi business. This extensive system is anticipated to increase the enforcement and monitoring capabilities of the industry regulators and operators of taxi fleets. The initiative is based on best international practices, and it provides an end-to-end process from guidelines to the penalty system to an effective public communications plan. Additionally, Suruhanjaya Pengangkutan Awam Darat (SPAD) will engage with car manufacturers for a leasing option to decrease the amount of down payment for car loans upon renewal of vehicle permits and continue to assist with maintenance and repair facilities [22].

4.4. Enhancing the Integration of the Urban Public Transport System

The most integrated transport systems are considered as the best ones. Integration of work stream undertakes several initiatives to integrate existing public transport facilities. The approach considers the integrated routes and transport facilities such as the mobility of commuters from source to destination and returning to the source. The integration of bus and rail services will encourage people to use public transport if stations are nearby or connected by bridges. Similarly, facilities such as Park-n-Ride will motivate the travelers to use trains or buses for long distances while parking their privately owned vehicle at the station. Such practices will result in less congestion on the roads [22].

4.5. Management of Travel Demand

About three million cars enter the Kuala Lumpur city centre every day causing congestion due to over-capacity. To overcome congestion, initiatives are to implement journey planners for public transport users and to enhance parking control and management. A journey planner will provide commuters an opportunity to plan their journey based on routes with the least amount of interchanges, fastest routes and routes with lesser walking distances. This initiative will be implemented by setting up an integrated one-stop journey planner across all modes by providing information such as routes, travel fare and schedule from the beginning of the proposed journey to the destination. The parking control and management initiative will review the parking rates in the Kuala Lumpur city centre to ensure a more competitive rate and to build effective pedestrian walkways [22].

5. The Era of Smart Mobility: Implementation in Malaysia

Smart mobility is a crucial component of smart cities implementation [24]. It encompasses several dimensions of the smart cities on numerous aspects comprising the citizens' quality of life [25]. The most imperative goals of smart mobility include reducing air pollution, traffic congestion, and noise pollution, increasing people safety, improving mobility speed and reducing cost. Nowadays, smart mobility is practiced via various methods such as carpooling, car sharing and cycling. Through carpooling also known as ridesharing, rides are shared with other to fill out the empty seats and to reduce the use of extra vehicles on roads. This facility helps in reducing traffic congestion and air pollution and improves mobility without adding new vehicles to the system. Cycling is considered as an eco-friendly method of mobility and use of it has been on the rise in Europe mainly in the cities with flat land and appropriate cycling infrastructure. For shorter distances such as few miles or less, cycling is often considered as the fastest mode of mobility to get to the workplace. Recently, car sharing has transformed the mobility. Companies such as Uber and Lyft have facilitated the commuters and drivers with smart technology that could be used to reserve transport services on individual or sharing basis. Using the applications drivers can access locations and customers using Global Positioning System (GPS) technology [26].

When it comes to the implementation of smart mobility facilities, there are significant barriers. For example, people are not willing to ride using carpool and car share due to emerging safety and security concerns. However, with the introduction of companies such as Grab and Uber, people are practicing carpooling as they feel safe since the companies can track the driver and travelers response in real-time. Cycling is not very common yet due to various reasons such as weather as the country stays hot throughout the year and another reason is lack of dedicated safe cycling lanes. However, the government is steadily working to make cycling as one of the mobility methods in Malaysia, Kuala Lumpur City Council went on to approve funds for two more cycling lanes in some parts of the city with a total budget of £765,000 [27]. oBike started a station-less smart cycle share system in Malaysia to promote cycling. From the current status, it could be said that Malaysia is steadily moving towards smart and eco-friendly mobility methods. With the improvements and developments, adoption could be further increased that would lead to the sustainable environment with reduced emission of air pollutants and improved health as well as lifestyle.

6. Conclusion and Future Work

European cities such as Bristol, Helsinki, and Stockholm have implemented green mobility patterns as discussed in Section 5. Malaysia is steadily adopting those practices, but there are significant barriers to overcome before full implementation. Currently, the biggest problem faced is a generation of traffic-related air pollutants that can be reduced mainly with reduction of privately owned vehicles. The increase in the use of public transport such as bus, train, taxi and carpooling services can result in lesser number of vehicles that means lower emissions of pollutants. The government has undertaken several initiatives to improve public transport, for example, nowadays infrastructure is connected using air-conditioned bridges, people have facilities such as Park-n-Ride, carpooling using Grab and Uber. However, there are still several improvements required regarding security, safety, journey planning applications and pricing policies. Efficient implementations are required from the government to compete with the smart mobility facilities of European cities with less traffic-pollution index. Apart from government efforts and initiatives, people must be educated via community engagement activities and gamification to realize the impact of their mobility method so that they could practice smart mobility by preference. This research mainly focused on data analysis of years

2014 and 2015, the next phase of this research would include an in-depth study with broad coverage of the past few years' data with emphasis on air pollution and healthcare.

7. References

[1] Almselati A, Rahmat R and Jaafar O 2011 An Overview of Urban Transport in Malaysia *The Social Sciences* **6** pp 24–33.

[2] UNEP 2001 The Role of the Transport Sector in Environmental Protection, Department of Economic and Social Affairs *United Nations Environment Programme Commission on Sustainable Development Ninth Session* (New York) pp 16 – 27.

[3] Aziz A and Amin N 2012 Transforming the Land Public Transport System in Malaysia: Transforming the Land Public Transport System in Malaysia *JOURNEYS* pp 30 – 37.

[4] MIROS, General Road Accident Data in Malaysia (1997–2016) https://www.miros.gov.my/1/page.php?id=17.

[5] Saxena A 2009 Greenhouse Gas Emissions: Estimation and Reduction *Asian Productivity Organization* (India).

[6] Olivier G, Janssens-Maenhout G, Muntean M and Peters J 2013 Trends in Global CO2 Emissions: 2013 Report *PBL Netherlands Environmental Assessment Agency* (The Hague).

[7] Salahudin S and Abdullah M 2013 Emissions: Sources, Policies and Development in Malaysia. *International Journal of Education and Research* **7** pp 1–12.

[8] Shahid S, Minhans A and Che O 2014 Assessment of Greenhouse Gas Emission Reduction Measures in Transportation Sector of Malaysia *Jurnal Teknologi* **4** pp 1–8.

[9] Kari 2008 Automobile Emissions and the Environment: The Malaysian Experience. Making Choices about Hydrogen, Transport Issues for Developing Countries http://www.idrc.ca/en/ev-132167-201-1-DO_TOPIC.html.

[10] Ahmad, Sany and Waleed 2013 Air Pollution Study of Vehicles Emission in High Volume Traffic: Selangor, Malaysia as a Case Study *WSEAS TRANSACTIONS on SYSTEMS* **12**.

[11] Department of Environment Malaysia, 2015 Pollution Sources Inventory https://enviro.doe.gov.my/ekmc/wp-content/uploads/2016/09/7-EQR-2015-Bab-6-1.pdf.

[12] Illinois Emission Control And Air Pollution http://mste.illinois.edu/tcd/ecology/student.html.

[13] Rozmi, Hesam and Rahim 2013. Passengers Preference and Satisfaction of Public Transport in Malaysia, Part II: A Comparative Analysis of Komuter and LRT Network *Research Journal of Applied Sciences, Engineering and Technology* **6** pp 1450-1456.

[14] Suria H 2012 Petunjuk Prestasi Bas Bandar dan Sumbangannya Terhadap Kemampanan Sistem Pengangkutan Awam PhD Thesis Universiti Sains Malaysia (Penang).

[15] Noor R and Yusfida A 2016 Theorizing the Concept of Urban Public Transportation Institutional Framework in Malaysia. *MATEC Web of Conferences*.

[16] Jayaraman K, Jacylin C, Shiau S and Soh K 2011 Robust Models for the Utilization of Public Bus Transport Services in Malaysia, *African Journal of Business Management*, **5** pp 10586–10600.

[17] Soh K, Wai W, Chu C and Yan H 2014 Improving Traffic Infrastructure in a Developing Country: An Investigation into the Usage of Public Bus Transport in Malaysia *Industrial Engineering & Management Systems* **13** pp 172–184.

[18] Smith M, Clarke, R 2000 Crime and Public Transport, *Crime and Justice* **27**.

[19] Ambak K, Atiq R, Ismail R 2009 Intelligent Transport System for Motorcycle Safety and Issues *European Journal of Scientific Research* **28** pp 601–612.

[20] Kamaruddin R, Osman I and Che C 2012 Customer Expectations and Its Relationship towards Public Transport in Klang Valley *Journal of Asian Behavioural Studies* **2** pp 29-38.

[21] Ibtishamiah N, Adji M and Karim M 2013 Public Transport Passengers' Perception and Demand Satisfaction: A Case Study at Petaling Jaya Municipal District, Malaysia *Proceedings of the Eastern Asia Society for Transportation Studies*.

[22] Pemandu GTP 2.0 Improving Urban Public Transport http://gtp.pemandu.gov.my/gtp/Improving_Urban_Public_Transport-@-GTP_2@0_Improving_Urban_Public_Transport.aspx.
[23] Bernama, MRT Built To Increase Usage Of Public Transportation, http://gtp.pemandu.gov.my/gtp/Media_Coverage-@-MRT_Built_To_Increase_Usage_Of_Public_Transportation.aspx.
[24] Clara B, Renata D and Beatrice D 2016 Smart Mobility in Smart City Action Taxonomy, ICT Intensity and Public Benefits, Empowering Organizations *Lecture Notes in Information Systems and Organisation*.
[25] Arena M, Cheli F, Zaninelli D, Capasso A, Lamedica R and Piccolo, A 2013 Smart mobility for Sustainability *AEIT Annual Conference 2013: Innovation and Scientific and Technical Culture for Development*.
[26] Peter V, Abhijit K, Tiffany F and William E 2015 Smart Mobility: Reducing Congestion and Fostering Faster, Greener, And Cheaper Transportation Options https://dupress.deloitte.com/dup-us-en/industry/public-sector/smart-mobility-trends.html.
[27] Ling L, 2015 How a crowd-sourced map changed Kuala Lumpur's ideas about cycling, https://www.theguardian.com/cities/2015/sep/18/how-crowd-sourced-map-kuala-lumpurs-ideas-cycling.

Acknowledgment
This research is funded by Taylor's University Malaysia. The work done is part of the research grant application ID (216422456) entitled as Smart Sustainable Cities: Connecting Bristol with Kuala Lumpur.

Modelling of indoor air pollutants dispersion: new tools

V Busini, S Favrin, G Nano, M Derudi

Politecnico di Milano, Department of Chemistry, Materials and Chemical Engineering "G. Natta"
E-mail: valentina.busini@polimi.it

Abstract. Ventilation systems are used for create a thermally comfortable environment and good indoor air quality. It is therefore essential to have adequate tools for predicting the performance of these systems. Among the various approachs, the computational fluid dynamics could be a useful tool for the design of the ventilation system. When dealing with pollutants dispersion problems, a steady state averaged simulation can be misleading because it is not able to properly predict and model peak concentrations, which can be relevant even if temporary. An interesting approach is the use of LES (Large Eddy Simulations) simulations to obtain a better description of concentrations oscillations. In this framework, the aim of this work is the validation of simulation carried out using the FDS (Fire Dynamic Simulator) software with an actual case study, already studied with a mock-up. Secondly, two new configurations of the ventilation system are proposed, in order to stress the capacity of the software to describe complex and different features, classical of HVAC (Heating, Ventilation and Air Conditioning) systems. Interesting conclusions about efficiency are drawn from the comparison, highlighting the potentiality of the software.

1. Introduction

Ventilation is used in buildings for create a thermally comfortable environment and good indoor air quality. It is therefore essential to have adequate tools for predicting the ventilation performance of buildings: analytical models [1]; empirical models [2]; small-scale models [3]; real-time models [4]; multi-zone models [5], [6]; CFD models (Computational Fluid Dynamics). The last provide more detailed information on ventilation performance, representing a much more sophisticated tool than other models. Hayashi et al. [7] conducted a CFD analysis to examine the characteristics of a ventilation system in a closed environment, with a two-dimensional model. As for the three-dimensional CFD models Kassomenos et al. [8] analysed the dispersion of vinyl chloride monomer into a polyvinyl chloride pipe manufacturing plant. Siddiqui et al. [9] have proposed a model for a risk analysis in case of chlorine releases in an industrial environment. Predictive capacity of CFD models require a proper validation process, typically performed by reproducing experimental evidence; for this reason, Lee et al. [10] have validated their model with the estimated the concentration of a tracer gas, isobutene, using a small-scale model; Deevy et al. [11] have predicted the dispersion of a mixture of sulphur hexafluoride and helium in a thermally stratified room; Lambert et al. [12] have studied the release of carbon dioxide into a ventilated space.

Due to the high computational cost required, CFD simulations are often performed as RANS (Reynolds Averaged Navier-Stokes) simulations. When dealing with pollutants dispersion problems, a steady state RANS simulation can be misleading because it is not able to properly predict and model peak concentrations, which can be relevant even if temporary. An interesting approach is the use of LES simulations in order to obtain a better description of concentrations oscillations.

IOP Conf. Series: Earth and Environmental Science **167** (2018) 012016 doi:10.1088/1755-1315/167/1/012016

In this framework, the aim of this work is the validation of simulation carried out using the FDS (Fire Dynamic Simulator) software with an actual case study, already studied with a mock-up. Secondly, a comparison between two different configurations is performed, in order to stress the capacity of the software to describe complex and different features, classical of HVAC (Heating, Ventilation and Air Conditioning) systems. Interesting conclusions about efficiency can be drawn from the comparison, highlighting the potentiality of the software.

2. Materials & Methods

In this work the version 6.5.3. of FDS was used. FDS is a CFD modelling software developed by the BFRL (Building and Fire Research Laboratory) and NIST (National Institute of Standards and Technology), for the simulation of turbulent reactive flows (i.e., fire effects) through the solution of the transport equations using a LES (Large Eddy Simulations) approach. It is a low Mach solver designed to investigate fire behaviour and perform safety assessments [13]. Main advantages of the software are the low computational cost and the robustness of the algorithm; while the main drawback is related to meshing limitation: the geometry description with a rectilinear mesh is the only way to guarantee a high computational performance. Other details about the simulations (e.g., grid and boundary conditions) are discussed in the next section.

3. Results and discussion

The case study simulated in this paper is that proposed by Rota et al. [14]; in which they have considered a ventilation system designed to reduce the risk to health in an industrial warehouse where calibration of thermocouples was carried out involving the use of alcohol baths (usually ethyl or methyl alcohol), which, by evaporation, dispersed within the working environment. The baths were placed in one half of the room, the rest was used for assembling and dismantling the thermocouples. In this type of working environment, it was not possible to apply a local ventilation system since it could have altered the calibration process.

Mock-up in scale 1:10, made of polymethyl acrylate, consisted of a base of 1.6 (L) x 0.8 (W) m with walls high 0.4 m. To create different airflow conditions within the room, the model had two openings on the rear wall, which represented windows of 0.1 (W) x 0.2 (H) m and one aperture, the door on the front wall of 0.2 (W) x 0.3 (H) m. At the centre of the ceiling there was a circular hole of 9 cm in diameter, connected to an intake system, which represents the ventilation system. The four diffusers simulating the thermocouple calibration counters, consisted of rectangular elements of 0.1 (L) x 0.2 (W) x 0.1 (H) m, each with two slots of 0.01 (L) x 0.16 (W) m placed on the upper base, through which tracer gas, ethane, was flowing.

The experimental tests have been carried out on three different cases, modifying the number of active pools and the opening of the windows, while the door on the right and the intake duct were always open; in the first case the two windows were closed and only two baths were in operation (Case 1), in the second case the other two tanks had been activated (Case 2), and in the third the windows had been opened (Case 3).

The concentration of ethane inside the mock-up has been measured at a constant height of 0.17 m, corresponding to the breathing height; the gas has been sampled by the use of long-needle semi-automatic injection syringes inserted into the simulation room by small sampling ports placed on the roof, arranged on two rows and numbered progressively, as in Figure 1, and then it has been analysed in a gas-chromatographic system Since the geometry of the room was symmetrical with respect to the vertical plane (from the door to the wall between the windows), the reference work contained ethane concentration measurements performed only in one half of the simulation chamber (as in Figure 1). Five measurements have been made for each sampling point.

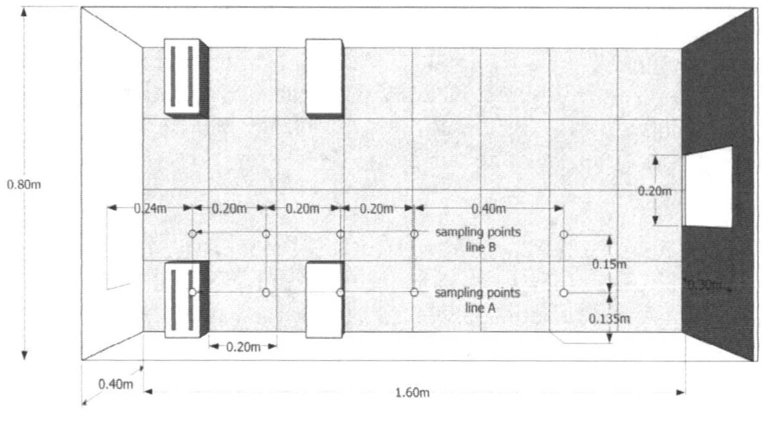

Figure 1. Position of samplings, top view [14].

The mock-up geometry was reproduced in FDS without any modification except: providing doors with a minimum thickness to allow the software to resolve the incoming streams from the opening, and making square the ventilation duct via the equivalent area to improve the solution through the square cells (as sketched in Figure 2). Uniform mesh was selected, with cubic cells of side 1 cm.
For the sake of brevity, only the Case 1 will be considered in this work; the relevant input data are reported in Table 1.

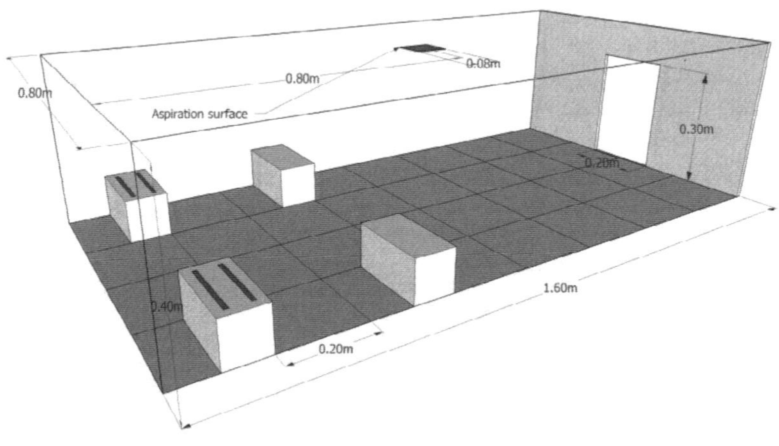

Figure 2. Geometry used in FDS for reproducing the mock-up Case 1.

Table 1 Input parameters of experimental test and simulations.

	Ethane mass flow rate [kg/s] per bath	Aspiration rate [m/s]	Aspiration volumetric flow rate [m^3/h]
Mock-up	1.1e-6	4.5	100
FDS	1.1e-6	4.5	100

Due to the high turbulence levels, a punctual monitor was supposed to be not suitable for a realistic description, thus in the simulations volumetric concentration monitors were used; this means that the simulation data are the mean, with the error bars, obtained among the values of the cells contained inside a volume centred in the experimental sample point, with a side of 4 cm (i.e., the mean on 64 cells).

To compare the results obtained from the simulations with the corresponding experimental ones, they were mediated on time, being the LES simulation transient. The mean time was obtained for each position, on the output values provided by the software, from the time the system reaches the stationary state, t_{ss},(around 100 s) until the simulation is completed (t_{fin} = 300 s).

Once obtained the mean value from FDS, the comparison with the experimental data was performed and sketched in Figure 3 and Figure 4.

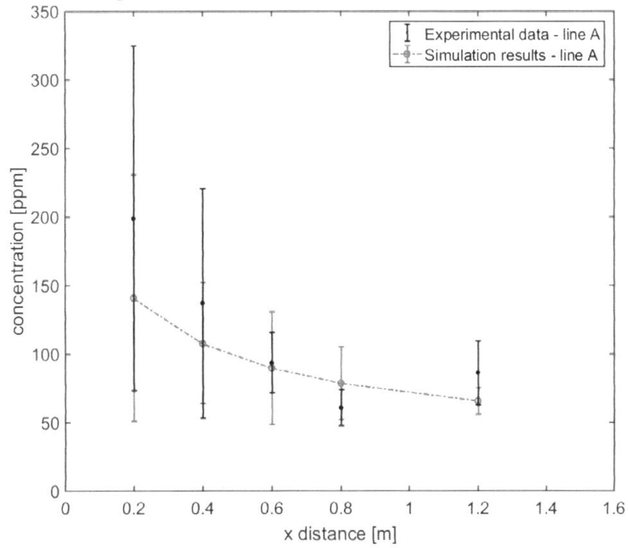

Figure 3. Comparison between experimental measurements and FDS predictions for the mock-up Case 1 at y = 0.135m (line A in Figure 1).

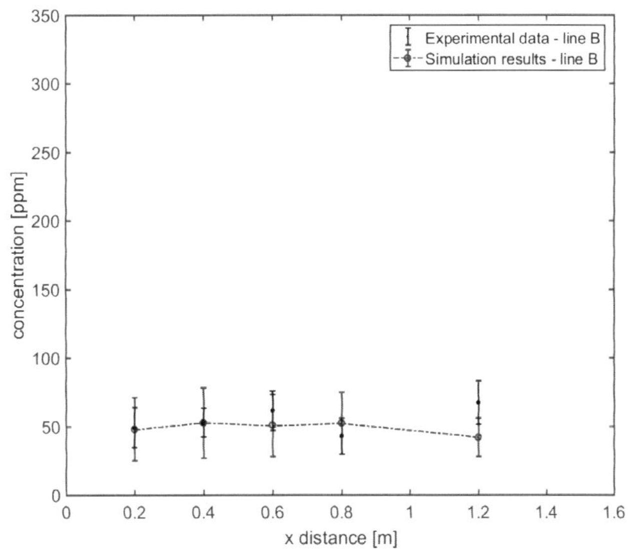

Figure 4. Comparison between experimental measurements and FDS predictions for the mock-up Case 1 at y = 0.285m (line B in Figure 1).

Good agreement between simulation predictions and experimental results was achieved, with a good consistency in both values and performance. Even better coherence is shown by the monitors located in the central part of the room (Line B of Figure 1) where lower concentrations were found.

It's worth spending a few words about the deviation of the results, both experimental and CFD, because highlight a very positive peculiarity of the software. It is possible to see that a really low reproducibility was obtained by the experimental tests even if all the precautions were taken in the measurement process. FDS results help to understand why measured values presented a high variability: the region of the mock-up in the tank area is strongly turbulent. This can be understood observing the deviation presented by the simulation results, which are of the same order of magnitude of the experimental deviations.
Moreover, the predicted concentration in the aspiration channel was validated with the experimental data. The comparison, reported in Table 2, is a further validation of the CFD simulations.

Table 2 Comparison between the concentration value of ethane at the intake system obtained during the experimental campaign [14] and the one obtained with FDS

	C_{exp}	C_{FDS}
Ethane concentration [ppm]	58	58.9

After the validation phase, two new configurations of the ventilation system were simulated. Keeping in mind that for the analysed case study (i.e., baths for calibration of thermocouples) it is not possible to position aspiration hoods directly upon the baths to avoid alterations to the calibration process, the two practical configurations, sketched in Figure 5, were studied and simulated.
In order to be able to compare the efficiency of the proposed solutions, the same tracer mass flow and the same number of ACH (Air Changes per Hour) were used in all the simulations, correcting the air flows.
The former (labelled as case N1) is characterized by a central channel of air emission and a certain number of exhaust hoods on the ceiling. This is a common solution applied often in huge environments, though to ensure a good air exchange; while in the latter (labelled as case N2) the air emission is obtained from a plenum posed on the ceiling and aspirations slits in the floor. On the other hand, this solution is frequently adopted when a strict control of pollutants is meant to be achieved; for example, in white and operating rooms.
These two configurations offer a presentation of various elements characteristics of a ventilation system such as air jets, hoods and plenum diffusers. The possibility offered by the software to model them should not be underestimated.
In Figure 6 an Figure 7 the comparison between the experimental data and the results obtained with the ventilation solutions N1 and N2 at the two sampling line are shown. Observing the results of N1, the higher levels of ethane calculated could be explained by the increased turbulence level in the room, which is generated by the jet air entries. This increased turbulence lead to an improved mixing in the volume, worsening the scenario by far. On the other hand, the second configuration (N2), exploiting a laminar airflow reduces the turbulence, preventing the pollutant mixing and allowing a timely aspiration of concentred puffs.
The N2 solution, with the laminar containment developed is absolutely more efficient. On the side of the modelling tool, the speed with which these scenarios can be analysed and understood are very well demonstrated. As an example, in Figure 8 are reported the visualization of the ethane concentrations for the three cases.

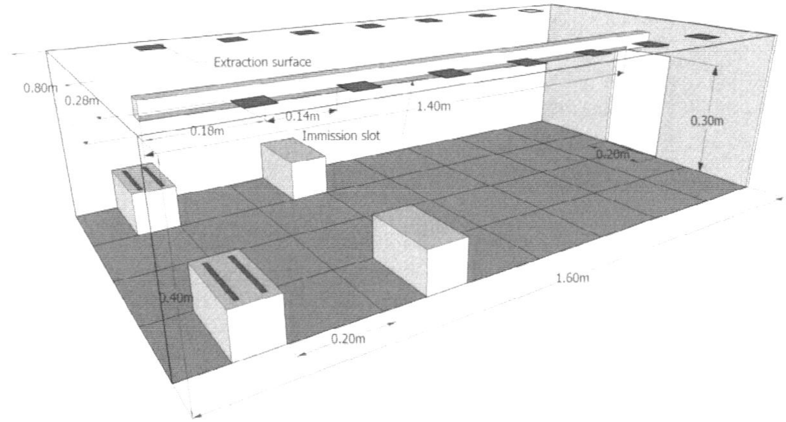

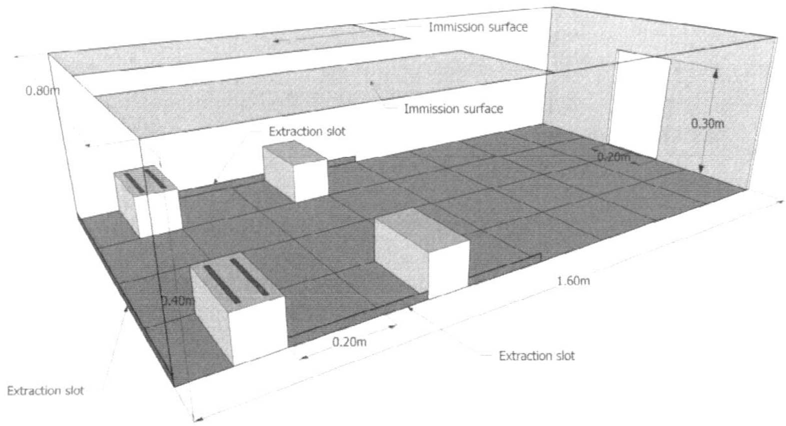

Figure 5. Sketches of the two new ventilation systems: N1 (top), N2 (bottom)

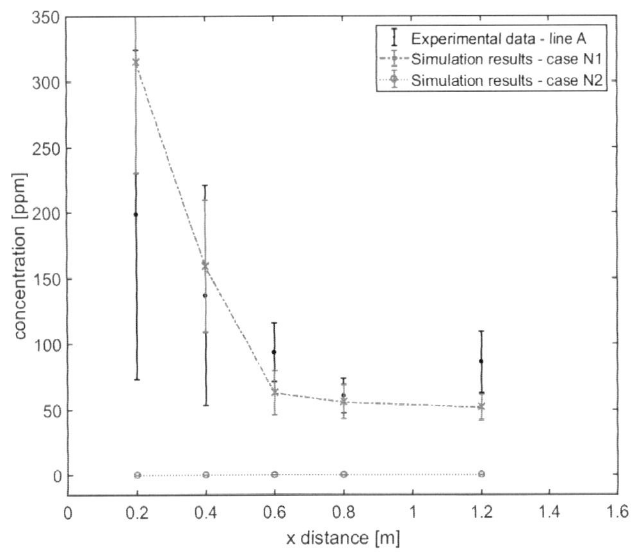

Figure 6. Comparison between experimental data and FDS predictions for the mock-up Case N1 and N2 at y = 0.135 m

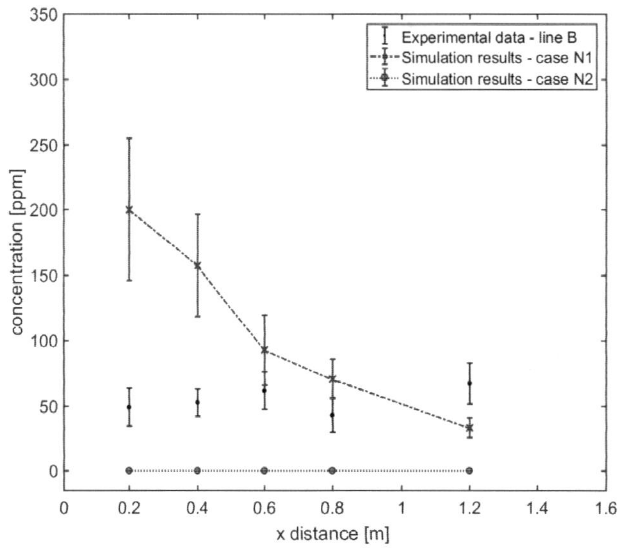

Figure 7. Comparison between experimental data and FDS predictions for the mock-up Case N1 and N2 at y = 0.285 m

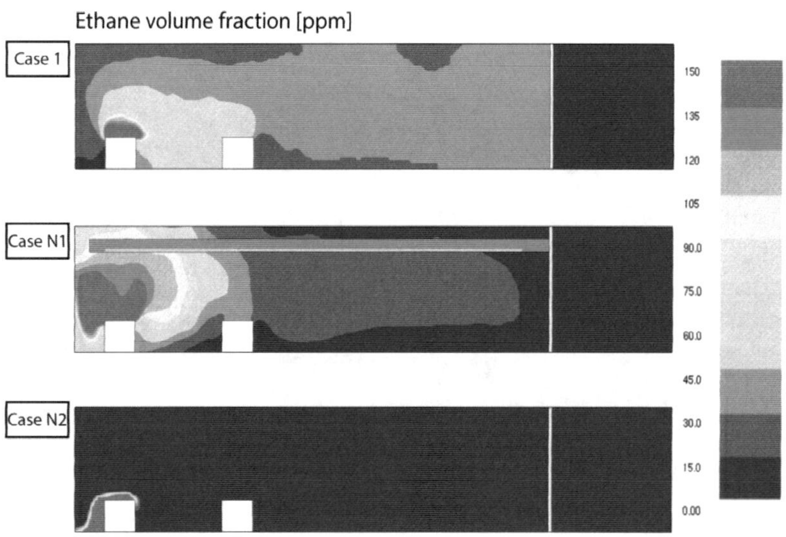

Figure 8. Ethane concentrations contours for the three cases: base case, N1 and N2

4. Conclusions

In the first part of this work a specific validation process was undertook with good results, highlighting that FDS, used in this work, undoubtedly represents an effective tool for analysing fluid-dynamic problems concerning the transport and dispersion of pollutants in confined environments.
The possibility offered to understand physical phenomena could be really useful in industrial practice to address some actual problems in ventilation design.
In the second part of the work, two alternative configurations of the ventilation systems were proposed; this has enabled to prove the effectiveness of FDS during the phase of the design of the plant as it allows easily evaluating the performance of any ventilation system and monitoring the most important parameters.

5. References

[1] Halios C H and Helmis C G 2007 On the estimation of characteristic indoor air quality parameters using analytical and numerical methods. *Sci Total Environ.***381**(1-3):222-32.

[2] Cho Y J, Awbi H B and Karimipanah T. 2008 Theoretical and experimental investigation of wall confluent jets ventilation and comparison with wall displacement ventilation. *Build Environ.***43**(6):1091-100.

[3] Yu H, Liao C M, Liang H M and Chiang K C. 2007 Scale model study of airflow performance in a ceiling slot-ventilated enclosure: Non-isothermal condition. *Build Environ.***42**(3):1142-50.

[4] Zhang Z, Zhai Z Q, Zhang W and Chen Q Y. 2007 Evaluation of various turbulence models in predicting airflow and turbulence in enclosed environments by CFD: Part 2-comparison with experimental data from literature. *Hvac&R Res.***13**(6):871-86.

[5] Axley J. 2007 Multizone airflow Modeling in buildings: History and theory. *Hvac&R Res.***13**(6):907-28.

[6] Yan D, Song F, Yang X, Jiang Y, Zhao B, Zhang X, et al. 2008 An integrated modeling tool for simultaneous analysis of thermal performance and indoor air quality in buildings. *Build Environ.***43**(3):287-93.

[7] Hayashi T, Ishizu Y, Kato S and Murakami S. 2002 CFD analysis on characteristics of contaminated indoor air ventilation and its application in the evaluation of the effects of contaminant inhalation by a human occupant. *Build Environ.***37**(3):219-30.

[8] Kassomenos P, Karayannis A, Panagopoulos I, Karakitsios S and Petrakis M. 2008 Modelling the dispersion of a toxic substance at a workplace. *Environ Modell Softw.***23**(1):82-9.

[9] Siddiqui M, Jayanti S and Swaminathan T. 2012 CFD analysis of dense gas dispersion in indoor environment for risk assessment and risk mitigation. *J Hazard Mater.***209**:177-85.

[10] Lee E, Feigley C E and Khan J. 2002 An investigation of air inlet velocity in simulating the dispersion of indoor contaminants via computational fluid dynamics. *Ann Occup Hyg.***46**(8):701-12.

[11] Deevy M, Stewart J R, Ren Z, Gobeau N and Saunders C J. 2008 A comparison of a range of models for dispersion in a partially stratified room. *Environ Modell Softw.***23**(4):511-9.

[12] Lambert AR, Lin CL, Mardorf E and O'Shaughnessy P. 2010 CFD Simulation of Contaminant Decay for High Reynolds Flow in a Controlled Environment. *Ann Occup Hyg.***54**(1):88-99.

[13] McGrattan K, McDermott R, Hostikka S and Floyd J. Fire Dynamics Simulator User's guide. 5 ed: NIST special pubblication 1019-5; 2010.

[14] Rota R, Canossa L and Nano G. 2001 Ventilation design of industrial premises through CFD modelling. *Can J Chem Eng.***79**(1):80-6.

IOP Conf. Series: Earth and Environmental Science **167** (2018) 012017 doi:10.1088/1755-1315/167/1/012017

Taking into account climate change adaptation in urban area through the CFD FLUENT simulation model example: an urban sector of the ORAN agglomeration

H BELHORMA[1], M CHACHOUA[2] and K MAHDI [3]

1 PhD Student in Climate Change and Adaptation, EGEAT Laboratory (Geographical and Territorial Planning Area), Faculty of Earth and Universe Science, Belgaïd Campus University of ORAN 2 Mohamed Ben Ahmed, AlGERIA
Email: hayatbelhorma@gmail.com
2 Senior university lectureer, Urbanist / Amenagist, Faculty Science of Earth and the Universe, Belgaïd Campus University of ORAN 2 Mohamed Ben Ahmed, ALGERIA
3 University Lectureer, IHFR (Hydrometeorological Institute of Training and Research), ORAN, ALGERIA

Abstract. Large cities cause microclimatic changes, the largest of which is known as the urban Heat Island Effect .The acceleration of urban development has caused variations in microclimatic conditions. Urban geometry and human activities in the urban environment interact with microclimatic parameters (wind speed and direction, air and surface temperature, solar radiation), also transforming the climate of our cities. It is for us to study the impact of climate change on urban environments and to understand the consequences of geometric and morphological factors on microclimatic conditions at the city scale, using tools called Computational Fluid Dynamics (CFD) model, we will try to show how urban geometry, in all its states interacts with microclimatic parameters (solar radiation, wind flows, air and surface temperature). Applied to the agglomeration of ORAN / ALGERIA on an existing urban sector (district HLM),we will try to show the strong correlation between failures and errors of urban design and the increase and generalization of Heat Island Effect in urban area.

1. Introduction

Actually, the debate on climate change has focused on all its physical aspects [1]. However, the effects of climate are felt by both populations and their living spaces [2].The large agglomerations of today are designated as environments conducive to the development of several microclimatic problems [3]. Urban development, which is accelerating more and more in its last decades under the process of metropolization, leads to the modification of microclimatic conditions [4].Climate change is now an issue and a challenge for climatologists, developers, architects and producers of urban environments that are increasingly adapted to the new climate[5]. We have adopted this topic, which consists in questioning the relationship between the urban form and various climatic parameters of the urban climate. The urban geometry of urban morphology as well as human activities in the urban environment interact with microclimatic parameters (solar radiation, wind flow, air and surface temperature), thus modifying the climate of our cities [6].

2. Methodology

IOP Conf. Series: Earth and Environmental Science **167** (2018) 012017 doi:10.1088/1755-1315/167/1/012017

The present research aims to study the relationship between urban morphology and microclimatic elements inscribed in a problem of adaptation of cities to climate change. To do this, we opted for an agglomeration: ORAN/ALGERIA. The observation of the phenomenon will take place on a portion of the urban space of ORAN agglomeration. It is in an existing neighborhood: HLM.

To answer our previously described problematic, we have used the simulation model CFD FLUENT to estimate the influence of certain meteorological elements (wind flow, temperature, and solar radiation) in urban area. CFD (Computational Fluid Dynamics) is a software used to simulate all fluid flows, compressible or incompressible, it is based on the method of finite volumes. CFD FLUENT is composed of three elements: The GAMBIT pre-processor is a 2D / 3D mesh maker that allows us to mesh the geometry domain. It generates msh*files for FLUENT. The FLUENT post processor allows us to solve and simulate fluid mechanics and heat transfer problems using the finite volume method. We have followed in the GAMBIT software the following steps: we have started by defining the geometry of the problem, then we realized the mesh and we checked it, in the end we defined the boundary conditions and definitions of the computational domains .The visualization of the results was realized by the postprocessor "FLUENT". [7]

3. Results

The simulation was carried out using the FLUENT software belonging to the CFD (Computational Fluid Dynamics) group, which is part of the eulerian numerical models [7]. In this section we have tried to show the importance of this tool for numerical modeling, particularly the results obtained in the area of building design and urban planning. Applied to these, the model requires taking into account local climate data.In our case, the basic climatological data are generally measured in the meteorological station of the IHFR stations (Hydrometeorological Training and Research Institute) and ONM (National Office of Meteorology) of ORAN.

3.1 Application of FLUENT software to Oran urban space

In this work, using this simulation model, it is for us to try to apply it to on a portion of the existing urban space HLM neighborhood Gambetta that is characterized by high densities and urban spreading to place ORAN among the most important cities in Algeria. It was chosen for its proximity to the IHFR so that the climatological data of the station are the closest to reality.

Figure 1: Satellite photography of the HLM city (Source: Google Earth-2014)

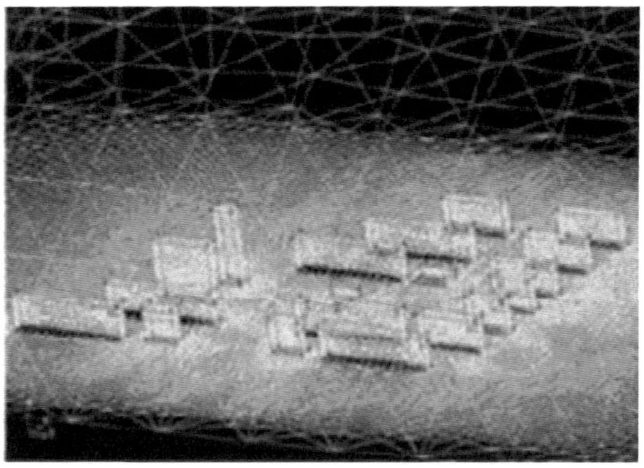

Figure 2: representation of the geometric design by GAMBIT 2.4 of the HLM district (ORAN) in3D

3.2 The effect of the wind

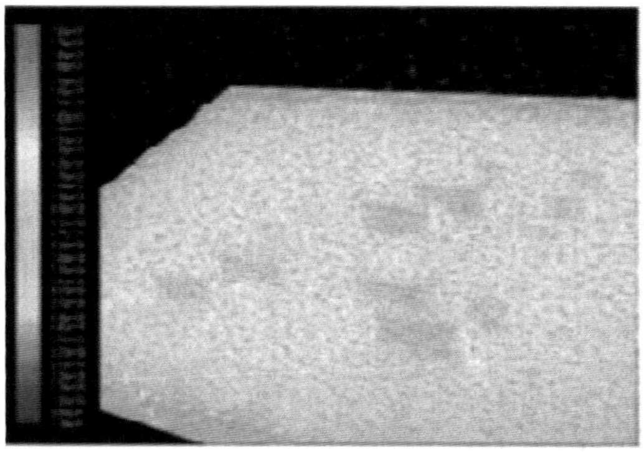

Figure 3: Illustration of the 3D velocity field with their mesh

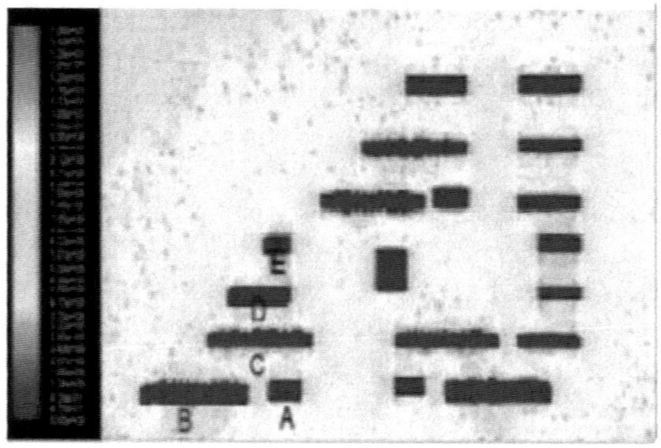

Figure 4: Wind contour ($m.s^{-1}$) at z = 3m

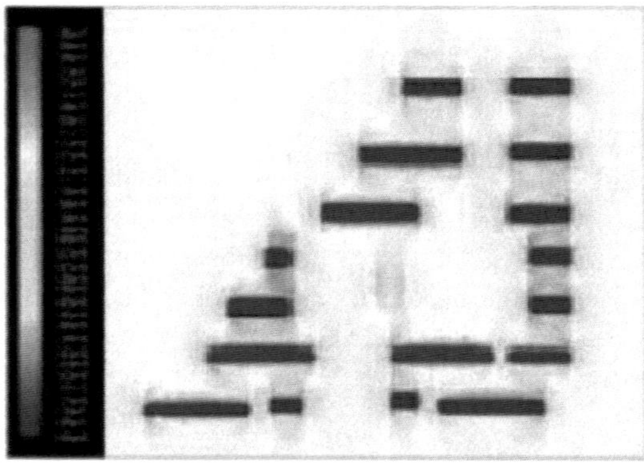

Figure 5: Wind contour ($m.s^{-1}$) at z = 10m

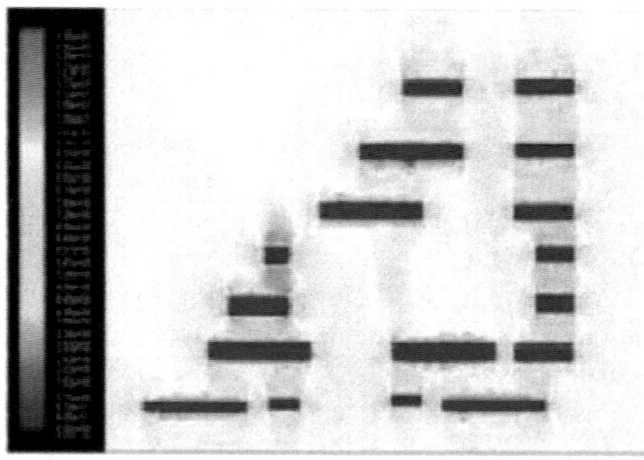

Figure 6: Wind contour ($m .s^{-1}$) at z = 20m

3.3 The effect of temperature

3.3.1. In winter

Figure 7: Static temperature contour (° C) for the 21/12/2014 at 09:00

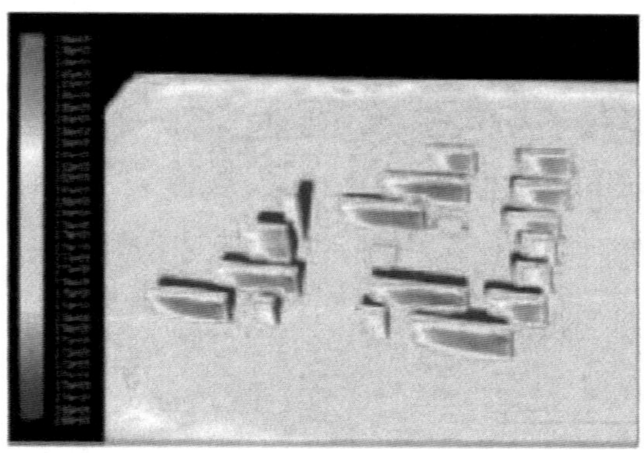

Figure 8: Static temperature contour (° C) for the 21/12/2014 at 12:00

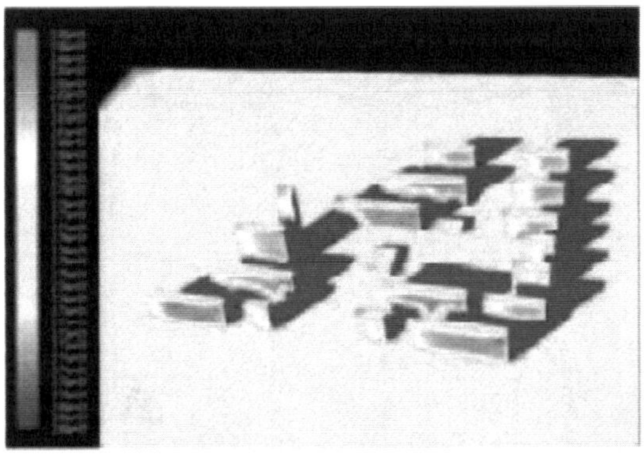

Figure 9: Static temperature contour (° C) for the 21/12/2014 at 16:00

Figure 10: Static temperature contour for the 21/12/2014 à 17:00

3.3.2. In summer

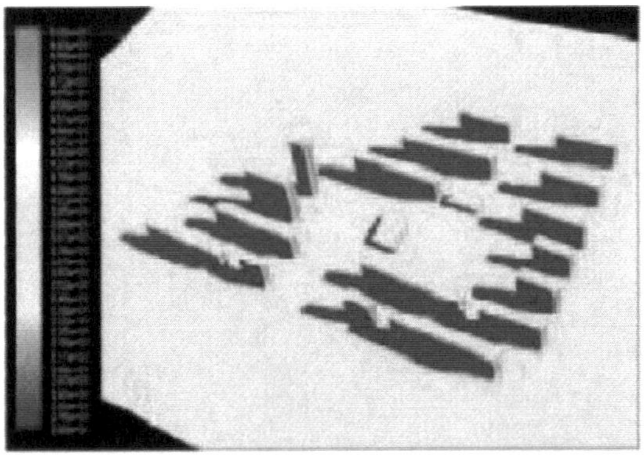

Figure 11: Static temperature contour (° C) for the 06/21/2014 at 08:00

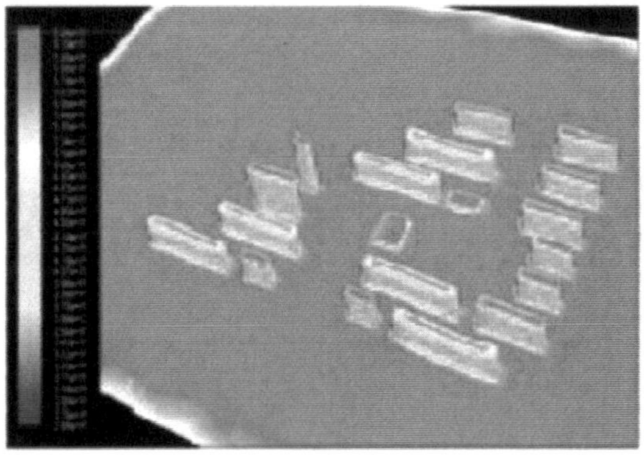

Figure 12: Static temperature contour (° C) for the 06/21/2014 at12:00

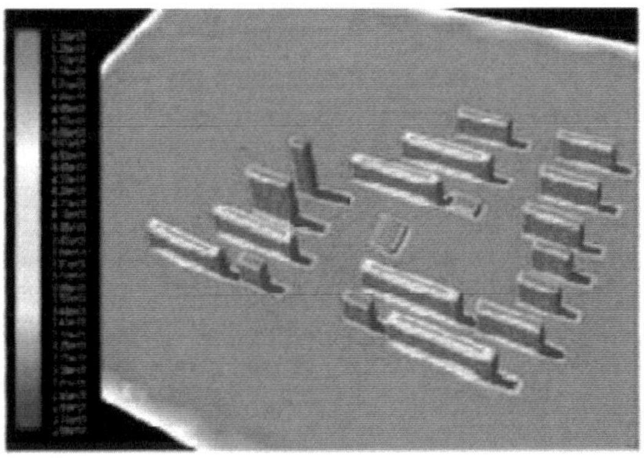

Figure 13: Static temperature contour (° C) for the 06/21/2014 at 16:00

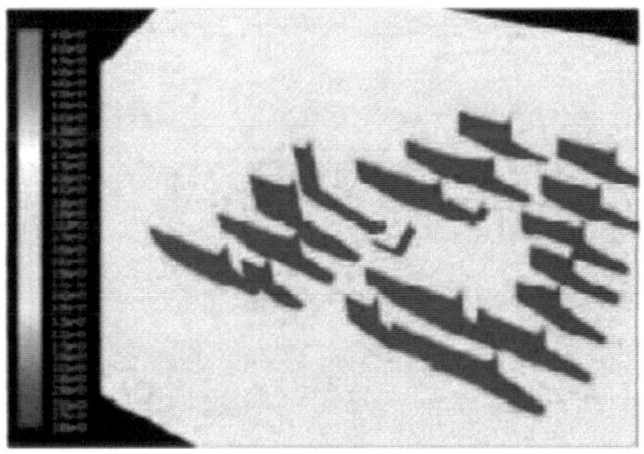

Figure 14: Static temperature contour (° C) for the 06/21/2014 at 18:00

3.4 The solar radiation effect

3.4.1. In winter

Figure 15: Solar Flux Contour (w.m^{-2}) for the 12/21/2014 at 09:00

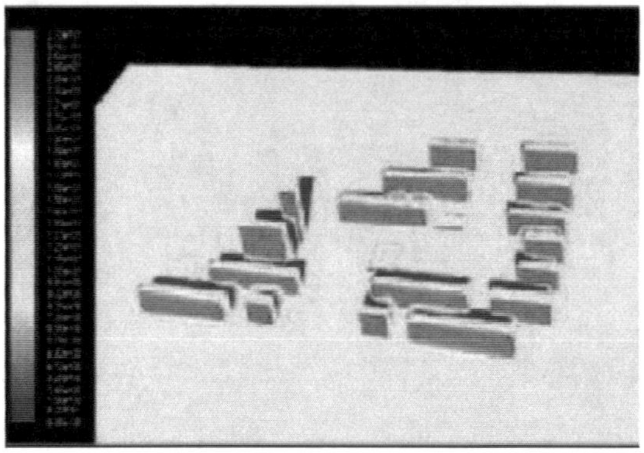

Figure 16: Solar Flux Contour (w.m^{-2}) for the 12/21/2014 at 12:00

Figure 17: Solar Flux Contour (w.m^{-2}) for the 12/21/2014 at 16:00

Figure 18: Solar Flux Contour (W.m^{-2}) for the 21/12/2014 at 17:00

3.4.2. In summer

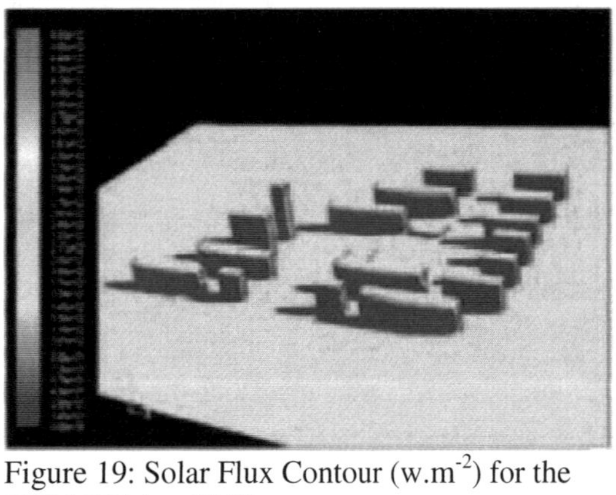

Figure 19: Solar Flux Contour (w.m^{-2}) for the 06/21/2014 at 08:00

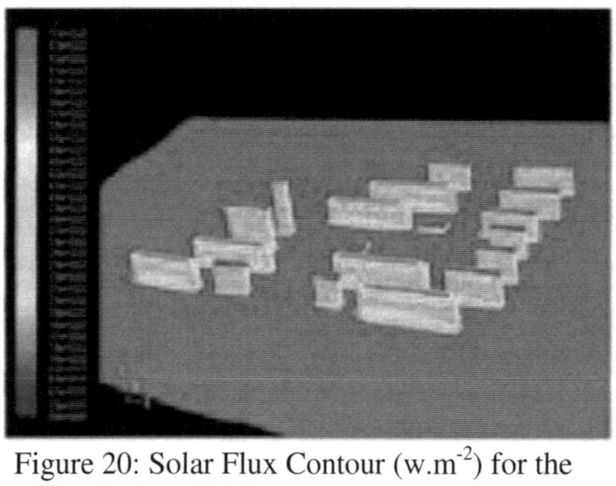

Figure 20: Solar Flux Contour (w.m^{-2}) for the 06/21/2014 at 12:00

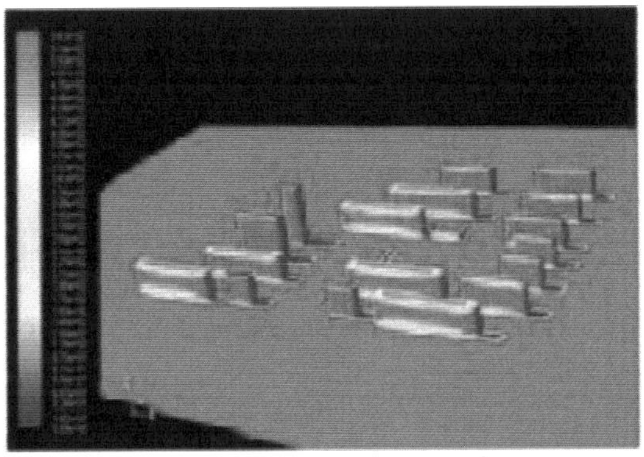

Figure 21: Solar Flux Contour ($w.m^{-2}$) for the 06/21/2014 at 16:00

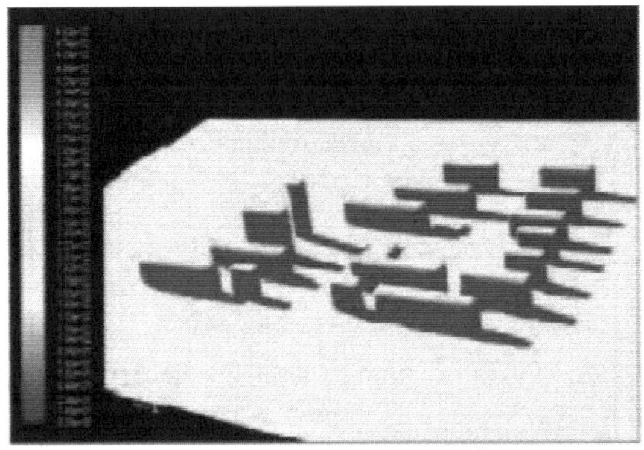

Figure 22: Solar Flux Contour ($w.m^{-2}$) for the 21/06/2014 at 18:00

4. Discussion

In our work we have chosen to use CFD simulation model for the parametric study of air movements around buildings. The Figure 1 represents the location of the neighborhood studied HLM ORAN, obtained by Google Earth, in Figure 2, we represent the 3 D geometry of the HLM city by Gambit 2.4.The figure 3 illustrates the speed field and their mesh in 3 D.

The Figures 4, 5, 6 show the results of our simulation carried out using the FLUENT software of the wind flow around buildings raised to different levels (Z = 3m, Z = 10m, Z = 20m) in the middle of the district HLM.The series of air speeds are presented in blue (very low speeds) and in red (very high speeds).

It can be concluded that in Figures 4, 5, 6 : between buildings D and E (these two buildings are on the left in the figures), we notice relatively high wind speeds, there is a circulation of air in this area which implies the presence of the canyon effect [8] , in our example (the height of buildings D = 45m, E = 36m, the width of the street = 18m, the ratio H / W = 2.5),it has a positive effect in the summer and it can refresh the street.

The Figures 7, 8, 9, 10 show the results of simulation for the winter season for the day of December 21st, 2014, expressing the static temperature contour on the HLM district at 9h, 12h, 16h and 17h.

The x-axis; represents the direction East. The y-axis; shows the direction North .The highest values are in red, the lowest values in blue.

At 9h, the temperature of East walls being high, at 12h, the South walls of the buildings are warmer because the position of the sun is in front of these walls, at 16h and17h, we notice the increase of the shade of the buildings because of the decrease of the solar height which has repercussions on the decrease of the temperature of the district.

In conclusion: the orientation of the buildings shows that, in winter, the South and West facing walls remain warmer than those oriented towards the North.

The Figures 11,12,13,14 show the results of our simulation for the summer season for the day of June 21st ,2014 ,expressing the static temperature contour (° C) on the district HLM at 8h, 12h,16h and 18h.we find that: at 8h, the temperature of the East walls is higher than the West due to the sun. From 12h to 16h, the temperature of the walls, terraces and the surface of the soil are very high in this period because the duration of sunshine is long and the height of the sun is maximum. At 18h, it is the West walls that are warm.

The Figures 15,16,17,18 represent the results of our simulation for the winter season for the day of December 21st, 2014 expressing the variation of solar radiation on the HLM district at 9h, 12h, 16h, and 17h. We note that: at 9h the facades exposed to the East receive more solar energy. At 12h, the facades exposed to the South are those that capture the most solar energy, at 16h and 17h, the West facing façades receive the maximum of solar energy. The buildings A, B form a solar mask for buildings C and D they prevent solar radiation from reaching other surfaces, so it is difficult to capture solar radiation in these spaces because of shading created by buildings who are neighbors.

The Figures 19,20,21,22 show the results of our simulation for the summer season for the day of June 21st,2014 explaining the evolution of solar radiation on the district HLM at 8h, 12h, 16h and 18 h According to the results, we find that: The simulations carried out at 8h show that the facades exposed to the East receive a large amount of solar energy, at 12h and 16h the terrace of the buildings are the areas where the energy collected is very high, at 18h the facades exposed to the West receive more energy.

In conclusion; the buildings that mask the sun are generally considered unpleasant because they bring a little shading in summer but become a real obstacles to sunshine in winter. So during the winter season the most favorable direction in the morning is the **South** because it allows to take advantage of solar energy and at night it is the **West** direction that it also allows to take advantage of solar energy and therefore reducing the use of heaters with low temperatures. In summer, the West direction allows to take advantage of the natural ventilation (West wind) during the day where (the temperatures are high) and subsequently the reduction of use of air conditioners. Given these elements, it appears that the **South-West** direction would have better suited to the city of HLM instead of the **North-South** direction that was favored during the design of the city.

5. Conclusion

The results we present, from the CFD FLUENT simulation model make it possible to provide elements of response to architects and urban planners in the design phase of an urban project. FLUENT presented itself as a simplified general code, essential for intensive use in urban design.

This code is part of the movement of works executed in thermal building. In this study, we wanted to illustrate the thermal behaviour of an urban space in the face of radiative and dynamic demands. We have highlighted the creation of a microclimate specific to the urban system, the surface temperature field being a function of the characteristics of the outside climate, but also of the configuration of the studied system.

The simulation relating to the air temperature made it possible to illustrate the phenomenon of Heat Island generated by a set of buildings. We have highlighted significant temperature differences between the system and the external environment. The FLUENT software allows by its modular and multi-model composition, the progression of the mathematical models proposed. However, additional research is needed to experimentally validate this calculation code.

6. References

[1] Achour-Bouakkaz N, 2006. The relationship between the urban Heat Island, climate change phenomenon and the density of the built plan: case of the city of AlGERIERS.

[2] Stella T, December 2012 Relations between urban morphology, microclimate and pedestrian comfort: application to the case of Eco-districts.

[3] Athamena Kh, April 2013 Modeling and simulation of urban microclimates: Study of the impact of urban morphology on comfort in outdoor spaces. Case of Eco-districts.

[4] Fouzia B, 2005 Impact of geometry of urban canyons on comfort external heat "case of CONSTANTINE coudiat".

[5] Abdeldjouad M, Jun 2014 Impact of building morphology on the dispersion of atmospheric pollutants Case study: ANNABA city.

[6] Djaafri M, Jun 2014. Urban form, climate and energy which indicators and which tools?

[7] Matene E L, Presentation of the Fluent Code and the numerical method for the forecast (chapter 04 -PFE 2009 - numerical simulation of turbulent flow through flow conditioners.

[8] SIGRID R, Jun 2007. Development of methodological tools and techniques to help design quality urban environments to promote the sustainable development of cities

An Enhanced Seismic Activity Observed Due To Climate Change: Preliminary Results from Alaska

Adven Masih[1]

1 Ural Federal University, Ekaterinburg, Russia
E-mail: adven.masikh@urfu.ru

Abstract. The impact of human induced climate change on the rising temperature cannot be neglected. According to the Intergovernmental Panel on Climate Change (IPCC) 2012 report, the mean temperature roughly rises up to 3°C relative to 1990. Permafrost in Siberia and Alaska has started to thaw for the first time since it formed 11,000 years ago, has caused by the recent rise in temperature over the past six decades. The melting rate of glaciers has become significantly higher, causing a noticeable rise (0.19meters) in the sea level globally. Climate change can trigger catastrophes such as earthquakes, volcanic eruptions, tsunamis and landslides due to melting glaciers and rising in sea level. The melting of glaciers driven by global warming warns us of a seismically turbulent future. When glaciers melt, the massive weight on the Earth's crust reduces and the crust bounces back in what scientists call an "isostatic rebound". The process can reactivate faults and lift pressure on magma chambers that feed volcanoes, hence increases seismic activity. The paper discusses the correlation between rise in temperature due to global warming and earthquake frequency using Pearson's correlation coefficient and regression analysis based on a case study from Alaska.

1. Introduction

The global warming is not restricted to climate change, the entire planet is experiencing the consequences of disturbed energy balance. When we expand our context, it turns out that the dangers of global warming are far greater for humanity than implied by climate change. This disturbed energy balance could be in the form of an isostatic rebound, a volcanic activity or an earthquake.
The rise in mean surface temperature of air has become significant since at least the early 20th century globally, indicating the Earth surface is getting warmer [1]. According to NASA and NOAA most of the warming occurred in the past 35 years, with 15 out of the 16 warmest years on record have occurred since 2001 [2]. Since 1950 the warming rate of Alaska has reached to an alarming level of +5.3°F per century which is faster than any other state in United States of America, by keeping margin of almost a full degree-per-century with the second place holder Minnesota's +4.4°F [3]. The warming rate of Alaska is more than double against the warming seen in the rest of United States. Warming in the winter has increased by an average of 6°F [4] and has led to changes in ecosystems like, earlier breakup of river ice in the spring. As the climate continues to warm, average annual temperature in Alaska is projected to an additional 2 to 4°F further increase by the middle of 21st century reported by U.S. Global Change Research Program [4].
The consequences of climate change lead to a lower sea ice extent (or area of ocean covered by ice) on record. For example according to National Snow and Ice Data Center 2014 report sea ice extent recorded in September 2014 was nearly 700,000 square miles which was less than the historical 1979-2000 average for that month - the difference is more than twice the size of Texas [5]. According to US

EPA 2015 annual report decline in the thickness and age of sea ice throughout the Arctic has been witnessed, with recent measurements indicating a loss of 50% of sea ice since 1979.
An enhanced volcanic activity has been observed due to deglaciation in Arctic has led to a hypothesis that the unloading of the glaciated ice in a volcanic terrains can increase volcanism through decompression melting in the shallow mantle or storage time reduction in crustal magma. Recently, the enhanced deglacial volcanic activity in Southeast Alaska sourced from Mount Edgecumbe Volcanic field has been correlated with the rapid isostatic adjustment, occurred following a retreat of regional glaciers [6]. The finding is consistent with the hypothesis that isostatic rebound is associated with ice-unloading which increases the volcanism [6], that can possibly influence the frequency of the earthquakes as well.
A certain earthquake is related to climate change? Surely not, but the earthquake activity can be linked to climate change, at least in a mountainous region where glaciers form and later they melt. The process introduces isostatic uplift and then will be a feedback between these processes [7]. Glaciation of mountainous areas introduce extra load on Earth's crust, which introduces stress in the crust. Depending on the rate of the isostatic uplift due to melting glacier and the composition of the underlying strata, these stresses can be released as landslides, ductile deformation and also as earthquakes. It may look ridiculous that climate change can affect the tectonic plate moment beneath the crust, but scientists like McGuire argue that it cannot just intensify disastrous events like cyclones, volcanoes, tsunamis but can trigger earthquakes as well.

2. Methods and study area

The work includes Alaska's 42 years average temperature and earthquake data starting from 1974 to 2016 (Table 1). The annual temperature data in Table 1 is provided by *Alaska Climate Research Center*. The temperature data in Table 1 is an average of temperature data gathered from 250 different stations in Alaska. Whereas the earthquake data is obtained from USGS earthquake catalogue and have been categorized on the basis of magnitude (M_w) in Table 1.
The research adopts the following methodologies: a regional study; selecting the study region and information collection. For the purpose of this research, published reports and news articles, along with journal publications were the major sources of information. Initially *stacked line graph* was constructed using Microsoft Excel to look for trends and patterns in the earthquake data set, further, *Pearson correlation* was employed to verify the correlation between variables (temperature and earthquake frequency) according to the proposed hypothesis that the increased earthquake activity can be correlated to rising temperature. Pearson correlation is used to predict the causality between two variables based on correlation coefficient value r.
Alaska is located in the far northwest of North America. It shares border with Canada to the east, the Arctic Ocean to the north and Pacific Ocean to the south and west as shown in Figure 1. Alaska is known for its largely undeveloped land, mountains, glaciers, harsh climate, seismic activity and biodiversity. A recent research findings by Alaska Climate Research Center (ACRC) show that there's some additional warming recorded in Alaska after 1976, marking a phase change of pacific decadal oscillation from negative to positive phase (Figure 2) [8]. Taking a linear trend in figure 2, an annual average temperature variations can be seen over the last 6 decades.
According to Alaska Volcanic Observatory (AVO) Alaska makes up about 80% of all active volcanoes in United States. Since 1760, 30 Alaskan volcanoes have had more than 240 confirmed eruptions with an average of 1 event per year. However for the past 40 years in Alaska on average more than two eruptions per year have been recorded [9].
Unluckily Alaska is an earthquake prone region and has more coastline than any other state in U.S. Reports suggest that around 75% of all major events recorded in United States of magnitude 5 and above on Richter scale happen in Alaska [10]. In fact, magnitude 6 and 7 can nearly happen anywhere (figure 3), whereas the most destructive events – magnitude 8 on Richter Scale and above primarily occur in shallow part of the subduction zone, where dense oceanic rocks of the Pacific thrust sink under the continental rocks of the Alaska. Figure 3 presents the historical earthquake record from 1970 to 2012, while other examples of high magnitude earthquakes happened in Alaska before 1970 include the most powerful and worst earthquakes ever hit North America named "1964 Good Friday

Earthquake" of magnitude $9.2M_w$ and "Rat Islands Earthquake" of magnitude $8.7M_w$ in 1965, both have been categorized as second and eighth largest earthquakes respectively, ever recorded worldwide by USGS.

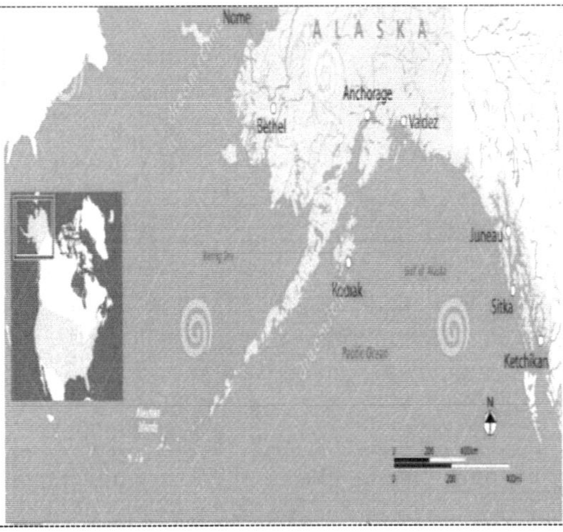

Figure 1. Regional map of Alaska

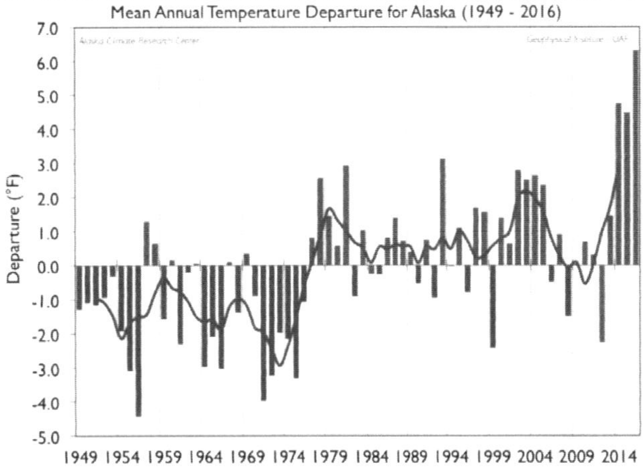

Figure 2. Mean annual temperature of Alaska Source: ACRC

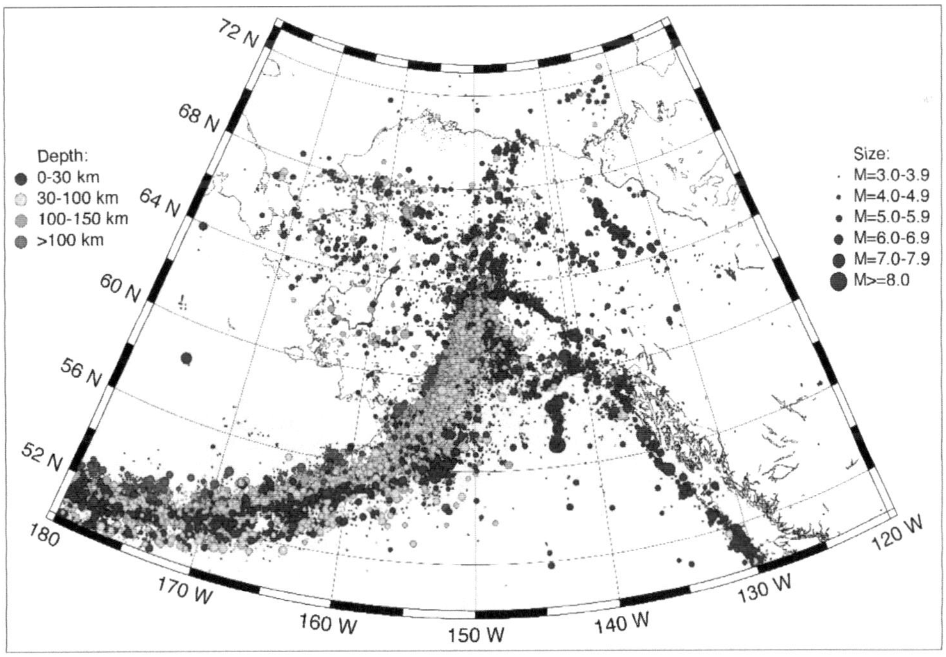

Figure 3: Historical map of earthquakes occurred in Alaska reported from 1970-2012, Source: AEC

3. Climate change and projected potential for seismic response

Climate change deserves a credit for its versatility. Global warming is not merely a fraction rise in the temperature over time, but these few degrees are and will continue affecting our human society and natural environment in the form of droughts, sea level rise, and floods if temperature continue to rise, however now seems it can influence another extreme class of events known as earthquakes.

U.S. Artic Research Commission reported [11] that the consistent rise in the temperature is a serious concern in Alaska and Arctic because often melting of glaciers and permafrost due to air and surface warming lead to ground destabilization as well as ecosystem changes. According to Sujib Kar [12] seismicity has been increasing due to global warming. Kar in an interview [13] said that "If you look at the number of earthquakes of magnitude 5 or more in Richter scale from 2001 to 2015 in earthquake-prone areas, you will be able to comprehend the dynamics. The rise of global temperature was phenomenal during this period". He further added that the total number of earthquakes recorded in 2001 with magnitude 5 or more on Richter scale across the globe were merely 157, whereas the numbers for same type of events occurred in 2015 were nearly ten times higher – 1556 [13]**.** The trend according to the proposed hypothesis shows that rising temperature due to global warming in Alaska might have contributed to the sharply increased cumulative earthquake frequency in recent decades.

Bill McGuire, a professor of geophysical and climate hazards at University College London, in his book entitled "Waking the Giant" argues that, if the present trend of climate change driven by human continue, the temperature and sea level will rise as compared to post-glacial period, it would be a surprise if some of the countless faults under the subsurface did not respond to the new distribution of global water certain to occur as world continue to heat up. He told India Climate Dialogue that "it's not a theory anymore, there is huge amount of evidence for the clear relationship between climate change and earthquakes particularly in Scandinavia and North America at the transition from the last Ice Age. The seismic response of ice unloading can clearly be seen in Alaska" [13]**.** McGuire [14] argues that when climate changed naturally in the past and planet emerged from an ice age, large ice sheets covering in much of the planet retreated. They were so heavy that the resulting release of pressure on the earth's crust caused it to '*bounce back*' triggering earthquakes, tremors and even volcanic activity [15] along preexisting fault lines. The last ice age occurred 20,000 years ago when temperature started rising caused large ice sheets to retreat, our earth is still responding to that ice age and the crustal bounce back rate has been modelled [16].

Earthquake and active volcanoes are interrelated for their same origin – tectonic disturbance. Generally when a volcano erupts it is very common for an earthquake to ensue. One can source the other, or both can occur at the same time because earthquakes are common around an active volcano and are usually triggered either by the moment of the magma beneath the earth surface or in response to strain exerted in an area of weak faults due to magma exclusion during volcanic activity. However such earthquakes in most cases are not of very high magnitude thus do not cause much damage directly.

4. Results and discussion

For earthquake data minimum earthquake magnitude picked is 2.5 on Richter scale, below 2.5 no events were considered. Table 1 witnesses a prominent temperature change of over +5°F during the study period in Alaska. In addition the average temperature during last 15 years (33.4°F – 0.77°C) has risen half a degree Celsius more than the temperature during first 15 years (32.5°F – 0.27°C) whereas for the years in-between (1989 –2001) the average temperature noted is 32.97°F – 0.54°C which confirms the alarming rate of consistent temperature rise in Alaska during last 42 years.

Due to the unexpected rapid melting of glaciers, Alaska has already lost almost a vertical kilometer of ice cover due to which stresses on the underneath rocks are being released [17]. This removal of load in return causes the earth crust to rise upward due to *isostatic rebound* [18], the released stresses could deform the earth surface [19] causing earthquakes. It is to be noted in Table 1 that temperature is consistently rising with over 13 times increased earthquake frequency (cumulative frequency) during the study period. Almost every sharp rise in temperature in Table 1 has consequently increased the total number of earthquakes recorded for the particular period.

Table 1. Total Number of earthquakes recorded based on the magnitude (M_w).

Years	Average Temp. (°F)	Minor EQs (2.5 – 4)	Moderate EQs (4 -5.9)	Major EQs (6 +)	Total Earthquakes	Cumulative Frequency
1974-77	30.02	259	224	0	483	161
1977-80	33.08	514	300	5	819	273
1980-83	32.6	373	306	3	682	227.3
1983-86	32.42	420	325	7	752	250.6
1986-89	34.36	646	424	9	1089	363
1989-92	33.3	1542	254	6	1802	600.6
1992-95	34.3	2415	304	4	2723	907.6
1995-98	32.2	2291	274	1	2566	855.3
1998-01	32.1	1896	302	9	2207	735.6
2001-04	34.96	3687	472	4	4163	1387.6
2004-07	33.43	3398	306	2	3706	1235.3
2007-10	32.2	3906	297	2	4205	1401.6
2010-13	31.4	4602	295	3	4900	1633.3
2013-16	35.03	5680	324	6	6010	2003.3

4.1. Earthquake frequency trend analysis

The stacked line graph plotted for the average frequencies of minor, moderate, major and cumulative frequency of earthquakes recorded allowing to easily identify and compare the trends in a time series data. It is understandable in Figure 4 that the average earthquake frequencies irrespective of the type of earthquakes follow an increasing trend. In fact minor, moderate and major earthquakes lines are almost overlapping, depict that frequencies of all types of earthquakes are indubitably increasing in a similar fashion during the study period.

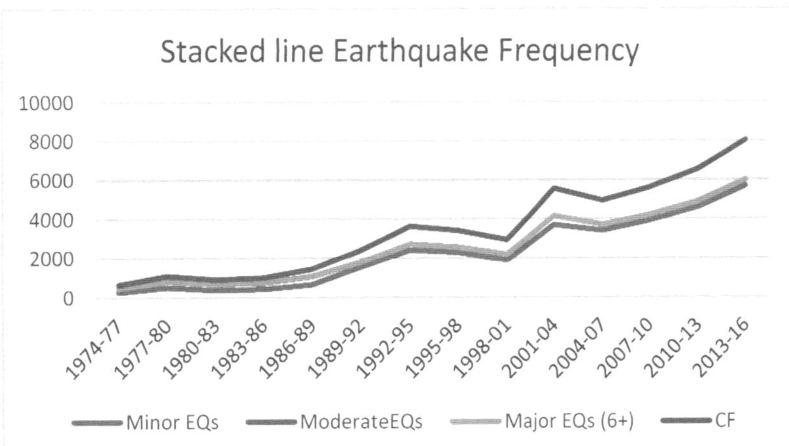

Figure 4. Stacked line graph for minor, moderate, major and cumulative earthquake frequencies (CF)

Deglaciation due to climate change promotes volcanism as discussed in section 1 which sources faults to reactivate. The effect of increased volcanic activity can be prominent in Alaska as most of the region is covered with glaciers melting at a high rate that can increase the frequency of earthquakes by reactivating pre-existing weak tectonic faults which makes the region sensitive to the changes like un-loading heavy ice cover. Thus a sharp increase in the frequency of earthquakes from 2001 – 2016 in figure 4 can be correlated to NASA reported 15 hottest years since 2001 [2] that has caused the glaciers and sea ice to melt at rapid pace and sea level to rise at a record rate of 85.5mm [20].

4.2. Correlation analysis

To get more insight into the data and to further verify the correlation between temperature and seismic activity Pearson's correlation coefficient was calculated at $P < 0.01$ i.e. with 99% confidence interval (chance of error is 1 out of 100) for minor, moderate, major and combined (moderate + major) earthquake data as shown in Table 2. As we are seeking a correlation between temperature and frequency of earthquake events, therefore, only temperature column is considered.

Table 2. Pearson's correlation coefficient (r), "r" is significant at P<0.01"*" and at P<0.12"**"

	Temperature	Minor EQs	Moderate EQs	Major EQs	Combined (Mod+Maj) EQs
Temperature	1				
Minor EQs	0.34	1			
Moderate EQs	**0.69***	0.17	1		
Major EQs	**0.44****	-0.15	0.42	1	
Combined (Mod+Maj) EQs	**0.70***	0.16	0.999	0.46	1

Earthquake behaviour is chaotic and skills for forecasting them are limited, however the rise in the frequency of small earthquakes is arguable as glaciers melt and sea level rises. Due to tremendous increase in the low magnitude earthquake records since 21st century beginning, the correlation coefficient between temperature and minor earthquakes was expected to be the strongest among all others presented in table 2, however interestingly, it is not even significant which reflects the technological progression of recent seismic station network expansion by USArray all over U.S. On the other hand, rise in temperature showed a strong positive correlation with moderate earthquake records, whereas its significant correlation with major earthquakes at $P < 0.12$ (medium confidence) was totally a surprise. Furthermore, a combined effect of both moderate and major earthquakes was also considered against the temperature variations that startlingly represented even greater correlation value r than its value for moderate earthquakes which warns that the rate of isostatic rebound driven by melting glaciers due to regional rise in temperature of Alaska has become high enough to influence the subsurface tectonic plates and can yield both moderate and major earthquakes by reactivating the subsurface faults.

To further validate our point and to testify the biasness of correlation coefficient, line graphs with linear regression trend for earthquakes of magnitude 4 and above versus temperature were plotted and set parallel in Figure 5 & 6 to visually examine the correlation between temperature and seismic activity in Alaska. Interestingly both presented analogous increasing trends. Although the earthquake regression line looks less steep as compared to temperatures', however the increase in the frequency of seismic records is evident. The sharp rise in average temperature of Alaska has not just escalated the frequency of earthquakes during particular periods such as 1986-89 and 2001-04, but each fluctuation either high or low in temperature graph seems influencing the earthquake frequency accordingly (Figure 5 & 6) is not merely a coincident, however, these frequent earthquakes most probably are the isostatic responses due to climate change in study region.

Adding carbon dioxide to the atmosphere inevitably causes the temperature to rise, and more you add, the greater the temperature rise [21]. No doubt GHG emission has made life possible, but the accumulation of greenhouse gases and particles such as inorganic black carbon soot due to enhanced anthropic activities has caused the Earth surface temperature to rise constantly according to IPCC 2014. According to USArray earthquake monitoring network –Transportable Array (TA) deployment of seismic stations expansion from 2004–2015 across 48 states including Alaska has enabled recording high quality data i.e. smaller subsurface movements are now recorded with precision, hence number of earthquakes recorded every year are increasing. Thus, technology advancement during recent decades might have contributed most likely to the low magnitude earthquake records in particular. The minor earthquakes listed in table-I confirms the sophistication of our capabilities of measuring earthquakes below magnitude 4 on Richter scale with time. Still, it seems that the technological progression perhaps have not affected the concerned seismic records of magnitude 4 and above because the tools used previously in 80s and 90s were able to record such events efficiently which table-I confirms.

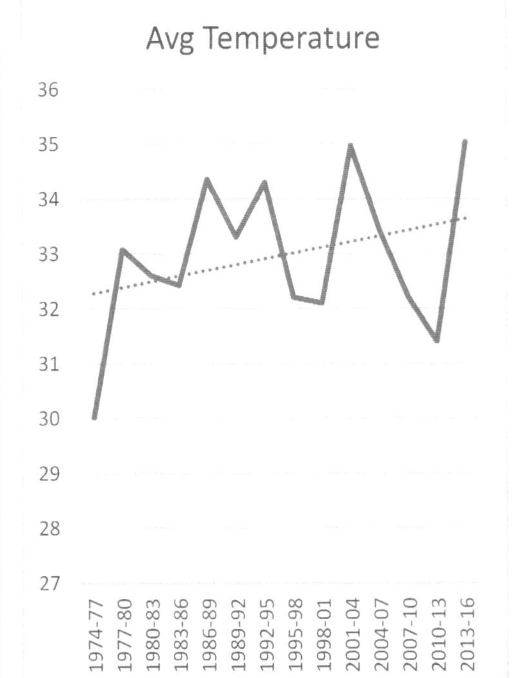

Figure 5. Average temperature graph

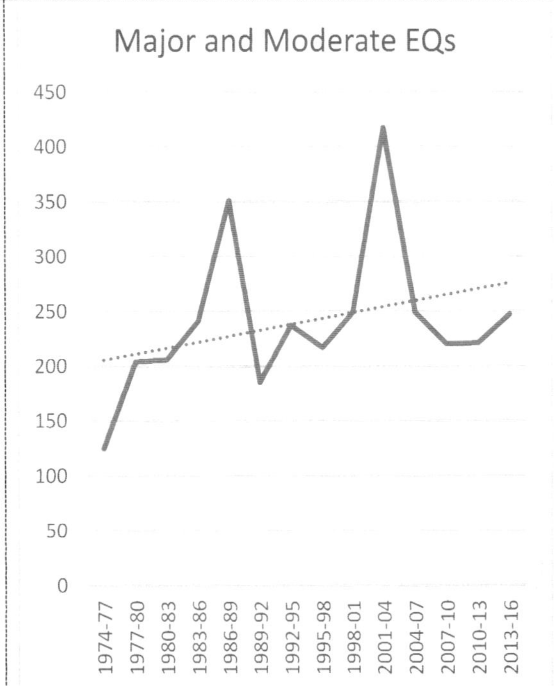

Figure 6. Combined (moderate and major) EQs

A total of 4468 earthquakes of magnitude above 4 on Richter scale have been reported from the study area including 61 major seismic events during the study period. The preliminary results by means of stacked and regression line graphs along with Pearson's correlation coefficient suggest that even though the postglacial stress due to rapid shrinkage of sea ice cover and melting glaciers has not yet started influencing the major earthquakes of magnitude 6 or more on Richter scale much, however the noticeable rise in the frequency of moderate earthquakes of magnitude ranging from 4 to 5.9 points to a seismically turbulent future, if the present trend of glaciers and sea ice melting due to rise in temperature driven by anthropogenic activity remains the same, soon many of the countless subsurface faults will start responding in the form of major earthquakes.

5. Conclusion

Temperature is rising globally due to global warming, particularly in Alaska it may result in a handful of benefits in terms of longer growing season for agricultural crops, tourism attraction and access to natural resources which are currently inaccessible due to the thick ice covers. However, the study conducted has used stacked line graph, Pearson's correlation coefficient and linear regression line graph to show that it can significantly influence the subsurface tectonic plate moments through accelerated seismic activity as well. It seems that rise in regional temperature due to global warming causing the glaciers to melt, which in turn depressurizing the underlying rocks, hence affecting the earth to rebound and faults to reactivate, therefore labeling the region seismically active with obvious increase in the frequency of volcanoes and earthquakes. The results presented are preliminary therefore further research will be conducted.

6. References

[1] Joseph F B, Matthew J, David D 2017 *Global and Planetary Change* **148** 55–71

[2] Brown D et al. 2 016 NASA NOAA Analyses Reveal Record-Shattering Global Warm Temperatures in 2015 https://www.nasa.gov/feature/goddard/2016/climate-trends-continue-to-break-records_Accessed 25 May 2017.
[3] Deke A, May 2016 Alaska: Last Frontier on the front lines, NOAA https://www.climate.gov/news-features/blogs/beyond-data/alaska-last-frontier-front-lines-climate-change_Accessed 25 May 2017.
[4] Chapin, F, Trainor S, Cochran P, Huntington H, Markon C, McCammon M, McGuire AD, and Serreze M, 2014: Ch. 22: Alaska. Climate Change Impacts in the United States: The Third National Climate Assessment, J. M. Melillo, Terese (T.C.) Richmond, and G. W. Yohe, Eds, U.S. Global Change Research Program, 514-536.
[5] A better year for the cryosphere 2013 Arctic Sea Ice News & Analysis, National Snow and Ice Data Center http://nsidc.org/arcticseaicenews/2013/10/a-better-year-for-the-cryosphere/ Accessed 17 April 2017.
[6] Praetorius S, Mix A, Jensen B, Prahl F 2016 *Earth and Planetary Science Letters* **452:**79-89
[7] Molnar P and England P 1990 *Nature* **346**, 29-34.
[8] Temperature changes in Alaska, The Alaska Climate Research Centre http://climate.gi.alaska.edu/ClimTrends/Change/TempChange.html Accessed 26 June 2017.
[9] Alaska's Volcanoes., Alaska Volcano Observatory. https://www.avo.alaska.edu/volcanoes/about.php_Accessed 26 May 2017.
[10] Why Earthquake happen in Alaska, Alaska Earthquake Center https://earthquake.alaska.edu/earthquakes/about Accessed 26 May 2017
[11] U.S. Arctic Research Commission Permafrost Task Force 2003 Climate Change, Permafrost, and Impacts on Civil Infrastructure. Special Report 01-03 (pp.72), *U.S. Arctic Research Commission*, Arlington, Virginia. Accessed 5 June 2017.
[12] Kar S 2013 *International Journal of Engineering Science Invention* ISSN (Online): 2319 – 6734, ISSN (Print): 2319 – 6726.
[13] Basu J, 2016 Climate change raises earthquake risk, Indian Climate Dialogue http://indiaclimatedialogue.net/
[14] McGuire B 2010 Climate forcing of geological and geomorphological hazards, *The Royal Society* (http://rsta.royalsocietypublishing.org/content/368/1919/2311.
[15] Sigvaldason G, Annertz K, Nilsson M 1992 *Bull Volcanol* **54**:385-392
[16] Paulson A, Zhong S, Wahr J, 2007, *Geophysical Journal International* **171**.
[17] McGuire B 2012 Science Weekly podcast: Will climate change unleash geological mayhem? *The Guardian* https://www.theguardian.com/science/audio/2012/feb/13/science-weekly-podcast-climate-change
[18] Wicander R, and Monroe S 2006 Historical Geology Chap 17: Cenozoic Geological History: The Pleistocene and Holocene Epochs. Belmont, CA: Thomson Brooks/Cole.
[19] Pagli C and Sigmundsson, F 2008 *Geophysical Research Letters* **35**:L09304
[20] Sea Level, NASA, (https://climate.nasa.gov/vital-signs/sea-level/_(Accessed August 20, 2017).
[21] Wadhams P 2017 A farewell to Ice: Report from the Arctic. *Oxford University Press*. 50-55.

IOP Conf. Series: Earth and Environmental Science **167** (2018) 012019 doi:10.1088/1755-1315/167/1/012019

Factors affecting greenhouse microclimate and its regulating techniques: A review

G Li[1], L Tang[1], X Zhang[2], J Dong[1] and M Xiao[1]

1 Department of Architecture and Built Environment, University of Nottingham Ningbo China, Ningbo, China
2 Solar Energy Research Center (SERC), Dalarna University, Falun, Sweden
E-mail: Guozhen.Li@nottingham.edu.cn

Abstract. This paper reviews factors affecting greenhouse microclimate and its regulating techniques towards upgrading the greenhouse applications in the area of southeast China which have little or very basic technology integration. The microclimate of greenhouse is apparently influenced by the shape and its orientation, the wind direction, the property of covering material, and the use of insect-proof screen as they eventually affect the total solar radiation, the thermal characterises, and the flow pattern inside. The natural ventilation and sun block are the most common method to cool the greenhouse, but more efficient evaporative cooling such as pad-fan system, misting/fogging system and roof sprinkler are required with extreme temperatures. The earth to air heat exchanger and the heat storage using phase change material may be used for heating or cooling throughout the year which are more economic and energy-saving than other traditional thermal technologies. The reviewed knowledge provides insights into upgrading greenhouse applications in Ningbo area towards more sustainable and efficient greenhouse farming.

1. Introduction

Greenhouse technology allows farmers to grow many kinds of crops in regions and seasons of adverse climatic conditions which brings high profit. It also allows people to enjoy almost any fruit or vegetable at local market at any time regardless of whether the food is out of season. The popularity of greenhouse farming techniques increased the crop production for the whole world and contributed to the reduction of world hunger problems. However, the success of greenhouse crop productions relies on the capacity to carefully regulate the microclimate inside the greenhouse in order to meet the requirement for crop production and quality [1]. It is reported that most plants grown in greenhouse prefer an optimal temperature between 17 °C and 27 °C, with extreme temperature of 10°C and 35°C [2] which is achieved by means of natural ventilation, shading and other cooling and heating techniques. Besides, the design parameters of greenhouse such as the shape, size, orientation, height, opening of side and roof vents, the wind direction outside the greenhouse, and the properties of covering material play an important role in influencing the microclimate of the greenhouse. This paper therefore reviews factors influencing the microclimate of the greenhouse and the latest technology for regulating the microclimate of the greenhouse, in particular, the cooling and heating techniques.

2. Factors influencing the microclimate of the greenhouse

2.1. Shape and Orientation

As mentioned, the air temperature inside of a passive greenhouse is directly dependent, besides the ambient air temperature and heat transfer confident of covering material, on the solar radiation. The solar radiation obtained by the greenhouse at a certain time and site is dependent on its shape and orientation which eventually regulate the inside air temperature with most of the direct solar radiation at being obtained at its floor which increases the air temperature inside the greenhouse. Besides, greenhouse also accepts diffusive and ground reflected radiation from roof and each wall, therefore the shape and orientation of the greenhouse have a crucial importance on the air temperature inside the greenhouse. The selection of orientation of a greenhouse (either along East-West or North-South orientations in longitudinal axis) can determine the total solar radiation received by the greenhouse , also can reduce the heating loads of the installed systems thereby saving a lot of cost. Sethi [3] compared five single span shapes of even-span, uneven-span, vinery, modified arch and Quonset type as shown in Figure 1 in terms of transmitted total solar radiation in India. Results show that uneven-span shape greenhouse receives the maximum and Quonset shape receives the minimum solar radiation during each month of the year at all latitudes. East-west orientation is the best suited for year-round greenhouse applications at all latitudes as this orientation receives greater total radiation in winter and less in summer except near the equator. Whist, Chandra et al [4] argued that north-south orientation results in better homogeneity of the microclimate for northern climatic conditions in India. While, Dragićević [5] concluded that, for uneven-span greenhouses orientated east-west and north-south (as comparison) at different latitudes at the northern hemisphere of Belgrade, Serbia, the east-west orientation of uneven-span solar greenhouse is the best suited during each month for all examined latitudes.

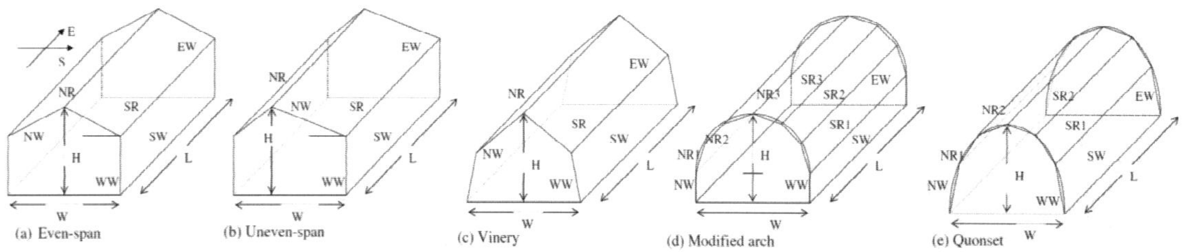

Figure 1. Greenhouse shapes in E-W orientation compared by Sethi [3]

2.2. *Wind direction*

The wind blows a greenhouse causing a direct positive pressure at the windward wall when the wind hits it. Whilst, the wind moves around the greenhouse and leaves the leeward wall with a negative pressure known as a sucking effect. If there are any vent openings on the windward and leeward walls of the greenhouse, fresh air will rush in the windward wall opening and exit the leeward wall opening to balance and relieve the pressures on the windward and leeward walls. The wind direction outside the greenhouse is an important factor influencing the flow of air and microclimate inside the greenhouse. Roy and Boulard [6] numerically studied the effects of wind direction on the climatic parameters inside a greenhouse which showed clearly the influence of the wind direction on the velocity, temperature and humidity distributions inside. Teitel et al. [7] found that when the wind blows perpendicularly to the greenhouse opening plane there is a cross-flow between the windward and leeward openings that results in gradients in the air and crop temperatures between windward and leeward sides. Gradients in temperature, humidity and air velocity are observed in the vertical direction as well. The temperature and humidity ratio were larger near the roof than near the crop while the air velocity was higher near the crop than near the roof. The gradients in the vertical direction were larger than those encountered in a horizontal plane in a direction parallel to the mean flow (from windward to leeward opening). The temperature difference between inside and outside and the turbulent fluxes were generally higher near the windward opening than near the leeward one. In addition, the ventilation rate increases linearly with wind speed. Khaoua et al. [8] numerically investigated the effect of wind speed and roof vent opening configuration on airflow and temperature patterns in a compartmentalised glasshouse. Results show that opening configurations combined with wind speeds strongly affect inside ventilation and microclimate parameters. Moreover, inside partition hinders the air circulation between different parts and causes significantly different internal microclimates within each compartment. Researchers also found that the wind direction affects the level of ventilation suggesting that a greenhouse with large vertical roof windows works better with a windward condition, whereas the multi-span greenhouse works better with a leeward condition [9].

2.3. Covering materials
It is important to choose suitable material for the covering of greenhouse as it will influence the radiations to plants, the diffuse light, the level of shading, and the temperature or humidity which are of concern for crop growth. Besides, factors such as the durability of material and its cost should be considered. So far, the most widely used covering are the flexible plastic films, such as polyethylene, and the rigid covering material such as plates of polycarbonate or the glass. The biggest advantages of the polyethylene are its malleability, easy installation and the cost-benefit ratio it offers in comparison with the other materials. The polycarbonate plates offer more durability and resistance to the impact, so they are often used in more extreme climates. Its main disadvantage, opposite to the polyethylene, is its high cost. Greenhouses with glass as covering material are often used in cold weather climates. They have advantages such as a total water tightness and greater insulation than simple plastic. Its principal disadvantages are its high cost in comparison with the rest, its fragility against the impacts and its high weight which needs a more resistant and adapted structure to this material. As an alternative, the use of ETFE (Ethyl Tetra Fluoro Ethylene) which is a type of plastic normally used in architecture in buildings, is starting to be studied. ETFE has high resistance to corrosion and load, supporting up to 400 times its weight. However, this material is very costly (up to 15 times more expensive than conventional plastic) and requires much more specialization in terms of the installation and structures adapted for it [10]. Montero et al. [11] suggested that diffusive films are preferred over clear films because they improve light uniformity and increase light interception by the crop. The U-V blocking films are a promising technique to reduce pest infestation. The ability of insects (whitefly and thrips) to move is associated with UV radiation; hence, by using plastic materials that absorb UV radiation, virus-disease transmission can be mitigated. Besides, they recommend multilayer rather than single-layer films since they allow addition of the positive properties of each of the components that form the film.

2.4. Insect-proof screen
In recent years crops have been grown under insect-proof screen to protect crop production from pest invasion. Greenhouses in the Mediterranean and southeastern Europe are often equipped with ventilation openings to offer good microclimate conditions for plant growth. However, these ventilation vents also serve as a port for pests to enter the greenhouse. Screens are characterized by their porosity which is the open area to its total area, their mesh size, the thread dimension, the light transmission/reflection and their resistance to airflow. In order to prevent very small pests to enter the greenhouse, screens with very fine mesh screens have to be used. However, the use of insect-roof screen would reduce the light transmission and hinder ventilation due to resistance to airflow. This is especially true duing warm and hot seasons when effective ventilation is most needed to avoid very hot and humid conditions for crop growth [11]. When insect-proof screen is used, it is important to consider the effect of screens on increasing the pressure drop on the ventilation openings which consequently reduce ventilation. The pressure drop on screens is mainly dependent on the screen porosity (ε) which can be calculated from the geometric dimensions of the screen by [11]:

$$\varepsilon = \frac{(l-d)(m-d)}{ml} \tag{1}$$

Where: l and m are the distance between the centres of the two-adjacent weft and warp threads respectively. d is the diameter of the threads. The temperature difference between greenhouse and ambient air with screens (ΔT_{sw}) and without screens (ΔT_w) can be related to the porosity of the screens by below correlation:

$$\Delta T_{sw} = \Delta T_w(5 - 4\varepsilon) \tag{2}$$

It should be borne in mind that equation 2 only give a rough estimate of the ΔT_{sw} as the temperature difference with and without a screen depends on greenhouse type, crops, weather etc. Considering the drawback of utilizing insect-proof screen, it is suggested that the ventilation openings are uncovered when there is no risk of pest invasion [12]. In addition to the effects on the ventilation rate, the screens reduce light transmission into the greenhouse by creating strips of shadow on the crop when they are installed on roof openings. In dusty regions the shadow effect may be worsen due to the accumulation of dust on the screens [11].

3. Greenhouse regulating technologies

3.1. Natural ventilation

Natural ventilation is a common way to cool the crop area in summer as it removes heat inside the greenhouse by a pressure difference created by wind and temperature. Besides temperature control, greenhouse ventilation also serves the imperative functions of humidity control, CO_2/oxygen replacement and air circulation. Sase [13] suggested that under the condition of a mild climate, proper design and good control of natural ventilation can ensure effective cooling and uniformity of the microclimate inside the greenhouse. Omer [14] investigated various designs of low energy greenhouse and stated that various change such as the size and shape of the vents can improve air movement over crops. Von Zabeltitz suggested that, if local wind speed is high enough, the volume to floor ratio of the greenhouse should be as large as possible to maintain favourable environment for crop growth. The side vents can be as large as possible, but roof vent opening should not be too large to avoid rain entering the greenhouse during the rainy season. The American Society of Agricultural and Biological Engineers (ASABE) standards [15] recommended that in tropical conditions the total sidewall vent area and the total ridge vent area should be the same with at least 15-20% of the floor area of the greenhouse. Other researchers reported different ratios of ventilation area to the surface roof area that were optimum to their specific applications. Those ratios reported are summarized in Table 1.

Table 1. Different ratios of ventilation area to the surface roof area reported by researchers

Researchers	Year	Ratio of vent to surface roof area	Conditions
(ASABE) standards [15]	2003	15-20%	Tropical conditions
Connellan [16]	2000	20%	Higher temperature regions
Kamaruddin et al. [17]	2000	40%	Tropical regions
Montero et al. [18]	2001	33%	-
Albright [19]	2002	10%	-
Campen [20]	2005	40.4%	Under Indonesian conditions
Hermanto et al. [21]	2006	60%	In humid tropics

Simone [22] argued that it is difficult to recommend the exact vent size for ideal ventilation. Generally, more smaller vents are better than one large one with the total venting area equal to about 1/6 the area of the south facing glazing. It is recommended to have at least one low vent and at least one high vent so that hot air will rise up and exit the upper vent(s) and draw fresh cooler air into the lower vents. Also, the vents should be positioned in a way which will prevent cold air from flowing directly across the plants.

In addition to natural ventilation, Artificial ventilation using induced draught and exhaust fans, blowers, helps in maintaining a greenhouse temperature closer to ambient due to higher rate of air change than that could be achieved through natural ventilation. In most greenhouse applications, fan induced ventilation is employed together with evaporative cooling system thus leading to pad-fan systems [23].

3.2. Pad-fan system

Mechanical cooling (for example, refrigeration-type air conditioning) is possible, but seldom affordable or practical for commercial greenhouses. Evaporative cooling techniques using pads & fans, fogs, mists, sprinklers can be very effective and affordable when used properly [24]. Evaporative cooling is a process that reduces air temperature by evaporation of water into the airstream. As water evaporates, energy is lost from the air causing its temperature to drop. Evaporative cooling is one of the most effective cooling method for controlling the temperature and humidity inside a greenhouse.

As shown in Figure 2a, pad-fan system applies a combination of evaporative pads and extractor fans to hermetic greenhouses. In this method, one of the walls is replaced with a wet pad, and there are fans on the opposite side wall to intake the air. The pad-fan system has a greater air saturation efficiency than that of the fog system which is also cheaper and less consumable in water and energy. But the method only works well with hermetic greenhouse structures which facilitates all incoming air passes through the evaporative pads on the side wall with powerful extractor fans being installed on the opposite wall to provide the suction effects [25].

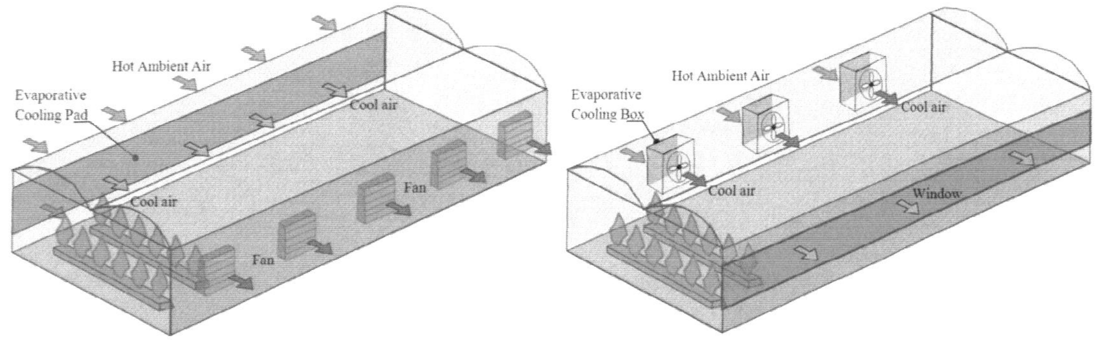

Figure 2. (a) Pad-fan evaporative cooling and (b) Evaporative cooling boxes (after Franco, 2014)

In addition to the pad-fan systems, the evaporative cooling boxes (Figure 2b) provide an alternative system which consist of an evaporative cooling unit with a cooling chamber, usually equipped with a plastic packing of a honeycomb mesh in order to allow air to come into contact with the water. The unit has a pump for water recycling and a fan for impelling the cooled air horizontally into the greenhouse. Evaporative cooling boxes do not require extremely hermetic structure of the greenhouse as the fan is an integral part of the cooling unit which impels the cool air into the greenhouse [25]. It worth mentioning that the suitability of evaporative cooling is restricted to particular region and climate. E.g. it is not suited to humid tropics due to high humidity levels.

3.3. Misting/fogging Systems

Greenhouse Misting Systems or fogging systems allow greenhouse temperatures to be reduced and humidity levels increased providing a growing environment for crops. The cooling is achieved by evaporation of small droplets of spraying water. The advantage of mist and fog systems over wet pad systems is the uniformity of conditions throughout the greenhouse, therefore eliminating the need for forced ventilation and airtight enclosure. Li and Willits [26] compared the cooling performance of a low-pressure (405 kPa) and a high-pressure fogging system (6.89 MPa) in two empty naturally ventilated greenhouses under summer conditions. The experiments showed that the evaporation efficiency of high-pressure system was at least 64% higher than the low-pressure system, however, the cooling efficiencies of the high-pressure system were only 28% higher than that of the low-pressure system. Katsoulas et.al. [27] suggested that mists, fogs, and sprinkler systems can be used with or without mechanical ventilation, but they still require controlled air exchange to maintain the correct humidity/air temperature balance.

3.4. Roof Sprinkler Cooling system

Roof Sprinkler Cooling systems provide cooling by sprinkling of water on the roof surface of greenhouse to form a thin layer of free water surface which evaporates and cools the surface. These systems are simple, and appear to be effective, but are not suited in areas where water is in short supply. Cohen et al. [28] carried an experimental comparison of evaporative cooling in a naturally ventilated glasshouse due to wetting the outer roof and inner crop soil surfaces. It was found in the glasshouse with a growing tomato crop, temperature differences between the air inside and outside of the glasshouse were affected by the wetting treatments. With growing plants, or wet soil surfaces, evaporative cooling of the roof surface appeared to have a minor effect on glasshouse air and foliage temperatures. These temperatures, however, were strongly affected by evaporative cooling of soil and foliage surfaces. Direct evaporation of a water droplet suspended in a spray mist in the air appeared to be the most effective site for evaporative cooling of the glasshouse.

3.5. Apply sun block

One of the best ways to avoid heat accumulation in the greenhouse is to block some of the solar energy from getting inside in the first place though this would also block light for photosynthesis. However, it can often seek a balance between the need for sufficient light for plant growth and the need to control potentially damaging temperatures by blocking, filtering, and reflecting incoming light on the outside of the greenhouse covering, or by using fixed or retractable shade materials inside the greenhouse. Plants are adapted for varying light levels and most species reach their maximum photosynthetic efficiency at rather low light intensities relative to full sunshine. Extremely high light levels can actually stop or slow down photosynthesis which directly reduces the available carbon dioxide required for photosynthesis. Therefore, providing some shade can often be beneficial (20-75% depending upon the species and the light intensity), resulting in a net gain in photosynthesis and continued healthy plant growth [24].

3.6. Earth to air heat exchangers (EAHES)

An earth-to-air heat exchanger draws air through buried ducts or tubes for cooling the greenhouse in summer and heating in winter. As the temperature of the ground below 3 meters is practically constant it therefore provides space conditioning throughout the year, with the incoming air being heated in the winter and cooled in the summer by means of earth coupling. Systems can be driven by natural stack ventilation, but usually require mechanical ventilation. As a new sustainable development technology, the greenhouse earth to air heat exchanger is more economic and energy-saving than other traditional thermal technologies for greenhouse because of the high heat capacity of the shallow soil. The technique is suited to greenhouse with a moderate cooling demand, located in climates with a large temperature differential between summer and winter, and between day and night.

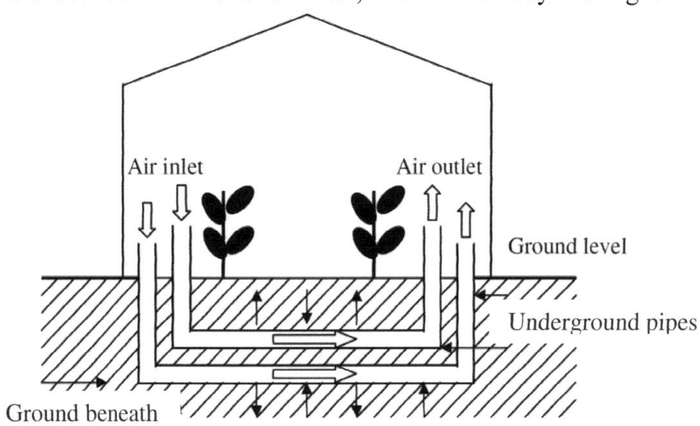

Figure 3 Coupling of a greenhouse with an earth-to-air heat exchanger system (after Sethi, 2008)

As depicted in Figure 3, the EAHES commonly consists of underground pipes and the airflow system, which forces the air through the pipes. Pipes usually run along the length of the greenhouse, with entrance and exit points of the circulating air at opposing sides. Cold air from inside the greenhouse is circulated through the underground pipes. Heat is transferred from the soil to the air stream and then returned to the greenhouse. This repeated circulation of cold greenhouse air causes the heating effect inside the greenhouse. The same system can also be used for cooling during the summer conditions [29]. Mavroyanopoulos et al. [30] studied the performance of a greenhouse heated by an earth-air heat exchanger consisting of 20 aluminium pipes with 15m length, 0.2m in diameter and 0.2mm thickness installed at a depth of 2m in a greenhouse of floor area 150 m^2. The greenhouse air temperature was adjusted by blowing the greenhouse air through the pipe system whenever the air temperature dropped below 12°C or exceeded 28°C. This resulted in a minimum mean night air temperature of 8.1°C inside the greenhouse with an absolute minimum of 7°C while the outside minimum mean and absolute minimum air temperatures were −0.8°C and −3.0°C, respectively, during the same period.

3.7. Phase change material storage

Due to the high cost of energy in greenhouse, the storage of heat is important for a greenhouse to provide optimum inside conditions during winter months. The strategy of passive heating system of greenhouse is to reduce the heat losses and at the same time to transfer excess heat from inside the greenhouse during the day to heat storage. This heat is used during the night to satisfy the heating needs of the greenhouse [31]. Besides other heating system such as water storage, rock bad storage, mulching, movable insulation, thermal curtain and earth to air heat exchanger, the use of phase change materials (PCM) is a promising alternative method. PCM are substances with a high latent heat which is capable of storing and releasing large amounts of energy when melting and solidifying at a certain temperature. In the heat storage cycle, PCM can store large amounts of heat in melting from solid to liquid at a constant temperature (phase transition temperature). In heat dissipation cycle, a circulating fluid air or water can extract heat from the storage unit causing the phase change material to solidify. Berroug et al. [31] developed a numerical thermal model to analyze the thermal performance of a greenhouse with a north wall made with the phase change material of $CaCl_2{\cdot}6H_2O$ as a storage medium. Results shows that with an equivalent to 32.4 kg of PCM per square meter of the greenhouse ground surface area, temperature of plants and inside air were found to be 6–12°C more at night time in winter period with less fluctuations. Relative humidity was found to be on average 10–15% lower at night time. Caprara et al. [32] compared a traditional sensible heat storage system of water tank storing unit with a latent heat storing unit based on PCMs. It was proven that PCMs materials are much more suitable for low temperature applications than sensible heat storing materials (water). In addition, PCMs unit shows a better response to the hourly energy fluctuations of solar collectors and greenhouse demand than water tank unit. Due to the limit of the paper, a more detailed review in terms of other heating technologies for agricultural greenhouse can be referred to the work of Sethi et al. [29].

4. Conclusions

The design parameters and the selection of microclimate control techniques are vital for sustainable greenhouse farming in the regions of adverse climatic conditions. This paper therefore reviewed the factors influencing the microclimate of the greenhouse and the latest advance in technology for regulating the microclimate of the greenhouse. The influence of greenhouse design parameters and its functional characteristics on microclimate of greenhouse was reviewed in terms of its shapes and orientations, the wind direction, the selection of covering materials and the effects of insect-proof screen. The importance of natural ventilation was stressed and the recommend location of opening vents and the ratio of ventilation area/surface roof area were summarized. The performance of evaporative cooling including pad-fan system, misting/fogging system and roof sprinkler system was discussed. The advantages of earth to air heat exchanger and the heat storage method using phase change material were introduced which can be used for heating or cooling throughout the year. The above knowledge offers insights for the design and selection of regulating techniques for advancing greenhouse applications in southeast China's Ningbo area with low technology structures. In the future, we will carry out further comparison and investigation of the reviewed microclimate regulating techniques to find out suitable solutions for local greenhouse farming.

5. References

[1] Benni S, Tassinari P, Bonora F, Barbaresi A and Torreggiani D 2016 Efficacy of greenhouse natural ventilation: Environmental monitoring and CFD simulations of a study case *Energ Buildings* **125** pp 276-86
[2] Campotti C A, Latini A, Matteo S and Viola C 2014 certificati bianchi guida operativa per la scheda 40E—Gli Incentivi Del DM 28 Dicembre 2012 Sui Sistemi Serra ENEA, Rome (2014)
[3] Sethi V P 2009 On the selection of shape and orientation of a greenhouse: Thermal modeling and experimental validation *Sol. Energy* **83**(1) pp 21-38

[4] Chandra P, Srivastava P C and Maheshwari R C 1982 Solar greenhouse design consideration, in ISAE XIX Annual Convention 1982 Udaipur.
[5] M. S and Dragićević 2011 Determining the optimum orientation of a greenhouse on the basis of the total solar radiation availability *Thermal science* **15**(1) pp 215-21
[6] Roy J C and Boulard T 2005 CFD prediction of the natural ventilation in a tunnel-type greenhouse: Influence of wind direction and sensibility to turbulence models. International Society for Horticultural Science (ISHS), 2005 Leuven, Belgium
[7] Teitel M, Liran O, Tanny J and Barak M 2008 Wind driven ventilation of a mono-span greenhouse with a rose crop and continuous screened side vents and its effect on flow patterns and microclimate *Biosyst Eng.***101**(1): pp 111-22
[8] Ould Khaoua S A, Bournet P E, Migeon C, Boulard T and Chassériaux G 2006 Analysis of Greenhouse Ventilation Efficiency based on Computational Fluid Dynamics *Biosyst Eng.* **95**(1) pp 83-98
[9] Rico-García E, Reyes-Araiza J L and Herrera-Ruiz G 2006 Simulations of the climate in two different greenhouses *Acta horticulturae* **719** pp 325-32
[10] Huete J 2017 Diferents covering materials for greenhouses. [cited 2017; Available from: http://www.jhuete.com/covering-materials-greenhouses/.
[11] Montero J I, Teitel M, Baeza E, Lopez J C and Kacira M 2013 Greenhouse Design and Covering Materials, in Chapter from formal publication, Food and Agriculture Organization of the United Nations (FAO).
[12] Ben-Yakir D and Chen M 2008 Studies of thrips migratory flights in Israel. Acta Phytopathologica et Entomologica Hungarica **43**(2) pp 243-8
[13] Sase S 2006 Air movement and climate uniformity in ventilated greenhouses *Acta horticulturae* **719** pp 313-24
[14] Omer A M 2010 Constructions, applications and the environment of greenhouses *Afr. J. Biotechnol.* **825**(25) pp 7205-27
[15] ASABE 2003 Standards for Heating, Ventilating and Cooling Greenhouses, 2003 USA
[16] Connellan G J 2000 Australian Flower Conference: Greenhouse Design for a Healthy Cut Flower Crop 2000 Australia
[17] Kamaruddin R, Bailey B J and Douglas M P 2000 Physical properties of covering materials for naturally ventilated tropical greenhouse *J. Trop. Agric. Food Sci.* **28**(1) pp 55-69
[18] Montero J I, Anton A, Kamaruddin R and Bailey B J 2001 Analysis of thermally driven ventilation in tunnel greenhouse using small scale models. *J Agr Eng Res* **79**(2) pp 213-22
[19] Albright L D 2002 Controlling greenhouse environments, International Society for Horticultural Science (ISHS), Leuven, Belgium, pp 121–25
[20] Campen J B 2005 Greenhouse design applying cfd for indonesian conditions. 2005 International Society for Horticultural Science (ISHS), Leuven, Belgium. pp 419-24
[21] Harmanto M, Tantau H J and Salokhe V M 2006 Optimization of ventilation opening area of a naturally ventilated net greenhouse in a humid tropical environment. 2006: International Society for Horticultural Science (ISHS), Leuven, Belgium pp 165-71
[22] Rudge S 2014 Greenhouses for the Northern Climate [cited 2017 10.25] Available from: http://www.emr.gov.yk.ca/agriculture/pdf/Greenhouses_for_the_Northern_Climate.pdf.
[23] Ganguly A and Ghosh S 2011 A Review of Ventilation and Cooling Technologies in Agricultural Greenhouse Application. *Iranica Journal of Energy & Environment* **2**(1) pp 32-46
[24] Argus 2017 Greenhouse Cooling Application Note [cited 2018 01.02] Available from: http://arguscontrols.com/resources/Greenhouse-Cooling-Argus-Application-Note.pdf.
[25] Franco A, Valera D L and Peña A 2014 Energy Efficiency in Greenhouse Evaporative Cooling Techniques: Cooling Boxes versus Cellulose Pads *Energies* **7** pp 1427-47
[26] Li S and Willits D H 2008 Comparing low-pressure and high-pressure fogging systems in naturally ventilated greenhouses. *Biosyst Eng.***101**(1) pp 69-77
[27] Katsoulas N, Baille A and Kittas C 2001 Effect of misting on transpiration and conductances of a greenhouse rose canopy *Agr Forest Meteorol* **106**(3) pp 233-47

[28] Cohen Y, Stanhill G and Fuchs M 1983 An experimental comparison of evaporative cooling in a naturally ventilated glasshouse due to wetting the outer roof and inner crop soil surfaces *Agric Meteorol.* **28**(3) pp 239-51
[29] Sethi V P and Sharma S K 2008 Survey and evaluation of heating technologies for worldwide agricultural greenhouse applications *Sol. Energy* **82**(9) pp 832-59
[30] Mavroyanopoulos G N and Kyritsis S 1986 The performance of a greenhouse heated by an earth-air heat exchanger *Agric For Meteorol.* **36**(3) pp 263-8
[31] Berroug F, Lakhal E K, El Omari M, Faraji M and El Qarnia H 2011 Thermal performance of a greenhouse with a phase change material north wall *Energ Buildings* **43**(11) pp 3027-35
[32] Caprara C and Stoppiello G 2009 Potential use of phase change materials in greenhouses heating: Comparison with a traditional system *J Agr Eng* **3** pp 25-32

Acknowledgments
The authors wish to acknowledge the below grants: Ningbo Enrich People Project (2016C10035), Ningbo Science & Technology Bureau Innovation Team project (2015B11011) and Ningbo Innovation Team Project on indoor air pollution control technology (2017C510001).

Institutional barriers for the implementation of climate change adaptation actions in the Mexican coastal zones

Norma Patricia Muñoz-Sevilla[1], Isaac Azuz-Adeath[2] and Maxime Le Bail[1]

1 Centro Interdisciplinario de Investigaciones y Estudios sobre Medio Ambiente y Desarrollo, Instituto Politécnico Nacional, Mexico
2 Centro de Innovación y Diseño, Colegio de Ingeniería, Centro de Enseñanza Técnica y Superior, Campus Ensenada, Mexico.
E-mail: nmunozs@ipn.mx

Abstract. This paper has as its object of study the coastal and marine areas of Mexico; specifically, it focuses on the analysis of existing barriers for the implementation of actions tending to the adaptation to climate change of these fragile and dynamic regions of the national territory. The study makes a general review of the main meetings, agreements and international entities related to the global climate that have served as the basis for the XXI Conference of the Parties, from which emanated the "Paris Agreement". After analyzing some implications of the aforementioned agreement and presenting evidence of the current situation, the main legal and normative instruments of the national marine and coastal areas are reviewed; the main risks in these areas of the national territory associated with climate change are evaluated; the main existing adaptation measures are analyzed; and, the most important socioeconomic and political barriers that exist for their correct implementation are established. Finally, some lines of action are proposed to improve the current vulnerability of the coastal areas and increase their resilience.

1. Introduction

Interest in long-term changes in climate has been the subject of scientific attention at least since the beginning of the last century [1], as well as concern for its impact on human development [2]. It is during the World Conference on Climate of 1979, when researchers of the climatic phenomenon show for the first-time signs of alarm about the effect that human activity was producing on the global climate.

This concern of the international scientific community for the climate system, prompted in 1988 the United Nations Environmental Program (UNEP) and the World Meteorological Organization (WMO) to create the Intergovernmental Panel on Climate Change (IPCC) and, years later, the United Nations Framework Convention on Climate Change (UNFCCC). This last one establishes as its main purpose: to evaluate exhaustively, objectively, openly and transparently, scientific, technical and socioeconomic information the risk of changes in the climate induced by man, its potential impacts and adaptation and mitigation options [3]-[5]. One of the main instrument of the UNFCCC, the Kyoto Protocol, seeks that the signatory countries achieve a reduction in greenhouse gas emissions.

The Conference on Environment and Development, held by the United Nations in Rio de Janeiro, Brazil, in June 1992, generated some of the global proposals and guidelines for action that have had the greatest impact on the historical development of contemporary civilization.

The UNFCCC recognizes since its appearance "that human activities have substantially increased the concentrations of greenhouse gases in the atmosphere and that this produces an increase in the global temperature of the earth, which could adversely affect ecosystems and humanity", and clearly states as its main objective: "the stabilization of concentrations of greenhouse gases in the atmosphere at a level that allows to prevent anthropogenic interference that is dangerous for the climate system". The signatories of the UNFCCC have been holding regular meetings (COP), and the one held in Paris, France in 2015 (COP 21), resulted into the "Paris Agreement", whose 29 articles' ultima goal for the signatory countries is to stop increasing and if possible reduce the greenhouse gases concentration in the atmosphere. Other sections address the importance of national and regional strategies for communication, capacity building, risk reduction, technological development and, in particular, adaptation and mitigation measures and possible sources of financing for actions and of collaboration strategies.

Although the "Paris Agreement" has been signed by many nations, it is a political document that lacks clearly established quantifiable goals, except for subparagraph 1(a) of Article 2: "Maintain the increase in the world average temperature at maximum 2°C above the levels of the pre-industrial era and strive to limit the increase in temperature to 1.5°C above that level ". To achieve this, the document proposes in Article 4 that countries reach as soon as possible the maximum peak of greenhouse gas emissions and seek their rapid reduction to achieve a balance between emissions and reductions during the second half of this century.

The difficulty of this task is that there is no scientific evidence showing an improvement in global climate conditions or that they can be optimistic about the reduction of greenhouse gas emissions into the atmosphere. Figure 1 shows the historical concentration of CO_2 in the atmosphere recorded in the observatory of Mauna Loa [6]. From the 18th century to some recent relevant events.

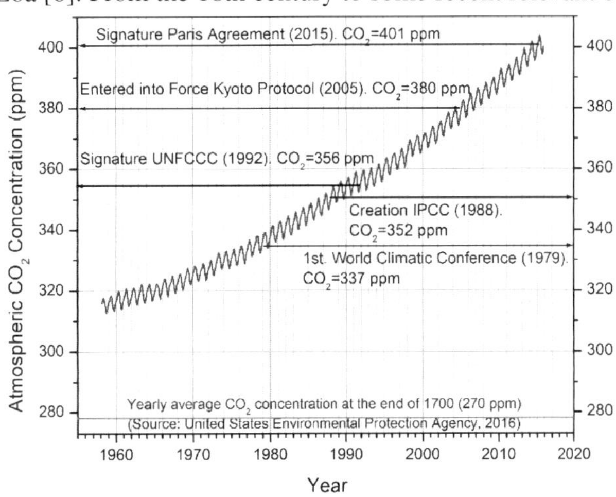

Figure 1. Atmospheric CO_2 concentration recorded by the Mauna Loa observatory (based on data from: Tans, P. and R. Keeling, 2016, NOAA / ESRL and Scripps Institution of Oceanography).

Because reliable records of the global temperature of the planet began after 1880 and that the IPCC establishes there is a low level of confidence existing in the measurements of temperatures previous to the year 1880 and during the industrial revolution [7], there would be no precise data of the maximum permissible global temperature not be exceeded. Despite this lack of definition and the non-technical nature of the "Paris Agreement", greenhouse gases emitted into the atmosphere continue to increase; and it is clear that the human being is affecting the global climate system and having an impact on natural and human systems.

Some of the most important impacts of climate change that are particularly important for this study, which focuses on the coastal states of Mexico, are: an increase in atmospheric temperature and upper layers of the oceans, an accelerated rise in the average sea level, occurrence with greater frequency and intensity of extreme weather events, acidification of the ocean, changes in natural biogeochemical cycles, increase in the incidence and spread of diseases caused by vectors, alterations in agroclimatic systems and fishing.

2. Case study

Mexico has a territorial area of approximately 1.96 million km^2, a maritime area of 3.15 million km^2 (61.5% of the national territory) and 11,122 km of coastline, shared by 17 coastal states and around 155 municipalities with seafront. The latitudinal extension of the country, its physiography and its opening to the Pacific Ocean, the Gulf of Cortés, the Gulf of Mexico and the Caribbean Sea, has allowed the existence of a varied and rich biological and landscape diversity.

According to the last census in 2010, the total population of the coastal states was 51.9 million people (46% of the national total), of which 17.5 million lived in municipalities with seafront. Between the years 2005 and 2010, 94% of the existing coastal municipalities reported an increase in their population, while only 6% had a decrease in it. This trend of population and urban growth in the coastal areas of Mexico will continue to increase the pressure and deterioration of coastal and marine spaces and resources; generating great challenges in terms of creation and maintenance of infrastructure, services, transportation and communication networks and, finally, increasing the risk and vulnerability to the phenomena related to climate change.

The 2013 National Inventory of Greenhouse Gases shows that mobile sources, electric power generation and industrial processes contribute with 416 Mt of CO2e of emissions to the atmosphere, which represents 62.5% of the total national. The coastal states house approximately 40% of the national vehicle fleet, 53% of the national manufacturing companies, and 63% of the national energy total is generated there, which makes them key CO_2e generators. Additionally, it is important to consider the 75,000 marine fishing vessels that operate on the Mexican coast, of which 68,245 have a capacity lower than one ton and, therefore, regularly lack of an adequate preventive maintenance (from the emission perspective). SEMARNAT [8] reported for 2008 that 48% of CO emissions from fixed sources, 55% from mobile sources and 67% from area sources, were generated in coastal states.

The natural and human systems present in the coastal areas of Mexico are especially sensitive to the impacts of climate change. These areas are intrinsically dynamic spaces for which strategies that favour their sustainable development must be sought. In this sense, this paper will address the following issues: first, it will analyze the relevant legal and regulatory instruments for the coastal and marine areas; then, it will analyze the main risks associated with climate change and variability and, finally, it will review existing and necessary adaptation measures and analyze the main barriers to the implementation of climate change adaptation actions.

3. Legal instruments for climate change and coastal zones

The National Policy for Seas and Coasts of Mexico establishes the need for coastal municipalities to count with an "Atlas of Risk" in which particular attention must be paid to the effects and impacts of climate change (section 1.4.1). On the other hand, strategy 1.5 proposes "to promote the elaboration, conclusion and execution of the Climate Change State Programs for the 17 coastal states". This policy for seas and coasts is contemporary with the General Law of Climate Change [9], which had as main antecedents the creation of the Inter-Ministerial Commission on Climate Change, the appearance of the National Strategy for Climate Change and the National Climate Change Program 2008-2012, published in 2009.

The General Law of Climate Change (LGCC), in addition to creating the National Institute of Ecology and Climate Change (INECC) and the National Council on Climate Change, establishes among its general provisions the elaboration and application of public policies for the adaptation to climate change and for the mitigation of emissions of gases and compounds of greenhouse effect. It defines the attributions of the federation, states and municipalities in the matter of climate change; the principles of the National Climate Change Policy and its evaluation mechanisms, as well as the guiding principles of the National Policies for Adaptation and Mitigation of Climate Change. It also specifies the guidelines of the National Climate Change System, of the CICC, of the National Council on Climate Change and creates the following planning instruments: the National Climate Change Strategy 10-20-40, the Special Program on Climate Change and the Climate Change State Programs as the main normative instrument of government actions on climate change, the LGCC is a solid, coherent instrument with the necessary instruments for the country to adequately address the challenges posed by climate change.

From the perspective of the environmental sector, the Climate Change Strategy for Natural Protected Areas [10] and the recent appearance of the Strategy towards 2040 [11], highlight the importance and commitment of these ecological policy instruments [12] with the conservation of the ecosystems and the environmental services that they provide, as a substantive axis in mitigating the impacts of climate change. For this purpose, it establishes as part of its vision for the Natural Protected Areas (NPAs) and the Priority Regions for Conservation (CPRs): "Contribute to increase the resilience of human communities, ecosystems and their environmental services, as well as to the mitigation of climate change in the NPA, CPR and other conservation modalities". It should be remembered that approximately 1.6% of the country's marine territory is part of the ANP network and that 70 marine and coastal RPCs have been identified, delimited and characterized [13].

To implement climate change adaptation and mitigation actions, different state (PEACC) and municipal (PACMUN) level action plans and programs for climate change action have been developed. According to the 5th National Communication to the UNFCCC [14], there were 8 PEACC completed in the country, of which 3 belonged to coastal states, 23 were in development (13 coastal) and 1 in planning corresponding to the state of Colima. After 2012, only Morelos has decreed its PEAAC (non-coastal state). Regarding the PACMUN in the reference year, only 16 of the 156 coastal municipalities had the official document.

4. Climate change and risk in the Mexican coastal zones

Mexico's coastal and marine zones are highly vulnerable to the direct impacts and side effects of climate change. In the opinion of the authors, the adaptation and mitigation measures at the local level (coastal States and Municipalities) have not been developed and implemented with the adequate urgency, mainly due to the lack of economic resources, human capacities and sensitivity of the decision makers to the phenomenon of climate change.

The following list shows some of the most important climate change impacts in Mexico´s coastal and marine zones:

1. Increase in the atmospheric temperature and in the upper layers of the ocean: a) the coastal states of Mexico are the main agricultural (66%), bovine (66%) and fishing (98%) producers of the country. Global warming, climatic variability, droughts and heat waves impact the food system, while in the oceanic part capture zones are shifting; b) the thermal expansion of the oceans and the increase of fresh water from the melting of glaciers and rain, have produced an accelerated rise in sea level, which puts at risk low-level-living and floodable coastal areas; c) the increase in the temperature of the surface layers of the ocean leads to an increase in the intensity and frequency of tropical systems (torrential rains, floods, landslides, intense winds, coastal erosion, etc.) also affecting public infrastructure, private companies and putting human lives at risk; d) the synergistic and often non-linear effect produced by the interaction of these phenomena can substantially increase the vulnerability of coastal areas through an intensification of the erosive processes on sandy beaches. This combination of phenomena can also affect the water quality and availability by increasing the salinization in groundwater reservoirs, and e) the increase in evaporation, cloudiness and rainfall, as a consequence of the increase in temperatures, favors the appearance, faster propagation and expansion to new areas of vectors transmitting diseases in humans (e.g. dengue) and associated gastrointestinal diseases due to the decomposition or deficient conservation of food, as well as agricultural pests.
2. Increase in the CO_2 levels in the atmosphere: a) the increase in atmospheric CO_2 levels favors a greater uptake of this compound by the ocean, decreasing its hydrogen potential (pH) and producing its acidification. The impact on the biogeochemical cycles of these and other processes associated with climate change are still under discussion, and b) ocean acidification has severe impacts on marine and coastal ecosystems. Sea level rise, ocean acidification, climatic variability (e.g. *El Niño* phenomenon) and changes in sea temperatures have led to the "whitening" of coral reefs at unprecedented levels in history [15]. Marine animals provided with calcareous shells are taking longer to reach their historical sizes. Microorganisms and

algae - important to the medical industry - are being affected by the acidification of the oceans and coastal bodies of water.

More information on climate change impacts on the Mexican coastal and marine zones is afforded by Muñoz-Sevilla y Le Bail [16], Escobar Briones [17], Azuz-Adeath [18], Aguilar et al. [19], Agüero [20] and Yañez-Arancibia [21].

In summary, it can be said with high confidence that the main recipients of the effects associated with climate change will be the coastal states of the country and, in particular, the municipalities with coastal fronts that depend on the services and functions of the marine and coastal natural ecosystems, as well as the infrastructure and services derived from the use of these valuable areas.

5. Adaptation measures and actions

The National Strategy for Climate Change Vision 10-20-40 proposes 3 strategic axes for adaptation: 1) to reduce vulnerability and increase the resilience of the social sector in the face of the effects of climate change; 2) to reduce vulnerability and increase the resilience of strategic infrastructure and productive sectors in the face of the effects of climate change, and 3) to conserve and use ecosystems sustainably and maintain the environmental services they provide.

41 lines of action emerge from these strategic axes, of which many have a clear impact on the coastal areas where a large part of the strategic (e.g. energy, communications and transport, military) and productive infrastructure (e.g. tourism, port, livestock and agriculture) is located.

However, despite the clear lines of action proposed, the country lacks the economic capacity to adequately address the problem of climate change and particularly to implement effective long-term adaptation measures.

To mitigate the impacts of climate change in coastal zones, at least the following requirements must be met: a) to establish mandatory vehicle verification mechanisms in all states and municipalities of the country in parallel with a development of public transport, b) to encourage the transition to low carbon economies at the municipal level, evaluating energy efficiency and implementing the use of clean energy, starting with government facilities, industries and public services (e.g. public lighting generated by wind or solar energy), and c) to maintain and seek to restore the natural vegetation percentage of the coastal municipalities.

To adapt to climate change in the coastal areas of the country, there are not many alternatives, but we could mention: a) the relocation of settlements and infrastructure located in risk areas, b) the modifications to the "traditional" types of crops, adapting new products or seasonality to changes in climatic conditions, c) the implementation of engineering measures to protect from the impacts of climate change (e.g., protection infrastructure for coastal areas, elevation of facilities of economic or strategic importance, communication routes and infrastructure of services), and d) the technological innovation to create large-scale infrastructure that "sequesters" carbon from the atmosphere and oceans.

Regardless of the selected adaptation option, the economic cost is very high and an important long-term planning is required. To the extent that this issue is not urgently addressed, Mexico's coastal zones will continue to degrade and suffer the incremental and uninterrupted impacts from climate change.

6. Barriers for the implementation of adaptation measures

From the local level (coastal towns), the main barriers to the implementation of adaptation actions to face climate change are: lack of economic resources; lack of long-term vision or regulatory instruments for long-term projects; technical capacities and planning often limited; prioritization of what is urgent and important; ignorance of the phenomenon of climate change; scarce social participation in decision-making; disengagement between government agencies and sectors of plans and programs; lack of instances to formalize mutually beneficial associations with the private sector, academic and international organizations, and often political differences associated with the partisan origin of the mayors and state governors.

In the case of coastal states, it is important to note that the State Development Plans (PEDs) of all coastal states include elements related to the importance of climate change. However, in this State strategic planning instruments, there are few concrete strategies to contribute to mitigation or adaptation. Within the specific measures identified, one can generally speak of the following: a) to use clean energy and resilient infrastructure, b) to establish adaptation and mitigation measures for climate change for the use of natural resources, cultivation areas and the reduction of the impact of natural disasters, c) to make proposals for social awareness, technical training, scientific research and the use of geographic information systems as support elements for decision making, and d) to make proposals in the legal field such as: the creation of the State Councils on Climate Change, initiatives on the Law on State Climate Change and preparation of State Programs of Action in the face of Climate Change.
However, despite the existence of this type of proposal in the State PEDs, in most cases no concrete instruments are presented to make them operational, nor there is talk of the need to clearly establish integral strategies and programs, binding with other sectors (e.g. business, academic, social) to address climate change issues.

7. Conclusions

To achieve the objectives set out in the "Paris Agreement" regarding the maximum desirable increase in global temperature, there are few options available: to stop increasing and, if possible, to reduce greenhouse gas emissions.
In Mexico, the main generating sectors of this type of gases have been identified and quantified; legal and normative instruments have been established; high-level technical and consultative bodies have been created related to the knowledge of the phenomenon, which have created technically very well elaborated strategies and programs.
The marine and coastal areas of Mexico are undoubtedly the most vulnerable regions of the country and with the greatest risk to the current and potential effects of climate change. Additionally, these regions have high rates of population growth, which implies, among others, a greater demand for energy, means of transport, generation of waste and, in general, sources of greenhouse gas production. On the other hand, considering the importance of the agricultural, livestock, energy and manufacturing sectors (main GHG generators) of the coastal states in the national context, it is clear that the coastal and marine zones of Mexico should be one of the spaces of greater attention in the policies of adaptation and mitigation to climate change.
However, the adoption and implementation of many policies and actions at the regional and local level, in addition to the lack of economic, human and technical resources and official municipal management times (3 year-terms), the lack of long-term vision in local planning and development instruments make coastal municipalities the weakest link in the structure of national attention to climate change.
The costs of adapting to climate change in the coastal areas of the country are high but, in the mid-term, the costs generated by the losses could be greater. Mitigation actions may be less costly from the economic point of view but carry political risks for the governors and decision makers. However, in the long term, the only option to achieve the sustainability of the coastal and marine areas of the country in the context of climate change is to take intelligent and effective measures based on scientific and technical knowledge.

8. References

[1] Ward R D (1906). Changes of Climate. *Popular Science Monthly*, Vol. **LXIX**, October, p. 458-470.
[2] Huntingtong E (1917). Changes of Climate and History. *American Historical Review*, Vol. **XVIII**, No. 2, January, p. 213-232.
[3] IPCC (2014). *Climate Change 2014: Synthesis Report*. Contribution of Working Groups I, II and III to the Fifth Assessment Report of the Intergovernmental Panel on Climate Change [Core Writing Team, R.K. Pachauri and L.A. Meyer (eds.)]. IPCC, Geneva, Switzerland, 151 pp.
[4] Hartmann D L *et al.* (2013). Observations: Atmosphere and Surface. In: *Climate Change 2013: The Physical Science Basis. Contribution of Working Group I to the Fifth Assessment Report of*

the Intergovernmental Panel on Climate Change [Stocker, T.F., *et al.* (eds.)]. Cambridge University Press, Cambridge, United Kingdom and New York, NY, USA

[5] Wong P P *et al.* (2014). Coastal systems and low-lying areas. In: *Climate Change 2014: Impacts,Adaptation, and Vulnerability. Part A: Global and Sectoral Aspects. Contribution of Working Group II to the Fifth Assessment Report of the Intergovernmental Panel on Climate Change* [Field, C.B., *et al.* (eds.)]. Cambridge University Press, Cambridge, United Kingdom and New York, NY, USA, pp. 361-409.

[6] EO/NOAA (2016). Earth Observatory, National Oceanic and Atmospheric Administration (EO/NOAA). Visited on November 2017 at: http://earthobservatory.nasa.gov/Features/WorldOfChange/decadaltemp.php

[7] Floud R and McCloskey D (1981). The industrial revolution 1780-1860: a survey. In: *The economic history of Britain since 1700.* Cambridge University Press, pp. 103-127.

[8] SEMARNAT (2014). *El Ambiente en Números. Selección de estadísticas ambientales para consulta rápida*. Secretaría de Medio Ambiente y Recursos Naturales (SEMARNAT), 59 pp.

[9] DOF México (2012). Ley General de Cambio Climático. *Diario Oficial de la Federación* del 06-06-2012.

[10] CONANP (2010). *Estrategia de Cambio Climático para Áreas Protegidas*. Comisión Nacional de Áreas Naturales Protegidas (CONANP), Secretaría de Medio Ambiente y Recursos Naturales (SEMARNAT), México, 23 pp.

[11] CONANP (2014). *Estrategia hacia 2040: una orientación para la conservación de las áreas naturales protegidas de México*. Comisión Nacional de Áreas Naturales Protegidas (CONANP), Secretaría de Medio Ambiente y Recursos Naturales (SEMARNAT), México, 80 pp.

[12] DOF México (2015). Ley General del Equilibrio Ecológico y Protección al Ambiente. *Diario Oficial de la Federación* 28-01-1998 (original). Diario Oficial de la Federación 09-01-2015 (texto vigente).

[13] Arriaga Cabrera L *et al.* (Coords) (1998). *Regiones marinas prioritarias de México*. Comisión Nacional para el Conocimiento y uso de la Biodiversidad. México. Azuz-Adeath I (2012). The role of global oscillations and teleconnections within atmosphere and ocean in regional coastal management. *Ocean & Coastal Management* **69**, p. 78-91.

[14] SEMARNAT-INECC (2012). *Quinta Comunicación Nacional ante la Convención Marco de las Naciones Unidas sobre el Cambio Climático*. Secretaría de Medio Ambiente y Recursos Naturales (SEMARNAT) e Instituto Nacional de Ecología y Cambio Climático (INECC), 400 pp.

[15] Normile D (2016). El Niño´s warmth devasting reefs worlwide. *Science*, Vol. **352**, Issue 6281, pp. 15-16.

[16] Muñoz-Sevilla N P and Le Bail M (2015). Sistemas costeros y zonas inundables. In: Ortiz Espejel B, Muñoz-Sevilla N P and Le Bail M (Coords). (2015). *Reporte Mexicano de Cambio Climático. Volumen 2: Impactos, Vulnerabilidad y Adaptación*. Universidad Nacional Autónoma De México/Programa de Investigación en Cambio Climático. ISBN Volumen: 978-607-02-7370-4

[17] Escobar-Briones E (2015). Sistemas Oceánicos. In: Ortiz Espejel B, Muñoz-Sevilla N P and Le Bail M (Coords) (2015). *Reporte Mexicano de Cambio Climático. Volumen 2: Impactos, Vulnerabilidad y Adaptación*. Universidad Nacional Autónoma De México/Programa de Investigación en Cambio Climático. ISBN Volumen: 978-607-02-7370-4

[18] Azuz-Adeath I (2012). The role of global oscillations and teleconnections within atmosphere and ocean in regional coastal management. *Ocean & Coastal Management* **69**. Pp 78-91.

[19] Aguilar A, Sánchez A and Martínez B (2012). Economic Impacts of Climate Change on Two Mexican Coastal Fisheries: Implications to Food Security. *Economics E-Journal*, No. **64**. 20 pp.

[20] Agüero J M (2014). Long-Term Effect of Climate Change on Health: Evidence form heat waves in Mexico. Inter-American Development Bank (IDB), *IDB Working Papers Series* No. **481**, 26 pp.

[21] Yañez-Arancibia A (2015). *Cambio Climático. Adaptación y Mitigación hacia agendas siglo XXI*. AGT Editor, S.A., 254 pp.

Construction of Revised TPB Model of Customer Green Behavior: Environmental Protection Purpose and Ecological Values Perspectives

Ting Liu[1,2], Shenglang Jin[1], Hongbing Zhu[1], Xinzheng Qi[1]

1 College of Tourism, Huangshan University, Huangshan 245021, China
2 Faculty of International Tourism and Management, City University of Macau, Macau 999078, China
E-mail: 13855913660@163.com

Abstract. This study constructs a comprehensive conceptual model of consumer behavioral intentions in green hotels. For the purpose of protecting the environment, based on Schwartz's values theory, this study identifies ecological values that influence consumers' green hotel visit intention. This study aims to develop an extended Theory of Planned Behavior (TPB) research model which includes environmental concern and perceived moral obligation to predict consumers' intention to visit green hotels. The green hotel behavioral intentions will provide insights to better understand the green consumer and to better manage the green hotel.

1. Introduction

Environmental issues have accompanied the development of human society and also exert different impacts on human beings. Since the onset of the New Environmental Paradigm [1] in the late 1970s, increasing attention has been placed on the interface between society and the natural environment, including the importance of environmental consciousness as a factor that can influence human behavior [2]. Public concern about environmental issues has been on the increase. Many consumers are aware that their buying behavior may do harm to the environment and start to search and purchase eco-friendly products [3]. This increased awareness and interest in sustainable consumption is expected to influence consumer purchase decisions [4]. Moreover, sustainable consumption has drawn more attention from corporate decision-makers due to stricter environmental regulation and growing stakeholder pressures focused on preserving the environment [5]. With more people demanding green travel, this environmental concern has changed into ecologically favorable behaviors and eco-friendly purchasing decisions [6]. Green consumption has become an important force, which can protect the environment. Green consumerism, and one facet of such behavior, began to emerge eventually across different industries and markets.

Energy saving, ecology, science and technology is the development trend of Green Hotel. It is not only the progress and the development of the social, but also the need of human beings to seek harmonious coexistence with nature. Figure 1 shows the hotel industry has to make good cooperation with economic development and environment, make rational utilization of resources, and take environmental protection as the center, therefore, the industry could obtain the sustainable development.

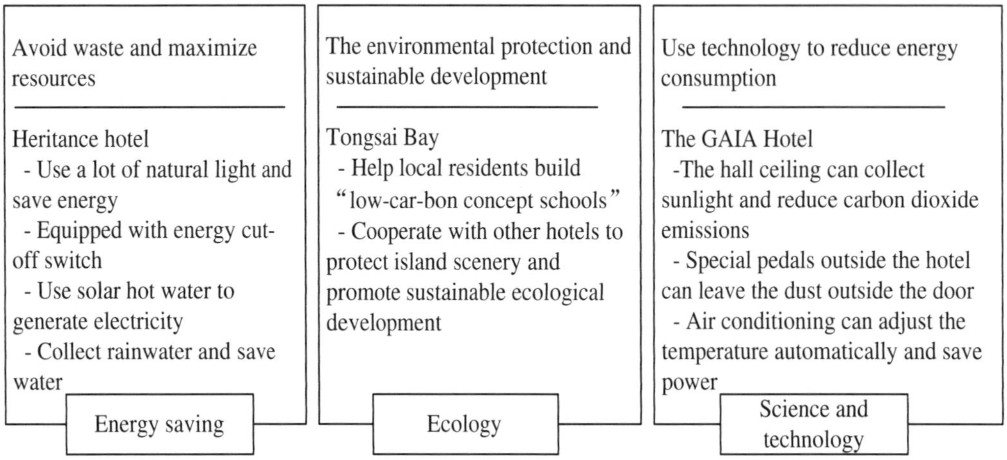

Figure 1. The development trend of Green Hotel

The study has four objectives:

- To understand the application of TPB in the hotel industry.
- To construct TPB framework indicators for green hotels.
- To identify which TPB framework indicators are suitable for use in the hotel industry.
- To provide practical and research implications for future green hotel development.

2. Literature Review

A narrative literature review methodology was adopted to enable synthesis of the current knowledge on the topic. Such an approach will allow for a broad perspective to be presented on a particular area. This narrative review of the research literature presents a summary about Green Consumerism \ Green Hotel, ecological values, theory of Planned Behavior model. The aim of this review was to consolidate research of the topic that may lead to better outcomes for the Green Hotel.

2.1. Green Consumerism and Green Hotel

Hotels that are nature-oriented, environmentally focused and environment friendly (i.e. eco-friendly) in use of energy, water, and materials that do not exacerbate impacts on the earth and environment. Green hotels refer to "the hotels that are managed based on the concept of sustainable development, which means they adhere to cleaner production, advocate green consumption, protect the ecological environment and make rational use of resources".[7]Green hotels are committed to the standards of 3Rs (reduce, reuse, recycle) and 2Es (energy, efficiency), whose core content consists of economical principles applied in the energy consumption within the hotels, proper guidance over the consumption of guests, sensible choices of the environment and its influence, re-understanding of employees, re-sponsibility and commitment to the community, and concerns of the hotel proprietors and managers over the course of nature and history.

Green consumerism can be explicated as a particular form of socially conscious or socially responsible consumer behavior, and hence may be referred to as environmentally concerned consumption. Green consumers normally engage in the purchase of products that are considered green, environment friend-ly, proenvironmental, or sustainable. [8]

2.2. Ecological Values

Values are distinguished from one another by means of the nature of motivational goal they articulate. Schwartz proposed a broad taxonomy of 56 values. These values might vary in their importance on a personal and cultural level. However, the core of these values is considered universal. In fact, studies undertaken across countries and diverse cultures established two broad facets of these values—self-transcendence versus self-enhancement and openness to change versus conservation [9]. Self-transcendent (i.e., altruistic or biospheric) compared to self-enhancement (i.e., egoistic) dimension is especially connected to environmental beliefs and behaviors, as environmental behavior generally entails a clash between short-term personal benefits and long-term collective concerns.

Environmental values are enduring, individually defined moral obligations to protect the environment. Both the supply and demand sides of the tourism industry of purchasing and consumption have been shown to be affected by environmental values [10]. Mair's study on tourists participating in carbon-offset programs found travelers that purchased offsets had positive environmental values [11]. Dolnicar (2010) found environmental values, or a moral obligation to protect the environment, was a strong predictor of environmentally-friendly tourist behavior and environmental behavior at home [12]. Ham and Han (2012) show that purchase-related loyalty, measured as intentions to visit, acceptance of price premiums, and providing recommendations, were amplified by consumers with positive environmental values [13]. However, Millar and Baloglu (2011) suggest values and beliefs may not be consistent with actions when traveling because people go into a different mode, and may be less concerned with the long-term impact of their decision-making [14]. The environmental values are positively related to consumers' motivations to select a hotel based on its sustainability practices [15].

2.3. Theoretical Base- Theory of Planned Behavior model

The Theory of Planned Behavior model, as proposed by Ajzen in 1991[16]. According to Ajzen, intention is predicted by three factors: attitude towards the behavior, perception of social pressure to perform the behavior (subjective norms), and control over performance of the behavior (perceived behavioral control). Consumers' behavior when staying at a green hotel differs from others in terms of their attitude, subjective norm, perceived behavior control, and intention.

The main basic assumption of the TPB model is that most behavior people engage in is under their own control and is rational. Moreover, the decision factor in a person's actual behavior is the tendency to behave, that is, behavioral intention. In addition, a person's personality, age, occupation, gender, etc. have no direct impact on his/her behavioral intention. In fact, these variables can only affect behavioral intention indirectly through attitude and subjective norms [16]. According to the TPB model, an individual's performance of a specific behavior is determined by his/her behavioral intention to perform the behavior. This behavioral intention is in turn determined by three factors related to the behavior: the person's attitude, subjective norms, and perceived behavioral control [17]. In view of the fact that perceived moral obligation as an indispensable construct of pro-environmental behavior, this study incorporated perceived moral obligation into the TPB model to form the extended TPB model. As for the antecedent of the components of the extended TPB model, this study pointed that environmental concern would exert some influences on consumers' pro-environmental behavioral intentions through situation-specific beliefs while believing that environmental concern would have some direct impacts on the components of the extended TPB model. [18]

3. TPB Model Construction Framework

Many tourists prefer destinations that practice environmental protection. This preference translates into tourists' purchasing and booking lodgings at locations that practice environmental protection and support related local charities are practiced. This is driven by a rise in consumers' environmental awareness [19]. Green Hotel has a high recognition in the user community. Safety, health and environmental protection is another name of Green Hotel. With the improvement of tourists' awareness of environmental protection, the requirements of the tourists to the Green Hotel have also changed. (As shown in Figure 2)

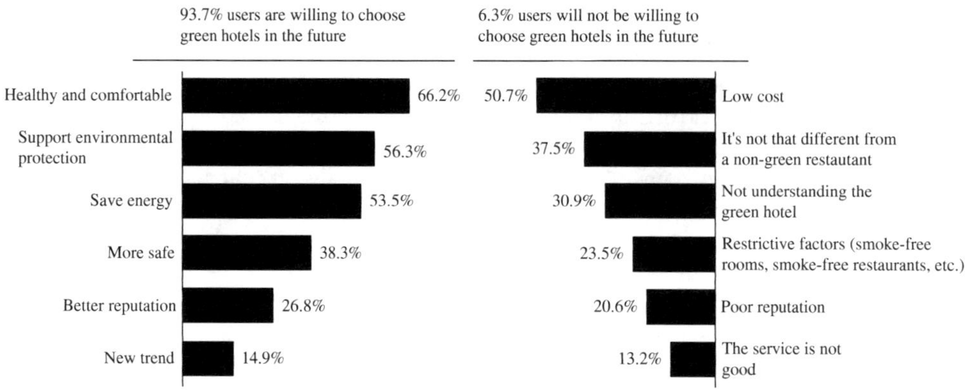

Figure 2. The intention and reason of Chinese tourists choose the Green Hotel [20]

In summary, a customer can purchase a night at a green hotel, sacrifice some convenience, luxury, and/or quality, and even pay a premium price to stay at a green hotel. This consumer behavior seems to result from their eco-centric attitudes, which in turn ensue from their biospheric values.
Based on the discussions of the concepts and theory, the research framework of this study is depicted in Figure 3.This revised TPB model integrates the environmental concern, ecological values and the moral obligation into the framework, which describes the relationships and influences on consumers' intention to visit green hotel and purchase behavioral intentions in the future (As shown in Figure 3).

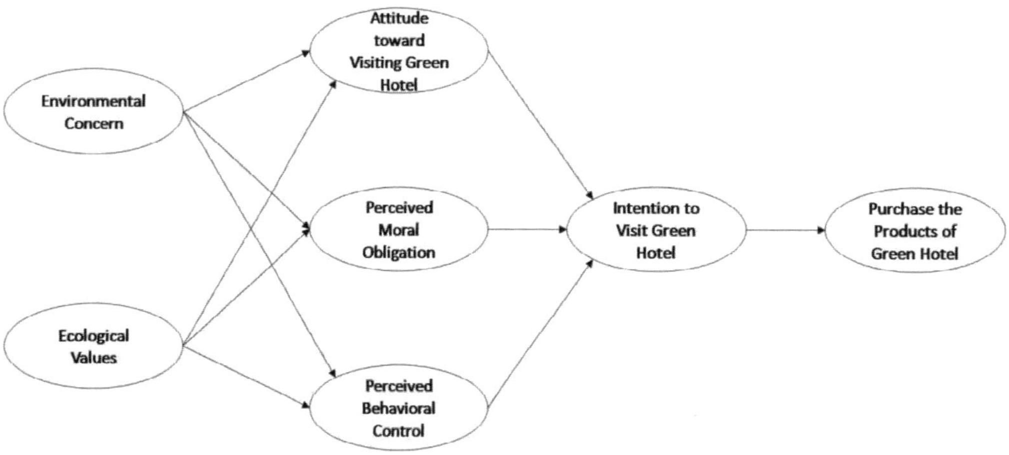

Figure 3. Proposed theoretical framework of consumers' visit intention of green hotel.

4. Implications

4.1. Practical Implications

The model might provide paramount implications for practitioners. Support for a robust effect of biospheric values on green hotel behavioral intentions will make a strong case for involving customers in the environmental management process. Green initiatives such as recycle receptacles, signs encouraging green behaviors, and linen and terry reuse programs, can prove to be beneficial for the practitioners not only financially, but also for enhancing guest satisfaction and loyalty. It is imperative for practitioners to look beyond these practices and think about innovative ways of involving the guests in the environmental management process as most of these practices are increasingly becoming commonplace.

4.2. Research Implications

The model we constructed integrates a number of theories such as values theory, value-attitude-behavioral hierarchy, ecocentrism and anthropocentrism ethic, and green signaling. Our model proposes that this unique combination of theories can be used to predict green hotel specific behavioral intentions. Empirical validation will show which particular values and attitudes can influence green consumer behavior on a green lodging context. As a result, providing all our hypotheses are supported, it will become possible to attract all types of consumers to the green hotel. Every consumer has a unique combination of core values and it is possible to strategically trigger each core set of values through various environmental actions. By understanding the prevalent theories and the relationships that our model proposed, the practitioner can certainly understand the hotel guest better. The goal of our study was not to propose new theory but to build a comprehensive model of green hotel consumer behavior based on relevant theories in psychology and mainstream consumer behavior research literature.

5. Conclusions and Future Directions

To help hotel make marketing decisions, this study builds upon prior literature on green consumer behavior and makes two contributions: 1) the paper extends the study of environmental values and environmental self-efficacy to a tourism context, 2) the paper constructs the TPB model of the customer green behavior from the ecological values for the environmental protection purpose.

The future research direction is that the TPB model could be applied to the study of hotel customer green behavior intention. The reliability and validity of the model are further verified by the data obtained from the questionnaire.

6. References

[1] Catton W R and Dunlap R E 1978 Environmental sociology: A new paradigm *Am.Sociol.* **13** p 41–9
[2] Dunlap R E and Catton W R 1979 Environmental sociology *Annu. Rev. Sociol* **5** p 243–73
[3] Lee J S, Hsu L T, Han H and Kim Y 2010 Understanding how consumers view green hotels: how a hotel's green image can influence behavioral intentions *J. Sustain. Tour* **18** p 901–14
[4] Dean M, Raats M M and Shepherd R 2012 The role of self-identity, past behaviour and their interaction in predicting intention to purchase fresh and processed organic food *J. Appl. Soc. Psychol* **42** p 669–88
[5] Hult G T M 2011 Market-focused sustainability: market orientation plus *J. Acad. Mark* **39** p 1–6
[6] Yao-Hsu T, Chung-Tai W and Ta-Ming W 2014 Attitude towards green hotel by hoteliers and travel agency managers in Taiwan *Asia. Pac. J. Tour. Res* **19** p 1091–109
[7] Butler J 2008 The compelling "hard case'' for "green" hotel development *Cornell. Hosp. Q* **49** p 234-44
[8] Imran R and Dennis R 2017 The influence of values and attitudes on green consumer behavior: A conceptual model of green hotel patronage *Int. J. Hospit. Tourism. Admin* **9** p 1-28
[9] Schwartz S H, Melech G, Lehman A, Burgess S, Harris M and Owens V 2001 Extending the cross cultural validity of the theory of basic human values with a different method of measurement *J. Cross. Cult. Psychol* 32 p 519–42
[10] Holden A 2005 Achieving a sustainable relationship between common pool resources and tourism: The role of environmental ethics *J. Sustain. Tour* **13** p 339-52
[11] Mair J 2011 Exploring air travellers' voluntary carbon-offsetting behaviour *J. Sustain. Tour* **19** p 215-30
[12] Dolnicar S 2010 Identifying tourists with smaller environmental footprints *J. Sustain. Tour* **18** p 717-34
[13] Ham S and Han H 2012 Role of perceived fit with hotels' green practices in the formation of customer loyalty: Impact of environmental concerns *Asia. Pac. J. Tour. Res* **18** p 731-48
[14] Millar M and Baloglu S 2011 Hotel guests' preferences for green guest room attributes *Cornell. Hosp. Q* **52** p 302-11
[15] Akehurst G, Afonso C and Martins G H 2012 Re-examining green purchase behaviour and the green consumer profile: New evidences *Manage Decis* **50** p 972-88

[16] Ajzen I 1991 The theory of planned behavior *Organ. Behav. Hum. Dec* **50** p 179–211
[17] Justin P, Ashwin M and Jayesh P 2016 Predicting green product consumption using theory of planned behavior and reasoned action *J. Retailing. Consum. Serv* **29** p 123–34
[18] Mei-Fang C and Pei-Ju T 2014 Developing an extended Theory of Planned Behavior model to predict consumers' intention to visit green hotels *Int. J. Hosp. Manag* **36** p 221– 30
[19] Teng-Yuan H, Chung-Ming C, Nae-Wen K and Sally M Y 2014 Establishing attributes of an environmental management system for green hotel evaluation *Int. J. Hosp. Manag* **36** p 197–208
[20] White paper on green development of Chinese lodging industry 2016 http:// www.data.iresearch.com.cn

Acknowledgment
This work is supported by Humanities and Social Sciences Research Project of Anhui Provincial Department of Education (NO.SKHS2015B06), the Project of Cultivating Tourism Youth Professionals of Anhui Province (NO.AHLYZJ201613), Anhui Social Science Key Project (NO.SK2015A526), Humanities and Social Sciences Research Project of Anhui Provincial Department of Education (NO. SK2012B460).

IOP Conf. Series: Earth and Environmental Science **167** (2018) 012022 doi:10.1088/1755-1315/167/1/012022

Implication to Environmental Education of Indigenous Knowledge and the Ecosystem of Upland Farmers in Aklan, Philippines

E Arangote

Associate Professor, Aklan State University, Banga, Aklan, Philippines
E-mail: emily.m.arangote@gmail.com

Abstract. This paper defined the association between the indigenous knowledge, cultural practices and the ecosystem its implication to the environmental education to the farmers. Farmers recognize the need for sustainability of the ecosystem they inhabit. The cultural practices of farmers on use of indigenous pest control, use of insect-repellant plants, soil management practices that suppress diseases and harmful pests and conserve soil moisture are deemed to be ecologically-friendly. Indigenous plant materials that were more drought- and pest-resistant were grown. Crop rotation was implemented with various crop seeds to increase their disease resistance. Multi-cropping, planting of perennial crops, categorization of soil and planting of appropriate crops, planting of appropriate and leguminous crops, allotting land as watershed, and preserving traditional palay seed varieties were found to be beneficial in preserving the environment. The study also found that indigenous knowledge about crops are still relevant and useful to the current generation. This ensured the sustainability of our environment and incumbent on policy makers and educators to support and preserve for generations yet to come.

1. Introduction

Today the world is facing the reality of the inevitability of climate change brought about by the use of modern technology. Much of this modern technology comes from the developed countries. However, its debilitating effect is felt by all of us, whether we come from a first world or third world country.

Indigenous knowledge, traditional, endogenous or classical knowledge [1] refers to knowledge accumulated through experience and transmitted from generation to generation. Farmers acknowledge that the sustainability of the ecosystem is the basis of their continued wellbeing.

Its debilitating effect is felt by all, whether we come from a first world or third world country. In the Philippines, all have felt natures destructive power- in Panay island during the passage of Typhoon Fengshen also termed locally as Typhoon Frank in June 2008 [2] and Typhoon Haiyan or locally known as Typhoon Yolanda [3] particularly, in the Visayas it left many lives and countless properties lost. Today, the Visayans are still in the process of rebuilding their lives, even more than a few year after Frank and Yolandas wrath.

Humanity is confronted by complicated problems such as diminishing natural resources and deteriorating ecosystems services because of threats arising from unprecedented growth and consumerism [4]. People cannot deny the fact that plants, animals and microorganisms support life on earth, in the form of supporting, regulating, provisioning and cultural services [5], [6]. Climate change has already been found to affect the functioning, appearance, composition and structure of ecosystems. First world and developing countries need to be cognizant and responsive to this threat to our environment. To address these threats, natural and social sciences have contributed through discovering generating and applying knowledge about ecosystem conservation and restoration and by creating stronger policies and practices of sustainable development [2], [7]. Communities have shown [7] in the past and increasingly today that they can collaborate for long-term resource management. It is important that the knowledge-practice-belief complex of indigenous peoples regarding biodiversity conservation is fully acknowledged and utilized in order to effectively manage ecosystems and biodiversity [8].

In Scott's paper [9] on sixteenth-century Visayan food and farming, he found that Visayan's lived in permanent settlements, an evidence that their swiddening techniques were not destructive. It is a favourable balance between their numbers and the land available to them permitted new swiddens to be made in the secondary growth on abandoned swiddens without cutting into virgin woodlands. In the paper it was pointed out that an environment which was able to carry a sparse Visayan population that burned off a small portion of its cover every year. In the paper of [10], also observed that farmers in Central Panay have high respect for nature. They do not go about wantonly destroying their environment. They even went to as far as hiring a shaman to perform the rituals that would appease spiritual beings.

The Trousdale thesis paper [11], uses an innovative and promising approach for sustainable tourism planning based on the multiple objective decision analysis in Guimaras Island, in Western Visayas. The paper of Steward [12] referred to this concept as cultural ecology. They stated that different environmental contexts shape the cultural development of a community. Also mentioned in the paper of Stewards [13]. Upland people develop a unique social and economic beliefs and practices or culture that is different from the lowlanders. Rewarding Upland Poor for Environmental Services (RUPES), as mentioned in the paper of Villamor [14], is a program operating on its third year in the Philippines. It has a two-pronged goal: to conserve the environment and at the same time, to alleviate poverty in the community. The program and its partners tested and developed environmental service reward (ESR) mechanisms in various areas of the country. These include strategies such as carbon sequestration, watershed and biodiversity protection, and landscape beauty in (a) Baticulan, Negros Occidental; (b) Sibuyan, Romblon; (c) Bakun, Benguet; and (d) Kalahan, Nueva Vizcaya. Please areas depend highly on their existing forests and would benefit in knowing about environmental services (ES).

The people inhabiting the twenty-first century could learn much lesson from our ancestors on how indigenous knowledge could help preserve the environment. Their practices were sustainable such that it ensured that our ecosystem could provide for the needs of todays and future generations. The research [2] found that local knowledge systems contribute to biodiversity conservation and preservation of ecosystems services, tropical ecological and biocultural rehabilitation introduced in [15], sustainable water management, genetic resource conservation and management of other natural resources, as well as in ecosystem restoration and adaptive management.

In Aklan, kaingin a slash and burn farming method farmers have been observing traditional beliefs and practices which are a reflection of an environmental world view rooted in their culture [16]. But, increasing population pressure and higher social cost due to higher social expectation could eventually make people adopt technology that ultimately degrade the environment. The upland farmer's closer contact with lowlanders seems to have influenced them to adopt the ways of the lowland farmers in applying fertilizer and pesticide to increase palay yield. Social scientists have already proven that fertilizer and pesticide use in farming contribute to pollution, which in turn are part of the drivers of climate change.

This paper is intended to identify the different indigenous knowledge used by upland farmers in relation to environmental education in the province of Aklan. Determine if it's ecologically - friendly and ecologically-destructive. Also to find how Indigenous Knowledge (IK) could be integrated in teaching students to improve their awareness on environmental conservation and protection.

The paper is organized as follows: methodology is discussed in section II, results and discussion in section III, and we conclude our work in section IV and give recommendations in section V.

2. Methodology

2.1. Research Design
A descriptive research design using a qualitative approach thru direct observation and interview is used. Qualitative research enables to describe the phenomena of interest in great detail, in the original language or context of the research participants and come up with generalized ideas through gathered data.

2.2. Topological Area
The study was conducted in the Province of Aklan in the Philippines where majority of its people are engaged in agriculture particularly in farming. The province is subdivided into 17 municipalities; but on the basis of this study the researcher only limits the farmer-respondents being chosen from six (6) municipalities who solely practiced upland farming using traditional knowledge. The said municipalities were Ibajay, Altavas, Balete, Libacao, Madalag and Tangalan. Figure 1 shows the location of the study.

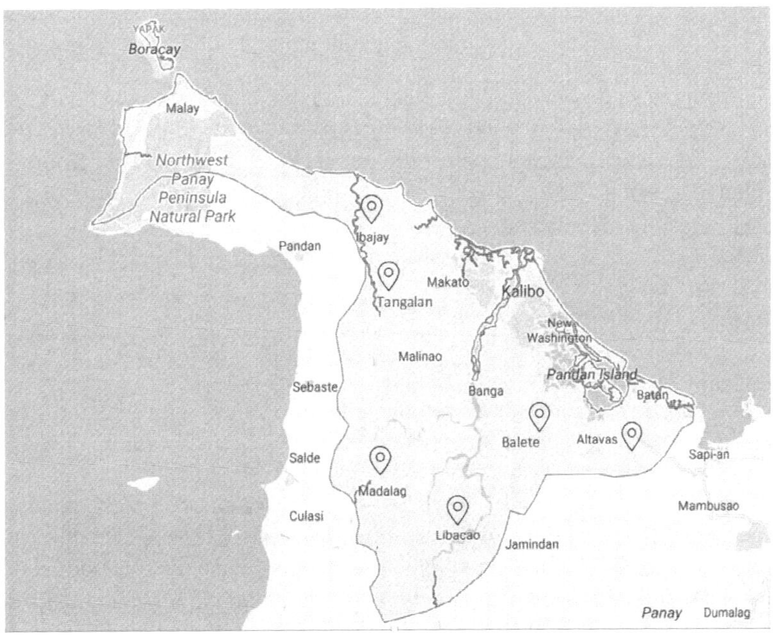

Figure 1. Location map of the study, with position symbol are the location of respondents.

2.3. Sampling Design
This study used the purposive sampling technique. It is one of the most common sampling strategies in qualitative research that determines the group of participants according to preselected criteria relevant to a particular research question. Thus, this technique is appropriate for the study because it focuses on the indigenous knowledge practiced by the local people in Aklan and their actual experiences in upland farming.

2.4. Respondents of the study
Out of eighteen (18) farmers interviewed thirteen (13) were males and five (5) were females, as shown in Table 1. The waning interest of the youth in taking up farming as a vocation and employment has resulted in the drop of the number of farmers in farming areas of Aklan. It is reported that among the basic sectors, farmers, fishermen and children belonging to families with income below the official poverty threshold or poor families posted the highest poverty incidences in 2015 at 34.3%, 34.0% and

31.4%, respectively. These sectors consistently registered as the three sectors with the highest poverty incidence in 2006, 2009, 2012 and 2015 [17], [18], [19].

Table 1. Distribution of farmer-respondents according gender.

Barangay	Number of Respondents	
	Male	Female
Monlaque, Ibajay	4	2
Aparicio, Ibajay	2	1
Cabugao, Altavas	1	0
Cortes, Balete	0	1
Ogsip, Libacao	2	0
Alas-as, Madalag	1	1
Tamokoe, Tangalan	3	0
Total	18	

2.5. Statistical Design
The data gathered from the study were documented and analysed qualitatively analysis on the basis of answering the research objectives. Extensive field notes which were subsequently coded and analysed in a variety of ways based from the responses of the respondents using the Key Informant Interview (KII) and In-depth interviews.

3. Results and discussions

3.1. Indigenous knowledge that are ecologically-friendly

3.1.1. Use of Indigenous Pest Control
Upland farmers used *eagpit* a type of mousetrap made from bamboo with *lingkad* (dried copra) as bait to control rats. To drive away the *maya* a native birds in the locality, farmers hanged *papalid* a fern genus or *banog nga bunot* a mock eagle made from coconut husk. Also tin cans or anything that make sounds, such as biscuit cans and tin plates, were tied to a string and controlled by a man from a *payag* or *bugawan* a makeshift hut while shooing the birds. The locals referred to this as *uylog* since the noise and the swaying movement of the string drove the birds away from their rice field.

Another means was the *tawo-tawo* a scarecrow made from *liay it niyog* made up of dried coconut leaves and dressed with worn-out clothes of a farmer. This is usually installed during night time by the farmer himself by walking directly to the spot without looking behind; otherwise, the wild birds would notice the surrogate farmer and they would not get scared. When plants were on the panicle-bearing stage, farmers hanged spoiled fish or any seafood that was fishy smelling on a pole posted around the *kaingin*. These attracted the pests, which the farmers then collected and burned. An upland farmer in Madalag said that before he practiced *panueo it tanangaw* using torch light to get rid of the bug. Upland farmers always made rounds on their field every day. When rice stalks were observed to be turning gray, they were pulled off and brought home to be hanged above their *dapog* a cooking area in the house with the belief that this will drive away evil which cast doom on the rice plants.

3.1.2. Spirit of *cooperation*
The spirit of cooperation was very much visible among most of today's traditional farmers. When a *kumpare* a name-call to a friend who served as godfather to the baptism or wedding of one's child schedules his farm preparation, neighbors would come to extend their help for free. Some called this *sagibin*, others considered it as *dagyaw*. With a *sag-ob nga tuba* mean a volume of tuba or coconut wine inside a bamboo pole and *sumsuman* a dish prepared to eaten with *tuba*, heavy work became

lighter through this cooperative effort. A somewhat likeness was noted by [9] of farmers in the sixteenth century. He found that farmers, by exchange labor, took turns in doing the farm work from planting to harvesting, with the owner feeding them all. However, this form of mutual help system was slowly being eroded because the market economy had reached the uplands.

3.1.3. High *respect* for nature and unseen spirits

Aklanon upland farmers had high respect for nature and unseen spirits. A *medico* (quack doctor) was hired to perform a ritual to negotiate with unseen spirits who are believed to be owners of the land to relocate somewhere else. A *sagda* an offering in the form of a butchered pig or chicken was offered in exchange for the negotiation. Farmers respected nature because of its power over them. Nature's wrath would be unleashed to those who failed to observe them. The modern phenomenon of flash floods or landslides may mean God's wrath for the *kaingin* farmers.

3.1.4. Respect *for* rice grains

A key informant said "*ga sunggod ro paray*" which meant that rice grains would leave the farmer and move to a neighbour's *tambobo* a grain house if the farmer's household would not observe the traditional way of harvesting and postharvest care to the *palay* or rice grains. An informant said that rice grains could get angry and run away, hence, must be treated with respect because palay possess *dunggan* a consciousness or spirit. This belief ensured that care must be observed in handling the palay grains. Use of insect-repellant plants. In the preplanting rituals, insect-repellant plants, such as *lampunaya* (mayana), *tabako* (tobacco), *tangead* (lemon grass), and *euy-a* (ginger) were planted ahead of the palay grains. These plants' smell repelled pests.

3.1.5. Multi-cropping and inter-cropping

Aside from palay, their main crop, farmers also planted corn, *gabi*, beans, squash, cassava, banana and some other perennial plants. In fact, corn was planted ahead of the palay as corn could be harvested earlier, which answered their need for a substitute for rice. Coconut and hemp were also planted to augment the family's income as prices of these products were high. The farmer also ensured not to leave the land "bald" or without vegetation for they were aware of the destructive effect of heavy rains on the top soil. Legumes, such as mongo, string beans and the like were believed to enhance the fertility of the soil as well as being a source of food and or income.

3.1.6. Living in harmony with nature

Kaingin farmers believed that *bata nga taeon* a virgin forest lands should not be converted to kaingin yet. Upland areas that had been used as *kaingin* for about three to five farming seasons need to be vacated for vegetation to regrow. Once abandoned, this area can again be cultivated after seven years. This is in accordance to time and season. According to the paper [20], "to ensure fertility of the land, a season of rest was given to the earth. It was a season of waiting until the land was ready and nature could nourish it."

3.1.7. Categorization of soil and planting of appropriate crops

According to a farmer, crops were not just planted anywhere. Skilled farmers could easily identify the type of soil (e.g., *bukaeay, napo, tikud*) and the appropriate crop suited to the type of soil. For instance, cassava grew well in a *tikud* type of soil and that frequent planting of corn would make the soil *tikud*. Farmers also left the rice stalks during the *gaeo* (harvest) to decompose. From the paper [20], it is mentioned that "planting of legumes was a way of nourishing the earth and so was the utilization of farm, plant and animal wastes as natural fertilizers. Some used leaves from madre de cacao, threshed rice stalks, rice hulls, corn cobs and chicken dung."

3.1.8. Preserving traditional varieties of bisaya and other upland varieties

In the absence of modern technology of germ-plasm banking, the upland community had demonstrated that they were the original keeper of many rice varieties. In the upland areas of Libacao, farmers identified twenty-seven rice varieties while farmers in Ibajay added nine more.

IOP Conf. Series: Earth and Environmental Science **167** (2018) 012022 doi:10.1088/1755-1315/167/1/012022

3.2. Ecologically-destructive practices

3.2.1. Cutting of trees
In general, the under brushing and consequent burning of the kaingin meant the cessation of the flora and fauna in the site. Although the area can naturally regenerate, the increasing number of households practicing kaingin farming will, in aggregate, destroy a large area of forest. The Aklan upland is part of a biodiversity conservation program of the United Nations Development Program, thus, there is a need to integrate this concern with the upland communities.

3.2.2. Use of modern fertilizers and pesticides
Upland farmers have started adopting the practices of the lowland farmers. An informant observed that pests attacking upland rice crops were the same with paddy rice. Unknowingly, they were applying ecologically destructive pest management method.

3.2.3. Short observance of the fallow period
Most upland farmers were applying three to four years of fallow period. Thus, lands used as *sulit* after a shorter period of rest reduced fertility and consequently resulted to poor harvest. The land may eventually become cogon land. Inheritance practice fragmented the land. Hence, sons would have a smaller size to cultivate. This land constraint contributed to a shorter observance of the fallow period. Watershed Destruction. Headwaters were observed to have run dry upon transforming the area to kaingin. In general, farmers had a limited understanding of the function of watersheds.

3.2.4. Planting fewer varieties of bisaya
The preferences of consumers drive the farmers to plant the varieties that were in high demand. This increasingly market-driven production of *bisaya* resulted in the preservation of few varieties that had markets in the lowland and the extinction of other varieties unless a mitigating measure is set up in place by the government.

3.3. Indigenous knowledge and its implication to environmental education
Using cultural ecology as a framework, upland indigenous knowledge is understood as the accumulated product of the people's ways to adapt to their upland environment. This has served them well in the past and will continue into the future. Indigenous knowledge that are ecologically-friendly need to be preserved and transmitted to our students. They should be taught to value and appreciate the rich culture of our upland communities. Indigenous knowledge may yet be our answer to prevent the continuous degradation of the environment, and hopefully, mitigate the debilitating effects of climate change.

4. Conclusions
The upland farmers have been observing traditional beliefs and practices which are a reflection of an environmental world view rooted in their culture. They possess an ecologically sound farming practices, such as being the community-based gene bank of rare rice varieties. Their indigenous knowledge need to be preserved since these practices have helped maintain our ecosystem until now. It must be noted that our ancestors have been doing kaingin farming as evidenced by the writings of the early Spanish missionaries in the sixteenth century. This proves that upland farming is sustainable. However, these practices are gradually overtaken by socio-economic factors like the effect of market forces, demographic changes, rising cost of living and higher social expectation. Such forces will ultimately kill the desirable traditional practices of upland farmers.

5. Recommendations
The national and local government units should recognize, support, and preserve the indigenous ecological knowledge of the upland communities and ensure that these are transmitted to the present

and future generations through the formal and informal school systems. The researchers recommend the documentation of indigenous knowledge and its use in natural resource management. The upland communities should be given a chance to participate in these activities hand-in-hand with government agencies like the Department of Environment and Natural Resources, Department of Agriculture, Department of Agrarian Reform and the local government units. Folk ecological knowledge and wisdom need to be maintained and to create new contexts for their continued practice with the help of educational institutions, teachers, students and non-government organizations working in collaboration with each other. Develop clear instructional materials on traditional knowledge systems to be used in schools and in extension programs to impart information on the value and threats to indigenous knowledge systems. Craft curriculum and methods to integrate traditional knowledge into formal education systems of agencies, researchers and practitioners who work together with communities. Enhance the indigenous and spontaneous community spirit and leadership of the tribal elders in the upland communities. The government needs to support measures to mitigate the effects of kaingin by designing educational and technical programs for upland farmers and their children while preserving the unique aspects of upland culture and practices.

6. References

[1] Fenstad E J 2002 Science and traditional knowledge *Report from the ICSU study group on science and traditional knowledge*, Paris: International Council for Science

[2] Lum T and Margesson R 2014 Typhoon Haiyan (Yolanda): US and international response to Philippines disaster *Current Politics and Economics of South, Southeastern, and Central Asia* 2014 Apr 1;23(2):209

[3] Yumul G P, Servando N T, Suerte L O, Magarzo M Y, Juguan LV and Dimalanta C B 2012 Tropical cyclonesouthwest monsoon interaction and the 2008 floods and landslides in Panay island, central Philippines: meteorological and geological factors *Natural hazards* 2012 Jul 1;62(3):827-40

[4] Pandey D N 2002 Traditional knowledge systems for biodiversity conservation, *Organization of the United Nations (FAO) Forestry Paper* FAO, Rome Italy 2002:22-41

[5] Tauli-Corpuz V, De Chavez R, Baldo-Soriano E, Magata H, Golocan C, Bugtong M V, Enkiwe-Abayao L and Carino J *2009 Guide on climate change and indigenous peoples* Tebtebba Baguio City Philippines

[6] Yamin F and Depledge J 2004 *The international climate change regime: a guide to rules, institutions and procedures* Cambridge University Press

[7] Pretty J 2003 Social capital and the collective management of resources *Science* 302(5652):1912-4

[8] Gadgil M, Berkes F C, and Folke 1993 Indigenous knowledge for biodiversity conservation *Ambio* 151-6

[9] Scott W H 1992 Looking for the prehispanic Filipino: and other essays in Philippine history, Cellar Book Shop

[10] Funtecha H F 2001 Traditional farming in Central Panay: A microcosmic world in environmental and cultural awareness *Palayag CWVS Monograph Series* Center for West Visayan Studies, College of Arts and Sciences UP in the Visayas Iloilo City Philippines

[11] Trousdale W J 1996 Sustainable tourism planning with multiple objective decision analysis: a case study of the Guimaras farm tourism pilot project MA Thesis School of Community and Regional Planning The Colorado College BC

[12] Ember C R, Ember M and Peregrine P N 2002 Anthropology, Singapore: Pearson Education Asia Pte Ltd

[13] Steward J H 2006 The concept and method of cultural ecology, The Environment in Anthropology: A Reader in Ecology *Culture and Sustainable Living* 5-9

[14] Villamor G B and Lasco R D 2009 Rewarding upland people for forest conservation: experience and lessons learned from case studies in the Philippines *Journal of Sustainable Forestry* 28(3-5):304-21

[15] Allen W H 1988 Biocultural restoration of a tropical forest *BioScience* 38(3):156-61

[16] Hilario C R, Rico I A, Biray E T and Gonzales E I 2012 Traditional knowledge in upland farming towards ecological restoration *Restrospect* Vol. 1 Number College of Teacher Education, Aklan State University Banga Aklan Philippines
[17] Carino C M 2013 Aging Filipino Farmers to Affect Food Security *The Manila Times*
[18] Bersales L G 2017 Farmers, Fishermen and Children consistently posted the highest poverty incidence among basic sectors, Philippine Statistics Authority, Reference No.:2017-150.
[19] Acosta-Michlik L and Espaldon V 2008 Assessing vulnerability of selected farming communities in the Philippines based on a behavioural model of agent's adaptation to global environmental change. *Global Environmental Change*, *18*(4), 554-563.
[20] Dator-Bercilla J A 2004 Indigenous rice farming technology and the survival of farming communities in the age of globalization *Palayag (CWVS Monograph Series)* Center for West Visayan Studies College of Arts and Sciences UP in the Visayas Iloilo City Philippines

Acknowledgments
The author would also like to thank the anonymous referees for their valuable comments and helpful suggestions. Dr. Arangote acknowledges the Aklan State University, for support through the Research and Development Services Office.

Heavy Metal Contamination Assessment in Near Surface Soils: A Case Study from Subarctic Region of Russia

A Masih[1] and A P Sergeev[2]

1 Ural Federal University, Ekaterinburg, Russia
2 Institute of Industrial Ecology Ural Branch Russian Academy of Science, Russia
E-mail: adven.masikh@urfu.ru; alexanderpsergeev@gmail.com

Abstract: Under the initiative of a routine ecological screening taken by the Russian state territories, soil surface assessment of heavy metals such as Cr, Cu, Ni, Mn, etc. was conducted in a city called Noyabrsk located in a subarctic region of Yamalo-Nenets Autonomous Okrug (YNAO) Russia. During survey a total of 321 soil samples were collected from different parts of the region including residential (237 samples), industrial (80 samples) and from background locations (4 samples) of the city. In order to determine the general ecological conditions of the area chemical analysis was conducted. Geostatistical tools along with other statistical techniques have been adopted to explore, analyse and map the obtained concentrations of the heavy metals. Results drawn have revealed some moderate – high anomalies of As, Cr and oil concentrations in industrial as well as in residential regions that can be a threat to public health. The aim of the paper is to assess the pollution status of urban soil and discuss the soil contamination sources for the future planning and management of Noyabrsk city.

1. Introduction

Heavy metal assessment is an important part of state's responsibilities because it ensures the environmental safety of its residents. It also brings sagacity of investigating the prime emission sources of heavy metals as their impact on the surrounding environment can be significant. Various studies have been conducted on soil pollution as a result of numerous industrial activities in many parts of the World, however there are very few investigations to author's knowledge that have been made in subarctic region of Russia assessing the ecological conditions of urban areas for pollution risk. Essential environmental components such as soil, biota, snow, bottom sediments, etc. are generally the places of accumulation for pollutants from different sources, therefore, surfaces containing pollutants are possibly be the best sources to investigate the nature and characteristics of the pollution [1], [2]. Soil ecological screening is vital because the composition of soils are intervened most of the times by rocks, water and air [3]. These interventions get deposited in the form of soil layers which preserve pollutants record in it for decades either carried by sediments, water, and air or caused by anthropogenic activities. The chemical variability of the parent rock or soils should be considered while determining the origin of the soil surface contamination [3]. Field data gathered mainly depend on the type and intensity of emission source as well as on other factors such as hydrological and meteorological conditions and climatic variabilities [3]. Such irregularities while collecting data sometimes lead to spatial heterogeneity that can seriously affect the anomalies generated due to tainted data to predict the environmental pollution.
Interpolation a geostatistical tool has extensively been used in recent years in soil pollution estimation. It performs the pollution assessment to study the spatial variations and ensures the certainty of its estimates. Moreover it evaluates spatial pattern and distribution of pollutants and enables spatial

mapping [4]. Recent geostatistical studies [3], [5] have focused on urban soils and have discussed the usefulness of adopting interpolation techniques to address spatial pattern and uncertainties in soil concentration data.
The paper discusses near surface heavy metal contamination in subarctic region of Russia. Spatial interpolation is the main method employed assessing the heavy metal pollution in area by using ordinary and indicator kriging maps presenting the concentrations of heavy metals in the study area.

2. Material and methods

The study conducted was a part of a routine ecological screening survey structured by the authorities of the YNAO in August 2006. Noyabrsk is the largest city in the southern YNAO district located in the middle of arctic zone of Western Siberian plains covering an area of 38.84 square kilometres as shown in Figure 1. The local urban soils are mainly consist of illuvial-humus podzol with little content of organic components [3], [6] whereas its composition is 100% sand. The economy of the region majorly based on hydrocarbon production with a fact that the place has always considered home ground to one of the biggest oil and gas company "Gazprom". Due to the fact the major pollutants in the urban soils were expected to be phenol and mineral oil against the other high concentrations of heavy metals revealed such as Arsenic and Chromium.

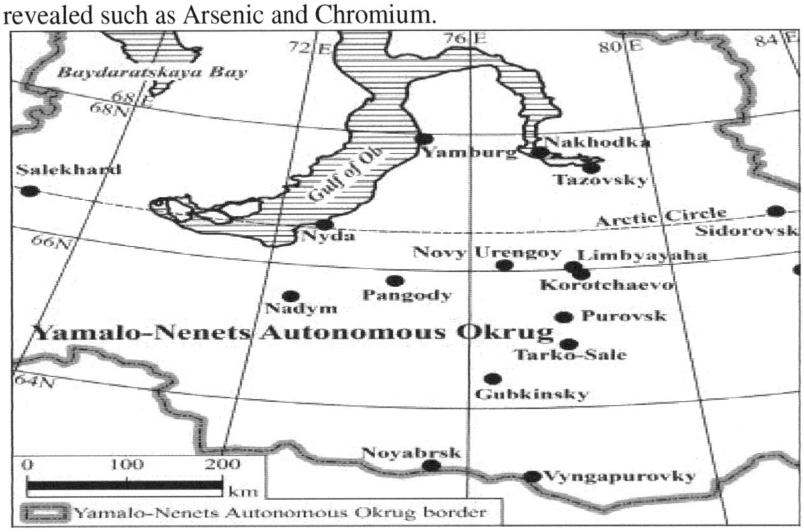

Figure 1. Location map of the study area

During the survey according to the extent of the region a total of 321 soil samples of different spatial scales on a squared-grid sampling scheme at a depth 0-5cm were collected from residential, industrial and background locations (typically acquired to compare values) of the city. An increased sampling interval in the industrial zone was adopted in comparison with the samples collected from residential area. At each point, 5 soil samples were taken, 4 of which making a square geometry of side 1m around the 5th located in the middle. Geographical coordinates of each sampling point were recorded using GPS, and thereafter soil samples were stored in polythene bags each weighed about 1.5kg prior to chemical analysis. Preparation of all samples and their chemical analyses with the type of methods employed in laboratory have already been discussed in detail in our previous work [1]. Chemical analysis was aimed at determining the hydrogen ion exponents of the following agents with the soluble form of F, Cr, Co, Ni, Cu, Zn, and the total form of Al, Cr, Mn, Fe, Co, Cu, Ni, Zn, As, Cd, Sb, Hg, Pb, phenol and oil concentrations in the topsoil.
Geostatistical tool of variography was employed to analyse and interpret the data, generally based on a theory of localize variable distributed in space showing a spatial correlation such that closer the samples higher the correlation and vice versa [7]. Variography is used to measure the spatial variability and dependency of a regionalized variable by calculating variogram/semi-variogram. It provides the input information for the spatial interpolation. The variogram function is expressed as:

$$\gamma(h) = \frac{1}{2N(h)} \sum_{i=1}^{N(h)} (Z(x_i) - Z(x_i + h))^2$$

where $\gamma(h)$ is the semivariance (variogram), $Z(x_i)$is the value of the variable Z at location of x_i, and $N(h)$is the number of pairs of samples separated by the lag distance h.

Information generated through variography was used to calculate sample weighting factors for spatial interpolation by ordinary and indicator kriging.

The anisotropic spatial variability was also considered in accordance with the symmetrical property of variogram function for all significant variables. Concerning the modelling of experimental variograms to get the information about the spatial structure and input parameters for kriging, different models i.e. exponential and Gaussian models were employed. The information acquired was later used for calculating weighting factors for spatial interpolation by the ordinary and indicator kriging.

The obtained concentration of Cr, As and oil are mapped and interpreted using Geostatistical Analyst tool in ESRI ArcGIS software with ordinary and indicator kriging as interpolation techniques. Other routine statistical analyses such as probability distributions and Spearman rank correlation coefficients were performed by using Statistica software.

3. Results and discussion

The ministry of Natural Resources and Ecology of Russian Federation controls the total content of nine heavy metals [8]. There are different types of approaches used in Russia as permissible standards such as maximum permissible concentrations (MPC), others include provisional permissible concentrations (PPC) and background values comparison when there are no standards considered for the concentrations estimated. In Noyabrsk soil data set, for metalloid Arsenic (As) both MPC and PPC approaches were considered, Chromium (Cr) values were compared with the standards set by PPC and also with the world soil standards as well, whereas for oil concentrations background location values were estimated.

Table 1. Descriptive statistics of heavy metals in study area.

Elements	Mean	Median	Minimum	Maximum	Std.Dev.	Coef.Var.	Skewness
Al	11210.7	10270	101	32940	4764.8	0.43	1.27
Cr	**64.56**	**61.41**	**16.57**	**140**	**22.81**	**0.35**	**0.74**
Mn	141.63	130.45	61.81	529.1	54.21	0.38	3.4
Fe	13220.05	12785	101.	28270	4000.2	0.3	0.6
Co	4.69	4.26	2.00	101	5.589	1.19	16.25
Ni	11.76	11.01	3.58	101	6.652	0.57	8.18
Cu	18.85	14.67	5.01	341.49	25.3	1.34	9.29
Zn	17.31	14.14	5	133.63	12.5	0.73	4.14
As	**1.6**	**1.42**	**0.14**	**5.2**	**0.79**	**0.51**	**0.96**
Pb	6.33	4.56	1	120.01	10.09	1.59	7.99
Oil	**24.48**	**6.88**	**0.93**	**829.4**	**75.13**	**3.07**	**7.51**

The statistical summary of the data set for heavy metals with other soil variables including oil is given in Table 1. Interestingly, almost all chemical elements assessed were fallen within the permissible concentration limits set under the environmental law of Russia, expect, the arithmetic mean value for oil concentrations which is very high as compared to its median values with high skewness, confirming some high anomalies in the study area. It is to be noted that the mean value for the concentration of oil in the topsoil is 4-times higher than the median value. The concentration of oil has ranged from 0.93 to 829.40mg/kg with a mean value of 24.48mg/kg. Also the mean value for oil calculated separately for industrial zone (37.66mg/kg) was almost 2-times higher than residential zones (19.7mg/kg), while both are way higher than the average value (2.50mg/kg) of 4 background location concentrations collected.

The general toxicological GOST 17.4.1.02-83 operated in Russia divides heavy metals/metalloids into three classes based on the degree of hazardousness, according to which Arsenic makes into the list of elements considered the most hazardous with other elements include Cd, Hg, Se, Pb, Zn [9]. In Table 1, the mean value of As (1.6mg/kg) is slightly higher than its median value, which is nearly equal to the permissible limits set under MPC and PPC i.e. 2mg/kg for the land with composition sand or loamy sand [10] by the Russian Federation. The As metal concentrations in studied area showed that around 90 sample points i.e. 30% of the total sampling points of As concentrations were higher than the strict soil guidelines set for As by MPC and PPC.

Chromium (Cr) currently categorized in the moderately hazardous element class according to toxicological GOST. Table 1 shows that the mean value of Cr is 64.56mg/kg which is slightly higher than its median value with average skewness can join oil and As in the list of elements that can be a threat to public health due to its high concentration values in study area. The average concentration of Cr appeared in the selected region falls within the reference value set under environmental law with only few Cr sample points crossing the Russia's permissible clarke limit of 122mg/kg [11], [12]. However, in an international perspective the distribution of the Cr in study area still remains a point of interest, because the permissible chromium clarke in world soils is 59.5 mg/kg [11], [13]. Keeping that as a reference value, only 38% sample points in our case had a Cr concentration less than the guideline set for world soils whereas the remaining 62% have surpassed the reference value.

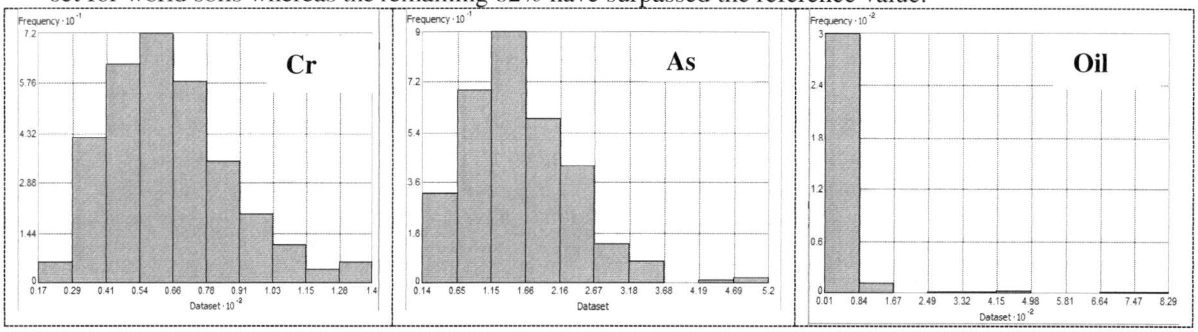

Figure 2: Frequency distribution for Chromium, Arsenic and Oil

Table 2. Spearman Rank Correlation.

	Al	Cr	Mn	Fe	Co	Ni	Cu	Zn	As	Pb	Oil
Al	1.0										
Cr	-0.02	1.00									
Mn	0.39	**0.51**	1.00								
Fe	0.13	**0.78**	**0.65**	1.00							
Co	0.06	**0.58**	**0.56**	**0.57**	1.00						
Ni	-0.05	**0.71**	**0.55**	**0.7**	**0.66**	1.00					
Cu	-0.18	**0.75**	**0.45**	**0.72**	**0.54**	**0.78**	1.00				
Zn	0.30	0.05	0.35	0.08	0.20	0.30	0.11	1.00			
As	0.23	**0.27**	**0.39**	**0.38**	**0.29**	0.25	0.25	0.09	1.00		
Pb	0.35	0.09	0.21	-0.09	0.03	0.18	0.06	**0.54**	0.09	1.00	
Oil	-0.09	-0.01	0.05	-0.08	0.16	0.30	0.09	**0.48**	-0.13	**0.36**	1.00

*Highlighted numbers are significant at P<.05

According to Table 1 and Figure 2 the distribution of oil concentrations in Noyabrsk topsoil is highly positively skewed in comparison with the moderately skewed and way normally distributed concentration values of As and Cr. The coefficient of skewness of Cr, As and oil clarkes distributions were computed using Kolmogorove-Smirnov test, knowing to the fact that normal distribution is required in linear geostatistics [14], because abnormal distribution can sometimes seriously affect the kriging results. Coefficient of variation (CV) values calculated for the data set suggested that among As, Cr and oil concentrations, oil showed the greater variation against the two metals.

Not all heavy metals entering surface soil are considered pollutants. Therefore, to get more insight into data acquired, Spearman Rank Correlation was performed at $P < 0.05$ to investigate the relationship among elements (Table 2). Chromium (Cr) presented a strong positive correlation with Mn, Fe, Co, Ni, Cu, and As. There existed no significant correlation between oil concentrations and Cr according to Table 2, rather it showed a negative correlation with As, which suggest that the concentration values of Cr and As are most probably not sourced by the high consumption of oil in the industrial zones of the city, however the strong correlation of Cr and As with one another as well as with the top half of the elements presented in Table 2, indicates the presence of single emission source.

3.1. Geostatistical and anisotropy analysis

To make better decisions variography analysis has been performed using topsoil parameters. Also for creating surfaces the data was thoroughly explored for obvious errors that can drastically affect the output prediction surfaces. Before creating surfaces, distribution of data was examined, global trends were identified and removed, and spatial autocorrelation and directional influences were understood in Geostatistical Analyst. Before picking the best model experimental variograms were tested with different theoretical models such as Spherical/Gaussian/Exponential functions and their results were compared. The selection criteria of each model was based on which best fitted to experimental variogram behaviour (table 3).

Table 3: Experimental variograms

Variable	Model	Nugget	Sill	Spatial Dependency
Cr	Exponential	326.6	575.9	Moderate
As	Exponential	0.13	0.63	Strong
Oil	Gaussian	3.2	3202.8	Strong

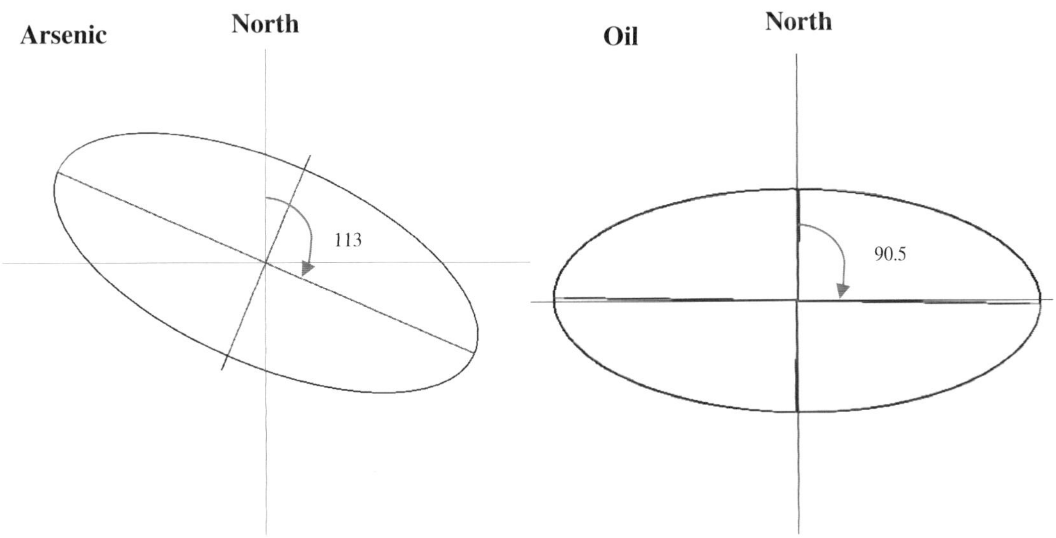

Figure 3. Anisotropic effect for As and Oil estimated clockwise from North by using geostatistical analyst determines their distribution along major axis with prependicular minor axis.

For to the degree of spatial dependency the work by Cambardella [15] is followed, according to which if the ratio of nugget to sill multiplied by 100 is less than 25% then the variable is considered strongly spatially dependent, if the ratio ranges from 25% to 75% the variable is moderately spatially dependent and in case the ratio crosses 75% then data is weakly spatially dependent.

Following the rule, the experimental variograms of As and Oil concentrations appeared to be strongly spatially dependent in comparison with moderately spatially dependent variogram of Cr. The anisotropic effect of elements shown in figure 3 clearly separates oil from As anomaly that could have been the possible emission source due to its minimum presence in industrial zones. Anisotropic analysis represents that the major oil anomaly appeared in study area is mainly distributed along E-W direction making a right angle with North, whereas residential region bounded As anomaly lies in NW-SE direction (Figure 3).

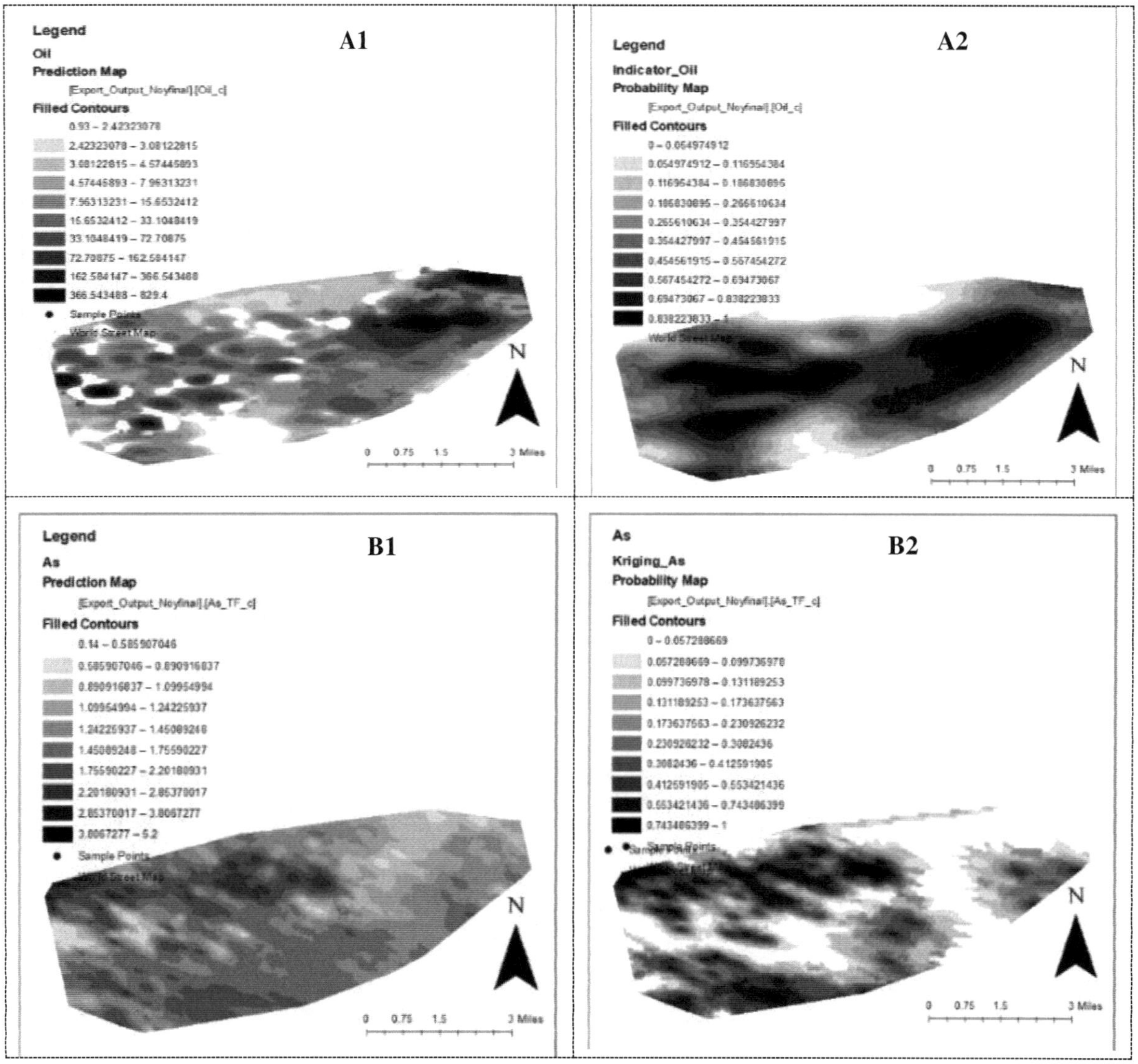

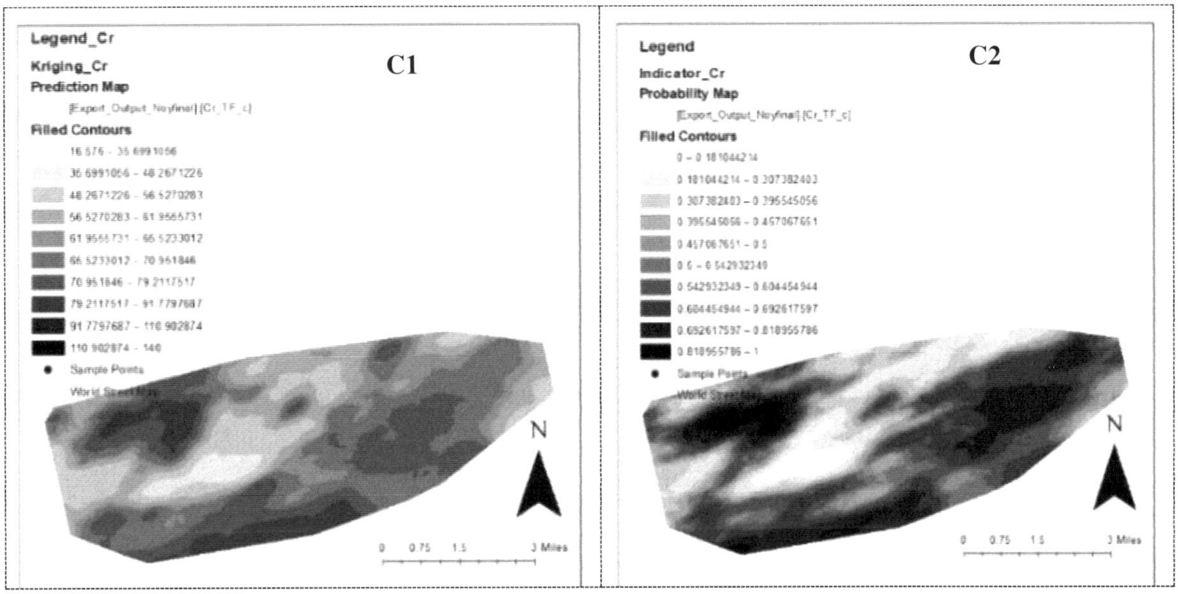

Figure 4. Kriged maps of oil, As, Cr (A1, B1, C1) clarkes using ordinary kriging and probability maps for the estimated concentrations (mg kg−1) of oil, As, Cr (A2, B2, C2) using Indicator kriging.

To precisely mark the probable polluted zones that have violated the guidelines set by the State, the indicator kriging technique adopted requires a threshold value for each element– oil, As and Cr. The threshold values assigned to; oil concentrations are compared with background values, Arsenic was assigned value set by MPC and PPC, and Cr has been compared with the its median value (61.41mg/kg) which is nearly equal to reference guide set for world soils. It is quite obvious in Figure 4 that the concentration of oil is significant in industrial zones with its few major spots in residential zone, whereas the high concentrations of As rather appeared in settlement zones. On contrary, kriged map of Cr have shown a parallel existence in both residential and industrial regions, however, its anomaly in residential area is comparatively higher than in the industrial zone of the city. Furthermore, the kriged maps confirm the correlation between oil, As and Cr (Table 2) i.e. the association of both metals – As and Cr with oil is either negative or not significant to be correlated. Moreover, the absence of As in the industrial zones is evident that the concentration of As metal (most hazardous) in residential is probably not been sourced by industrial activities but perhaps region has been contaminated by human-induced activities in the urban soil.

It wouldn't be justified declaring that the sampling area is polluted, however, a considerable portion i.e.30% of the metalloid (As) the most hazardous to local standards and 62% of Chromium (Cr) values according to international guideline set for world soils have crossed the permissible boundaries which can be a threat to public health, therefore, seek local government intention.

4. Conclusion

The geostatistical approach adopted for the study has examined and interpreted the topsoil heavy metal concentrations in Noyabrsk a city situated in subarctic region of Russia. Although the region is well known for its large amount of hydrocarbon extraction yet surprisingly, almost all chemical elements assessed appeared within the local permissible ecological limits expect the moderate anomalies of As, Cr and oil particularly in residential zones. The results drawn indicate that around 30% sampling points of Arsenic a most hazardous and strictly monitored metalloid with a permissible concentrations of just 2mg/kg in Russia, have violated the ecological limits. On the other hand a very few samples of Chromium a heavy metal have surpassed the local reference values, while in an international perspective 62% sampling points of the same metal (Chromium) have violated the limits set for world soils. Although oil clarke variations revealed in industrial zone of urban topsoil are two-times higher as compared to the settlement zones, however, the study conducted showed that it doesn't share a strong correlation with the high concentrations of Arsenic and Chromium metals to be the soil contamination source. It seems as anthropogenic activities are possibly the emission source of moderate anomalies of Arsenic and Chromium within settlement region especially, hence responsible of contributing into the urban soil pollution of the city.

5. References

[1] Sergeev A P, Baglaeva E M and Shichkin A V 2010 Atmospheric Pollution Research **1** 44 - 49.
[2] Saet J E, Revich B A, Yanin E P, Environment Geochemistry, Russia, 1990, pp 84-108.
[3] Sergeev A P, Baglaeva E M, Antonov K, Medvedev A, Rakhmatova A 2015 *15th international multidisciplinary scientific geoconference SGEM* **II** 27-34.
[4] Komnitsas K and Modis K 2006 *Sci.Total Environ.* **371**: 190–196.
[5] Rodriguez, J A, Nanos N, Grau J M, Gil L and Lopez A M. 2008 *Chemosphere.* **70**: 1085–1096
[6] Dobrovolsky G V, Urusevskaya I S 2004, *Geography of soils Russia*, 195-200.
[7] Pannatier, Y. 1996. Variowin: Software for Spatial Data Analysis in 2D. Springer-Verlag, New York.
[8] Federal Report on the State of the Environment in the Russian Federation in 2007, Minprirody RF, Moscow, 2008 (in Russian).
[9] GOST 17.4.1.02-83 Environmental Protection, Classification of Chemical Substances for Pollution Control Standard, Moscow (1983) (in Russian)
[10] Chernova O V, Beketskaya O V, *Eurasian Soil Science*, 2011, Vol. **44 (9)**, 1008–1017.
[11] Vodyanitskii N Yu 2016 *annals of agrarian science* **14** 257 – 263.
[12] Greenwood N N and Earnshaw A. 1997 *Chemistry of the Elements* (second ed.), Butterworth-Heinemann, Oxford.
[13] Kabata-Pendias A 2011 Trace Elements in Soils and Plants (fourth ed.), CRS Press, Boca Raton, USA.
[14] McGrath D, Zhang C and Carton C T, 2004 *Environ. Pollut.* **127(2)**: 239–248.
[15] Cambardella, C A, Moorman T B, Novak J M, Parkin T B, Karlen, D L, Turco R F and Konopka A E 1994. *Soil Sci. Soc. Am. J.* **58**: 1501–1511.

Monitoring Lead in Soil Water System Using Complex Permittivity

S Al Martini[1] and J Q Shang[2]

1 Associate Professor, PhD, P.Eng., MASCE, College of Engineering, Abu Dhabi University, Abu Dhabi, UAE
2 Professor, PhD, P.Eng., Western University, London, Ontario, Canada
E-mail: samer.almartini@adu.ac.ae, jqshang@uwo.ca

Abstract. The complex permittivity was investigated to detect lead contamination in soil. The system used consists of an Automatic Network Analyzer (ANA), a sample holder, coaxial cables and a personal computer for data processing. Five samples of Halton Till soil were prepared and mixed with five different concentrations of hybrid blends of lead salts ($PbSO_4$, $PbCl_2$ and $Pb(NO_3)_2$). The objective was to investigate the capability of complex permittivity to detect lead, when introduced to the soil in hybrid blend salts. Chemical analysis was conducted on the pore water of the six soil samples to investigate lead transport and the potential risk of lead in groundwater sources. The results show that the complex permittivity is sensitive to the change in lead concentration in soil as the real part decreases and the imaginary part increases with the increase of lead concentration. The real part decreased from 23 to 17 when the concentration of the lead increased from 0 to 1800 ppm. The imaginary part increased from 16 to 24 as the lead concentration in soil increased from 0 to 1800 ppm. The chemical analysis of the pore water showed that the electrolytes increased as lead concentrations increased in soil. The results displayed that the complex permittivity could be used as a non-destructive tool to monitor lead contamination in soil.

1. Introduction

Heavy metals (lead, mercury, cadmium and arsenic) are toxic and can cause contamination in the environment. The release of heavy metals to the environment affects human health, wildlife and the ecosystems. Many hazardous substances and heavy metals such as Pb, As, Ba and Cd are released to the subsurface as a result of industrial activities. Improper disposal of these substances has a significant adverse impact on the environment [1].

Many industries all over the world, and especially in the developed countries, generate wastewater daily which contains high levels of metals. In the past, at the start of the industrial revolution, any damage that heavy metals could have caused to the environment was not known. Many industries discharged their wastewaters directly to rivers and lakes causing pollution of water systems. Eventually, the contamination is transferred to humans, wildlife and vegetation. The environmental problem has become an important issue in the world. Strict environmental regulations have been established in order to protect the environment from pollution.

Lead in soil usually results from leaded gasoline, pulverized lead in paint, and from waste disposal. The lead emitted to the atmosphere from various sources accumulates in the soil. Its accumulation depends on the rate of its disposal from the atmosphere and the magnitude of lead released to the environment. Inorganic forms of lead such as lead oxide and lead chloride are the most widely released forms of lead to soil. The concentration of lead in natural soils is usually less than 50 ppm [2].

Lead in soil in populated areas has resulted in an increase of lead levels in the blood of both children and adults. Lead in soil causes severe health problems to children engaged in hand to mouth lead transfer. Therefore, soil should be addressed as a potential lead pathway to people. People should be aware of its danger and the hazard that it represents, so that they can take the required precautions. Regulators and the public health community should acknowledge this as a very serious matter, so that strict regulations can be issued to protect people, animals and the environment from lead contamination [3].

The vast industrial revolution of the last century brought environmental contamination issues to the forefront of mankind attention, which raised environmental awareness. In the last decade environmental contamination became one of the most important global concerns. More recently, soil contamination problems have attracted a lot of research endeavours and resulted in a wealth of pertinent published literature. In particular, it has been recognized that more efficient methods to study and monitor soil contamination process need to be developed.

The conventional monitoring methods involve soil and water sampling and a chemical laboratory analysis. Drawbacks of these methods are that they are expensive and time consuming. It is of great importance to find an alternative, in-situ, fast and non-destructive method, that is able to delineate soil contamination. The study of the interaction between the soil-water system and the incident electromagnetic waves is known as microwave aquametry [4]. Using electromagnetic waves to characterize soil properties and to identify soil contamination is one of the alternative methods used to monitor soil contamination.

The response of the soil-water system to electromagnetic waves can be characterized in terms of conduction and polarization. Polarization represents the ability of material to store electrical energy, while conduction is the energy loss due to polarization and ionic conduction. Both terms are combined in one term known as the soil complex permittivity (ε^*). Theoretical and experimental studies have been carried out and results have been reported in the literature on the investigation of soil water complex permittivity [5]-[7]. It has been found that the complex permittivity reflects the physical and chemical properties of a soil-water system, making it feasible to study changes in soil contamination.

In the course of the current research, the soil complex permittivity method is investigated and evaluated for detecting lead contamination in soil. If the soil complex permittivity measurement is found to be sensitive to the chemical change in the soil pore fluid, the complex permittivity method will be a promising tool in lead site assessment. The system used to measure the soil complex permittivity is the transmission line technique, where the soil sample is placed in a coaxial holder to measure its complex permittivity. Because several researchers have had success when using the coaxial holder to characterize soil properties [8], [9]. More details about the used system can be found elsewhere [10].

This paper presents results of complex permittivity measurements in lead contaminated soils. Hybrid blends of three lead salts ($PbCl_2$, $PbSO_4$, $Pb(NO_3)_2$) at different concentrations were mixed with soil. The lead concentrations are selected based on Ontario Ministry of Environment and Energy (MOEE, 1996) for soil (200 ppm – 1000ppm).

The objective in this chapter is to investigate the relationship between soil complex permittivity and lead concentrations in hybrid lead salts, which could be used as an indicator to detect soil contamination with lead. The main objectives of the current research work can be summarized as follows:

- Study the relationships between lead concentrations in soil contaminated by lead and the complex permittivity.
- Investigate the capability of complex permittivity to detect lead in soil. This was done by measuring the complex permittivity of soils mixed with artificial solutions of three lead salts ($PbSO_4$, $PbCl_2$ and $Pb(NO_3)_2$).
- Investigate lead transport and the potential risk of lead in groundwater sources.

2. Preparation of Soil Samples Contaminated with Lead Solutions

The soil used in this investigation is Halton Till. The clay minerals in this soil are identified as illite, chlorite and smectite (trace), and the non-clay minerals are quartz, feldspar and carbonates including calcite (12.7%) and dolomite (3.5%) [11]. Multiple salt solutions were prepared using three lead salts (i.e. lead nitrate ($Pb(NO_3)_2$), lead chloride ($PbCl_2$), and lead sulfate ($PbSO_4$)) to study the possibility of tracing lead contamination in complex forms. The lead concentrations used in the investigation were based on Ontario Ministry of Environment (MOE). The lowest concentration were chosen according to the MOE Guideline (MOE 1996) for residential areas, while the highest concentration were based on the allowable lead concentration for industrial areas. Table 1 summarizes the solutions used to prepare soil specimens.

Table 1. The solutions used to prepare soil specimens

Chemical	Sample #	Contaminant Concentration in Soil (ppm)		
		$Pb(NO_3)$	$PbCl_2$	$PbSO_4$
$Pb(NO_3)_2$ + $Pb\ Cl_2$ + $Pb\ (NO_3)_2$	HT1	0	0	0
	HT2	48.76	44.6	51
	HT3	97.5	89.3	102
	HT4	274	251	287
	HT5	487	447	511
	HT6	867	795	907

Since lead chloride and lead sulfate have low solubility in water, they are sieved through a sieve #200. The lead that passed through this sieve was weighed and mixed with water. After oven drying the soil in the oven, the lead salt solutions were mixed with the soil; the soil was put in bags, sealed and left overnight for the soil moisture to reach equilibrium. The soil mixed with the lead solutions was then compacted in an apparatus designed by Scholte [9]. The soil samples were then mixed with the lead salt solutions and left overnight to reach equilibrium. The pore water was obtained by squeezing the soil sample using a Pneumatic Pore Water Squeezer (Serial No. GRC-3) which generated a maximum stress of 25 MPa on the soil samples. The time needed to squeeze each sample was about six hours.

3. Chemical Properties of Soil Pore Water

The concentrations of the major anions and cations were measured from the soil pore fluid. This was done to detect the chemical composition change in the pore fluid due to lead contamination. The pH and electrical conductivity of the pore fluid were measured using pH and conductivity meters. The anion concentrations (chloride, sulfate, nitrate) in the pore water were measured using High Liquid Performance Chromatography (515 HLsPC), which was capable of measuring the three anions at the same time. Cations, such as Pb^{+2}, Na^{+}, Ca^{2+}, K^{+} and Mg^{2+} were measured using a Philips Atomic Absorption Spectrometer PU9100X. The pH of the squeezed liquid was measured using an ORION Ag/AgCI SURE-FLOWTM Electrode (Model No. 9165BN).

Table 2 shows the chemical analyses for all soil samples mixed with lead solutions with different concentrations. As shown in the table, there was no trace of lead in the pore fluid of all five-soil specimens. The concentrations of calcium, magnesium, potassium, chloride, sulfate and nitrate increased when the concentration of lead in the soil increased (Table 2). Sodium and potassium did not show any significant increase. The pH of the pore fluids was remained in a very narrow range with an average of 7.4. This pH value is close to the value reported by Xei [12], which could be attributed to the carbonate in the soil that controlled the pH for all soil samples mixed with lead nitrate.

The analysis of the pore water shows significant amounts of hydrocarbonate (HCO^{3-}) Because of the high carbonate content, the pore water remains alkaline. Therefore, the precipitated lead carbonate remains insoluble due the high carbonate, which explains why lead is absent in pore waters of the soil samples tested. It is concluded that the alkaline pH of the soil plays an important role in the attenuation of lead. As indicated by Harter [13]), the presence of calcium (Ca) in soil plays an effective role in the absorption of lead. Because of the exchange reaction between calcium and lead, calcium desorbs while Pb adsorbs to the clayey soil [14]. Since calcium is situated at the edges of clay particles, it is easily exchanged with lead [15]. Moreover, the maximum cationic exchange occurs when soil pH is higher than seven [16]. It can be seen the increase in sulfate concentration with higher lead concentration introduced into the soil. It can also be attributed to the exchange reaction between sulfate and lead, sulfate desorbs while Pb adsorbs to the clayey soil.

Table 2. Summary of pore water analysis

Chemical	P SO_4 + Pb Cl_{2+} Pb $(NO_3)_2$					
Sample	HT1	HT2	HT3	HT4	HT5	HT6
Conductivity (mS/cm)	2.3	2.45	2.62	3.83	4.65	6.12
Pb^{2+}(mg/L)	ND*	ND*	ND*	ND*	ND*	ND*
Na^+ (mg/L)	245	230	185	215	205	245
Mg^{2+} (mg/L)	130	160	156	226	250	310
Ca^{2+} (mg/L)	760	860	906	1348	1784	2380
K^+ (mg/L)	12	13.5	7	5.5	4.75	8.75
Cl^- (mg/L)	179	87	54	260	470	766
SO_4^{2-} (mg/L)	1800	1087	1820	1800	1820	2030
NO_3^- (mg/L)	165	213	200	1530	1700	2500
HCO3 -(mg/L)	114	175	135	279.6	180	371

*ND: Non detectable

4. Complex permittivity for soil mixed with $Pb(NO_3)_2$+$PbCl_2$+$PbSO_4$ solutions

Figure 1a presents the relationship between lead concentration and the real relative permittivity at the dispersion frequency (320MHz). The real relative permittivities decreases when salt concentrations increase. The relationship is linear with R^2 = 0.812. Figure 1b presents the relationship between the imaginary part and lead concentrations at frequency 200MHz. They are also linearly related. The imaginary part increases with lead salt increase with R^2= 0.64. Figure 1c shows the relationship between the imaginary and relative real permittivity. Which are linearly related with R^2= 0.814.

After investigating the feasibility of measuring the complex permittivity to detect lead in soil when lead presents in the seven abovementioned forms, it is concluded that the complex permittivity is sensitive to the changes in lead concentrations in the soil.

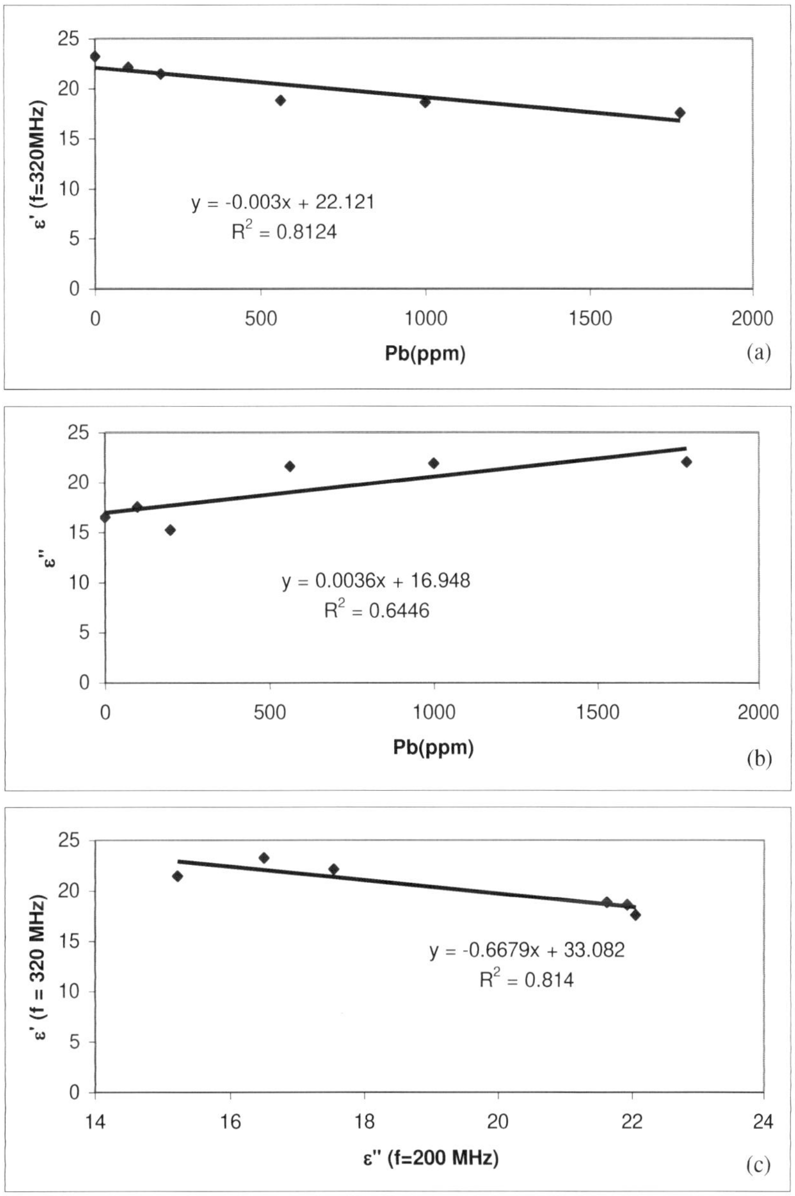

Figure 1. Halton Till mixed with lead chloride, lead nitrate and lead sulfate $PbCl_2$+$Pb(NO_3)_2$+$PbSO_4$) complex permittivity relationships

(a) ε' at dielectric dispersion (f=320 MHz) versus concentration
(b) ε" at (f=200 MHz) versus concentration
(c) ε' at dielectric dispersion (f=320 MHz) versus ε" at (f=200 MHz)

5. Complex permittivity of pore water for soil mixed with lead chloride, lead nitrate and lead sulfate

Figure 2a presents the relationship between lead concentration and the real relative permittivity along with the deionized distilled water at 2GHz. This figure shows that the real relative permittivities decrease when salt concentrations increase.

a

b

c

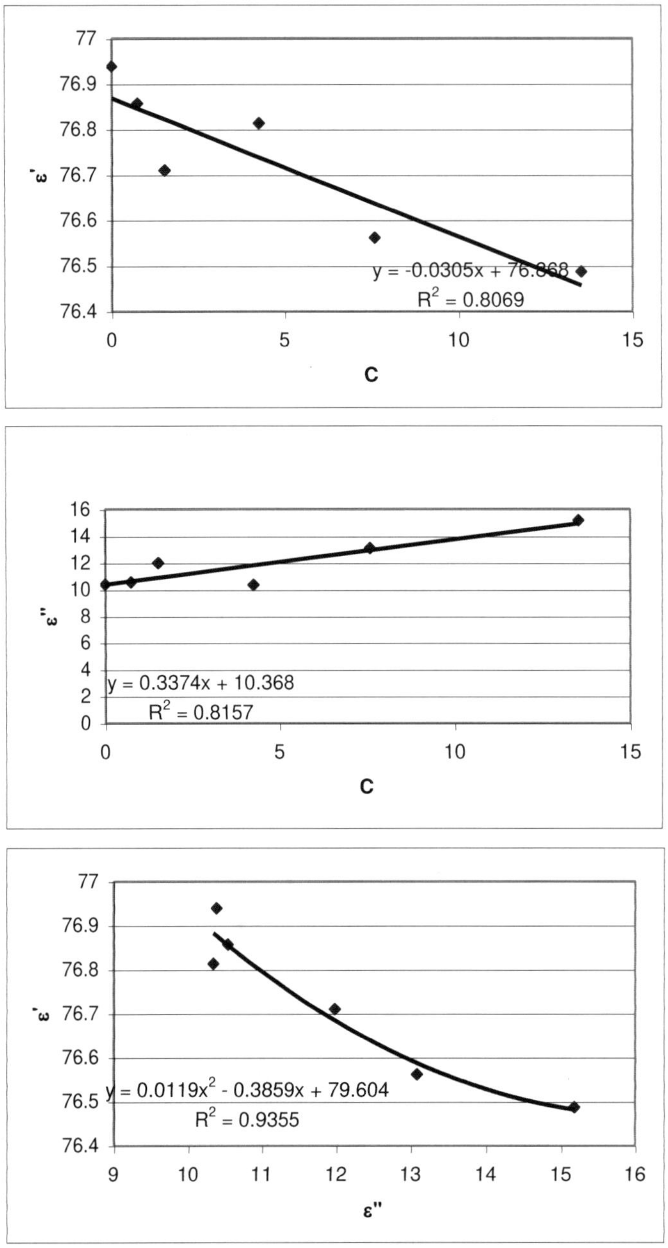

Figure 2. Complex permittivity of pore water squeezed from Halton Till soil mixed with hybrid lead salts (lead chloride, lead nitrate and lead sulfate) at f=2 GHz.

a. ε' versus concentration
b. ε" versus concentration
c. ε' versus ε"

The relationship is linear with R^2 =0.8. Figure 2b presents the linearly relationship between the imaginary part and lead concentrations at 2GHz. The imaginary part increases when lead salt increases with R^2=0.8152. The changes of the complex permittivities of the pore fluids are due to changes in static conductivities and ion concentrations of the soil pore fluids when soil samples are mixed with lead nitrate solutions. This indicates that complex permittivity measurements of pore fluid are sensitive to changes in lead nitrate and lead chloride and lead sulfate concentrations in soil. Figure 2c shows the second order relationship between the imaginary and relative real permittivity with R^2=0.93.

6. Conclusions

The complex permittivity system was investigated to detect lead in soil. Six soil samples were prepared with six different concentrations of ternary bends of lead salts (($PbSO_4$, $PbCl_2$ and $Pb(NO_3)_2$)). The soil samples were squeezed and the chemicals and complex permittivity of the pore water were measured. The following conclusions can be drawn:

- The complex permittivity of soil is sensitive to changes in lead concentration, the real part decreases and the imaginary part increases as lead concentration increases. The real part decreased from 23 to 17 when the concentration of the lead increased from 0 to 1800 ppm. The imaginary part increased from 16 to 24 as the lead concentration in soil increased from 0 to 1800 ppm.
- The chemical analysis showed that there is no lead in all pore water samples and lead was retained on soil.
- The results showed that the complex permittivity method can be used to detect lead in soil from a contaminated site where the form of lead and its concentrations are not known.

7. References

[1] Hutchinson T C and Meema K M, (1987), *Lead, Mercury, Cadmium and Arsenic in the Environment*, John Willy& Sons, Inc.

[2] Kabata-Pendias A, (2001), *Trace Elements in Soils and Plants*, (3rd Ed). CRC Press LLC.

[3] Mercier G, Duchesne J and Gibergues C A, (2002), *A Simple and Fast Screening Test to Detect Soils Polluted by Lead. Environmental Pollution*, V. 118, pp. 285-296.

[4] Kraszewski A, (1997), *Microwave Aquametry*, IEEE Press, New York.

[5] Al Martini S and Shang J, (2016), Complex Permittivity to Detect Lead in Soil with Various Salt Forms and Combinations, *1st International Symposium on Civil and Environmental Engineering (ISCEE 2016)*, Maleka, Malaysia.

[6] Al-Martini S and Shang Q J (2013), Complex permittivity of Lead in Contaminated Soils, *in the Proceeding of GeoMontreal 2013*, Montreal, Canada.

[7] Peplinski N R, Ulaby F T and Dobson M C, (1995), Dielectric Properties of Soils in the 0.3-1.3 GHz. Range, *IEEE Transactions on Geoscience and Remote Science*, V.33, No.3, pp.803-807.

[8] Umana J A, (1998), *Measurement of Dielectric Permittivity of Pavement Materials*, M.E.Sc. Thesis, Civil and Environmental Department, UWO.

[9] Scholte J W, (1999), *The Complex Permittivity of Compacted Halton Till*, M.E.Sc. Thesis, Civil and Environmental Engineering, the University of Western Ontario.

[10] Al Martini S and Shang Q J (2011), Complex Permittivity Measurement System for Detection of Soil Contamination, *Alhosn University Journal of Engineering and Applied Science* , Vol. 4, No. 1.

[11] Rowe R K, Shang J Q and Xei Y, (2002), Effect of Permeating Solutions on Complex Permittivity of Compacted Clay, *Canadian Geotechnical Journal*, Vol.39, pp1016-1025.

[12] Xei Y, (1999), *Complex Permittivity of Clayey Till Permeated by Aqueous Ionic Solution*, M.E.Sc. Thesis, UWO.

[13] Harter R D, (1979), Adsorption of Copper and Lead by Ap and B2 Horizons of Several Northeastern United States Soils, *Soil Science Society of America Journal*, Vol. 43, pp.679-683.

[14] Grim E R, (1953), *Clay Mineralogy*, McGraw- Hill Book Company, Inc.

[15] Olphen V H, (1963), *Clay Colloid Chemistry*, Interscience Publishers a division of John Wiley & Sons.

[16] Mclean E J and Bledsoe E B, (1992), *Behavior of Metal in Soil*, EPA/540/S-92.

Ecological Assessment of University Campus' Waterfront Landscape: *With Guangzhou Higher Education Mega Center as a Case*

Lin Zhu[1] and Shao Ping Guan[*]

1 Postgraduate Student, South China University of Technology, Guangzhou, China
*Corresponding author, Professor, South China University of Technology, Guangzhou, China
E-mail: 573079958@qq.com, 274392154@qq.com

Abstract. In recently years, the establishment of university campus has been on a rise, but the campus ecology is confronted with unprecedented challenges. Campus waterscape is one of the most important part of campus ecological system. The implementation of ecological design principle has been an important issue of planning programming. Besides, the current situation of campus ecology is worth being focused and studied. The paper has studied and analysed central waterscape of 5 universities in Guangzhou through field investigation. Based on the theory of water ecosystem service and Analytic hierarchy process (AHP), the paper aims to make a comprehensive comment on ecological environment of Guangzhou university town, whose problems are concluded. Corresponding policies are made to establish sustainable waterscape with good ecological environment, beautiful landscape and suitable activity venues, which will achieve the sustainable landscape effect for campus waterscape design.

1. Introduction

As an important natural resource, water is the material basis on which mankind depends, important factor influencing ecological balance and part of natural and urban environment. The university campus is like a small society in cities. Waterscape is a basis of forming good campus living environment and a substrate bearing outdoor activities and teaching. For university campus landscape, waterscape is both the reflection of campus ecology and an important channel for teachers and students to appreciate features and ecological culture of university waterscape [1]. There is a large waterscape near the school building and the author can see it when having a rest and looking far into the distance. There are few people staying there and the water is yellow where aquatic plants cannot grow. It is obviously that the ecological situation has been on an abnormal state. The reason for it is to be settled immediately through thinking, investigation and research.

2. Features and functions of campus waterfront landscape

2.1. An Introduction to Ecosystem of Campus Waterfront Landscape

The waterfront in campus is the junction between aquatic ecosystem and terrestrial ecosystem, including animals and plants, microbes, soils, air, rivers in the junction. The waterfront area is influenced by the two ecosystems. Its national conditions are good to the proliferation and development of biomes. Numerous biological species have the features of amphibious and the food chains are also complex. It is closely related to the overall ecosystem of cities.

2.2. Features of Ecosystem of Campus Waterfront Landscape
Ecological waterscape refers to a kind of waterscape type with sustainable ecological environment, good ecological image and abilities to achieve self-renewal and stability and harmonious existence between people and nature. After literature review, the author thinks that campus waterscape ecosystem can provide potential services. According to the Millennium Ecosystem Assessment Report of United Nations [2], the features mainly cover such four types as regulatory service (regulating ecological environment), supply service (supplying products for people), supportive service (supporting the overall ecosystem) and cultural service (providing spiritual feeling, entertainment, aesthetic experience or cultural education) [3].

2.3. Functions of Ecosystem of Campus Waterfront Landscape
As main open space in campus, campus waterfront space plays a role in ecology, leisure, landscape, culture, etc. The landscape can create micro climate, improve air quality, regulate humidity, beautify environment, increase diversity of species, protect and improve environment [4].

3. Investigation on overall evaluation of ecosystem of campus waterfront landscape

3.1. Research Objects
The paper has studied and analysed the central waterscape of Guangzhou university town through field investigation. Guangzhou Higher Education Mega Centre with an area of 34.4 square kilometres is located in Xin Zao Town, Pan Yu District, Guangzhou. It was constructed on 1st September, 2004, including such ten universities as Sun Yat-Sen University, South China University of Technology, South China Normal University, Guangzhou University, Guangdong University of Foreign Studies, Guangzhou University of Chinese Medicine, Guangdong College of Pharmacy, Guangdong University of Technology, Guangzhou Academy of Fine Arts, Xing Hai Conservatory of Music. As shown in "Figure 1". The waterscape in university is investigated and the degree of being concerned is studied. As shown in "Figure 2-Figure 4", the attention of South China University of Technology, South China Normal University, Guangzhou University of Chinese Medicine, Xing Hai Conservatory of Music, Guangzhou University of Technology is the highest. The paper does a further research on the five universities.

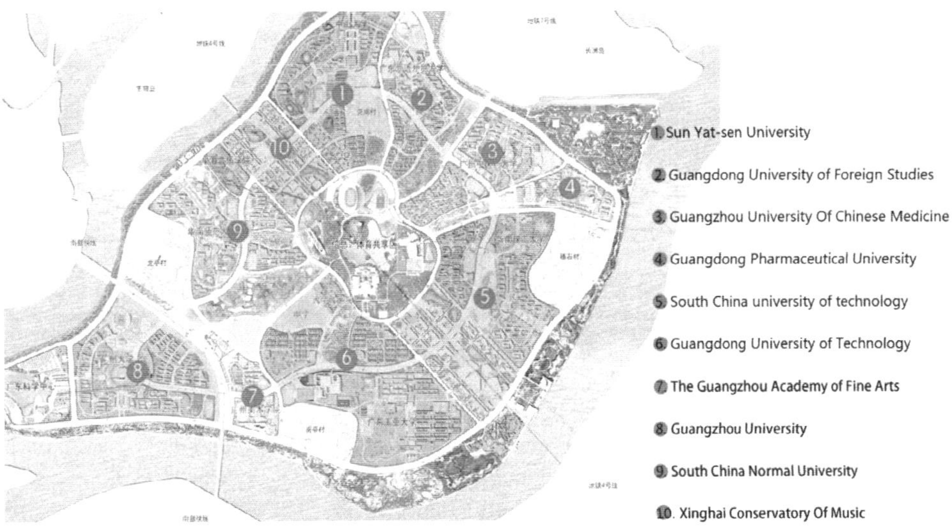

Figure 1. Guangzhou Higher Education Mega Centre

Figure 2. South China University of Technology & Guangzhou University of Chinese Medicine

Figure 3. Xing Hai Conservatory of Music & South China Normal University

Figure 4. Guangzhou University of Technology

3.2. Evaluation methods

3.2.1. Evaluation Factors System

Evaluation of ecosystem of campus waterfront landscape involves many indicators. A set of evaluation factors are defined according to data from the investigation, the theory of water ecology system, evaluation methods for international waterscape, systematic, objective and feasible principles and with Likert scale. Likert scale is the most common scale of summated rating scale, which was improved in 1932 by Likert (an American social psychologist) on the basis of original sum scale [5]. Primary factors showing the ecological situation of waterscape are selected [6], including regulatory service, supply service, supportive service and cultural service. Secondary factors include sewage interception, runoff control, self-purification of water, maintaining the diversity of waterfront plants, maintaining wildlife diversity, improving campus aesthetics and increasing opportunities of campus recreation. Tertiary factors include 22 indicators. It aims at constructing a comprehensive evaluation index system of ecological status of waterfront landscape.

3.2.2. Evaluation Factors Weight and Assignment

This template was designed for two affiliations. Evaluation factor weight is the key of overall evaluation and level analysis is an effective way to define weight coefficient. It is especially suitable to complicated problems that are difficult to analyse with quantitative indicators [7]. It makes the factors clear through dividing them into connected orderly levels. According to fuzzy judgement for reality, quantitative indicators are given on the basis of relative importance of such levels. Then mathematics methods are used to define weight coefficient of all elements' order of relative importance [7]. The steps of level analysis are as follow: A set of comparative values: The general comparison value is 1,3,5,7,9, As see from "Table.1". Standard group of numerical setting.

Table 1. Standard group of numerical setting

A pair of comparative value	Make up your mind
1	equal importance
3	weak importance
5	strong importance
7	very importance
9	absolute importance

Calculate the weight of the first grade evaluation factor, as see from "Table. 2".

Table 2. Weight of the first grade evaluation factor

a pair of comparative value	Supportive service	Regulatory service	Cultural service	Supply service	The geometric average	The weight
Supportive service	1	3	5	5	2.943	0.54570
Regulatory service	1/3	1	3	5	1.495	0.27721
Cultural service	1/5	1/3	1	1	0.508	0.09419
Supply service	1/5	1/5	1	1	0.447	0.08288
In Total					5.393	1

Finally, the two and three grade evaluation factors were compared. As see from "Table 3", the evaluation index set of waterfront landscape ecological condition was summarized. A comprehensive evaluation index system of ecological status of waterfront landscape.

Table 3. A comprehensive evaluation index system of ecological status of waterfront landscape

Primary factors	Secondary factors	Tertiary factors
Regulatory service (0.28)	Sewage interception (0.3)	Water volume (0.010)
		Water quality level (0.024)
		Sewage flowing automatically (0.030)
		rainwater collection (0.030)
	Runoff control (0.2)	Drainage method (0.036)
		Water permeability of soil (0.020)
	self-purification of water (0.5)	Water bottom material (0.040)
		Aquatic plant coverage (0.060)
		Waterscape plant coverage (0.040)
Supply service (0.08)	Fresh water supplementation (1.0)	Water cycle (0.080)

Primary factors	Secondary factors	Tertiary factors
Supportive service (0.55)	Maintaining the diversity of waterfront plant (0.55)	Width of waterfront planting belt (0.090)
		Reasonable structure of revetment pore (0.070)
		Arboreal coverage (0.080)
		Species of aquatic plants (0.063)
	Maintaining the diversity of wildlife (0.45)	Invasion of alien species (0.090)
		Growing of aquatic plants (0.078)
		Natural wetland rate (0.080)
Cultural service (0.09)	Improving campus aesthetics (0.4)	Coast shape (0.020)
		Water aeration (0.018)
	Increasing opportunities of campus recreation (0.6)	Usage (0.020)
		Population threat (0.014)
		Hydrophilic (0.020)

3.2.3. Evaluation Score Calculation and Classification

Cumulative summation formula is used to calculate comprehensive score of each campus waterscape investigated. Natural breakpoint classification of ArcGIS is used to classify the five universities according to the waterscape ecological situation into good, middle and bad situation.

The result shows that one is "good" with a proportion of 20% of all the investigated landscapes and 21.8% of the total area. Three are "middle" accounting for 60% of all investigated universities and 63.12% of the investigated area. One is "bad" accounting for 20% of the investigated waterscapes and 15.06% of the investigated area, as see from "Table.4".

Table 4. Ecological status evaluation level

Number	Key investigation school name	Composite scores	Ecological status evaluation level
1	South China University of Technology	4.806	Bad
2	Guangzhou University of Chinese Medicine	5.303	Middle
3	Xinghai Conservatory of Music	5.601	Middle
4	South China Normal University	5.419	Middle
5	Guangzhou University of Technology	6.195	Good

4. Overall Evaluation

The investigation and analysis has shown that the ecological situation of waterscape in Guangzhou university town is serious and few universities are "good" in this aspect. An obvious problem is that usage rate of large area is low, so people's activities has light influence on it. The "middle" situation accounts for the largest portion, which constitutes the main body of waterscape in Guangzhou university town. The development and utilization for it is unreasonable, changing and even degenerating some aquatic features and posing threat on sustainable development of water. The "bad" situation accounts for the least portion, but some waterscapes have lost basic ecological functions. Main problems can be listed as follow:

4.1. Problems in regulatory service
The waterscape systems in South China University of Technology and Xing Hai Conservatory of Music are the type of container with weak water circulation, low self-purification capacity, degenerating capacity of regulating runoff. There is no connected campus aquatic ecosystem. The two universities use a lot of ground floor, causing large-area impermeable and low soil immersion rate. Waterscape revetment focuses on hard retaining wall, pursuing one-side formality and causing running water to become dead water, which has exerted negative influence on biological migration.

4.2. Problems in supply service
The water cycle management systems in the five universities are not efficient and have a long water cycle period. The fresh water resource (rainwater, etc.) is not effectively collected and used.

4.3. Problems in supportive service
The biological diversity in the five areas has decreased with few waterfront plants, imperfect species, which is related to the treatment of revetment. Waterfront plants don't connect water with hard landscape (pavement and path) to form a transition. The arboreal coverage is low and few aquatic plants are used in campus waterscape.

4.4. Problems in cultural service
The usage rates of the five waterscapes are not high and few hydrophilic platforms are established in the five universities. Though aquatic area is large enough, there exist many dead angles. Accessorial building and incomplete leisure facilities bring inconvenience to the using of waterscape.

5. Conclusions and Recommendations
According to previous investigation and documentation, the paper comes up with an idea that as to waterscape in Guangdong university town, perfect ecological protection and management system should be established and campus ecological protection and management capabilities should be improved. In waterscape ecology, water quality should be improved and multi-level filter strip should be widened and constructed to raise the water self-purification rate. At the same time, reuse of water circulation system can be established. Waterfront landscape designed to protect biological diversity should be built; vegetation types and landscape configuration should be enriched; landscape wetlands of internal ecological water treatment should be built; space should be opened to enrich the spatial morphology of vegetation communities and stratification of landscapes. Besides, opportunities of recreation can be increased by covering such topics as popular science education, public participation, leisure activities through three levels-watching and staying, entering and participating, exploring and interacting [8].

In a word, water is the source of life. People have an innate yearning for water and the environment of people is closely related to water environment. Waterscape in campus is an integrated natural system and landscape, providing a continuous leisure space and environmental cognitive space for campus life. The paper aims at emphasizing the idea of protecting ecology in terms of waterscape design. If we do good to water, campus waterscape will bring more pleasant appreciation and huger public profit.

6. References
[1] Mei W. 2006 *J.* Based on the concept of ecological planning of the campus planning. *Journal of Shan xi building* **2** 1-2 of 2
[2] Mea M 2005 *Ecosystems and human well-being: Wetlands and Water Synthesis* (Washington DC: World resources institute)
[3] Jieqiong W and Qi Z. 2015 *J GIS-based eco-services evaluation model for evaluating the physical form of urban green infrastructure. Landscape architecture* **7** 109-17

[4] Lingyun X and yuncai W 2015 *J. From the landscape urbanism to ecological urbanism. Journal of Chinese urban forestry.* **13**(6) 23-26
[5] Baidu encyclopaedia 2007 *https://baike.baidu.com/*
[6] Binyi L and Jieqiong W 2007 *J. Eco-analysis-based regional landscape planning--research and application of analysis model of dominant eco-factors and modified eco-factors. Landscape architecture* **2** 82-7
[7] JunHua Z M 2005 *Planning and design of the survey analysis and practical* (Beijing: China building industry press) p169-186
[8] Echols S and Pennypacker E 2008 *J. From storm water management to artful rainwater design. Landscape Journal* **27**(2) 268 – 290

Acknowledgement
The work is upported by "the Fundamental Research Funds for the Central Universities" (project number: XMS17).

Integrating the concept of ecosystem services and values in Land use planning

M H Mohammadi Ashnani[1*], A Danehkar[1], M Makhdoum[1] and V Majed[2]

1 Faculty of Natural Resources, University of Tehran, Iran
2 Faculty of Economics, University of Tehran, Iran
*E-mail: ashnani@ut.ac.ir

Abstract. Many high-profile efforts conducted recently have proposed integrating ecosystem service values in the decisions made about land usage; however, there are not many evidences of using such approach in practice. The objective of the present study is to give an overview of the obstacles in the way of using ecosystem service assessment and valuation in managing the environment. In addition, a few solutions to achieve a comprehensive and practical framework are proposed. When it comes to land planning, the results should be comprehensive to facilitate comparison with other alternatives. One approach to this end is to aggregate the results to a manageable set through weighting methods. Here, the way of utilizing weighting method in making decisions about land usage is examined and potential areas of development are proposed. Moreover, the differences in the available patterns of use are examined, the cause and consequences of the differences are discussed, and necessity of the found differences is investigated.

1. Introduction

Adding environmental aspects to planning, policymaking, and programming activities had drawn an increasing attention from decision-makers in the both public and private sectors [1].

To realize this goal several analysts and practitioners had demanded an improved and more systematic use of analytical tools to comprehend and evaluate economic, technological, and environmental systems in an attempt to add pertinent and structured knowledge to decision-making processes [2], [3]. In fact, there are some analyzing tools for land use/landscape planning for different systems available that are widely used.

The farmers have been challenged by citizens and firms to add more functions along with food production to land/landscape use. This means that the landscape needs to carry out several functions simultaneously.

The objective is to achieve landscape performance in the context of sustainable development; that is, utilizing the natural resources based on a comprehensive consideration of all ecological, social, and economic functions while preserving the potentials for providing goods and services to future generations [4]-[6]. In the case that these concerns are realized for landscape/land development purposes, the decisions about modifying landscapes need to be made after taking into account the economic, social, and ecological functions and values. Under sustainable landscape development, man changes the landscape only to improve its performance and achieve extra value. Thereby, science needs to link the physical structure and function of the landscape to the economic, sociocultural, and ecological values needed by the users [7]. Since scientific knowledge influences landscape-development processes, the desired landscape values, at first, should be linked to the intended changes in the structure and functioning of the physical landscape [8].

The top challenge as to biomass-landscape is to make decision about the optimal allocation and management of a large number of land use options. Landscape functions and services represent a key concept in policy making given that decision-makers need to handle a clear demand for landscape services by a wide range of stakeholders [9]-[12].

Given the above introduction, the present study is an attempt to offer an overview of the challenges in the way of using ecosystem service assessment and valuation for environmental management and introduce a number solutions to achieve a comprehensive and practical framework.

2. Land use planning

A systematic assessment of land and water potential is called land-use planning, which examines different options for land use based on economic and social conditions to find the best possible land-use option. The objective is to choose and actualize the land use policies that will realize the needs of the public in the best possible manner, while the resources are preserved for the future. What powers the attempt for planning is the need for change, the need for better management, or the need for a completely different pattern of land use that is demanded by changes in the circumstances.

All types of rural land use are under consideration: agriculture, pastoralism, forestry, wildlife conservation, and tourism. In addition, planning guides us in the face of conflict between rural and use and urban or industrial expansion through determining which regions are the most suitable for rural use [13].

One probable mistake in the early stages of land-use planning is to focus only on land resources and neglecting the way they can be used. Decisions for land use are not only based on land suitability and the demand for products and criticality of using the land for a specific purpose is also taken into account. Information about suitability of the land, the demands for alternative products or uses and chances to fulfill the demands for the available land now and in the future need to be taken into account [13]. There is no two land use planning projects completely similar as the local circumstances and objectives can be varied notably and each project demands a different treatment. Still, we can introduce ten steps as a guide (Figure 1) [13].

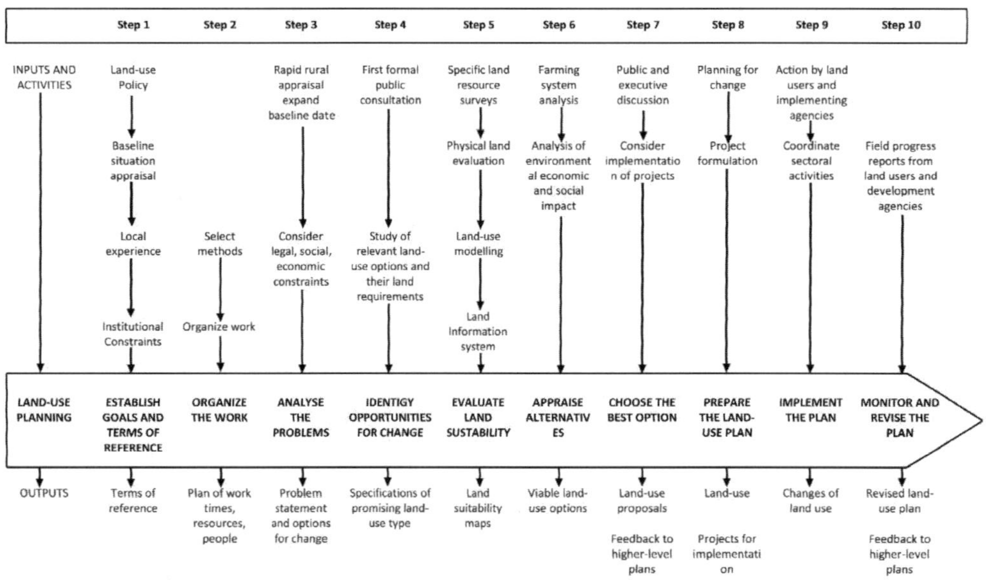

Figure 1. Steps in land-use planning

Each step needs a specific activity or a set of activities and the information generated in the past step is used in the next step.

By adopting a broader point view, the steps can be categorized into the following logical sequence [13]:

- Identifying the problems (Steps 1-3)
- Determining what are the alternative solutions (steps 4-6)
- Deciding about the best alternative and develop a plan (steps 7, 8)
- Implementing the plan and finding how it functions and learning from the experience (steps 9, 10)

3. Weighting methods

The environmental systems analysis tools discussed in this paper have their roots in different scientific disciplines. The proposed set of weighting methods adopted as the tools are different and regardless of their large overlapping, the categorizations adopted here are based on different logics and none cover all types of methods.

3.1. Monetary methods

3.1.1. Different types of economic values. Environmental economists distinguish different types of economic values [14]. The first recognized difference is between use and non-use value of environmental goods and services. Furthermore, use values are categorized into direct (e.g. the timber value of a forest) and indirect use values (e.g. the value of carbon fixation of a forest). A non-use value refers to the value that people attach to an amenity without taking into account if they use it or not [15].

3.1.2. Market prices. A marginal valuation of goods and services at the available level of supply and demand is provided by market prices. The market value is used for computing damage costs, loss of production, and loss of capital. Still, it is notable that the market values only cover the direct use values, and they are not a reliable measure of the total economic value [15].

3.1.3. Revealed willingness to pay. Relying on the information about people's expenditures on the related market goods, under the revealed preference method, the analyst tries to find the value of non-marketed goods [16]. The value of lost property and the value of production losses both are taken into account in the market-related damage. The willingness to pay (WTP) for environmental quality, in travel cost study, is determined based on differences in expenditures for visiting recreational sites with different quality. The value of commodity such as landed property, under hedonic pricing method, is determined as a set of characteristics and one or more of them can be environmental [16]. The key problems in these methods are limited applications and difficulties to link the market data to the environmental amenities.

3.1.4. Stated willingness to pay. Estimates of the value a person perceives for environmental quality, under stated preference methods, are explained by surveying where they are asked to express their preferences. Among the available methods are contingent valuation and choosing the modeling. Given that the approach covers both use and non-use values, it is considered as the most comprehensive one. The mentioned preference method might be influenced by a variety of biases such as link to the survey design, the interview situation, information given to respondents, and strategic behavior [17].

3.1.5. Imputed willingness to pay. Imputed WTP covers many cost methods including damage cost avoided, replacement cost, and substitute cost method [18]. These methods assume that if people incur costs to avoid damages caused by lost ecosystem services or to replace the services of ecosystems, then, such service needs to be worth at least what is paid to replace them. However, there are problems in using cost approaches as costs are not linked to the extent of the damage and they are not pertinent to the felt severity of the problem. One may say that they are the lower limit to the benefits when only economic issues are under consideration. They are the best options when avoidance measures are determined; that is, there is a clear will to pay the costs [19].

3.1.6. Political willingness to pay. Finnveden et al. (2002) argued that societal value might be different from the sum of individual values and given that, it may sound reasonable to deduce values from the behavior of government [20]. The costs for achieving established target can be construed as society's WTP with political process as the mediator. Determining a target for pollution level or the protection of an ecosystem demands an implicit valuation as the decision highlights the decision-makers' willingness to pay the value for reaching the target. In a similar manner, taxes are the tools to internalize the externalities created by consumption or production. Payable taxes of emission can be explained as a political valuation of the created externalities.

3.1.7. Avoidance costs. Valuation based on avoidance costs refers to the estimation of the cost of imposing limitation on, for instance, emissions [21]. The difference between this method and costs-to-reach target estimates lies with the fact that the adopted limits are not decided by the actors but rather by analysts so that it represent no one's WTP and indicates a hypothetical situation. This can be used in greened economy GDP computations [21].

3.2. Non-monetary methods

3.2.1. Proxy methods. The easiest way to weigh different types of environmental impacts and compare them are proxy methods. These methods rely on countable quantitative measures that are assumed to represent the total environmental impacts [22].

3.2.2. Distance-to-target methods. Different categories of environmental impact based on the distance between current level of environmental pollution and future environmental target value are measured by these methods [23]. There will be no necessity for equally important targets when the targets are determined. Thereby, a sort of weighting between the impacts should be performed. Still, in many cases, the targets are considered to be of the same weight so that there is no inter-effect weighting [20].

3.2.3. Revealed willingness to pay. Discussing preferences in a panel to make judgement can be carried out in different ways. Experts, stakeholders, or lay people can participant in the panel and the process of discussing the issue can be classified into different ways[24]. Multi-criteria analysis is a group of methods designed to deal with the problems featured with several goals [15].

4. Integrating economic aspects in land use planning

Adopting a proper usage for an area of land is one of the most challenging part of land programing.
At the decision making point, the planners and decision-makers need to interact with each other.
At the point of decision, the roles of the planner and the decision-maker must interact.
Priority determining methods are either qualitative or quantitative and regardless of that, the proposed method can be categorized as follows[25]:

a. Comparative qualitative or quantitative methods.
b. Quality or quantitative methods to determine capacity of the area of land.
c. Quantitative methods of mathematical programming.

Through the simplest planning situation –i.e. new land settlement- it is possible to allocate land units to specific usages. Afterward, the settlers are brought in and they, at least at the beginning, need to practice the planned usages. In most of the cases however, the land is already occupied by settlers and used for cultivating, grazing, etc. purposes. Is such case, the objective of planning is to solve the current problems in the land-use systems; in addition, the land use cannot be allocated easily. It is possible to offer new land-use types for specific areas, which is done through developing services and providing inputs and services. In some cases, the options can be set out in a goal achievement matrix and ranked based on the chosen criteria.
The idea proposed here is to use analysis of environmental economy results as a weight in the seventh stage of land planning. Knowing that there is no consensus about these values in the society, it cannot be expected to have a generally accepted single valuation or weighting methodology. Still, it is

imperative to utilize multiple methods in parallel to show how the different values might influence the results of study.

5. Results and discussion

As shown by the review, there was a wide range of choices in land use and weighting methods for different environmental systems analysis tools. On the other hand, the way that the methods are used is different. It appears that these differences have to do with traditions and views about environmental economic valuation and weighting rather than applications. Nowadays, the lack of consistent environmental economic weighting/valuation sets is quite clear. To improve the quality and usefulness of these sets, more studies and development are needed. The results showed that the following points can be of interest for the future studies:

a. To compares land usages based on different methods of usage planning and comparing the results;
b. To use land planning methods featured with economic evaluation and comparing them with physical methods of land planning along with weighting based on environment economy;
c. To develop a new consistent environment economic valuation weighting/valuation sets; that is, using the same valuation method for all types of land uses;
d. To evaluate the employed methods and the resulted sets as to scientific quality, consistency, and data gaps and to test them in case studies to ensure reasonableness of the results
e. To develop several sets of value obtained from a large number of methods, using taxes, avoidance cost, cost-to-reach targets.

6. References

[1] Lenschow A, *Environmental policy integration: Greening sectoral policies in Europe*. 2002: Routledge.
[2] Thissen W A and Twaalfhoven P G, *Towards a conceptual structure for evaluating policy analytic activities.* European Journal of Operational Research, 2001. **129**(3): p. 627-649.
[3] Nilsson M, et al., *The use and non-use of policy appraisal tools in public policy making: an analysis of three European countries and the European Union.* Policy Sciences, 2008. **41**(4): p. 335-355.
[4] Brundtland G H, *World commission on environment and development (1987): Our common future.* World Commission for Environment and Development, 1987.
[5] IUCN, *Rio declaration on environment and development*, 1992, WWF: Switzerland.
[6] Linehan J R and Gross M, *Back to the future, back to basics: the social ecology of landscapes and the future of landscape planning.* Landscape and Urban Planning, 1998. **42**(2): p. 207-223.
[7] Haines-Young R, *Sustainable development and sustainable landscapes: Defining a new paradigm for landscape ecology.* Fennia, 2000. **178**(1): p. 7-14.
[8] Termorshuizen J W and Opdam P, *Landscape services as a bridge between landscape ecology and sustainable development.* Landscape ecology, 2009. **24**(8): p. 1037-1052.
[9] FAO, *Cultivation our futures*, in *FAO/Netherlands Conference on the Multifunctional Character of Agriculture and Land.* 1999, FAO/LNV,: The Netherlands.
[10] OECD, *Multifunctionality: towards an analytical framework.* 2001, Paris,: Organization for Economic.
[11] Bills N and Gross D, *Sustaining multifunctional agricultural landscapes: comparing stakeholder perspectives in New York (US) and England (UK).* Land Use Policy, 2005. **22**(4): p. 313-321.
[12] De Groot R S, et al., *Challenges in integrating the concept of ecosystem services and values in landscape planning, management and decision making.* Ecological complexity, 2010. **7**(3): p. 260-272.
[13] Agriculture Organization of the United Nations. Interdepartmental Working Group on Land Use Planning, *Guidelines for land-use planning.* Vol. 1. 1993: Food & Agriculture Org.
[14] Turner R K, Pearce D, and Bateman I, *Environmental economics: an elementary introduction.* 1994: Harvester Wheatsheaf.

[15] Ahlroth S, et al., *Weighting and valuation in selected environmental systems analysis tools–suggestions for further developments.* Journal of Cleaner Production, 2011. **19**(2): p. 145-156.
[16] Hanley N S and JF White B, *Environmental economics in theory and practice*. 2002: Palgrave macmillan.
[17] Carson R T, Flores N E, and Meade N F, *Contingent valuation: controversies and evidence.* Environmental and resource economics, 2001. **19**(2): p. 173-210.
[18] Mishra S, *Valuation of environmental goods and services.* , in *In: Singh, O.P. (Ed.), Environment and Natural Resources: Ecological and Economic Perspectives.* 2010, Regency Publications: New Delhi. p. pp. 34e54.
[19] Kopp R J, Krupnick A J, and Toman M A, *Cost-benefit analysis and regulatory reform: An assessment of the science and the art.* 1997.
[20] Finnveden G, et al., *Normalization, grouping and weighting in life cycle impact assessment.* Life cycle impact assessment: striving towards best practice. Society of Environmental Toxicology and Chemistry (SETAC), Pensacola, 2002.
[21] UN, *Integrated environmental and economic accounting*. 2003, European Commission, IMF, OECD and World Bank: New York, United Nations. p. 100-118.
[22] Huijbregts M A, et al., *Is cumulative fossil energy demand a useful indicator for the environmental performance of products?*, 2006, ACS Publications.
[23] Weiss M, et al., *Applying distance-to-target weighing methodology to evaluate the environmental performance of bio-based energy, fuels, and materials.* Resources, Conservation and Recycling, 2007. **50**(3): p. 260-281.
[24] Seppälä J, *Decision analysis as a tool for life cycle impact assessment*. 1999: Eco-Informa Press Bayreuth.
[25] Cocks K and Ive J, *Rural land use planning techniqes*. In Micro computer for lacal goverment planning and management. 1986, Sydney: Hargreen Pub. 112-122.

Investigating the effects of sea-level rise on morphodynamics in the western Giens tombolo, France

M T Vu[1,2], Y Lacroix[1,3] and V T Nguyen[2]

1 University of Toulon, Avenue G.Pompidou, La Valette du Var, 83162, France
2 University of Transport and Communications, Cau Giay, Hanoi, 100000, Vietnam
3 MEMOCS, Università Degli Studi dell'Aquila, Italy
E-mail: mvu@univ-tln.fr

Abstract. Rising sea level along with the occurrence of greater and more frequent storms would cause not only coastal flooding, but also beach erosion and shoreline retreat problems. The Almanarre beach along the western Giens tombolo is socio-economically and heavily vulnerable to accelerated sea level rise due to its high touristic value and low-lying topography. Therefore, it is necessary to quantify the impacts of sea level rise (SLR) on the morphodynamics in this area, e.g. to evaluate the relationship between the beach erosion and SLR. Coupled hydrodynamic and sediment transport numerical models are used to investigate the changes in current, wave and sediment dynamics when sea level rises. A total of 16 scenarios with and without SLR are simulated. The effectiveness of the coupled model is assessed by comparison of simulated values with available field measurements. The results presented in this work should be useful in the investigation of other coastal regions.

1. Introduction

The western branch of Giens double tombolo is located in Provence, South East of France (Figure 1). It has been the object of both accretion and erosion processes due to natural causes such as waves, winds and storms, or due to human interference. Courtaud [1] studied the shoreline evolution of Giens tombolo with the aid of aerial photographs and some field surveys, using imagery digital processing techniques. During the period from 1950 to 1998, the northern part of Almanarre beach, located between Landmark B01 to B23 (Figure 2), suffered an average retreat of -6.4 m/linear meter corresponding to an erosion area of -13,900 m^2, whereas the southern part between Landmark B23 and B46 was accreted with an average advance of +19.8 m/linear meter corresponding to an accretion area of +41,100 m^2. With the same technique, Than [2] estimated that the average annual rate of shoreline erosion is (-0.01 to -0.63) ± (0.27 to 1.82) m/year in the northern part and the shoreline accretion is (0.02 to 2.01) ± (0.14 to 5.1) m/year in the central and southern parts over the period from 1920 to 2012. By using numerical simulation, [3] indicated that the disappearance of Posidonia in Giens Gulf would cause a significant increase of current speed, significant wave height and sediment transport rate. In the other hand, by 2100, the global SLR due to climate change, forecasted to be around 44 cm, may have serious impacts on coastal regions [4]. According to Brunel [5], all the beaches of Provence underwent an average shoreline retreat of 5.8 (± 3.5) m due to a relative SLR of +11 cm between 1896 and 1998. If an increase in sea-level of about +44 cm is predicted in the end of this century, Brunel [5] also estimated an average retreat of -20 (±2) m, with values varying between -12 (±2) and -41 (±2) m. It suggests that SLR will exacerbate coastal the erosion problem in the study area. In this paper, the numerical models are used to present a quantitative analysis of the effects of SLR on hydrodynamics and sediment transport along Almanarre beach.

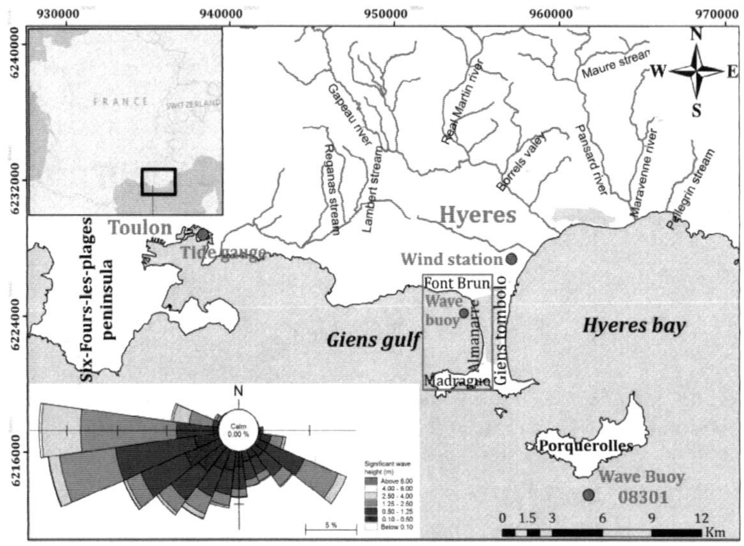

Figure 1. The study area and annual swell rose.

2. Data and Methods

2.1. Study area

Almanarre beach runs through the studied area with an overall direction of north-south and extends about 4.5 km along the Salt Road. Unlike a normal straight coast, the bathymetry in this area is highly complex due to the presence of submerged shoals and cross-shore deep troughs (Figure 2). The wind data of Hyères station (Figure 1) reveals that the west and southwest winds have strong influence on the wave agitation and the coastal morphology in Giens Gulf. These winds maintain the frequency of 25.66% of the observed time with the maximum speed of 21.47 m/s in the west direction. On the other hand, wave data of the buoy 08301 located approximately 1.8 km south of Porquerolles Island (Figure 1) reveals that the studied area is exposed to dominant west and southwest waves with respective frequencies of 36.92% and 28.84%. Waves are the main hydrodynamic process along the coast of Giens tombolo, as the maximum tidal variation is less than 0.3 m [6]. The seabed of the Gulf of Giens is covered by Posidonia seagrass - with an area of about 1800 ha [7], the most important endemic seagrass species in the Mediterranean Sea. It has a strong impact on the local morphodynamics, viz. dissipation of wave energy, reduction in current speed, promoting the accumulation of suspended sediment, and the reinforcement of the sandy soil [8]. The first mapping of the Posidonia was conducted in 1993 by the G.I.S. Posidonie. The results showed that the upper limit of Posidonia is on average around from -2 to -3 m depth and its lower limit from -30 to -35 m (Figure 3).

2.2. Numerical modeling

2.2.1. Model descriptions

In the present study, the MIKE21 coupled modules of Hydrodynamic (HD), Spectral Wave (SW), and Sand Transport (ST) developed by DHI [9] are applied to model hydrodynamics and sediment transport along the western Giens tombolo under the impact of sea level rise. Particularly, the HD module is used to compute surface elevation and currents which are induced by waves, whilst the SW module is used for transformation of wind-generated waves and swell from offshore and description of the detailed near-shore wave field. The ST module is used to estimate the sediment transport field and the morphological evolution [3]. The computational domain, about 9 km^2, has only one western open-sea boundary and a shoreline-land boundary enclosing remainders (Figure 2). The coupled model system is solved on a flexible 2D grid including 21,048 triangular elements. Accordingly, a coarser mesh with the largest grid cell area of 5000 m^2 is used for the offshore area and very fine mesh with the smallest one of about 40 m^2 in nearshore area along Almanarre beach (Figure 2).

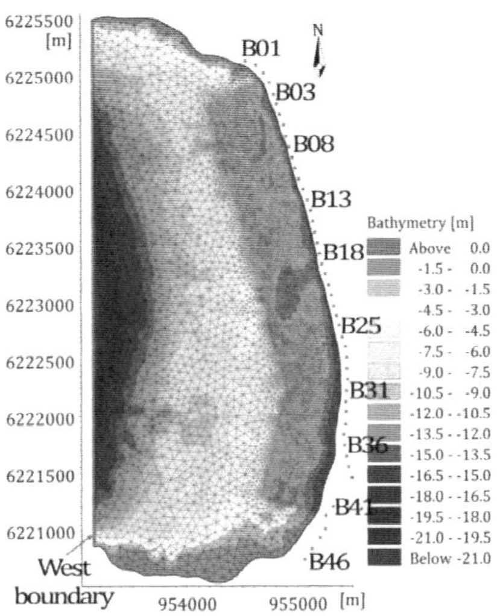

Figure 2. The computational mesh and bathymetry of the study area.

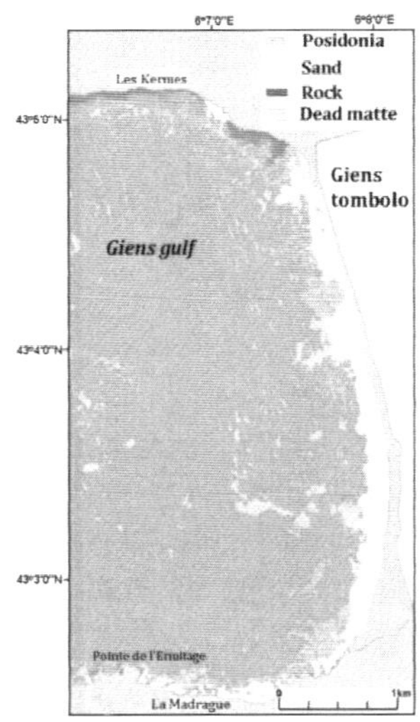

Figure 3. Distribution of Posidonia in Giens gulf. *(Source: Cartham-AAMP/Andromede, 2012).*

2.2.2. *Model calibration*

The model calibration was done by adjusting some main input parameters, viz. Manning's number, M, Nikuradse's roughness height, k_s, and the median grain size, D_{50} [3]. The simulated wave and current results are compared with the data measured at Almanarre beach from 31st October to 12th November 2000. The computed wave heights at the water depth of 3.5 m are reasonably close and show good agreement with measured values (Figure 4). The RMSE is about 0.12 m, corresponding to the difference SI of 22% (Table 1). Furthermore, the comparison of computed and measured current speed is displayed in Figure 5. The statistical scores for this comparison are R^2 = 0.64, RMSE = 0.02 m/s, and SI = 41%, respectively.

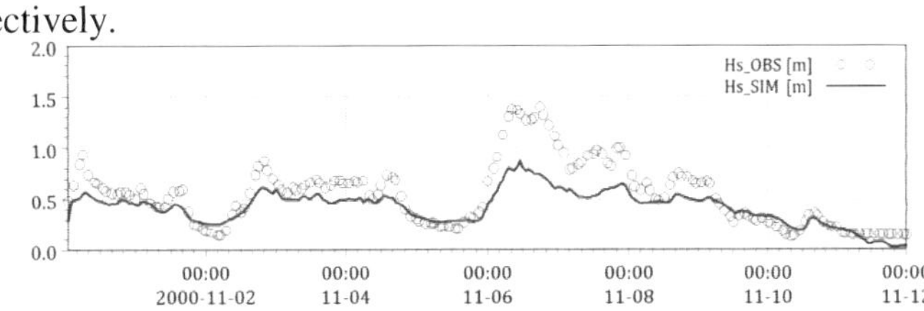

Figure 4. Comparison between the observed (OBS) and simulated (SIM) wave heights at Almanarre beach.

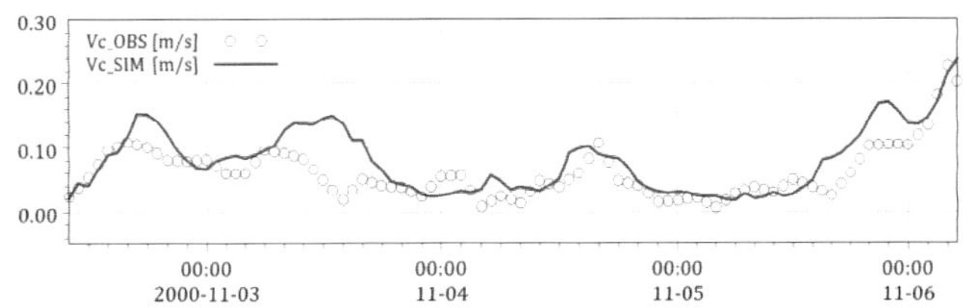

Figure 5. Comparison between the observed (OBS) and simulated (SIM) current speeds at Almanarre beach.

Table 1. Calibration of wave height and current speed for local scale (November, 2000).

Case	R^2	RMSE	SI
Hs_OBS vs Hs_SIM	0.86	0.12	0.22
Vc_OBS vs. Vc_SIM	0.64	0.02	0.41

The morphological evolution obtained from the ST model is compared with the bathymetry which was measured by E.O.L [10] in November 2008. Figure 6 shows the difference between the measured and simulated beach profile at Landmark B25 of Almanarre beach. The model setup has resulted in BSS of 0.39, which is on reasonable level according to Van-Rijn [11].

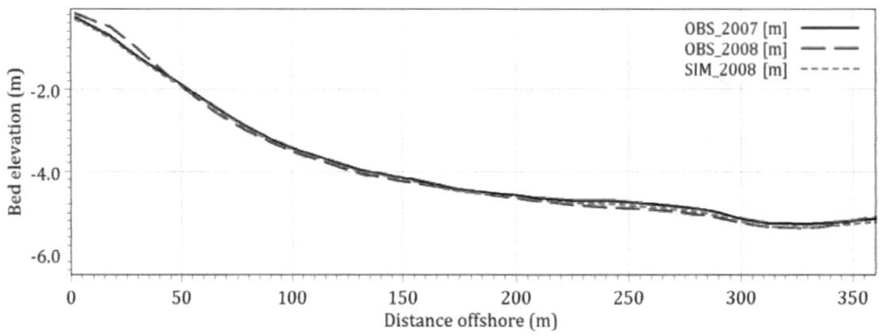

Figure 6. Comparison of observed (OBS) and simulated (SIM) profile at Landmark B25 of Almanarre beach.

2.3. Study scenarios
In order to clarify and quantify the impacts of SLR in the studied area, different wind directions and different levels of storms are modeled. The input data of the numerical simulations in SLR conditions are described in Table 2. For annual scenario, the sea level is about 0.5 m above CM which was determined by sum of average sea level over many years (+0.39 m) and the minimum rate of sea level variation at Marseille by 2040-2050 (0.11 m) [12]. On the other hand, sea levels in remainders are the same values as shown in present sea level conditions (No SLR) but adding 35 cm caused by the phenomenon of sea level rise between 2010 and 2060 [13]. Regarding the wave parameters, they were referred from the reports of ERAMM [14] and CEREMA [15].

Table 2. A summary of study scenarios for the western Giens tombolo.

No.	Scenarios	Sea level (m)		Wind		Offshore wave		
		No SLR	SLR	Speed (m/s)	Direction (°)	$H_{1/3}$ (m)	T_p (s)	MWD (°)
1	Annual	0.39	0.5	5.8	200 (S)	2.24	7.55	270
				7.5	230 (SW)			
				6.48	270 (W)			
				8.69	320 (NW)			
2	Decadal storm	0.95	1.3	12.55	230	6.56	9.12	225
3	Tri-decadal storm	1.00	1.35	19.83	230	7.1	10.3	225
4	Semi-centennial storm	1.15	1.5	29.59	230	7.34	11	225
5	Centennial storm	1.50	1.85	36.43	230	7.64	12	225

3. Results and Discussion

3.1. Effects of SLR on waves
The coastal morphology of the western Giens tombolo is influenced by four main wind directions, viz. northwest, west, southwest, and south. The change of wind direction accompanied with the SLR phenomenon can provoke the significant modification of wave field. It is clearly seen that the southwest winds cause the highest waves in the Almanarre beach, whereas the lowest waves are generated by the south winds (Figure 7), regardless of SLR. Particularly, the mean significant wave heights, H_s, found at the water depth of 3.5 m near Landmark B08 under the impacts of southwest and south winds are about 0.6 m and 0.57 m, respectively (Table 3). However, all the wave heights decrease gradually when they approach the shallow water areas and are dissipated by the bottom friction. In addition to generating the highest waves, the southwest winds also induce the largest radiation stresses i.e. S_{xx}, S_{yy} and S_{xy} (Table 3). When the sea level rises, deeper water areas enable larger waves to reach and break closer to the coastline, resulting in greater wave heights in nearshore zones. Indeed, the mean wave heights near Almanarre beach are increased by 1.22%-1.59% (Table 3). The southern winds induce the largest increase of wave height up to 1.59% at Almanarre beach, compared with the other wind directions. As sea level increases, the propagation of larger waves over

the shallow waters results in higher radiation stresses, boosted by 2.2%-5.1% at a water depth of 3.5 m in front of Almanarre beach when sea level rises about 11 cm (Table 3).

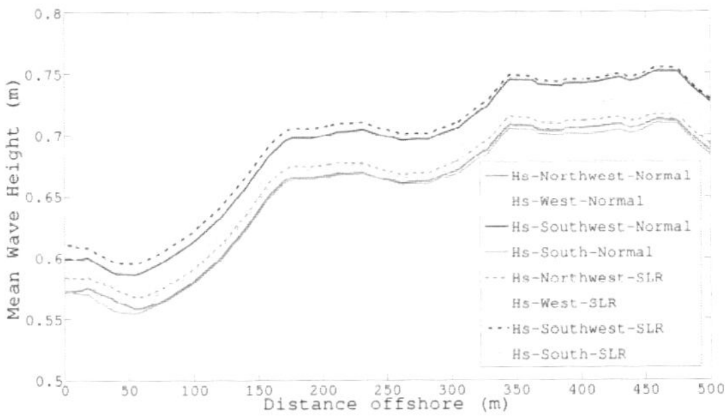

Figure 7. Modeled cross-shore variations in wave heights under the impact of SLR and variation of wind direction at Landmark B08.

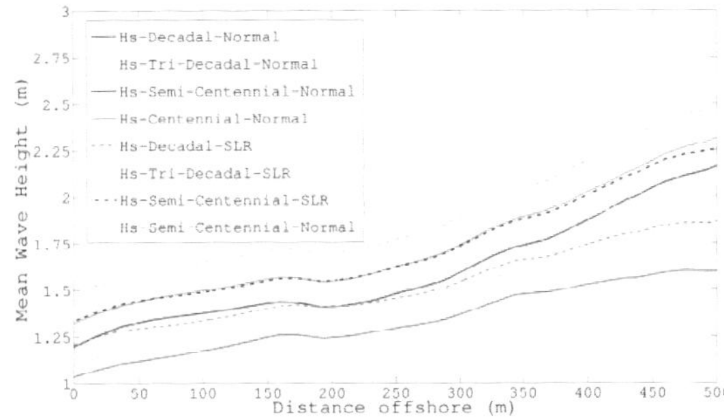

Figure 8. Modeled cross-shore variations in wave heights under the impact of SLR and variation of storm scale at Landmark B08.

Figure 8 depicts the simulated cross-shore variations in wave heights at Almanarre beach under the influence of different storms as well as SLR. It is noticeable that the higher level of storm causes the larger and stronger impacts on wave fields, regardless of SLR. Particularly, the mean significant wave height without SLR increases from 1.15 m in decadal storm to 1.46 m in centennial storm; while its value ranges from 1.31 m in decadal storm to 1.59 m in centennial storm (Table 3). Greater water depths due to storm surges with SLR reduce bottom friction and increase water depth relative to the wave height, resulting in larger and more energetic waves that could reach the beaches. As a result, the nearshore significant wave height is intensified by 8.91%-13.39% at the water depth of 3.5 m near Landmark B08 (Table 3). When SLR occurs along with storms, an increase in wave heights would generate a corresponding raise of radiation stresses. An increasing trend in wave heights is found in all the stormy scenarios, but the maximal increase occurs in the decadal storm and it reduces when the storm levels increase. The radiation stresses are boosted by 20.27%-32.88% at Almanarre beach, compared with those cases without SLR.

3.2. Effects of SLR on currents

In the Giens gulf, both the winds and waves induce the longshore currents [6]. Therefore, the variation of wind directions and the different wave climate were simulated in this work in order to investigate their effects on the current field. Figure 9 presents the modeled cross-shore variations in current speed (V_c) at Landmark B08 in Almanarre beach, caused by different wind directions and different sea levels. Even though the southwest winds induced the highest significant wave height in the studied area (Figure 7), the northwest winds have the strongest impacts on the current speed along the western Giens tombolo. This is valid for both no SLR case and SLR case. Specifically, the highest mean current speeds of 0.22 m/s and 0.23 m/s at the water depth of 3.5 m at Almanarre beach generated by the north-western winds are higher than other winds about 20%-36% in the nor SLR conditions and 14%-28% in the SLR conditions (Table 3). This paradox may come from the terrain of the studied area. In addition, the southern waves are blocked or limited by the presence of the rocky outcrop of Madrague (Figure 1). This paradox also confirms that the nearshore currents along Amanarre beach are mainly governed by the wind and the wave only plays a secondary role in the development of these currents. When sea level rises, the currents are stronger than those in the cases of no SLR conditions. For instance, the current speed at the water depth of 3.5 m near Landmark B08 is increased by 4.56%-16.1% (Table 3).

The impacts of storms along with SLR on the modelled cross-shore variation in current speed at Almanarre beach is illustrated in Figure 10. It is noted that the current speed is reduced from 150 m seaward of the shoreline to 350 m seaward of the shoreline before gradually increasing again to reach the coast. Similar to the annual conditions, this cross-shore evolution of current speed would be attributed to the seabed change. Without SLR scenarios, the higher level of storm usually causes the larger current speed at Almanarre beach, except for the centennial storm case. Specifically, the current speed of 0.2 m/s at the water depth of 3.5 m near Landmark B08 in the decadal storm condition is increased up to 0.42 m/s in the semi-centennial storm condition and significantly decreased to 0.39 m/s in the centennial case (Table 3). This sudden reduction would be explained by the fact that the sharp increase of 35 cm in the centennial storm could cause the local water volume expansion and the extensive submergence of low-lying areas. As sea-level rises, most of current speeds influenced by the storms are increased by 33.3%-86.7% at Almanarre beach. Apart from the cases of the semi-centennial storm, the current speed at the water depth of 3.5 m near Landmark B08 is reduced to 3.4% (Table 3).

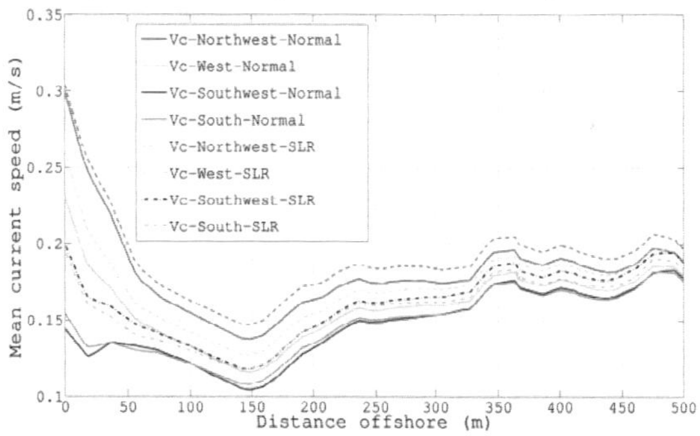

Figure 9. Modeled cross-shore variations in current speed under the impact of SLR and variation of wind direction at Landmark B08.

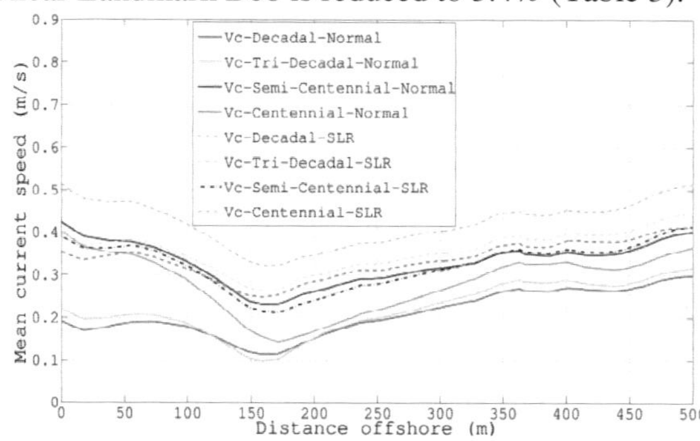

Figure 10. Modeled cross-shore variations in current speed under the impact of SLR and variation of storm scale at Landmark B08.

3.3. Effects of sea-level on sediment dynamics

The sediment transport in the study area is carried out by the current of longshore drift induced by the oblique waves striking the coast from the north to the south and from the west to the east [1]. Similar to the study of hydrodynamics, the effect of SLR on sediment transport rates (Q) was investigated by elevating the mean sea level (+0.5, +1.3, 1.35, 1.5, and 1.85 m). Figure 11 plots the modelled cross-shore variation in the sediment transport rates under the impacts of various winds along with SLR. It reveals that the higher wave-induced radiation stresses and current speeds (Figure 9) close to shore resulted in a narrow band of sediment transport concentrations along the shoreline. It is clearly seen that the northwest winds always have the strongest impact on the sediment transport rates, regardless of SLR. They induced the highest sediment transport rates of $4.5x10^{-5}$ $m^3/s/m$ at the cross-shore of Landmark B08 in Almanarre beach. By contrast, the lowest sediment transport rate of about $2.13x10^{-5}$ $m^3/s/m$ is generated in the south winds. It strongly confirms that the pattern of the sediment transport rates mostly depends on the current speed distribution (Figure 9). When taking into account the SLR, the sediment transport rates are intensified along Almanarre beach, regardless of wind direction. Particularly, at the water depth of 3.5 m near Landmark B08, the total load is increased by about 3%-23% probably due to the raise in nearshore current speed when taking into account the SLR phenomenon (Table 3). The maximal increase of 22.9% is observed in the southwest winds probably due to the highest waves along with sea level rise, while the minima increase of 3% is caused by the northwest wind.

Figure 12 shows the cross-shore variation in sediment transport rate under the impacts of storms with and without sea level rise. It is noticeable that the cross-shore distribution of sediment transport rates starts reducing from 350 m seaward of the shoreline to 150 m seaward of the shoreline at Almanarre beach, corresponding to the decrease of current speed (Figure 10), regardless of SLR. In addition, it verifies that the high storm level induces the largest total load of sediment. The maximum sediment transport rate of over $1.7x10^{-3}$ $m^3/s/m$ is reported in the centennial storm scenario. The current speed plays a decisive role in the sediment transport along the western Giens tombolo. Indeed, when sea level rises, the increase of current speed results in the boost in the total load at Almanarre beach. The total load is added by 13%-282% if SLR is taken into account (Table 3). The maximal increases are of 244.4% and 281.8% in the decadal and tri-decadal storm conditions, respectively. Whilst the minimal rise of 12.9% occurs under the impact of semi-centennial storm with SLR.

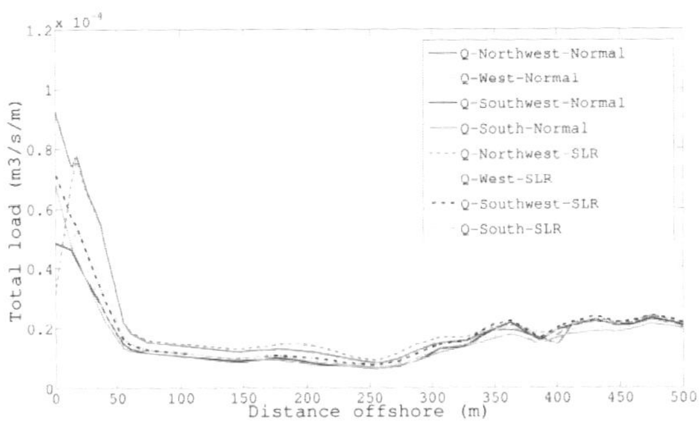

Figure 11. Modeled cross-shore variations in sediment transport rates under the impact of SLR and variation of wind direction at Landmark B08.

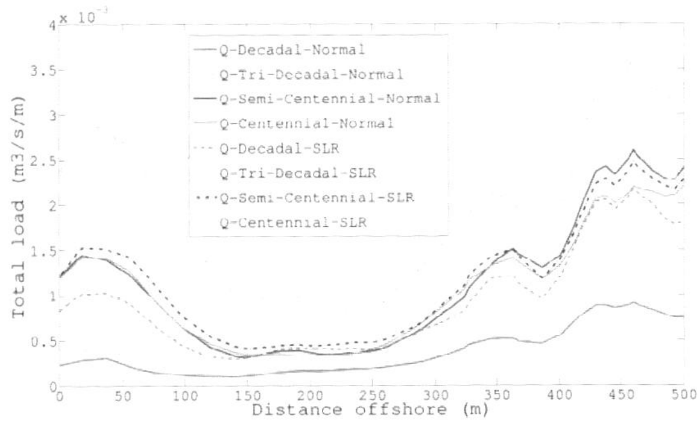

Figure 12. Modeled cross-shore variations in sediment transport rates under the impact of SLR and variation of storm scale at Landmark B08.

Table 3. Effect of sea level rise on hydrodynamic parameters and sediment transport in Almanarre beach.

Scenario		H_s (m)	S_{xx} (m^3/s^2)	S_{xy} (m^3/s^2)	S_{yy} (m^3/s^2)	V_c (m/s)	Q ($m^3/s/m$)
No SLR	Northwest	0.582	0.182	0.048	0.127	0.217	4.49E-05
	West	0.593	0.183	0.049	0.129	0.174	3.28E-05
	Southwest	0.603	0.184	0.050	0.131	0.141	2.40E-05
	South	0.570	0.165	0.045	0.118	0.139	2.13E-05
	Decadal	1.152	0.667	0.167	0.416	0.200	3.30E-04
	Tri-Decadal	1.246	0.789	0.189	0.473	0.222	4.34E-04
	Semi-Centennial	1.337	0.923	0.211	0.512	0.415	1.60E-03
	Centennial	1.457	1.107	0.251	0.613	0.389	1.72E-03
SLR	Northwest	0.591	0.188	0.050	0.130	0.227	4.62E-05
	West	0.600	0.187	0.051	0.132	0.187	3.47E-05
	Southwest	0.612	0.191	0.053	0.136	0.163	2.95E-05
	South	0.579	0.171	0.047	0.122	0.156	2.44E-05
	Decadal	1.307	0.881	0.222	0.540	0.373	1.14E-03
	Tri-Decadal	1.376	0.995	0.247	0.597	0.411	1.66E-03
	Semi-Centennial	1.456	1.110	0.270	0.651	0.401	1.80E-03
	Centennial	1.591	1.338	0.327	0.772	0.518	3.18E-03
Difference (%)	Northwest	1.57	2.88	3.62	2.95	4.56	2.96
	West	1.22	2.18	3.14	2.40	7.87	5.96
	Southwest	1.58	3.81	5.10	3.97	16.08	22.88
	South	1.59	3.35	4.42	3.45	11.73	14.25
	Decadal	13.39	31.93	32.88	29.80	86.68	244.36
	Tri-Decadal	10.38	26.07	30.34	26.05	85.60	281.80
	Semi-Centennial	8.91	20.27	27.89	27.31	-3.38	12.86
	Centennial	9.23	20.93	30.60	25.90	33.15	84.84

4. Conclusions

Numerical models have been developed successfully to quantify the impacts of SLR on hydrodynamic parameters and sediment transport in the Giens gulf. The results of this work strongly confirm that SLR leads to significantly increased wave heights, current speeds and sediment transport rates. The impact of SLR on morphodynamics in the stormy conditions is about ten times greater than that in the annual conditions. Particularly, the mean wave heights near Landmark B08 are increased by 1.22%-1.59% in annual conditions, whereas they are boosted by 8.91%-13.4% in the extreme events, comparing to those without SLR. In the stormy conditions, the nearshore current speed is intensified by 33.2%-86.7%, while it is only increased by 4.56%-16.1% in the no SLR scenarios. The raise in the current speed mainly leads to a corresponding increase of sediment transport rate by 2.96%-22.88% in the annual cases and by 12.86%-281.8% in the stormy cases. This would be explained by the fact that deeper water areas due to SLR along with storm surge allow larger waves to reach and break closer to the shoreline, resulting in more wave energy impacting the beach profile and taking sediment offshore.

5. References

[1] Courtaud J, *Dynamiques geomorphologiques et risques littoraux cas du tombolo de giens (Var, France méridionale)*. 2000, Université Aix-Marseille I. p. 263.

[2] Than V V, *Modélisation d'érosion côtière : application à la partie Ouest du tombolo de Giens*, in *LATP*. 2015, Aix Marseille Université: Marseille. p. 400.

[3] Vu M T, Lacroix Y and Nguyen V T, *Investigating the impacts of the regression of Posidonia oceanica on hydrodynamics and sediment transport in Giens Gulf.* Ocean Engineering, 2017. 146: p. 70-86.

[4] Paskoff R P, *Effects of sea-level rise on coastal cities and residential areas*, in *Climate Change, Human Systems and Policy*, A Yotova, Editor. 2009, Eolss Publishers Co. Ltd.,. p. 1-16.

[5] Brunel C and Sabatier F, *Potential influence of sea-level rise in controlling shoreline position on the French Mediterranean Coast.* Geomorphology, 2009. 107(1–2): p. 47-57.

[6] SOGREAH, *Etudes sédimentologiques de la rade d Hyères. Littoral de port Pothuau à la Badine*. 1988. p. 68 + annexes et cartes.

[7] Sinnassamy J M and Pergent-Martini C, *Localisation et état de l'herbier de Posidonies sur le littoral P.A.C.A: Var*. 1990: D.R.A.E. & G.I.S. Posidonie.

[8] Boudouresque C F, Bernard G, Bonhomme P, Charbonnel E, Diviacco G, Meinesz A, Pergent G, Pergent-Martini C, Ruitton S and Tunesi L, *Protection and conservation of Posidonia oceanica meadows*. 2012: RAMOGE and RAC/SPA. 202.

[9] DHI, *MIKE 21/3 Coupled Model FM*. 2014.

[10] E.O.L, *Suivi de l évolution des plages de la commune Hyères-les-palmiers*. 2010: Commune de Heres-Les-Palmiers.

[11] Van-Rijn L C, Walstra D J R, Grasmeijer B, Sutherland J, Pan S, and Sierra J P, *The predictability of cross-shore bed evolution of sandy beaches at the time scale of storms and seasons using process-based profile models.* Coastal Engineering, 2003. 47(3): p. 295-327.

[12] Gomis D, Tsimplis M, Marcos M, Fenoglio-Marc L, Pérez B, Raicich F, Vilibić I, Wöppelmann G, Monserrat S, and Jordà G, *Sea Level Rise and its Forcing in the Mediterranean Sea.* MedCLIVAR – Mediterranean CLImate VARiability, 2011.

[13] Brunel C, *Evolution séculaire de l'avant-côte de la Méditerranée française, impact de l'élévation du niveau de la mer et des tempêtes*. 2010, UNIVERSITE AIX-MARSEILLE I.

[14] ERAMM, *Etude sur la protection de la partie Nord du tombolo Ouest de Giens*. 2001.

[15] CEREMA, *Details of the campaign 08301 and 08302-Porquerolles*. 2014.

Acknowledgements

This work was financially supported by the 911 Project of Vietnam International Education Development, Ministry of Education and Training, Vietnam (Grant No. 911). The authors are grateful to EOL, CETMEF, CEREMA, SHOM, and REFMAR for sharing and providing the in situ measurement data. The authors also sincerely thank the DHI for allowing a free license for using MIKE software.

IOP Conf. Series: Earth and Environmental Science **167** (2018) 012028 doi:10.1088/1755-1315/167/1/012028

N[1]-(3-trimethoxysilylpropyl)diethylenetriamine grafted ordered mesoporous silica KIT-6 for CO_2 adsorption

D J Mei and J W Wei

College of Environmental Science and Engineering, Guilin University of Technology, Guilin 541004, China
E-mail: 59138522@qq.com

Abstract. Pure KIT-6 was synthesized by using tri-block copolymer P123 as the structure directing agent and tetraethyl orthosilicate (TEOS) as the silica source. It was functionalized with N^1-(3-trimethoxysilylpropyl)diethylenetriamine (denoted as 3N) by grafting in solvent at 80 °C. The materials were characterized by X-ray powder diffraction (XRD) and nitrogen adsorption/desorption and thermal gravimetric analysis (TGA). The CO_2 adsorption capacity was studied in a gravimetric analyzer. The resulted showed that the pore structure of KIT-6 was well kept after grafting of amine groups while surface area, pore size and pore volume were decreased. The CO_2 adsorption capacities of 3N-KIT-6 increased first and then decreased with the increase of 3N loading. The maximum adsorption capacity is 3.017mmol/g.

1. Introduction

Greenhouse effect has become a threat across the globe especially on the marine ecosystem. Carbon dioxide is the main greenhouse gas that causes global warming and the capture and sequestration of CO_2 is an effective method to reduce CO_2 emissions[1]. Compared with the traditional liquid amine scrubbing technology, adsorption approach has the advantages of low energy penalties and high selectivity[2]. Therefore, more and more researchers begin to focus on solid adsorbents. In the recent years amine functionalized silica have earned much attention for CO_2 capture. Various kinds of silica, such as MCM-series, KIT-series, and SBA-series[3], were employed as the support because of their high surface area, large and uniform pores and tunable pore sizes. The impregnation and grafting are two common ways to introduce amine into the solid supports[4]. Amines can be evenly distributed on the surface of the support by grafting method[5-6]. Compared with grafting method, impregnation can make more amines loaded on support, but their distribution is inferior because the amine groups often conglomerate[7]. Nowadays, more and more literatures reported on amine grafted MCM series and SBA series for CO_2 adsorption[8]. But only few of them reported on the CO_2 adsorption behavior in amine grafted KIT-6.

In this paper, we synthesized KIT-6 and grafted it with different volumes of N^1-(3-trimethoxysilylpropyl)diethylenetriamine (denoted as 3N). Characteristics of grafted KIT-6 were investigated through various techniques. The CO_2 adsorption performance was tested in the thermal gravimetric analyzer at 60 °C.

2. Experimental

2.1. The Synthesis and Modification of KIT-6

Pure KIT-6 was synthesized according to the previous report[9]. KIT-6 was grafted by using N^1-(3-trimethoxysilylpropyl)diethylenetriamine (denoted as 3N) as modifier. In a typical grafting process, 2 g of KIT-6 was dispersed in a 60 ml of toluene in a flask. The given amount of 3N was added in the mixture and refluxed at 80 °C for 12 h. Finally the modified materials were prepared and recorded as 3N-KIT-6-X, where X(X=0.5,1,1.5,2) is the 3N volume.

2.2. Characterization of Materials

The powdered X-ray diffraction (XRD) patterns were obtained on a X'Pert PRO powder diffractometer in the 2 θ range of 0.5–8°. N_2 adsorption/desorption isotherms were measured at liquid N_2 temperature. The samples were degassed at 80 °C for 10 h prior to measurement. The surface areas of samples were calculated using the Brunauer–Emmett–Teller (BET) equation and the pore size were determined by the Barrett–Joyner–Halenda (BJH) method.Thermal gravimetric analysis (TGA) was carried out under N_2 atmosphere from 25 to 750 °C with a heating rate of 5 °C/min.

2.3. Adsorption of carbon dioxide

CO_2 adsorption experiments were carried out by thermal gravimetric analyzer (TGA) in the mixed gases (15 vol.% CO_2 and 85 vol.% N_2) under a flow of 60 ml/min at 60 °C. Prior to the adsorption of CO_2, the samples were heated in N_2 with a flow rate of 100 ml/min at 110 °C for 1 h and then cooled to 60 °C.

3. Results and Discussion

3.1. XRD characterization

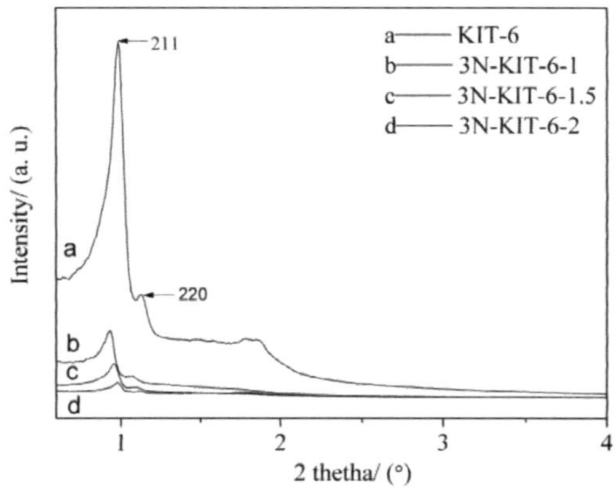

Figure 1. XRD patterns of the KIT-6 and 3N-KIT-6-X

As shown in Figure 1, the XRD pattern of KIT-6 has the clear diffraction peaks at (211) and (220) planes, indicating that KIT-6 has the double continuous cubic structure[9]. The major peaks of 3N-KIT-6 were retained but the intensity of these peaks were diminished with increasing amounts of grafting 3N. This indicated that the mesoporous structure of KIT-6 is preserved and 3N successfully grafted onto KIT-6.

3.2. N_2 Adsorption/Desorption

The N_2 adsorption/desorption isotherms of KIT-6 before and after modification with 3N are displayed in Figure 2. The surface area, pore volume and pore diameter of materials are summarized in Table 1. It can seen that all samples displayed type IV isotherms, which is the features of mesoporous materials. The surface area and pore volume of KIT-6 are 646.57 m^2/g and 1.4015 cm^3/g, respectively. With the increase in the 3N amount, the pore channels of materials were blocked and the surface area and the pore volume decreased rapidly. These features are in harmony with the results of XRD characterization which confirmed that 3N were successfully grafted onto KIT-6.

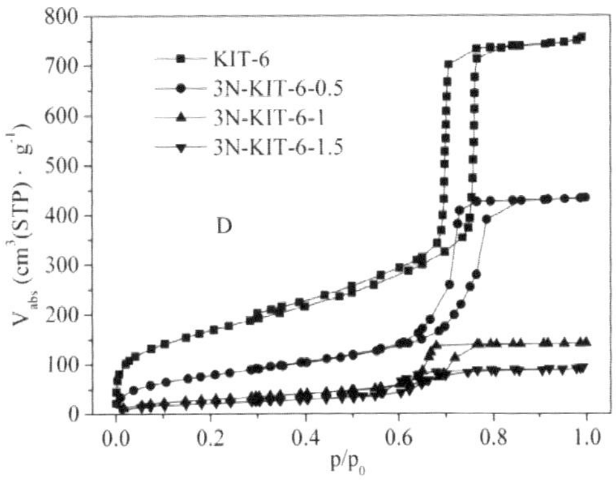

Figure 2. N_2 adsorption/desorption isotherms of samples

Table 1. Physicochemical properties of samples

Sample	Surface area ($m^2.g^{-1}$)	Pore volume ($cm^3.g^{-1}$)	Pore diameter (nm)
KIT-6	646.57	1.4015	4.61
3N-KIT-6-0.5	292.65	0.6892	4.61
3N-KIT-6-1	115.96	0.2313	3.53
3N-KIT-6-1.5	83.06	0.1514	2.71

3.3. TG studies

Figure 3 shows the TG curves of KIT-6 and 3N-KIT-6-X. For all samples, the weight loss below 100 °C was attributed to the impurity gas and physically adsorbed water. From 100 °C to 750 °C, KIT-6 has none obviously mass loss. As for 3N modified KIT-6, the weight loss between 200 °C and 500 °C was assigned to the thermal decomposition of 3N and the surfactant. Above the temperature of 200 °C, 3N-KIT-6-X shows different mass loss due to the decomposition of 3N with different ratios grafted onto KIT-6. These results indicate that 3N was successfully grafted onto KIT-6 and all samples have good thermal stability below 200 °C.

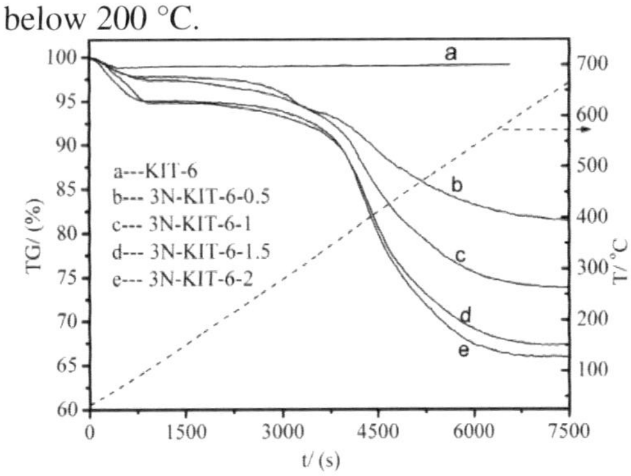

Figure 3. TG curves of KIT-6 and 3N-KIT-6-X

3.4. CO_2 Adsorption

Figure 4 depicts the CO_2 adsorption capacities of 3N-KIT-6-X. The CO_2 adsorption capacity of adsorbent was depends on pore size, pore volume and amine loading. For pure KIT-6, the CO_2 adsorption capacity was 0.642mmol/g. The CO_2 adsorption capacities of 3N-KIT-6-X increased first and then down with the increasing of the amount of 3N grafted. The sample shows an adsorption capacity of 1.474, 1.797, 3.017 and 1.882 mmol/g when it is modified with 3N of 0.5, 1, 1.5 and 2 ml, respectively. Among them, 3N-KIT-6-1.5 has a best adsorption capacity. The adsorption capacity increased was due to the chemisorption between amine functional group grafted on the surface and CO_2. But in higher 3N concentration,the adsorption capacity decreased, indicating that extra amine groups were covered on the outer surface of KIT-6 and limited the spread of CO_2.

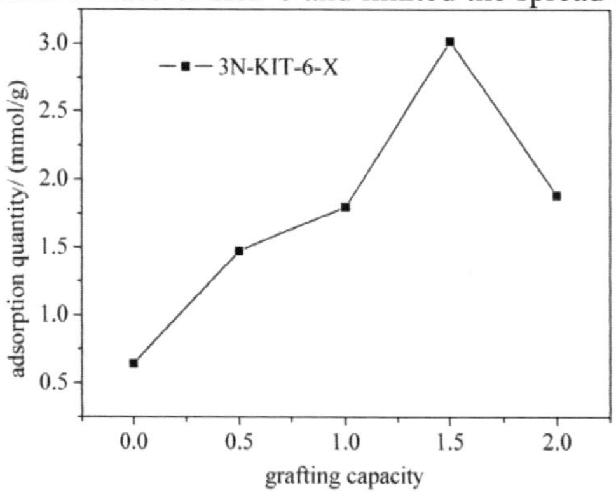

Figure 4. CO_2 adsorption performance of samples

4. Conclusions

In this study, a novel 3N grafted KIT-6 material (3N-KIT-6-X) for CO_2 adsorption was synthesized. Characteristics of 3N grafted KIT-6 and the adsorption performance were tested. The XRD and N_2 adsorption/desorption results indicate that after grafting the pore structure keeps well. TG analysisconfirmedthat 3N was successfully grafted onto KIT-6 and all samples have good thermal stability below 200 °C. The adsorption results indicate that 3N-KIT-6-X has larger CO_2 adsorption capacity. Under similar conditions, 3N-KIT-6-1.5 has a best adsorption capacity of 3.017 mmol/g.

5. References

[1] Macdowell N, Florin N, Buchard A, Hallett J, Galindo A, Jackson G, Adjiman C S, Williams C K, Shah N and Fennell P 2010 *Energy Environ. Sci* **3** 1645-69
[2] Piet M 2014 *Angew. Chem. Int. Ed. Engl.* **52** 11269-72
[3] Patil U, Fihri A, Emwas A and Polshettiwar V 2012 *Chem Sci* **3** 2224-9
[4] Han K K, Zhou Y, Wen J J and Zhu J H 2011 *Adv Mat Res* **287-290** 3081-4
[5] Kassab H, Maksoud M, Aguado S, PeraTitus M and Bonneviot L 2012 *RSC Adv* **2** 2508-16
[6] Li Y, Sun N, Li L, Zhao N, Xiao F, Wei W, Sun Y and Huang W 2013 *Materials* **6** 981
[7] Shi J J, Liu Y M, Chen J, Zhang Y and Shi Y 2010 *Acta Phys. Chim. Sin.* **26** 3023-9
[8] Yan X, Zhang L, Zhang Y, Yang G and Yan Z 2011 *Ind. Eng. Chem. Res* **50** 3220-6
[9] Kleitz F, Choi S H and Ryoo R 2003 *J. Chem. Soc. Chem. Commun.* **9** 2136

Acknowledgements

This research was funded byNational Natural Science Foundation of China (No.51566003) and National Natural Science Foundation of Guangxi (No.2015GXNSFAA139231).

Theoretical studies on bond dissociation enthalpies for model compounds of typical plastic polymers

J B Huang[1], G S Zeng[2], X S Li[3], X C Cheng[1] and H Tong[1]

1 School of Mechatronics Engineering, Guizhou Minzu University, Guiyang 550025, China
2 Key Laboratory of Jiangxi Province for Persistent Pollutants Control and Resources Recycle, Nanchang Hangkong University, Nanchang 330063, China
3 School of Resources, Environment and Materials, Guangxi University, Nanning 530004, China
E-mail: zgs77@163.com, lixinshengkm@126.com, tonghong63@126.com.

Abstract. The polyethylene (PE), polypropylene (PP), polystyrene (PS), polyvinyl chloride (PVC), polyethylene terephthalate (PET), bisphenol A polycarbcconate (PC) comprise above 80% of plastic polymers. The bond dissociation enthalpies for model compounds of the six typical plastic polymers were calculated by using density functional theory methods, and the possible degradation mechanisms of these plastic polymers were proposed. The degradation of PE and PP is mainly main-chain random scission mechanism; the PS degradation is mainly zipper scission and main-chain random scission mechanism; the PVC degradation is mainly branched-chain scission mechanism; the PET and PC decompose mainly through C-O homolytic cleavage, C-C homolytic cleavage and concerted reaction.

1. Introduction

Plastic plays an important role in improving people's living standards over the past several decades. It is a key of innovation of many products in various sectors such as packaging, electronic, construction, automotive, healthcare and others [1]. However, the applications of these plastic goods have continuously given rise to plastic waste pollution problems around the world. At present, most of these plastic wastes are disposed of through dumping into landfills or burning in incinerators along with other solid wastes [2]. But these conventional methods will be phased out in the near future due to the serious environmental pollution problems they lead to. There is a great need to deploy and develop new environmentally friendly and cost-effective waste plastic management technologies as currently used disposal options become less viable. Among these, thermochemical conversion to energy resources has been a significant way to use such waste plastics effectively and to meet with the growing demand from the energy sector [3].

Pyrolysis, the thermal decomposition of organic materials in the absence of oxygen at a mediate temperature [4], is considered as an environmentally friendly and widely used method and producing renewable energy from biomass [5]-[8]. Extensive researches were performed worldwide for the production of alternative fuels or chemicals from waste plastics through the pyrolysis [9]-[14]. The previous studies on mechanism of plastics thermal decomposition mainly focused on experimental studies to explore the distribution of products in different pyrolysis conditions. However, there are few researches about chemical reactions and formation mechanisms of the main products during plastics thermal decomposition. In the process of thermal decomposition, the free radical reaction is considered to be one of the main pyrolysis mechanisms [15], [16]. Bond dissociation enthalpy is an important thermodynamic quantity that can exemplify chemical activity of free radicals reaction, and the smaller the bond dissociation enthalpy is, the more easily the bond is broken [17]. So the theoretical studies on bond dissociation enthalpies contribute to the further understanding of thermal decomposition mechanism of plastic polymers. In this study, the bond dissociation enthalpies (BDE) for model compounds of six typical plastic polymers (as shown in Figure 1) were calculated by using density functional theory (DFT) methods, and the possible formation pathways of main products in the pyrolysis processes were analysed.

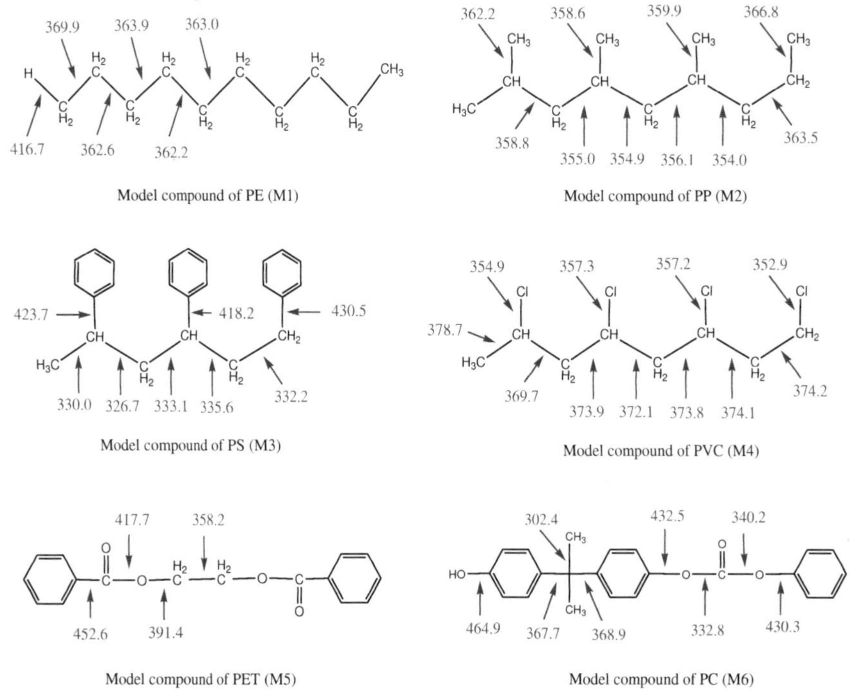

Figure 1. Model compounds of six typical plastic polymers (M1~M6) and the BDE values marked in blue, calculated by using M06-2X/6-31++G(d,p) (kJ/mol).

2. Calculation methods

All calculations were completed using Gaussian 09 suite of programs [18]. The literatures [19], [20] reported that the M06-2X and B3P86 functionals was superior to other functionals for the reproduction of BDE. So the equilibrium geometries of reactants, intermediates, transition states and products were optimized by employing density functional theory methods M06-2X [21] and B3P86 [22] with the 6-31++G(d,p) basis set. Standard thermodynamic parameters at 298 K, including zero-point energy correction (ZPE), were obtained by vibrational frequency calculations. Bond dissociation enthalpy is calculated as [23]:

$$\mathrm{BDE}_{298}(\mathrm{R\text{-}X}) = H_{298}(\mathrm{R}) + H_{298}(\mathrm{X}) - H_{298}(\mathrm{R\text{-}X}) \tag{1}$$

where H_{298} is the enthalpies of formation of the molecule and radical species, which is calculated using the following equation [24]:

$$H_{298} = E + ZPE + H_{trans} + H_{rot} + H_{vib} + RT \tag{2}$$

where ZPE is the zero point energy. H_{trans}, H_{rot} and H_{vib} are the standard temperature correction term calculated using the equilibrium statistical mechanics with harmonic oscillator and rigid rotor approximations.

3. Results and discussion

Calculations of bond dissociation enthalpies contribute to the understanding of free radicals reaction processes. Figure 1 marks the BDE values of the major bonds of model compounds of six typical plastic polymers (M1~M6), calculated by using M06-2X with the 6-31++G(d,p) basis set. Table 1 lists the BDE values of the major bonds of M1~M6, calculated at M06-2X/6-31++G(d,p) level and at B3P86/6-31++G(d,p) level. The average BDE value of main-chain C-C bond for model compound of polyethylene (PE) at M06-2X/6-31++G(d,p) level is 364.3 kJ/mol, and that of main-chain C-C bond for M1 at B3P86/6-31++G(d,p) level is 350.9 kJ/mol. The average BDE value of main-chain C-C bond for model compound of polypropylene (PP) at M06-2X/6-31++G(d,p) level is 357.1 kJ/mol, and that of main-chain C-C bond for M2 at B3P86/6-31++G(d,p) level is 329.5 kJ/mol; but the average BDE value of branched-chain C-CH3 bond for model compound of polypropylene (PP) at M06-2X/6-31++G(d,p) level is 361.9 kJ/mol, and that of branched-chain C-CH3 bond for M2 at B3P86/6-31++G(d,p) level is 342.6 kJ/mol. The average BDE value of main-chain C-C bond for model compound of polystyrene (PS) at M06-2X/6-31++G(d,p) level is 331.5 kJ/mol, and that of main-chain C-C bond for M3 at B3P86/6-31++G(d,p) level is 291.7 kJ/mol; but the average BDE value of branched-chain C-Caromatic bond for model compound of polystyrene (PS) at M06-2X/6-31++G(d,p) level is 424.1 kJ/mol, and that of branched-chain C-Caromatic bond for M3 at B3P86/6-31++G(d,p) level is 395.9 kJ/mol. So the homolytic cleavage of main-chain C-C bond in the PP and PS pyrolysis is the main degradation pathway. The average BDE value of main-chain C-C bond for model compound of polyvinyl chloride (PVC) at M06-2X/6-31++G(d,p) level is 373.8 kJ/mol, and that of main-chain C-C bond for M4 at B3P86/6-31++G(d,p) level is 345.8 kJ/mol; but the average BDE value of branched-chain C-Cl bond for M4 at M06-2X/6-31++G(d,p) level is 355.6 kJ/mol, and that of branched-chain C-Cl bond for M4 at B3P86/6-31++G(d,p) level is 343.7 kJ/mol. By comparing the BDE values, the order of the BDE value of the weakest bond for M1~M4 is as follows: PS < PVC < PP < PE. In the pyrolysis of PE, PP and PS, the free radical reaction mechanism is considered to be the main pyrolysis mechanism. For free-radical reactions, the bond dissociation enthalpies may serve as approximations to the activation energies [6]. In the PVC pyrolysis, PVC was initially dehydrochlorinated to generate HCl by concerted reaction mechanism [25], and the activation energy in concerted reaction is significantly lower than that in free radical reaction. Therefore, we can infer that the order of the thermal stability is as follows: PVC < PS < PP < PE, which is in accordance with the related experimental result [26].

Table 1. The BDE values of the major bonds of model compounds of six typical plastic polymers (M1~M6)

		BDE									
M1	M062X	416.7	369.9	362.6	363.9	362.2	363.0				
	B3P86	422.6	359.2	346.6	349.4	348.4	351.0				
M2	M062X	366.8	359.9	358.6	362.2	363.5	354.0	356.1	354.9	355.0	358.8
	B3P86	353.7	337.4	336.8	342.4	341.9	325.7	327.3	325.9	326.5	329.8
M3	M062X	430.5	418.2	423.7	332.2	335.6	333.1	326.7	330.0		
	B3P86	405.3	392.0	390.4	295.6	288.4	288.2	285.6	300.7		
M4	M062X	352.9	357.2	357.3	354.9	374.2	374.1	373.8	372.1	373.9	369.7
	B3P86	346.4	343.1	343.2	342.2	351.2	343.6	342.4	341.6	342.8	340.6
M5	M062X	452.6	417.7	391.4	358.2						
	B3P86	438.0	392.5	342.1	338.5						
M6	M062X	430.3	340.2	332.8	432.5	368.9	367.7	302.4	464.9		
	B3P86	385.4	307.9	302.3	389.2	329.2	328.9	267.1	462.0		

For the model compound of polyethylene terephthalate (PET), the BDE value of C-C bond is the lowest, 358.2 kJ/mol at M06-2X/6-31++G(d,p) level (338.5 kJ/mol at B3P86/6-31++G(d,p) level), and the BDE value of O-C bond is the second lowest, 391.4 kJ/mol at M06-2X/6-31++G(d,p) level (342.1 kJ/mol at B3P86/6-31++G(d,p) level). For the model compound of bisphenol A polycarbonate (PC), the BDE value of C-CH_c bond is the lowest, 302.4 kJ/mol at M06-2X/6-31++G(d,p) level (267.1 kJ/mol at B3P86/6-31++G(d,p) level), and the BDE value of O-Ccarbonyl bond is the second lowest, 332.8 kJ/mol at M06-2X/6-31++G(d,p) level (302.3 kJ/mol at B3P86/6-31++G(d,p) level). For PET and PC, it is clear that the C-C bond dissociation is easiest, followed by C-O dissociation of carbonate group. However, the carbonate group is reactive easily with water or alcohol, so the degradation may not exactly follow the bond dissociation enthalpies [10].

The BDE values calculated at M06-2X/6-31++G(d,p) level are higher than those at B3P86/6-31++G(d,p) level, but the variation trends of BDE values at two levels are similar. Comparing with the method B3P86, M06-2X is more accurate according to many other studies about carbohydrates systems [19], [20].

Based on related experimental results and experiences, we propose various possible thermal degradation pathways. Figure 2 depicts the proposed degradation reaction pathways in the pyrolysis process of PE and PP. The PE and PP degradation mechanism is mainly main-chain random scission, namely macromolecular main-chain cracks randomly to generate various free radicals. These free radicals further decompose into various products through intramolecular or intermolecular hydrogen transfer reaction. As shown in Figure 2, PE or PP can decompose into the radicals R1 and R2 through random cleavage of the C-C bond. The radicals R1 and R2 can further decompose into the smaller radicals and products such as ethylene, propylene, butane, ethane, propane and others.

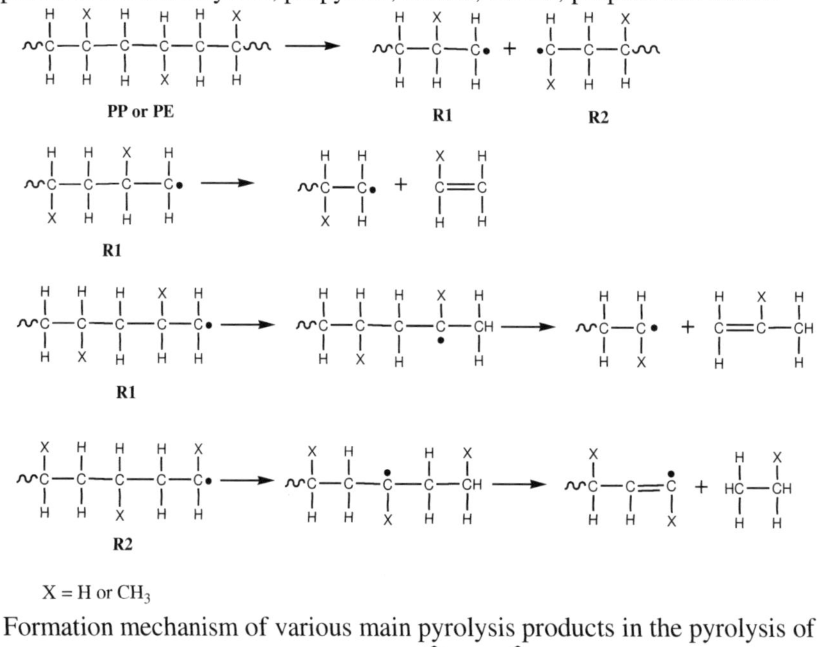

Figure 2. Formation mechanism of various main pyrolysis products in the pyrolysis of PE and PP

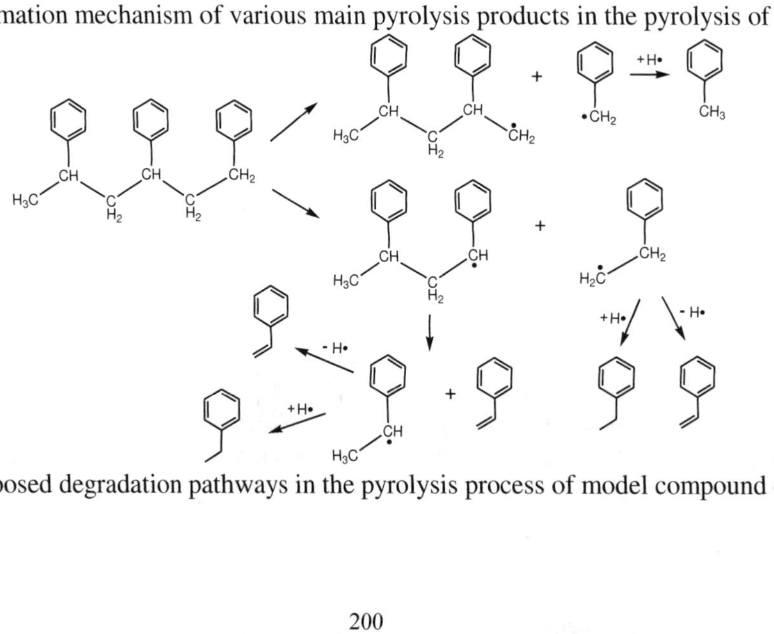

Figure 3. Proposed degradation pathways in the pyrolysis process of model compound of PS (M3)

IOP Conf. Series: Earth and Environmental Science **167** (2018) 012029 doi:10.1088/1755-1315/167/1/012029

Figure 3 describes the proposed degradation reaction pathways in the pyrolysis process of model compound of PS. The PS degradation mechanism is mainly zipper scission and main-chain random scission, and the PS pyrolysis produced a very high yield of monomer styrene due to zipper scission reaction. As shown in Figure 3, the model compound of polystyrene (M3) can decompose to form a free radical and benzyl, and the benzyl can be further converted into methylbenzene through a hydrogenation reaction. M3 can also decompose to form a free radical and phenyl ethyl, and the free radical can further decompose to form styrene and a smaller radical. The phenyl ethyl can be further converted into phenylethane through a hydrogenation reaction, or into styrene through a dehydrogenation reaction.

Figure 4 depicts the proposed degradation pathways in the pyrolysis process of model compound of PVC. The PVC degradation mechanism is mainly branched-chain scission, namely the elimination reaction of HCl occurs and then the main-chain is converted into the conjugated dienes chain, which can be further converted into benzene compound, often accompanied by crosslinking reaction.

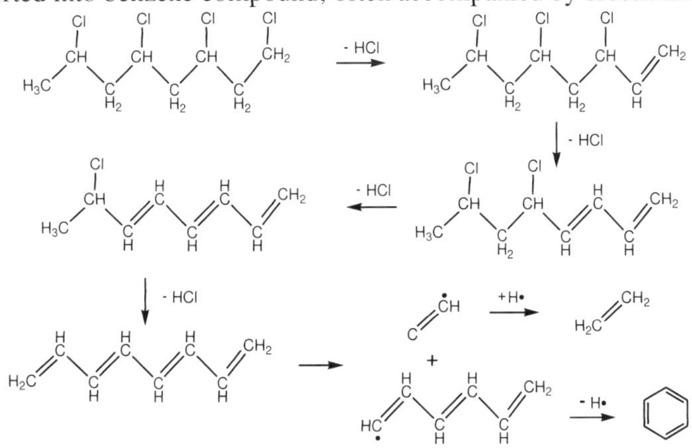

Figure 4. Proposed degradation pathways in the pyrolysis process of model compound of PVC (M4)

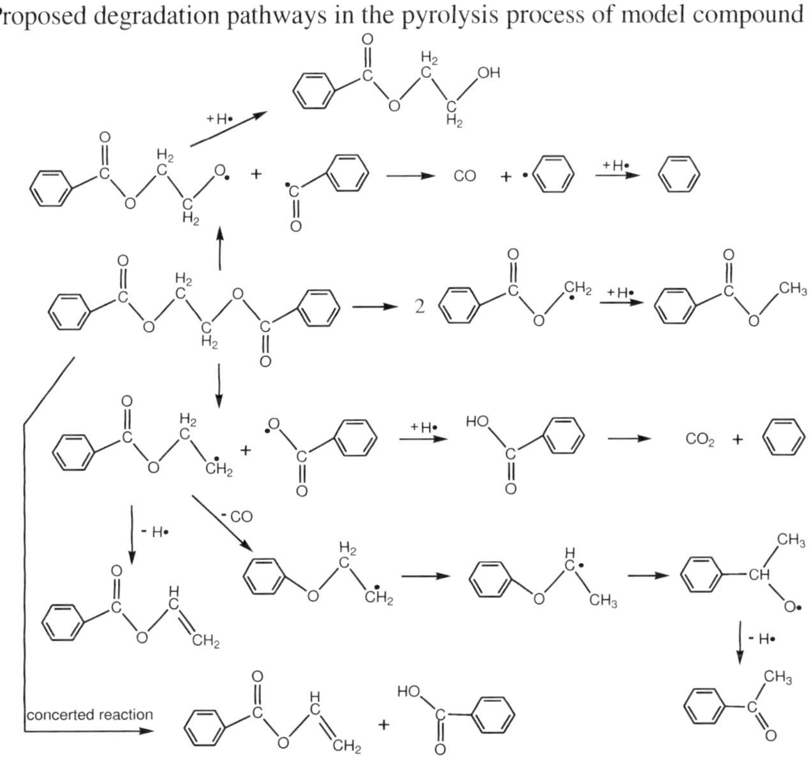

Figure 5. Proposed degradation pathways in the pyrolysis process of model compound of PET (M5)

Figure 5 depicts the proposed degradation pathways in the pyrolysis process of model compound of PET (M5). M5 can decompose into two same radicals through C-C homolytic cleavage, and the radical can be further converted into benzoic acid methyl ester. M5 can also decompose into the various radicals through C-O homolytic cleavage, and these radicals can further decompose into various products such as CO_2, CO, benzene, benzoic acid, acetophenone, 1-hydroxyl ethyl benzoate, vinyl benzoate and others. M5 can also decompose into benzoic acid and vinyl benzoate through concerted reaction, which is similar to the dehydration reaction.
Figure 6 describes the proposed degradation reaction pathways in the pyrolysis process of model compound of PC (M6). M6 can decompose into two radicals through C-O homolytic cleavage, and the radicals can further decompose into CO_2, CO and bisphenol A. The calculation results of the BDE show that methyl scissions occur easily in the isopropylidene linkage, and 1,1-bis(4-hydroxyphenyl)-ethane can be easily formed through C-CH_3 homolytic cleavage. M6 can also directly decompose into bisphenol A and carbonate through hydrolysis reaction. The carbonate can further decompose to form CO_2 and phenol by a decarboxylation reaction.

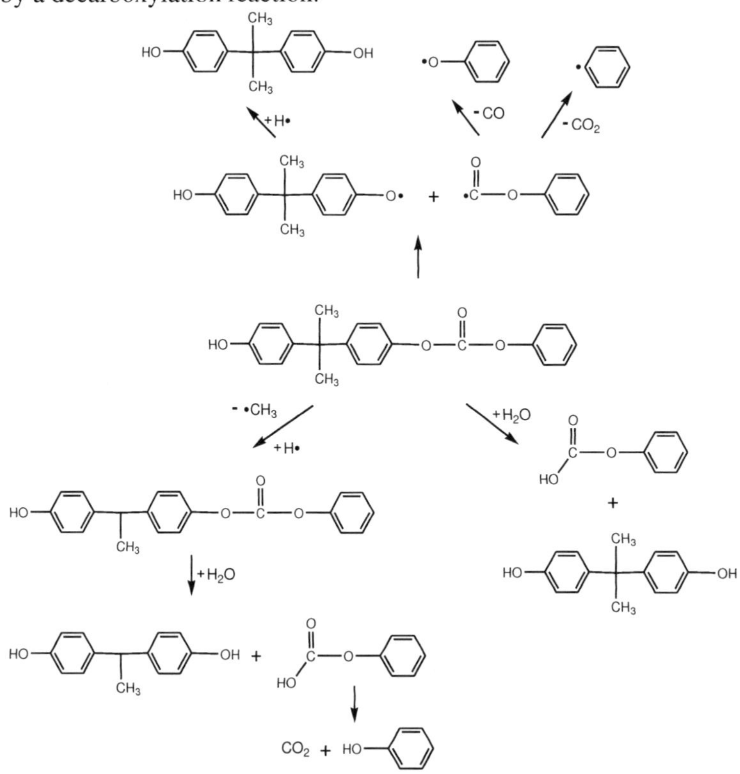

Figure 6. Proposed degradation pathways in the pyrolysis process of model compound of PC (M6)

4. Conclusions

In the present study, density functional theory methods M062X and B3P86 were used to investigate the bond dissociation enthalpies for model compounds of the six typical plastic polymers. The BDE values calculated at M06-2X/6-31++G(d,p) level are higher than those at B3P86/6-31++G(d,p) level, and M06-2X is more accurate according to many other studies about carbohydrates systems. On the basis of computational results, the order of the thermal stability for four kinds of polyolefins is as follows: PVC < PS < PP < PE. The possible formation pathways of main products in the pyrolysis of six plastic polymers were proposed. PE and PP degradation mechanism is mainly main-chain random scission; PS degradation mechanism is mainly zipper scission and main-chain random scission, and the PS degradation produced a very high yield of monomer styrene; PVC degradation mechanism is mainly branched-chain scission, in which the elimination of HCl occurs and then the conjugated dienes chain is formed; PE and PP pyrolysis processes is more complicated, and cPET and PC decompose mainly through C-O and C-C homolytic cleavage and concerted reaction.

5. Reference

[1] Sharuddin S D A, Abnisa F, Daud W M A W and Aroua M K 2016 *Energ. Convers. Manage.* **115** 308.
[2] Song H S and Hyun J C 1999 *Korean J. Chem. Eng.* **16** 316.
[3] Ahmad I, Khan M I, Khan H, Ishaq M, Tariq R. 2015 *J. Appl. Polym. Sci.* DOI: 10.1002/APP.41221.
[4] Liu W, Tian K, Jiang H and Yu H 2016 *J. Hazard. Mater.* **310** 217.
[5] Mckendry P 2002 *Bioresource Technol.* **83** 37.
[6] Huang J, Wu D, Tong H and Li W 2013 *J. Energy Inst.* **86** 189.
[7] Mohan D, Pittman C U and Steele P H 2006 *Energy Fuel* **20** 848.
[8] Huang J, He C, Wu L and Tong H 2016 *Chem. Phys. Lett.* **658** 114.
[9] Panda A K, Singh R K and Mishra D K 2010 *Renew. Sust. Energ. Rev.* **14** 233.
[10] Nam Jang B and Wilkie C A 2004 *Polym. Degrad. Stab.* **86** 419.
[11] Liu Y, Qian J and Wang J 2000 *Fuel Process. Technol.* **63** 45.
[12] Martín-Gullón I, Esperanza M and Font R 2001 *J. Anal. Appl. Pyrol.* **58-59** 635.
[13] Kumar Singh R and Ruj B 2016 *Fuel* **174** 164.
[14] Miskolczi N, Bartha L and Angyal A 2009 *Energ. Fuel.* **23** 2743.
[15] Britt P F, Buchanan III A C, Cooney M J and Martineau D R 2000 *J. Org. Chem.* **65** 1376.
[16] Huang J, Wu S, Cheng H, Lei M, Liang J and Tong H 2015 *J. Fuel. Chem. Technol.* **43** 429.
[17] Younker J M, Beste A and Buchanan III A C 2012 *Chem. Phys. Lett.* **545** 100.
[18] Frisch M J, Trucks G W, Schlegel H B, Scuseria G E, Robb M A, et al 2013 *Gaussian 09, Revision D.01*, Gaussian, Inc., Wallingford CT.
[19] Huang J, He C, Wu L and Tong H 2017 *J. Energy Inst.* **90** 372.
[20] Johnson E R, Clarkin O J and DiLabio G A 2003 *J. Phys. Chem. A* **107** 9953.
[21] Zhao Y and Truhlar D G 2008 *Acc. Chem. Res.* **41** 157.
[22] Perdew J P 1986 *Phys. Rev . B* **33** 8822.c
[23] Huang J, Liu C, Jin Q, Tong H, Li W 2014 *J. Renew. Sustain. Ener.* **6** 033116.
[24] Foresman J B and A. Frisch 2002 *Exploring chemistry with electronic structure method*, Gaussian, Inc., Pittsburgh, PA.c
[25] Shen Y, Zhao R, Wang J, Chen X, Ge X, et al 2016 *Waste Management* **49** 287.
[26] Dong P, Yin S, Bie R 2006 *Journal of Harbin Institute of Technology* **38** 1959.

Acknowledgments
This work was supported by the National Natural Science Foundation of China (No. 51266002) and the Key Project for Innovation Research Groups of Guizhou Provincial Department of Education (No. KY[2016]028) and the Open Research Funds of Key Laboratory of Jiangxi Province for Persistent Pollutants Control and Resources Recycle (Nanchang Hangkong University) (No. ES201880055).

Carbon Dioxide Minimum Miscibility Pressure with Nanopore Confinement in Tight Oil Reservoirs

R S Mohammad[1, *], S Zhang[1], E Haq[2], X Zhao[3] and S Lu[1]

1. Oil and Gas Field Development Engineering Department, China University of Petroleum-Beijing, Changping, China
2. Geological and Geosciences Department, China University of Petroleum-Beijing, Changping, China
3. Centre of Geoscience Computing, School of Earth and Environmental Science, University of Queensland, Brisbane, Australia
E-mail: 2846921094@qq.com

Abstract. CO_2-injection is one of the capable processes in EOR from low-permeable reservoirs and MMP determination is a key factor in estimating the displacement efficiency of the CO_2 in the EOR processes. The laboratory procedures for MMP determination recognized in the oil industry are slim-tube, rising-bubble, or vanishing interfacial tension (VIT). However, the presence of nanopores in tight formations influences phase equilibrium, causing reduction in MMP. Instead, the existing MMP correlations need to be modified for tight reservoirs that might result in reliable MMP. Therefore, MMP measurement is performed using WinProp to validate correlations for tight oil samples. This study presents MMP determination experimentally using VIT for tight oil samples in both recombined-oil and dead-oil conditions. Subsequently, results obtained from VIT are compared with slim-tube results and the relative error was found 4.86% for recombined-oil and 23.36% for dead-oil. A huge deviation between VIT and slim-tube is noted while measuring MMP for dead-oil, due to deficiency of multiple contacts miscibility and stabilization of heavier fractions. subsequently, an already incorporated correlation for MMP is utilized, considering the effect of nanopore confinement. This study provides an appropriate technique for predicting MMP considering the capillary pressure and solubility on well performance of tight reservoirs.

1. Introduction

Carbon dioxide enhanced oil recovery (EOR) is the main choice to execute CO_2 geological sequestration due to the extra economic advantage, prevailing infrastructure and acquaintance of petroleum and its operations. A CO_2 Miscible injection process have become commonly used procedure for the improved oil recovery worldwide. Thus, many CO_2 field applications have been recorded as successful [1], [2]. In miscible CO_2 injection process, the main task is to displace the trapped oil using miscible CO_2 injection process, which increases the displacement efficiency and enhance the oil recovery. minimum miscibility pressure (MMP) determination is a key parameter in estimating the capability of injected fluid in the EOR processes. The miscibility is achieved at the MMP, when interfacial tension among the CO_2 and reservoir fluid is approaching zero, which consequences in potential transfer of molecules across the interface leading to mutual miscibility and homogeneous fluid formation [3].

There are many experimental procedures for the determination of MMP such as Slim-tube (ST), vanishing interfacial tension (VIT) and raising bubble (RB). A number of studies presented the slim-tube for different gas-oil systems for calculation of MMP and formed various standards for its prediction. However, VIT experiment is achieving cohesion within the oil industry because it is requiring extremely short time and less expenses in estimating CO_2-Oil MMP while slim-tube needs extended time. The concept of VIT is to predict the miscibility situations by calculating the interfacial tension through the fluid phases against differing injectivity pressures or composition of the injected fluid and subsequently calculates the MMP by extrapolation [4]. The miscibility mechanism is achieved, as interfacial tension (IFT) among injected CO_2 and the reservoir fluid reaches zero and transmission of molecules take place between fluid bulks that results in miscibility development. Holm et al. introduced the MMP as oil recovery reaches 80% at gas breakthrough [5]. Waqar et al. conducted experimental work and compared VIT and ST test, however they did not consider the nanopore confinement effect in their study [6]. Rao et al. applied VIT to realize the miscibility conditions for a couple of gas–oil samples by altering the compositions of the injected fluid and injectivity pressure [7], [8]. Orr & Jessen have questioned the viability of this approach, but recent studies by Ayirala and Rao [9] showed that the experimental VIT MMP results were within 5 to 8 percent of the slim-tube MMP values [10]. Zhang et al. proposed correlation for CO_2-Oil MMP with nanopore confinement effect which lacks experimental analysis [11] in contrast to this study.

On the other hand, a diversity of corrections is in appreciation of CO_2-Oil MMP, which help engineers in making decisions to develop injectivity status and to design appropriate downstream facilities [12]. The central factors impacting CO_2-Oil MMP correlations are reservoir fluid compositions, reservoir temperature and CO_2 injectivity conditions [13]. In the previous studies, reservoir temperature was considered the most significant parameter influencing the CO_2-Oil MMP. However, few techniques have huge inconsistencies with the data obtained from laboratory work. It was noted that MMP is further affected by molecular weights of crude oil [14]. Subsequently, the prognosis of CO_2-Oil MMP is required to deem crude oil composition for computing MMP. Higher concentration of lighter components in the crude oil increases the MMP, while higher concentration of intermediate components reduces the MMP [15]. Most of the pores in tight reservoirs are on the order of a few nanometers varying from 3nm to 100nm and from 40nm to 1500nm for shales and sandstones, respectively [16].

Many researches pointed that the fluid phase behavior in these nanopores diverges notably then that of macro pores, which cause variations in Van der Waals forces, molecules orientation in such a nanoscale pore structure [17], as presented in Figure 1. The molecular orientation can be transformed due to nanopore confinement effect as proposed by Pitakbunkate et al [18]. They show that the conventional liquid-vapor equilibrium (VLE) calculations could be extended to unconventional reservoirs by including capillary pressure and shift in critical temperature and pressure in VLE calculations. A mathematical algorithm is used to calculate VIT by integrating the extended VLE calculation to any MMP algorithm, such as the multiple mixing cells (MMC) model of Ahmadi and Johns [19]. In this approach, incremental pressure increase as IFT approaches zero.

Nanopore confinement can cause fluid phase equilibrium alteration with shifting critical properties [20] resulting in different fluid behavior. Yangyang et al., observed that the CO_2 density near pore walls is higher than that in pore centres [21]. Since the crude oil phase equilibrium and fluid properties in a confined system deviate from those at their bulk state, CO_2-Oil MMP is decreased with confinement due to critical properties shift. Therefore, the consequence of nanopore confinement on CO_2-crude oil MMP is studied and incorporated into a new novel CO_2-Oil MMP correlation. The recommended screening technique is carried out by phase behavior module of CMG (Winprop) using tight reservoir crude oil samples.

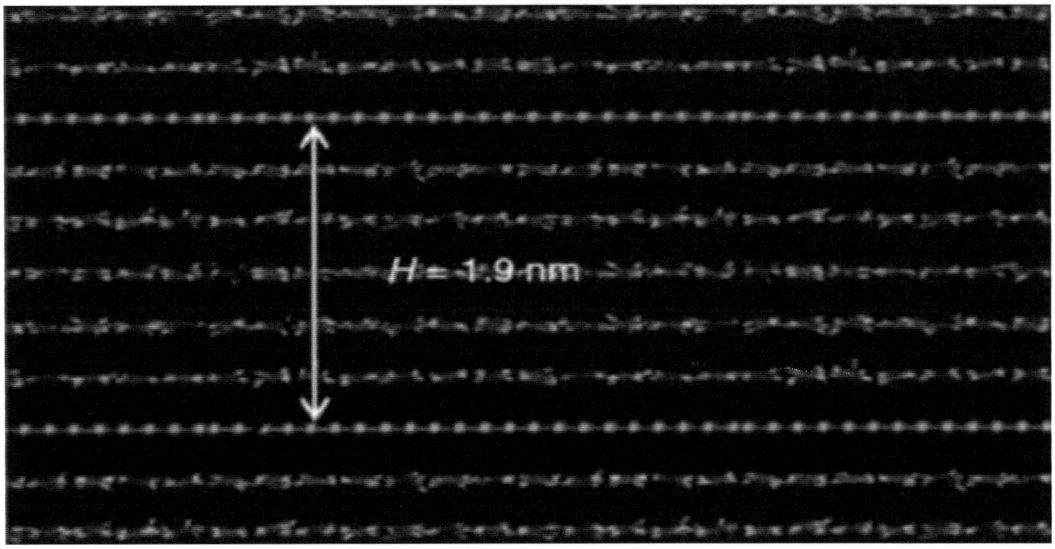

Figure 1. Molecular orientation in nanopore confined system [18].

As a result of the above-mentioned experimental MMP approaches, a study for MMP measurements using VIT experimental procedure is performed. The crude oil of tight reservoir is being utilized with CO_2 in both the stock tank oil (dead oil) and recombined oil (live oil) forms individually for displacement study. Furthermore, a comparative estimation of CO_2-crude oil MMP with confinement effect calculating correlations with VIT measured MMP was also carried out using CMG WinProp to find an appropriate correlation for precise MMP prediction.

2. Methodology

2.1. MMP Correlation

Many researchers developed correlations for computing crude oil- CO_2 MMP. Holm & Josendal (1974) modified correlation to predict the CO_2 MMP based on molecular weight of C_5+ and the desired temperature of the reservoir. Lee J. provided a model based on reservoir temperature and CO_2 vapor pressure for estimating MMP. Furthermore, procedure by Orr and Jensen was almost suitable correlation for predicting MMP of low temperature reservoirs. Yuan et al. presented correlation to predict CO_2-oil MMP using intermediates and heavier mole frictions of reservoir fluid and reservoir temperature. Shokir developed a new model for the prediction of both impure and pure CO_2 displacements. However, none of above-mentioned CO_2-crude oil MMP determining correlations might prove to be reliable MMP with confinements effect. Therefore, a novel correlation is proposed to take into account the effect of nanopore confinement in determining the CO_2-oil MMP.

The solubility parameters for hydrocarbon components can be measured by [22]:

$$\sigma = \frac{\alpha \delta^2 V^{0.33}}{A} \quad \text{Eq. (1)}$$

Where

δ is solubility parameters,
σ is interfacial tension,
V is molar volume,
α is constant and
A is coefficient for solubility parameters.

Furthermore, the oil solubility parameters can be determined from average molecular weights and reservoir temperature [23].

$$\delta_{oil} = 0.01M + 6.54 - 0.01(T - 25) \quad \text{Eq. (2)}$$

Where

T is reservoir temperature,

M is molecular weight.

The solubility parameters for CO_2 can be estimated by the following expression [24]

$$\delta_{CO_2} = bP_c^{0.5}\left(\frac{\rho_r}{\rho_{r(liq)}}\right) \quad \text{Eq. (3)}$$

Where

P_c is critical pressure,

α is constant,

ρ_r is reduced density and

$\rho_{r(liq)}$ is reduced density of gas compressed to a liquid state.

The pressure condition that corresponds to $|\delta_o - \delta_g| \approx 3.0\ (\text{cal/cm}^3)^{0.5}$ is the CO_2-crude oil MMP [23]. At reservoir temperature, the relationship between gas solubility and reduced density can be calculated by Equation (3). Then the CO_2 density can be obtained. According to Figure 2, the CO_2-oil MMP can be estimated.

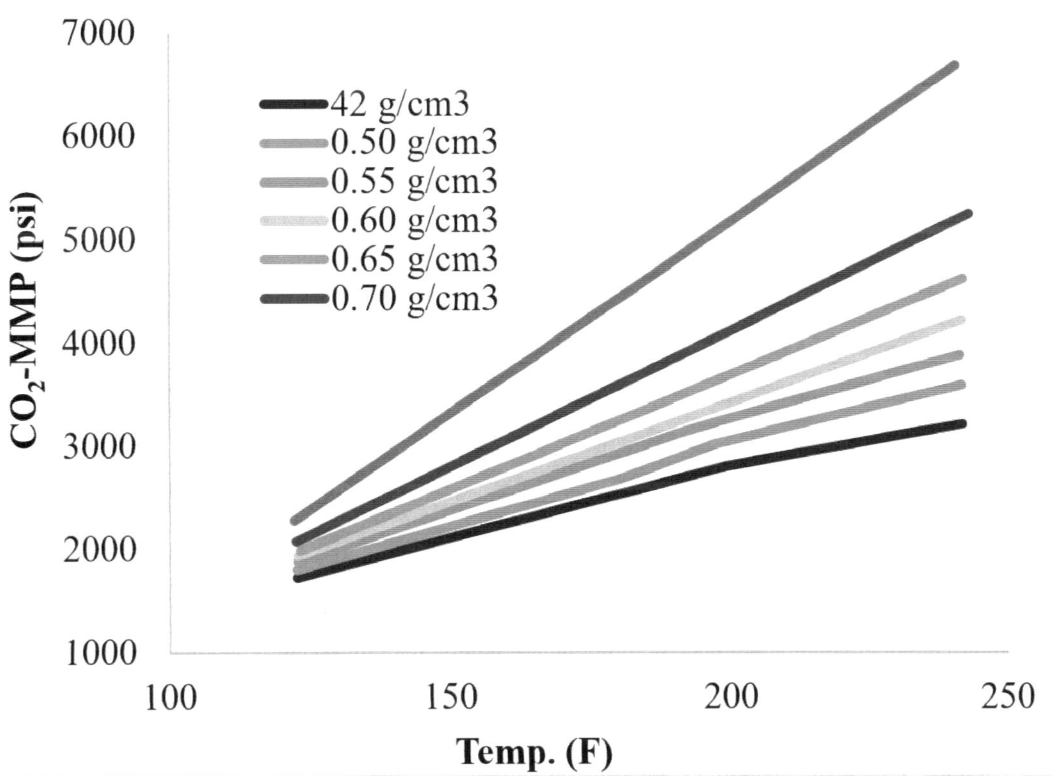

Figure 2. Density of CO_2 required for miscible displacement [17].

Taking the confinement effect into consideration, a critical property shift is given by the following equations [20]:

$$\Delta P_c = \frac{P_c - P_{cp}}{P_{cp}} = 0.9409\frac{\sigma}{r_p} - 0.2415\left(\frac{\sigma}{r_p}\right)^2 \quad \text{Eq. (4)}$$

$$\Delta T_c = \frac{T_c - T_{cp}}{T_{cp}} = 0.9409\frac{\sigma}{r_p} - 0.2415\left(\frac{\sigma}{r_p}\right)^2 \quad \text{Eq. (5)}$$

$$\sigma = 0.244\sqrt[3]{\frac{T_c}{P_c}} \quad \text{Eq. (6)}$$

Where ΔPc and ΔTc are the critical pressure and critical temperature shift due to nanopore confinement effect, respectively. Pc and Tc are the critical pressure and critical temperature, respectively. Pcp and Tcp are the pore critical pressure and pore critical temperature, respectively. Where, rp is the pore radius and σ is the Lennard-Jones size parameter.
Subsequent to the above steps, the CO_2 density can be obtained. Consequently, it requires to be modified by the following equation at frequent pore radius in 50 nm or less as proposed by Zhang et al [11]. Then the modified CO_2-Oil MMP corresponding to Figure 2 with the nanopore confinement effect can be achieved.

$$\frac{\rho_{confined}}{\rho} = -0.0003(lnr_p)^4 + 0.0088(lnr_p)^3 - 0.0973(lnr_p)^2 + 0.4578lnr_p + 0.2179 \quad \text{Eq. (7)}$$

2.2. Experimental Work

2.2.1. Recombination of Downstream Fluids

The live crude oil sample utilized in this study were collected from downstream system (gas from first-stage separator and dead oil from stock tank) and experimentally recombined at reservoir conditions based on the saturation pressure to reproduce a representative fluids of tight oil reservoir as validated in our previous studies [25], [26]. This study does not explain the recombination related processes, though it represents the reservoir fluid and recombined fluid compositions used in this work in Table 1. The carbon dioxide utilized in this study has a purity > 99.95%. The rest of PVT properties of tight reservoir fluid such as oil viscosity, density, gas-oil ratio and formation volume factor were measured using CMG WinProp as shown in Table 2 resulting in a good correlation with tight reservoir fluid data.

Table 1. Reservoir fluid and recombined fluid components.

Comp.	Pc *(psi)*	Tc *(^oK)*	Mol. Weight	Reservoir Fluid *(Mole %)*	Recombined Fluid *(Mole %)*	Average Absolute Error *(%)*
N_2	492.31	126.2	28.013	1.3643	1.3843	1.4691
CO_2	1069.86	304.2	44.010	6.7853	6.8817	1.4213
C_1H_4	667.19	190.6	16.043	27.034	27.426	1.4514
C_2H_6	708.34	305.4	30.070	5.5795	5.6450	1.1733
C_3H_8	615.76	369.8	44.097	5.7348	5.7831	0.8422
iC_4H_{10}	529.05	408.1	58.124	1.8855	1.8839	0.0828
nC_4H_{10}	551.09	425.2	58.124	3.3868	3.3806	0.1839
iC_5H_{12}	490.84	460.4	72.151	2.6760	2.6544	0.8066
nC_5H_{12}	489.37	469.6	72.151	3.7068	3.6693	1.0108
C_6H_{14}	477.03	507.5	86.010	5.1205	5.0579	1.2230
C_7+	145.69	905.1	274.00	36.755	36.233	1.4193
					AARE (%)	0.0828

Table 2. Comparison between PVT properties of tight oil reservoir and predicted properties using WinProp.

PVT Properties.	Sample	Model (Winprop)	Average Absolute Error (%)
Bubble Point (psi)	2652	2651.864	0.005130
Viscosity (cp)	0.270	0.2698	0.074070
Oil Density (lb/ft^3)	36.849	36.900	0.138402
Gas-Oil Ratio (scf/stb)	76.50	76.84	0.444444
Oil FVF (bbl/stb)	1.317	1.319	0.151860
		AARE (%)	0.162780

2.2.2. Vanishing Interfacial Tension Test

Vanishing interfacial tension (VIT) is a fast route to calculate the minimum miscibility pressure to economically recover oil during CO_2 injection process. VIT offers a method to directly measure the IFT between the two phases. The experimental schematic demonstration utilized in this work is illustrated in Figure 3. The apparatus comprises of two high pressure and temperature PVT cells and syringe pump (ISCO-260D). The upper cell contains recombined fluid phase and CO_2 phase is injected from syringe pump until a little head of free CO_2 is seen at the top, confirming that the recombined fluid is fully saturated with CO_2 phase. The measurement system is consisting of high pressure syringe pump to inject CO_2 as a solvent into the lower cell at stable desirable temperature. The PVT cell contains a moveable piston to manipulate the pressure inside the cell by pumping CO_2 on to the back of PVT cell as a pressurizing fluid. The valve between both the cells allows us to displace the system, while keeping the upper cell at a relatively higher pressure than the lower cell pressure. As the valve is opened the upper cell system is likely to pump the recombined fluid into the lower cell and once the droplet is formed the valve is closed or the moveable piston is utilized to pressurize and depressurize the system. The IFT measurements are conducted with few test droplets until the system have excess oil at the bottom of the lower cell, this is to achieve a stable droplet at the tip of the capillary tube. As the stabilized droplet of recombined fluid is introduced in the lower cell which is fully saturated with CO_2, in terms the CO_2 instantly swells the droplet which dissolves in the oil, thus significantly extending the volume of the oil phase, leading the lighter components of the oil phase to isolates into the vapor phase and achieve solubility through CO_2 phase. In the process one may achieve slight stability in terms of droplet. Through saturating/presaturating the oil phase with CO_2 (upper cell) and CO_2 phase with the oil (lower cell), some of that mass transfer upon the formation of the droplet at the tip of capillary is removed. However, the lower cell which has two lateral windows for optical observations; one of the lateral windows is utilized for lighting source and the other for camera to detect the shape of droplet and the magnified picture is digitized and stored in the operating system. Afterwards, the digitized droplet analyzes drop shapes through Laplace equation to fit the experimental drop shape for the determination of interfacial tension with the knowledge of density difference (density of the oil saturated with CO_2 at desired pressure and temperature and density of the CO_2 phase saturated with oil at desired pressure and temperature).

$$\Delta P = \gamma\left(\frac{1}{R_1} + \frac{1}{R_2}\right) \qquad \text{Eq. (8)}$$

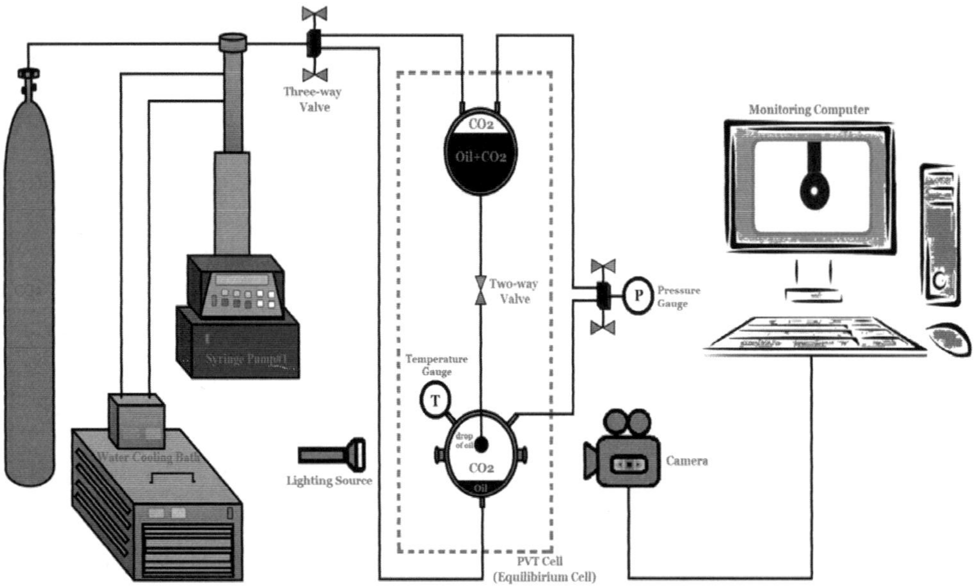

Figure 3. Vanishing Interfacial Tension measurement apparatus.

3. Results and Discussion

3.1. VIT measured MMP

MMP measurements using VIT technique are done in a high-pressure PVT cell. Crude oil is introduced as a drop phase into the chamber filled with the injected fluid. Pendant drop shape analyses are utilized to calculate the interfacial tension. The pressure is gradually increased by pumping more injection fluid on the back of PVT cell as pressurizing fluid. The interfacial tension is determined at different pressures at reservoir temperature. VIT technique for estimating MMP between reservoir fluid and injected fluid (CO_2) with the nanopore confinement effect of tight reservoir besides the slim-tube test data is validated at the reservoir temperature of 238 °F. The reservoir fluid compositions and critical properties are presented in Table 1. Vanishing interfacial tension measurement for minimum miscibility pressure data for both of live oil (recombined oil) and dead oil (stock tank oil) samples are illustrated in Figures. 4 and 5, respectively. However, it was observed that the live oil MMP estimated using VIT procedure is 3478 psi, which is comparatively higher in comparison with the calculated data of 3316 psi utilizing slim-tube technique with AARE of 4.86%. The close agreements designate that VIT procedure can precisely predict the MMP of the reservoir fluid in confined phase. The measurements are done at 9 different pressures, after which the line is extrapolated to zero IFT value. In contrast, stock tank oil (dead oil) MMP determination using VIT was 1821 psi, whereas ST measurement was 2376 psi with AARE of 23.36%. It is noted that a huge deviation between VIT and ST data occurred due to heavier fractions of dead oil. Thus, light fractions of dead oil are isolated into CO_2 gas phase, considered too low to be enriched for solubility with crude oil. In case of stock tank oil with high stabilized heavier fractions, this difference can be correlated to lack of multiple contact miscibility development between crude oil and gas phases compared to achieved miscibility in slim-tube case at respective injection pressure.

In order to consider the capillary pressure effect on the minimum miscibility pressure, the MMP is calculated with different pore sizes. The MMP for the pore size 5 nm is 2810 psi, whereas MMP increased to 3104 psi with the pore size of 10 nm. Therefore, the minimum miscibility pressure reduces as the pore size is reduced. Thus, the reservoir fluid and injected fluid (CO_2) will achieve miscibility at lower pressures and subsequently is helpful for CO_2 EOR. Figure 6 shows the determined IFTs and MMP considering capillary pressure effect for tight oil reservoirs.

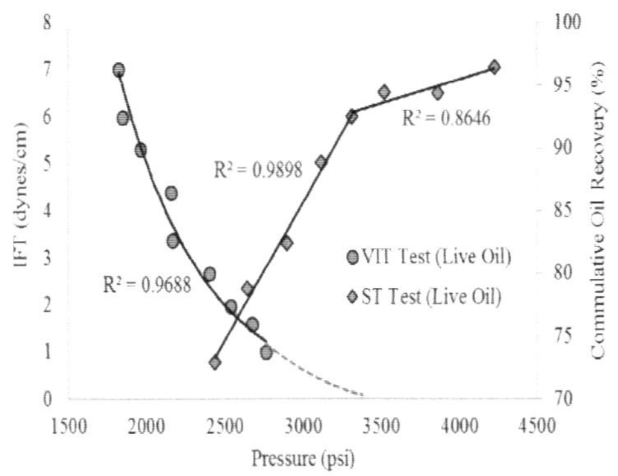

Figure 4. MMP calculation using VIT and comparison with ST test for live oil sample.

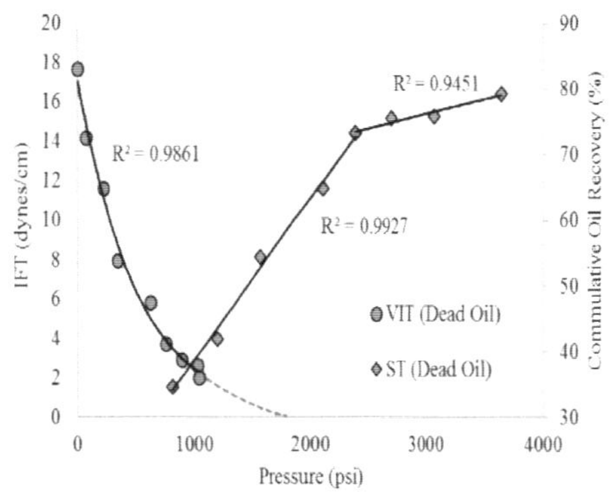

Figure 5. MMP calculation using VIT and comparison with ST test for dead oil sample.

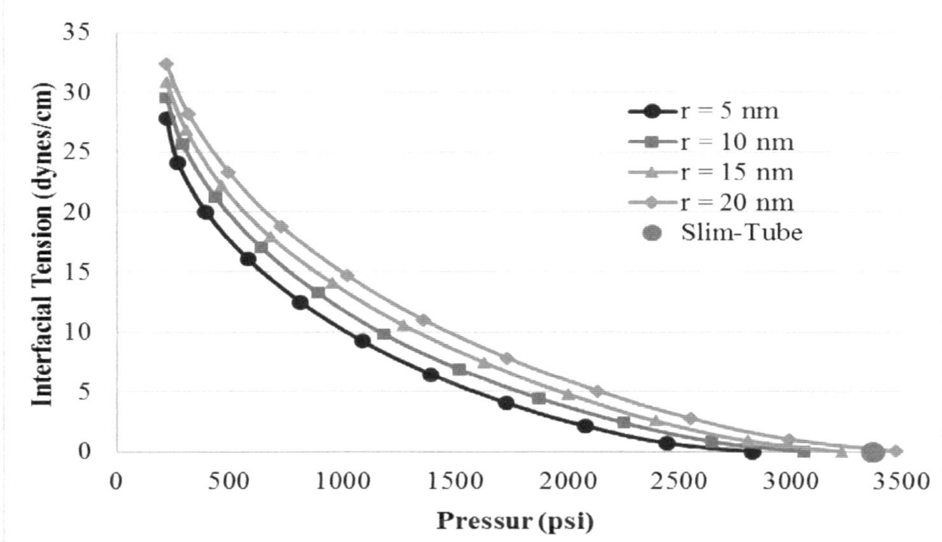

Figure 6. Determined IFTs and MMP with capillary pressure effect for live crude oil of tight reservoirs.

3.2. Correlation calculated MMP

A CO_2-Oil MMP can be measured in several methods and the correlation provides one of the easiest and attractive solutions to estimate MMP. Though, the correlations are used in certain conditions, however they are yet to be considered as an efficient method to roughly determine CO_2-Oil MMP. Mostly, the MMP correlations are function of the few parameters including reservoir temperature, molecular weights and mole fractions. Most of CO_2 MMP correlations are summarized in Table 3.

Table 3. Common CO_2 MMP Correlations.

Author	Correlation
Holm & Josendal	*Graphical correlation as shown in Figure 2.*
Cronquist	$MMP = 16T^{(0.744+0.0015X_{vol}+0.0011MW_{C_{5+}})}$

Orr & Jensen	$MMP = 0.101386exp\left[10.91 - \frac{2015}{255.372 + 0.5556(1.8T_R + 32)}\right]$
Emera & Sarma	$MMP = 5.0093 * 10^{-5}(1.8T_R + 32)^{1.164}(MW_{C_{5+}})^{1.2785}\left(\frac{X_{vol}}{X_{int}}\right)^{0.1073}$
Yelling &Metcalfe	$MMP = 12.6472 + 0.01553(1.8T_R + 32) + 1.24192 * 10^{-4(1.8T_R+32)^2} - \frac{716.9427}{(1.8T_R + 32)}$
Shokir	$MMP = -0.068616Z^3 + 0.31733Z^2 + 4.9804Z + 13.432$
Alston	$MMP = 0.000878T^{1.06}MW_{C_{5+}}^{1.78}\left[\frac{X_{vol}}{X_{int}}\right]^{0.136}$
Lee	$MMP = 7.3924 * 10^{2.772-\left[\frac{1519}{(492+1.8T_R)}\right]}$
Glaso	$MMP = 180 - 3.404MW_{C_{5+}} + \left(1.7 * 10^{-9}MW_{C_{7+}}^{3.730}e^{786.8MW_{C_{7+}}^{-1.058}}\right)T$
Shengli	$MMP = 675.01 - 6.9MW_{C_{6+}} + 1.7(T - T_{critical})$
Yuan	$MMP = a_1 + a_2MW_{C_{7+}} - a_3X_{int} + \left(a_4 + a_5MW_{C_{7+}} + a_6\frac{X_{int}}{MW_{C_{7+}}^2}\right)(1.8T_R + 32) + (a_7 + a_8MW_{C_{7+}} - a_9MW_{C_{7+}}^2 - a_{10}X_{int})(1.8T_R + 32)^2$

Recombined crude oil sample shown in Table 1 is used to validate the CO_2-Oil MMP correlations. Results in Table 4 are shown based on the correlation parameters. In general, the correlations involving molecular weights and temperature give a good prediction of CO_2-Oil MMP. The correlations only taking temperature into consideration significantly deviates from the results provided by the Winprop simulator. On the contrary results of MT and MTF correlations are similar but the equations become more complex if the crude oil composition is involved in the correlations.

Table 4. CO_2–Oil MMP correlation validation for each parameter.

	Recombined Fluid *(Live Oil)*	Stock Tank Oil *(Dead Oil)*
Temperature (MMP in psi)	2892.052	2240.832
Temp. MW (MMP in psi)	3250.295	2630.984
Temp., MW & MF (MMP in psi)	3150.219	2570.068
CMG Winprop (MMP in psi)	3319.913	2619.381

Afterwards, CO_2 MMP is predicted with a confinement effect at 20nm and 10nm pore radii as shown in Table 5. The correlation proposed by Zhang K., [11] gives a reasonable prediction of CO_2-Oil MMP, which also incorporates the alteration of CO_2–Oil MMP with nanoscale pore confinement. Comparison between experimental MMP and calculated MMP ITF values shown in Figure 7 suggest that the results from the correlation considering confinement effect have good coincidence which measured values.

Table 5. CO_2–Oil MMP correlation validation for nanopore confinement.

	Correlation (psi) r = 20 nm	CMG (psi) r = 20 nm	Correlation (psi) r = 10 nm	CMG (psi) r = 10 nm
Recombined Fluid	3343.118	3505.561	2968.154	3084.046
Stock Tank Oil	2214.726	2407.626	1849.760	2086.213

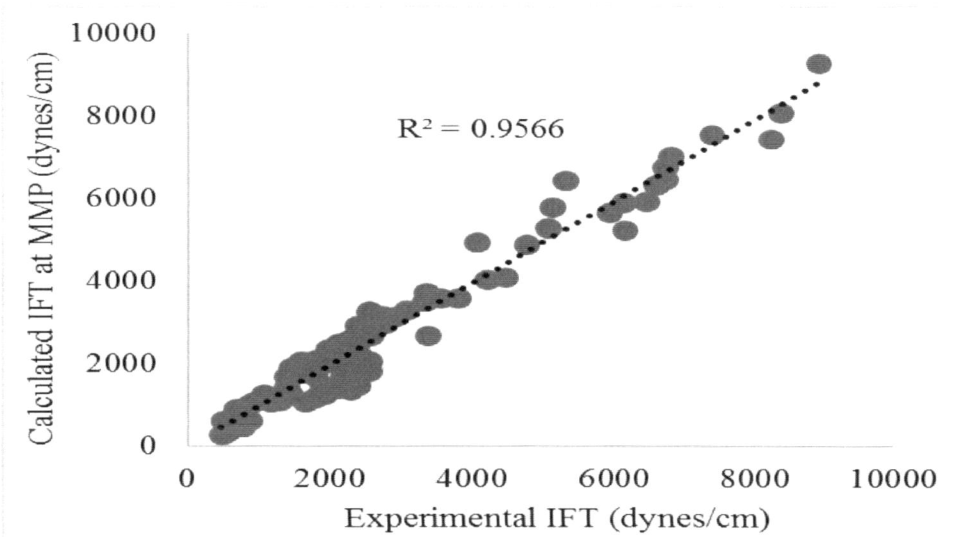

Figure 7. Calculated MMP IFT by confined correlation vs. experimental MMP IFT.

4. Conclusion

The VIT technique is efficient, less expensive and a reproducible technique in determining the MMP compared to irreproducible, most expensive and time consuming ST procedure. VIT test needs at most one day to estimate one MMP, whereas slim-tube requires a minimum of two weeks for one MMP determination. For recombined crude oil, VIT estimated MMP was found significantly good in contrast to slim-tube determined MMP that validates the multiple contact miscibility between recombined fluid and injected CO_2 gas. while in case of dead oil sample, higher deviation was noticed between ST and VIT estimated MMPs. This may be due to deficiency of multiple contacts miscibility and stabilization of crude oil heavier fractions.

With reference to common CO_2-Oil MMP correlations, it was observed that for both of the crude oil samples (recombined oil and dead oil samples), no correlation is able to calculate MMP in close consent with VIT and ST determined MMP. It emphasizes the correlations validity only for certain number of crude samples considered during correlations development. Therefore, most of the common CO_2-Oil MMP correlations are validated by Winprop simulator. Subsequently, a solubility depended technique is proposed and confirmed to predict the CO_2-Oil MMP considering the modifications by nanopore confinement. Commonly, the correlations with molecular weights and reservoir temperature can give a rough estimation of CO_2-Oil MMP. For tight oil reservoirs, taking confinement effect into account, CO_2-Oil MMP is reduced. This study entails the combination of solubility techniques and the nanopore confinement effect that provide an efficient method for MMP calculations.

5. References

[1] Sun J, Zou A and Schechter D (2016, April 11). Experimental and Numerical Studies of CO_2 EOR in Unconventional Liquid Reservoirs with Complex Fracture Networks. Society of Petroleum Engineers.

[2] Zhang K, Li Y, Hong A, Wu K, Jing G, Torsæter O and Chen Z (2015, October 20). Nanofluid Alternating Gas for Tight Oil Exploitation. Society of Petroleum Engineers.

[3] Lake L W, Enhanced Oil Recovery. 1984: Prentice-Hall Englewood Cliffs, NJ 234. ISBN: 0132816016

[4] Johnson J P and Pollin J S (1981, January 1). Measurement and Correlation of CO_2 Miscibility Pressures. Society of Petroleum Engineers.

[5] Holm L W and Josendal V A (1974, December 1). Mechanisms of Oil Displacement by Carbon Dioxide. Society of Petroleum Engineers.

[6] Waqar A, Gholamreza V, Al-Bemani A S, Al-Wahaibi Y, Experimental Determination of Minimum Miscibility Pressure, In Procedia Engineering, Volume 148, 2016, Pages 1191-1198, ISSN 1877-7058.

[7] Rao D N and Lee J I (2000, January 1). Evaluation of Minimum Miscibility Pressure and Composition for Terra Nova Offshore Project Using the New Vanishing Interfacial Tension Technique. Society of Petroleum Engineers.

[8] Dandina N R and Lee J I, Application of the new vanishing interfacial tension technique to evaluate miscibility conditions for the Terra Nova Offshore Project, In Journal of Petroleum Science and Engineering, Volume 35, Issues 3–4, 2002, Pages 247-262, ISSN 0920-4105,

[9] Ayirala S C and Rao D N (2011, September 1). Comparative Evaluation of a New Gas/Oil Miscibility-Determination Technique. Society of Petroleum Engineers.

[10] Franklin M O and Kristian J, An analysis of the vanishing interfacial tension technique for determination of minimum miscibility pressure, In Fluid Phase Equilibria, Volume 255, Issue 2, 2007, Pages 99-109, ISSN 0378-3812,

[11] Zhang K, Seetahal S, Alexander D, He R, Lv J, Wu K, … Chen Z (2016, June 13). Correlation for CO_2 Minimum Miscibility Pressure in Tight Oil Reservoirs. Society of Petroleum Engineers.

[12] Mohammed K E and Hemanta K S, Use of genetic algorithm to estimate CO_2–oil minimum miscibility pressure—a key parameter in design of CO_2 miscible flood, In Journal of Petroleum Science and Engineering, Volume 46, Issues 1–2, 2005, Pages 37-52, ISSN 0920-4105,

[13] Johnson J P and Pollin J S (1981, January 1). Measurement and Correlation of CO_2 Miscibility Pressures. Society of Petroleum Engineers.

[14] Yellig W F and Metcalfe R S (1980, January 1). Determination and Prediction of CO_2 Minimum Miscibility Pressures (includes associated paper 8876). Society of Petroleum Engineers.

[15] Emera M K, Javadpour F, & Sarma H K (2007, August 1). Genetic Algorithm (GA)-Based Correlations Offer More Reliable Prediction of Minimum Miscibility Pressures (MMP) Between Reservoir Oil and CO or Flue Gas. Petroleum Society of Canada.

[16] Nelson P H (2009). Pre-throat sizes in sandstones, tight sandstones, and shales. AAPG bulletin 93. 3: 329-340.

[17] Holm L W and Josendal V A (1982, February 1). Effect of Oil Composition on Miscible-Type Displacement by Carbon Dioxide. Society of Petroleum Engineers.

[18] Pitakbunkate T, Balbuena P B, Moridis G J and Blasingame T A (2016, April 1). Effect of Confinement on Pressure/Volume/Temperature Properties of Hydrocarbons in Shale Reservoirs. Society of Petroleum Engineers.

[19] Ahmadi K and Johns R T (2011, December 1). Multiple-Mixing-Cell Method for MMP Calculations. Society of Petroleum Engineers.

[20] Zhang K, Wang M, Liu Q, Wu K, Yu L, Zhang J and Chen S (2015, October 20). Effects of Adsorption and Confinement on Shale Gas Production Behavior. Society of Petroleum Engineers.

[21] Yangyang L and Jennifer W, Molecular simulation of CO_2 adsorption in micro- and mesoporous carbons with surface heterogeneity, In International Journal of Coal Geology, Volume 104, 2012, Pages 83-95, ISSN 0166-5162,

[22] Allan F M B, handbook of solubility parameters and other cohesion parameters, Second Edition. October 29, 1991 by CRC Press.

[23] Lange E A (1998, April 1). Correlation and Prediction of Residual Oil Saturation for Gas Injection EOR Processes. Society of Petroleum Engineers.

[24] Calvin G C, et al., (1968). High Pressure Gas Chromatography of Nonvolatile Species.

[25] Mohammad R S, Zhao X, Zhang S, et al. Arab J Sci Eng (2017) 42: 1633.

[26] Abu-Eishah S I and Mohammad R S (2016) Phase Behavior of a United Arab Emirates Stock-Tank Oil and Carbon Dioxide at Reservoir Conditions: Experiments and Thermodynamic Modelling. Open Journal of Yangtze Oil and Gas, 1, 1-22.

A comparison of the energy consumption for CO_2 compression process alternatives

S Jackson and E Brodal

Institute for Engineering & Safety, UiT Norges arktiske universitet, Norway
Email: steve.jackson@uit.no

Abstract. The efficient transportation of large volumes of CO_2 generally requires pipelines that will operate above the critical pressure of CO_2. Since most capture processes release CO_2 at low pressure, compression of CO_2 from the point of capture to pipeline will generally be required. The compression duty can be achieved using *conventional* multi-stage compressors or using newer shockwave type compressors. Pumping could also be used if CO_2 is condensed below its critical point. This paper presents a comparison the energy consumption associated with these compression process alternatives. The focus of the review is on the clarity of the comparison and the careful optimisation of each of the scheme considered. The main finding is that the performance advantages claimed for improved CO_2 compression process schemes are often optimistic because the based-line scheme compared against is not well optimized.

1. Introduction

The addition of carbon capture to a power station has a significant impact on efficiency. For a typical coal-fired plant with post combustion capture, the net efficiency could be expected to drop by as much as 7 to 12% due to the heat and power requirements of the capture process [1, 2].

Expressing the efficiency penalty in terms of equivalent lost electrical power production per unit CO_2 captured (kWh_e/tCO_2) is convenient for assessment purposes and this definition shall be used in this paper. Lucquiaud et al. [3] report that for an "nth of a kind CCS plant with current state-of-the-art solvent technology" as 250-300 kWh_e/tCO_2. Compression of CO_2 is expected to consume 90-120 kWh_e/tCO_2 of electrical power [4], making the contribution 30 to 50% of the total. As advances are made in capture technology, the importance of the energy consumption of the compression process will only become more important.

Multi-stage integrally geared type compressors have been used in high-pressure CO_2 applications for decades [5] and are identified in this article as the *conventional* approach to CO_2 compression. This type of compressor uses intercooling between each stage and typically has a pressure ratio around two [6]. Little useful heat is generated during compression and waste heat is normally rejected to a cooling utility.

A *conventional* eight-stage CO_2 compression process with pumping from above critical pressure is illustrated in Figure 1 and Figure 2. In this model, the compressors work isentropic, $\eta_{is} = 1$, and with a constant pressure ratio, $r_8 = 1.87$.

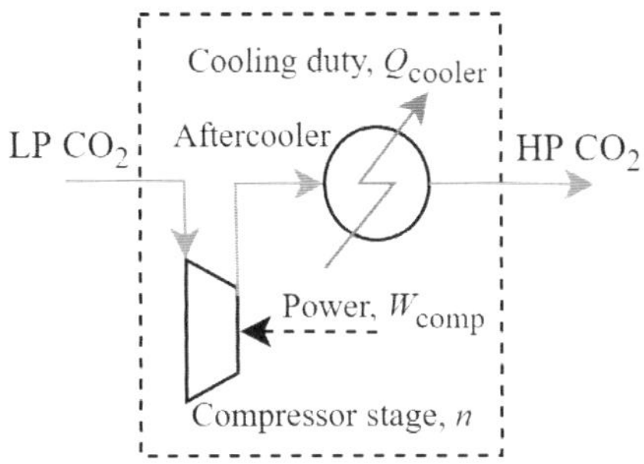

Figure 1. Illustration of a *conventional* compression process with n stages.

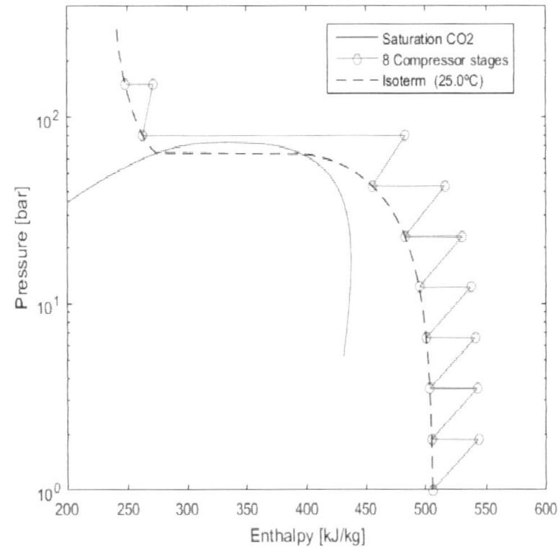

Figure 2. Illustration of a *conventional* eight-stage compressor process in a pressure-enthalpy diagram.

Above its critical pressure (7.38 MPa), CO_2 will behave with properties similar to that of a liquid at 25°C and in an eight-stage compressor process the final stage of compression can be replaced with a pump. To enable a process where CO_2 is pumped to the final pressure from below its critical pressure, refrigeration is needed to cool and condense CO_2 below ambient temperature.

Several studies have investigated compression schemes that utilize pumping of CO_2 from below its critical pressure where an external refrigerant is used to condense below ambient temperature CO_2 [7, 8]. Again, these studies have presented optimized schemes that demonstrate potentially significant CO_2 compressor power savings.

Figure 3 and Figure 4 show an illustration of a three-stage compression process with refrigeration and pumping, where the compression process uses isentropic compressors with pressure ratio, $r_8 = 1.87$.

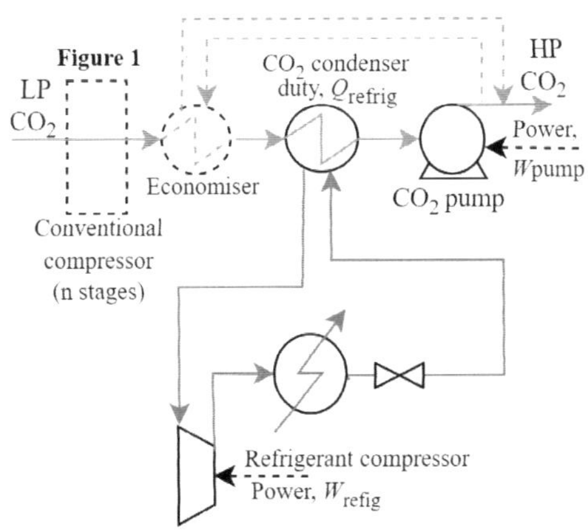

Figure 3. Illustration of a *conventional* compression process with refrigeration and pumping from below critical pressure.

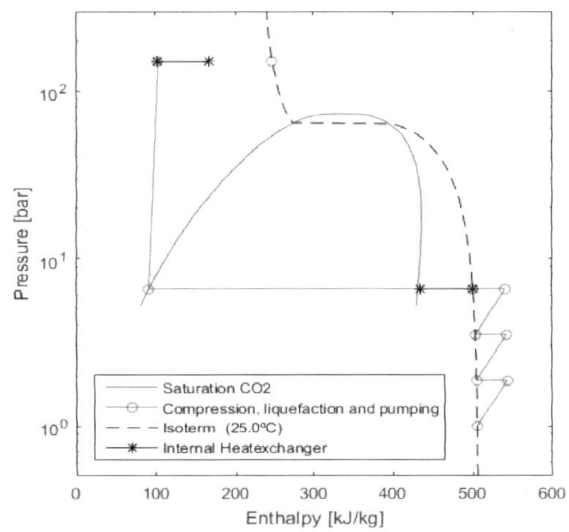

Figure 4. Process with three compressors (r=1.87), internal heat exchanger, refrigeration unit and pump.

Several studies have looked at the opportunity to utilize wasted heat from CO_2 compressor aftercoolers. Although some research has been conducted into the potential benefits of this approach when applied to *conventional*, low pressure ratio compressor designs [9], the focus of research activity has been *unconventional* compressor designs with high pressure ratios [10, 6, 11].

Figure 5 and Figure 6 illustrate a two-stage compression process where heat is recovered from the compressor aftercoolers in an ORC cycle. The ORC cycle is shown with an optional Internal Heat Exchanger (IHE), sometimes referred to as a recuperator.

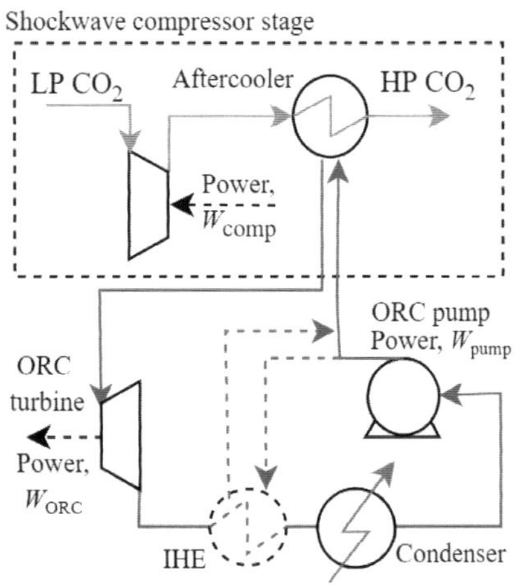

Figure 5. Illustration of a single-stage compression processing with heat recovery using an ORC.

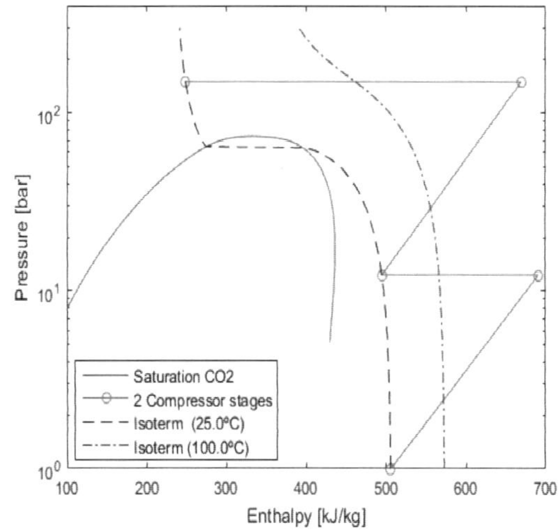

Figure 6. Two Compressor Stages (n = 2) and Pressure Ratio $r_2 = 12$.

Studies into heat recovery from CO_2 compressor aftercoolers have also generally found that a reduction in overall power is achievable. A recent focus for research and development work has been a shockwave type compression technology under developed [12]. This technology claims to offer several advantages over a *conventional* compressor design including "reduced capital costs, smaller footprint, and reduced parasitic plant impact" [13].

A less studied alternative of a pumped processes is one where the heat from compression is to run an absorption refrigeration process. If it is assumed that a bleed before the last compression stage is cooled down using absorption refrigeration process that is driven by the heat from the compressor, as shown in Figure 7, the amount of CO_2 which is liquefied is limited by the amount of useful heat generated in the compression stages and the efficiency of the absorption refrigeration process.

The comparison of the performance of different CO_2 compression schemes for real systems depends on a large number of parameters. For example, compressor efficiencies (per stage), equipment pressure losses, exchanger temperature approaches, etc., all need to be set on a consistent and realistic basis for a valid comparison. None of the studies mentioned have compared the full range of alternative CO_2 compression approaches listed above on a common basis and there is little consistency in the approached used to accounting for the relative value of waste heat and refrigeration duties.

The aim of this study is, therefore, to review all the main process alternatives for CO_2 compression on a simple and consistent basis so that a comparison can be made of the relative performance of these different approaches to CO_2 compression.

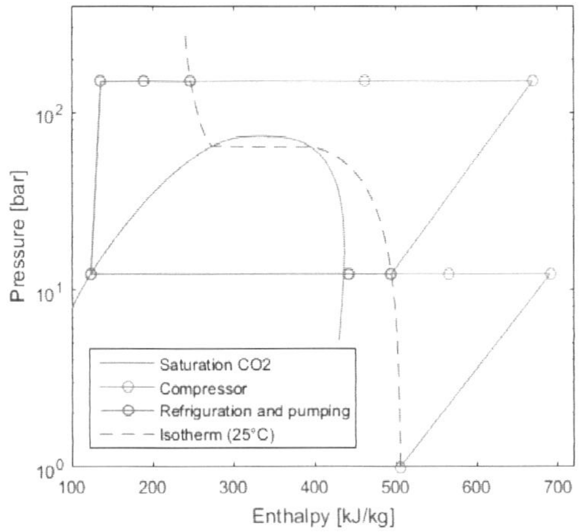

Figure 7. Illustration of a two-stage compression process where absorption refrigeration used to liquefy a CO_2-bleed using the heat generated in compression.

2. Method

The method adopted in this research work is to consider the review of the alternative compression processes in two parts: firstly, a theoretical review of minimum power consumption for a broad range of compressor system design cases based on the performance of idealized thermodynamic cycles; and subsequently, a review of the practically achievable minimum power consumption for a more focused selection of promising cases based on a set of assumed process equipment performance parameters. The first set of work is referred to in this report as *idealized modelling* the second *engineering design study*.

In both parts of the modelling work the feed stream to the compressor is consider to be pure CO_2, the feed pressure is 1 bara and the pipeline pressure 150 bara. Other modelling parameters and the tools used to construct the process models vary between the *idealized* and *engineering* parts of the study and are described under the following headings.

2.1. Idealized modelling cases

The performance of each of the idealized processes is modelled in Matlab based on enthalpy data from the CoolProp fluid package. Coolprop uses the Span & Wagner EOS [14] to calculate CO_2 properties which is implemented in NIST REFPROP and generally considered to be the benchmark for pure CO_2 property prediction.

- All machinery operates ideally with isentropic efficiency, $\eta_{is} = 1$
- Waste heat is rejected at a minimum temperature, $T_C = 25$ °C
- Heat exchangers have a temperature approach, $\Delta T_{min} = 0$ °C and pressure drop, $\Delta P = 0$ bar
- Compressors stage pressure ratio, r, is constant, i.e., $r_n = 150^{1/n}$, where n is the number of stages

2.1.1. Conventional compression *with refrigeration and pumping.* In a *conventional* CO_2 compression processes the pressure ratio per-stage, r, is usually limited to around 2.0 [6] and based on this, the minimum number of stages for these cases is set to 8, giving $r_8 = 1.87$. The total compressor power, W_{comp}, is the sum of the power for each stage, W_n:

$$W_{comp} = \sum_{1}^{n} W_n \tag{1}$$

When CO_2 is condensed above its critical pressure the refrigeration cooling duty, Q_{refrig}, and the refrigeration compressor power, W_{refrig}, reduces to zero. In all other cases, the refrigeration unit duty consumption is assumed to be the same as an ideal Carnot cycle, which represents the highest efficiency possible between two reservoirs with temperatures T_H and T_L.

$$COP_{refrig}^{ideal} = \frac{Q_{refrig}}{W_{refrig}} = \frac{T_L}{T_H - T_L} \tag{2}$$

The power consumption, W_o for refrigeration and pumping models, where the CO_2 is condensed below its critical pressure is the sum of CO_2 compression duty, W_{comp}, refrigeration duty, W_{refrig} and pumping duty, W_{pump}.

$$W_o = W_{comp} + W_{refrig} + W_{pump} \tag{3}$$

If the CO_2 exiting the pump at pipeline pressure is cold, so an economizer is used reduce the temperature before the refrigeration process to minimize the refrigeration duty, see Figure 3.

CO_2 has a triple point of -56.6 °C and 510 kPa, so if freezing is to be avoided in the liquefaction process, CO_2 must be condensed above -56.6 °C. In the *conventional* eight-stage compressor arrangement, the first opportunity to condense CO_2 above its freezing point is after Stage 3 at 650 kPa (see Figure 4).

2.1.2. Shockwave compression and heat recovery. The CO_2 compression is an energy intensive process with high stage outlet temperatures. Heat in the aftercoolers is available at a range of temperature levels. The total amount of heat available being:

$$Q_{HR} = \int_{25°C}^{\infty} q_{HR}\, dT \tag{4}$$

The maximum efficiency for a heat engine working between two reservoirs with temperatures T_H and T_L, is the Carnot efficiency:

$$\eta_{max} = 1 - \frac{T_L}{T_H} \tag{5}$$

The amount of work that can theoretical be recovered from the heat above temperature T_{HR}, using an infinite number of Carnot cycles and ambient temperature cooling T_C is therefore:

$$W_{HR}^{ideal} = \int_{T_{HR}}^{\infty} w_{HR}^{Carnot}\, dT = \int_{T_{HR}}^{\infty} q_{HR}\left(1 - \frac{T_C}{T}\right) dT \tag{6}$$

The total power consumption of the shockwave compression with heat recovery cases is calculated based on the compression power minus the maximum work that can be recovered in the heat recovery process, W_{HR}^{ideal}, as illustrated below.

$$W_o = W_{comp} - W_{HR}^{ideal} \tag{7}$$

2.1.3. Absorption refrigeration. Using heat at high temperature (T_H) for cooling at low temperature (T_L), the theoretical maximum coefficient of performance for an ideal cycle is [15]:

$$COP_{abs}^{ideal} = \frac{Q_L}{Q_H} = \frac{T_L}{T_H} \cdot \frac{T_H - T_C}{T_C - T_L}, \tag{8}$$

where T_C is the ambient temperature, e.g. cooling water.

Figure 7 shows a process where a CO_2 bleed stream is liquefied at $T_L \approx$ -40 °C. This absorption refrigeration process will have $COP_{abs}^{ideal} = 0.83$, assuming T_C = 20 °C and that all heat above T_H = 100 °C can be used in the refrigeration process.
A review of the literature for absorption refrigeration shows that the ideal COP is far from realistic when considering real systems. For example, the ideal coefficient of performance for an absorption refrigeration between T_H= 120 °C and T_C = 5 °C is $COP_{abs}^{ideal} = 4$, while a realistic achievable value is

$COP_{abs}^{real} \approx 1.0$ [16]. Because of this, the assessments made here consider a range of COP values from a maximum of COP_{abs}^{ideal} to a minimum of 0.4 (10% of the maximum, COP_{abs}^{ideal}).

2.2. *Engineering design study*

The results from the *idealized* modelling work identify both *conventional* refrigeration and pumping schemes with between 4 and 6 stages and shockwave compression schemes with 1 to 3 stages as having lowest power consumption. Accordingly, these cases were studies in more detail.

Compressor efficiency is a key parameter and must be selected on a reasonable basis for a fair comparison to be made. This study uses $\eta_{is} = 0.85$ for both *conventional* and shockwave type compressors based on performance claims made for shockwave compressors under development [17], indicating typical adiabatic efficiencies of 85% for both shockwave compressors with a pressure ratio of 9 and *conventional* centrifugal compressor stages operating with a pressure ratio of 2.

Other design parameters used in this part of the study work are summarized below.

- Isentropic efficiency for pumps and turbines, $\eta_{is} = 0.85$
- Waste heat is rejected at a minimum temperature, $T_{HR} = 25$ °C
- Heat exchangers have a temperature approach, $\Delta T_{min} = 5$ °C and pressure drop, $\Delta P = 30$ kPa
- Maximum stage pressure ratio for *conventional* compressors, $r = 2{,}0$
- Maximum stage pressure ratio for shockwave compressors, $r = 10{,}0$
- Minimum CO_2 pumping pressure = CO_2 critical pressure + 10 bar = 8,39 MPa
- Minimum sub-cooling prior to pumping, $\Delta T_{sub} = 10$ °C

Aspen HYSYS was used to build the process models required and energy flows were calculated using the Peng Robinson (PR) equation of state. Studies have confirmed that PR generally provides reasonable accuracy in predicting the relevant properties for pure CO_2 apart from region immediately around the critical point [18, 19].

Additional variables relevant to this part of the study work are the compressor stage pressure ratios; the ORC operating parameters; ORC working fluid and ORC process design. The selection of these parameters was the result of a set of optimization studies that are described under the following headings.

2.2.1. Optimisation of conventional compression cases. CO_2 compressor power is minimized when the stage pressure ratio is allowed to vary. Overall power is also minimized when CO_2 is pumped from immediately above its critical pressure. In the optimization of the *conventional* compressor cases the compressor stage pressure ratios were varied up to the 2.0 limit. Where this limit was reached in any single-stage, a new case was added to the optimization study where the pressure ratio limit was increased to 4.0 for the stage in question. In practical terms, this represents the addition of a new compressor wheel to the compressor design and not a new stage with after cooling. The CO_2 was pumped from 8.39 MPa in all cases.

Ammonia is a refrigerant that exhibits good energy efficiency over the temperature ranges of interest to this study. However, when heat exchanger pressure drops are introduced into the compressor design, the required refrigeration temperature after three stages of compression falls below -33 °C and an ammonia based system would be required to operate below atmospherics pressure. For this reason, the cases considered in this part of the study are for 4 and more stages of compression. In all cases a simple single-stage refrigeration cycle is used.

2.2.2. Optimisation of shockwave compression and heat recovery cases. In this study, heat is recovered using a single-stage ORC, mirroring the approach of Pei et al. [10] and Farajollahi et al. [9]. Selection of an optimum ORC working fluid and process scheme is a complex task in itself and in this study, and based on the work of Lai et al. [20] three working fluids and two process designs were identified for assessment: cyclo-pentane, n-pentane, n-butane, an ORC design with Internal Heat Exchanger (IHE) and an ORC design without IHE as illustrated in Figure 5.

In each of the ORC design cases the maximum pressure was set to 90% of the working fluid critical pressure and the cooling temperature set equal to that in the compression process. The remaining

ORC operating parameters and the CO_2 compressor stage pressure ration were optimized for minimum overall specific power, W_0. Results for all cases are presented in part 3.

2.2.3. Sensitivity cases. To ensure that the selected process design parameters do not favor one of the cases studied over the others, a set of sensitivity studies for key process parameters was conducted. The convention and shockwave compression schemes with lowest overall specific power consumption were used as the basis for the study and the parameters adjusted were as follows:

- Compressor, turbine & pump efficiency, $\eta_{is} = 0.75$ to 0.90
- Heat exchanger pressure drop, $\Delta P = 10$ to 50 kPa
- Cooling temperature, $T_C = 15$ to 45 °C

3. Results

3.1. Idealized modelling

3.1.1. Compression, liquefaction and pumping. Figure 8 and Figure 9 show how the energy usage changes with the number of compressor stages used in the process.

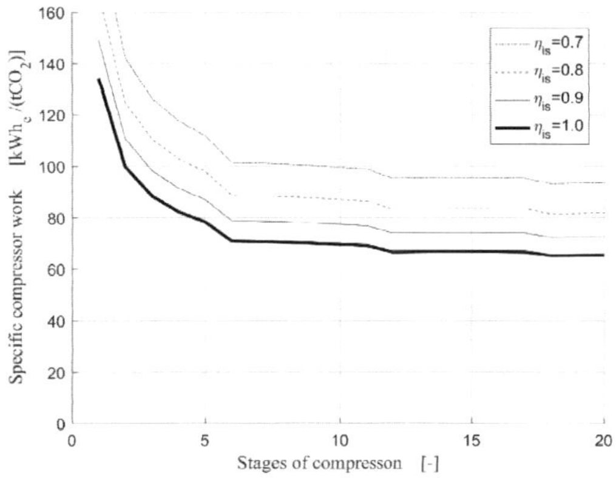

Figure 8. Impact of compressor stages on power for multi-stage compressor designs with equal pressure ration and isentropic efficiency between 0.7 and 1.0.

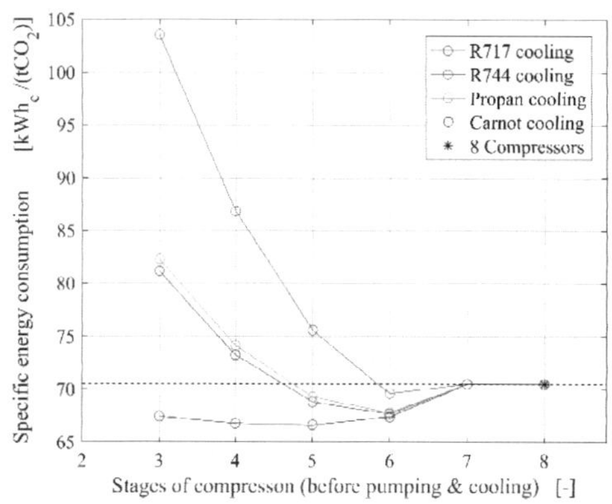

Figure 9. Energy costs in processes with n compressors (r = 1.87 pressure ratios), pump and ideal one-stage refrigeration cycles.

3.1.2. Compression with heat recovery. The amount of heat available at different temperatures is illustrated in Figure 10, where the total energy above 25 °C. The maximum work W_{HR}^{ideal} is illustrated at different temperatures in Figure 11. Figure 12 shows the overall specific energy consumption of the compression process with heat recovery at three different temperature levels.

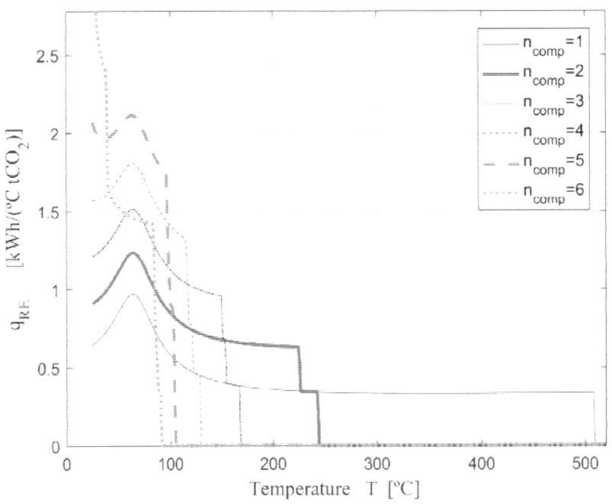

Figure 10. Waste heat from the CO_2 compression process at different temperatures.

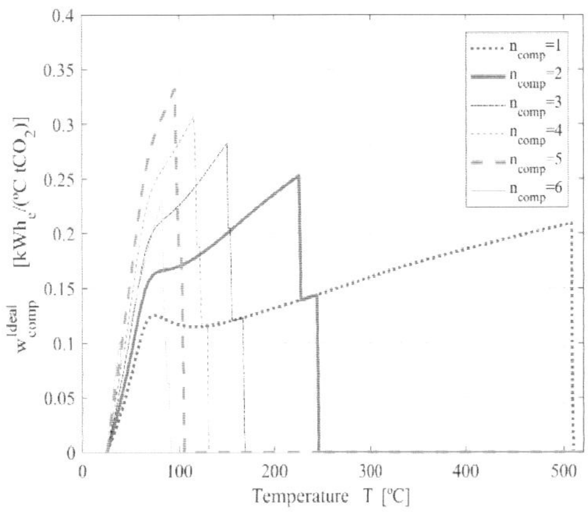

Figure 11. Maximum work which can be recovered by an ideal process from the waste heat produced in various CO_2 compression processes.

3.1.3. Shockwave compression and waste heat recovery (refrigeration). Figure 13 shows that this alternative is not more energy efficient than the *conventional* method, even when an ideal, reversible heat pump process is used as the basis for performance estimates.

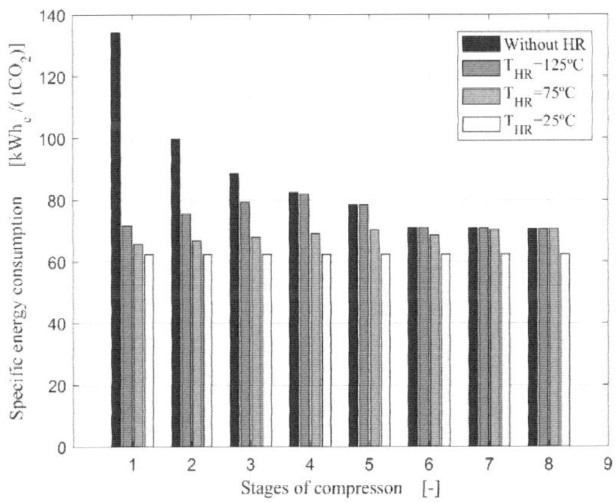

Figure 12. Specific energy consumption of the compression process with Carnot heat recovery above three different T_{HR} temperatures.

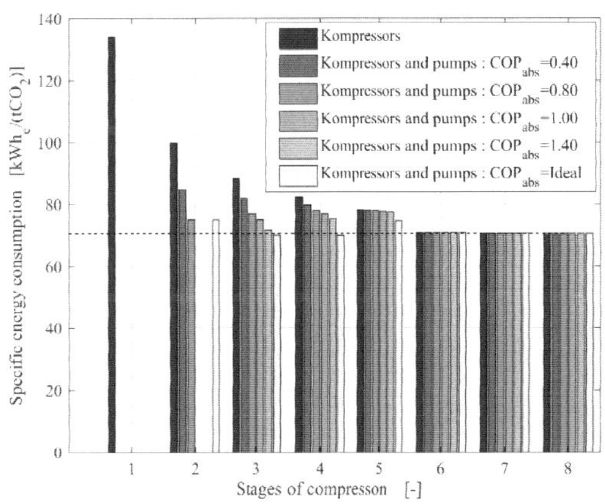

Figure 13. Spesific energy consumtion using absorption refrigeration and n compressors with pressure ratios $r = 150^{1/n}$. Heat from CO_2 above $T_{HR} = 100$ °C is used.

3.2. Engineering design study

The following heading present the results of the engineering design study work.

3.2.1. Conventional compression and pumping cases. Table 1 presents the results of the optimization study for the *conventional* compression cases.

Table 1. Optimisation parameters for *conventional* compression cases.

	Compressor pressure ratio, r			
	C1	C2[*]	C3	C4
Stage 1	2.00	3.98	3.78	3.78
Stage 2	2.00	2.00	2.45	2.46
Stage 3	2.00	2.00	2.00	2.01
Stage 4	2.00	1.92	1.83	1.83
Stage 5	2.00	1.72	1.66	1.66
Stage 6	2.00	1.52	1.45	1.45
Stage 7	1.86	1.24	1.21	1.19
Specific power, W_0 (kWh_e/tCO_2)	91.9	89.2	89.0	89.0

[*]Case C2 is taken to be the optimum case for the *conventional* compression process given the negligible performance gain achieved by adding one additional compressor wheel to this compressor design.

Table 2 presents some important performance results for the refrigeration and pumping cases based on the C2 compressor stage ratios shown in table 1.

Table 2. Summary of performance results for refrigeration cases.

Performance parameters	C2-4	C2-5	C2-6	C2-7
Number of stages before pumping (-)	4	5	6	7
Refrigerant temperature, T_{refrig} (°C)	-20.6	-0.60	10.0	-
Refrigeration compressor pressure ratio, r (-)	5,74	2,49	1,69	-
Refrigeration unit efficiency, COP (-)	3,78	7,80	14,0	-
Specific power, W_0 (kWh_e/tCO_2)	96,2	90,2	87.7	89,2

*for the seven-stage compressor design, the CO_2 condensing temperature is above 25°C

3.2.2. Optimization of shockwave compression cases. Three ORC working fluids and two cycle cases are considered as part of the optimization study for both two and three-stage compression cases. To simplify the presentation of the results, each of the cases is assigned a number which is summarized table 3.

Table 3. Summary of shockwave compression optimisation cases.

ORC fluid	two-stage compressor		three-stage compressor	
	without IHE	with IHE	without IHE	with IHE
Cyclopentane	S1-2	S4-2	S1-3	S4-3
nPentane	S2-2	S5-2	S2-3	S5-3
nButane	S3-2	S6-2	S3-3	S6-3

Table 4 presents the results from the optimization study conducted for the shockwave compression cases based on the case summary presented in table 3.

3.2.3. Overall performance comparisons. Figure 14 and Figure 15 present a beakdown of overall specific power contributions for optimum design cases identified in tables 2 and 3: C2, S3 and S9.

Table 4. Optimisation parameters for shockwave compression cases.

Performance parameters	S1-2	S2-2	S3-2[a]	S4-2	S5-2	S6-2	S1-3	S2-3	S3-3[a]	S4-3	S5-3	S6-3[b]
CO_2 compressor Stage 1, r (-)	10.0	10.0	10.0	10.0	10.0	10.0	10.0	9.20	6.60	9.59	10.0	-

CO_2 compressor Stage 2, r (-)	8.69	8.69	8.69	8.69	8.69	8.69	3.40	4.80	5.00	3.23	4.13	-
CO_2 compressor Stage 3, r (-)	-	-	-	-	-	-	3.40	4.61	4.90	5.07	3.79	-
ORC pressure, P_{max} (MPa)	1.20	3.04	3.53	1.33	3.04	3.53	1.83	3.04	3.53	2.10	2.60	-
ORC temperature, T_{max} (°C)	149	188	152	154	188	145	174	189	147	186	178	-
ORC condenser inlet, T_{cond} (°C)	83.0	90.7	61.0	39.3	35.2	32.0	82.6	90.7	48.3	39.3	35.2	-
ORC thermal efficiency, η_t (-)	0.17	0.19	0.18	0.18	0.21	0.17	0.19	0.19	0.17	0.21	0.21	-
Specific power, W_o (kWh_e/tCO_2)	93.0	87.2	86.1	97.4	92.4	88.5	98.6	94.0	87.7	101	95.3	-

[a]Cases S3 & S9 represents the optimum case for the shockwave compressor

[b] In case S9 the condenser inlet temperature is close to the cooling temperature (48,3°C cf. 25°C) and therefore, the IHE performs no useful function and the results for case S12 are omitted.

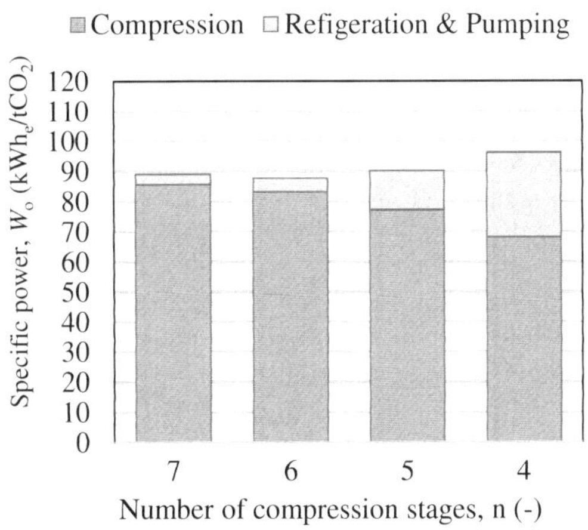

Figure 14. Breakdown of overall specific power contributions for the C2-4, 5, 6 and 7 design cases.

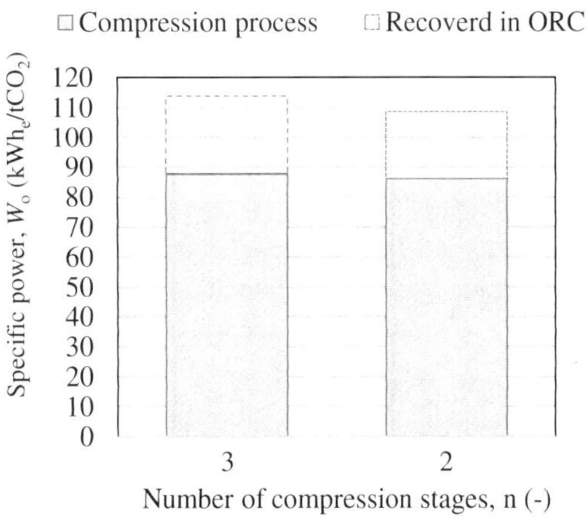

Figure 15. Breakdown of overall specific power contributions for the S3-2 and S3-3 design cases.

3.2.4. Sensitivity study. Figure 16-Figure 18 present variation in overall specific power with three key design parameters for the two best performaing *conventional* and shockwave compression cases: C2-6 and S3-2.

4. Discussion

Figure 8 shows that for a *conventional* compressor design with constant stage pressure ration the power consumption is within 10% of the theoretical minimum achievable power once the compressor has more than eight stages. Figure 9 shows that in the most optimistic case where CO_2 is condensed below critical pressure and pumped to the final pressure a maximum 10% saving in specific compression power is achievable, or that both cases have the same power given the most optimistic imaginable operating parameters.

Figure 9 shows that the minimum power for a scheme where refrigeration and pumping is employed should occur after around five stages of compression and provide a small overall saving in power compared to a *conventional* eight-stage design. The results presented in Figure 14 for the optimized support this, but also shown that the saving in overall power for an optimized scheme is almost negligible.

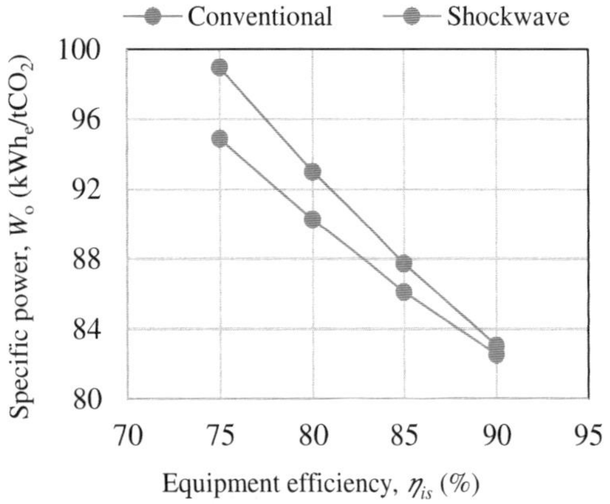

Figure 16. Variation in overall specific power for the *conventional* (C2), and shockwave (S3) design cases with varying equipment efficiency.

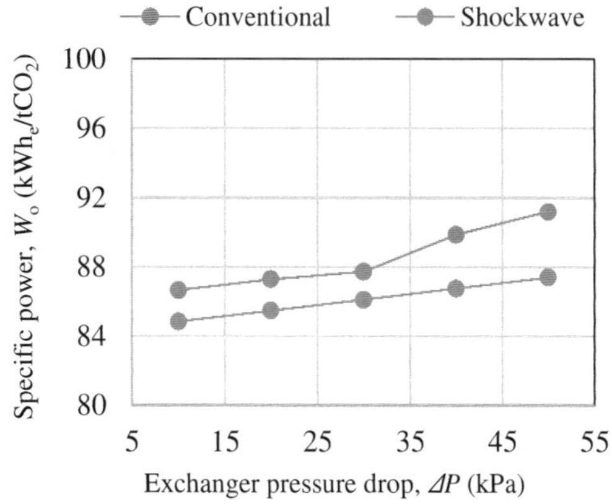

Figure 17. Variation in overall specific power for the *conventional* (C2), and shockwave (S3) design cases with varying exchanger pressure drop.

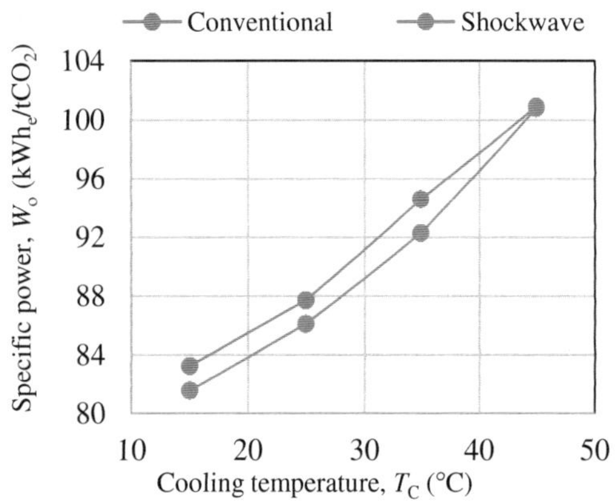

Figure 18. Variation in overall specific power for the *conventional* (C2), and shockwave (S3) design cases with varying cooling temperature.

The savings in overall power achievable in a scheme employing refrigeration and pumping are dependent on the efficiency of the refrigeration unit. In the extreme case, where refrigeration duty is freely available, e.g. at a location adjacent to an LNG import terminal, there is an obvious benefit for this type of a design. In the present study a single-stage ammonia refrigeration unit has been assumed. Table 2 shows that the COP for this refrigeration unit varies between 3.78 and 14.0, with the compressor pressure ratio varying from 5.7 to 1.7. Although a pressure ratio of 5.7 in case C2-4 potentially indicates that a two-stage refrigeration process may offer lower overall power, Figure 14 shows that the case C2-6 is provides lower overall power that cases C2-5 and C2-4, making further optimization of the C2-6 case redundant.

Figure 10 shows that only compressor designs with 3 or fewer stages generate a significant amount of waste heat above 100 °C. Figure 11 and Figure 12 show that, although heat below 100 °C can only be converted into work with low efficiency there is potential to recover some useful work from compression processes with up to 8 stages. This supports the findings of Farajollahi et al [9] who found that 3 MW_e could be recovered from a six-stage compressor with a duty of 13.51 MW_e. However, Figure 12 also shows that the overall specific power consumption for a compression process with heat integration is a minimum where the number of compression stages is also minimized and therefore that the energy saving reported by Farajollahi et al. [9] may be less significant if their design was compared to other, optimized, design alternatives.

The minimum practical number of stages achievable with a shockwave type compressor design is two. The optimization work carried out in this study for two and three-stage compression processes shows that pumping from above the critical pressure of CO_2 is still favorable for these cases. Figure 15 also shows that the optimum case is a two-stage compressor where the temperature level of the waste heat is highest and most beneficial use can be made in an ORC.

If the heat available in the shockwave compression cases can be used at a higher efficiency than that achieved in the present study the comparison of the *conventional* and shockwave results could be affected. The work of Bolland et al. [21] suggests that the use of this heat to replace steam extracted for boiler feed water heating at 200°C in a natural gas power plant could have a thermal efficiency up to, $\eta_t = 0.3$ and that below 80°C the efficiency will drop to less than $\eta_t = 0.1$. Since the average thermal efficiency shown from Table 4 for the optimized ORC schemes developed here is $\eta_t = 0.2$ and the average $T_{max} = 168$ °C, the potential gain from steam cycle integration can only be minor as would be the impact on the findings of this study.

The results of the sensitivity study presented in Figure 16-Figure 18 do not show any bias towards one of the two design cases considered here. They do, however illustrate the importance of compressor efficiency. If both compressors have a 90% efficiency the overall specific power consumption is almost equal and, over the range of efficiencies considered, the difference in efficiency separating the two compressors for any given overall power consumptions is only 2.5%. If the uncertainty in compressor efficiency is any greater than this, neither of the two cases has a clear advantage.

5. Conclusions

The results presented in this paper show that given equal compressor efficiency for both high pressure-ratio shockwave compressors and *conventional*, low pressure-ratio compressors, the minimum achievable power is essentially the same if both schemes are carefully optimized. The main conclusion of this study is, therefore, that the reduction in power consumption often claimed for CO_2 compression processes where either energy is recovered from waste heat in the compressor aftercoolers, or where CO_2 is condensed against a refrigerant and pumped to pipeline pressure from below its critical pressure is often optimistic. However, the result of the sensitivity study show that if one of these two alternatives have an advantage in terms of compressor efficiency, it is likely that this will translate into an advantage in terms of overall compression power requirements. Likewise, if a scheme using refrigeration and pumping has access to an external source of cooling, or if a shockwave compression process has a particular opportunity to use waste heat efficiently, this will play an important role in determining which process will be most efficient.

6. Highlights

- Comparative study of conventional and shockwave CO_2 compression
- Clear and unbiased comparison based on careful optimization of each case
- Concise presentation of results including assessment of the sensitivity to design parameters

7. Reference

[1] Supekar S D and Skerlos S J 2015 Reassessing the efficiency penalty from carbon capture in coal-fired power plants *Environmental Science & Technology* **49** 12576–84

[2] Rubin E S, Mantripragada H, Marks A, Versteeg P and Kitchin J 2012 The outlook for improved carbon capture technology *Progress in Energy and Combustion Science* **38** 630–71

[3] Lucquiaud M, Liang X, Errey O, Chalmers H and Gibbins J 2013 Addressing technology uncertainties in power plants with *Energy Procedia* **l37** 2359–68

[4] Jordal K and Aspelund A 2007 Gas conditioning–The interface between CO_2 capture and transport *International Journal of Greenhouse Gas Control* **1** 343–54

[5] Dittmer R and Strube R 2015 Integrally-geared compressors as state-of-the-art technology *Carbon Capture Journal* **47** 15–18

[6] Romeo L M, Bolea I, Lara Y and Escosa J M 2009 Optimization of intercooling compression in CO_2 capture systems *Applied Thermal Engineering* **29** 1744–51

[7] Duan L, Chen X and Yang Y 2013 Study on a novel process for CO_2 compression and liquefaction integrated with the refrigeration process *International Journal of Energy Research* **37** 1453–64
[8] Alabdulkarem A, Hwang Y and Radermacher R 2012 Development of CO_2 liquefaction cycles for CO_2 sequestration *Applied Thermal Engineering* **33** 144–56
[9] Farajollahi H and Hossainpour S 2017 Application of organic rankine cycle in integration of thermal power *Energy* **118** 927-36
[10] Pei P, Barse K, Gil A J and Nasah J 2014 Waste heat recovery in CO_2 compression *International Journal of Greenhouse Gas Control* **30** 86–96
[11] Harkina T, Hoadley A and Hooper B 2011 A comparison of the process integration of shockwave CO_2 compression with conventional turbo machinery into PCC power station design *Energy Procedia* **4** 1339–46
[12] Kuzdzal M 2015 Compression solutions for large-scale *Carbon Capture Journal* **47** 13-14
[13] DOE 2015 DOE Selects eight projects to receive funding for reducing the cost of CO_2 capture and compression *DOE* webpage http://wwwenergygov/fe/articles/doe-selects-eight-projects-receive-funding-reducing-cost-co2-capture-and-compression
[14] Wagner W and Span R 1996 A new equation of state for carbon dioxide covering the fluid region from the tripple-point *Journal of Physical and Chemical Reference Data* **25** 1509-96
[15] ASHRAE 2009 ASHRAE handbook of fundamentals *ASHRAE*
[16] Srikhirin P, Aphornratana S and Chungpaibulpatana S 2001 A review of absorption refrigeration technologies *Renewable and Sustainable Energy Reviews* **5** 343–72
[17] Ramgen Power Systems 2008 The rampressor competitive advantages *Ramgen Power Systems* Webpage http://wwwramgencom/tech_shock_rampressor.html
[18] Mazzoccoli M, Bosio B, Arato E and Brandani S 2014 Comparison of equations-of-state with P-v-T experimental data of binary mixtures rich in CO_2 under the conditions of pipeline transport *The Journal of Supercritical Fluids* **95** 474–90
[19] Wilhelmsena Ø, Skaugena G, Jørstadb O and Hailong L 2012 Evaluation of SPUNG# and other equations of state for use in carbon capture and storage modelling *Energy Procedia* **23** 236–45
[20] Lai A N, Wendland M and Fischer J 2011 Working fluids for high-temperature organic rankine cycles *Energy* **36** 199–211
[21] Bolland O and Undrum H A 2003 novel methodology for comparing CO_2 capture options for natural gas-fired combined cycle plants *Advances in Environmental Research* **7** 901–11

IOP Conf. Series: Earth and Environmental Science **167** (2018) 012032 doi:10.1088/1755-1315/167/1/012032

Bioremediation Process for Oil Contaminated Underground Water

M Imani[1], M Ebrahimnejad Miangoleh[2] and A Mokhtarpoor Bazmin Abadi[3]

1 Azarnegar Sari company, Shahabi street, Sari, Mazandaran, Iran
2 Bargh Araye Shomal company, khazar steert, Sari, Mazandaran, Iran
3 Gardening department, Islamic Azad University of science and research unit, Tehran, Iran
E-mail: mehdi.imani.mi@gmail.com

Abstract. The petroleum crude is a pollutant that penetrates from surface soils and reaches to the underground water sources. An experimental setup was used for validating bioremediation method that prevents pollutant dispersion in the underground water sources. For this purpose, diluted Autotrophicus bacterium the water was injected into a reservoir and recovered from a well at downstream. Furthermore, a numerical model based on diffusion of the pollutant in the porous media and a disappearing reaction rate was used for validation of existing models in proposed method. Finally, experimental and model results show that injection rate of microorganism has an optimum. In present work, bioremediation process in the experimental setup that has a recycling pattern was introduced for tracing of contaminant dispersion, and the results of a numerical modeling was validated versus experiment data. Moreover, biodegradation reaction cause to reduction of pollutant dispersion flux that shows better microorganism performance in the aquifer.

1. Introduction

Groundwater contamination is considered to be one of the most significant ecological and health problems, and it greatly affects the general environmental features of large regions and drinking water supply conditions in many countries [l]. The protection of public water supplies is of paramount importance. Groundwater, the main source of drinking water for about 50% of the world's total population, is said to be contaminated if some hazardous compounds in the water meet or exceed U.S. EPA standards [2].

The several research studies have been done for providing valuable experimental data in places that possibility of underground contamination exists. These studies help to taking strategies for preventing or controlling of pollutant distribution. For example, Liu et al. and AI-Awadi et al. reported field data of different places, otherwise, Lucianoa et al. and Ayotamuno et al. reported laboratory data [3-6].

Hes et al. studied bioremediation of a diesel fuel contaminated aquifer in laboratory aquifer [7]. Likewise Chen et al. [8], with application of enhanced bioremediation system from the spill area to the down gradient area (located approximately 300 m from the source area) found that the percentage of total hydrocarbon removal increased [9].

Contaminants can adhere to soil and subsequent rainfall can leach the adhered contaminant into groundwater. Recently, Clinton et al., Mirbagheri and Hashemi Monfared, and Du and Liang reported that state using numerical modeling has benefits and it could be used for anticipating of contaminant transportation in the porous media [10-12]. In addition, Suarez and Rifai studied about kinetic model of bioremediation and reported reaction rate parameters for several in situ cases [13].

2. Modeling

2.1. Flow in variably saturated porous media
Richards' equation models flow in variably saturated porous media. With variably saturated flow, hydraulic properties change as fluids move through the medium, filling some pores and draining others.

$$\left(\frac{c}{\rho g}+s_e s\right)\frac{\partial p}{\partial t}+\nabla\cdot\left[-\frac{k_s}{\mu}k_r\nabla(p+\rho g h)\right]=Q_s \tag{1}$$

where pressure, p, is the dependent variable. C represents the specific capacity, Se denotes the effective saturation, S is the storage coefficient, Ks gives the hydraulic permeability, μ is the fluid dynamic viscosity, k, denotes the relative permeability, p is the fluid density, g is acceleration of gravity, h represents the vertical coordinate, and Q_s is the fluid source (positive) or sink (negative). Solving Richards' equation for a dependent variable of pressure (p) provides features for specifying the values of the hydraulic head or pressure head on the boundaries of a model.
The fluid velocity across the faces of an infinitesimally small surface is:

$$u=-\frac{k_s}{\mu}k_r\nabla(p+\rho g h) \tag{2}$$

Dividing u by the volume liquid fraction, Θ characterize the average fluid velocity in the pores (u_a = u/Θ). As following, the saturation was defined with the van Genuchten equations:

$$\theta=\begin{cases}\theta_r+s_e(\theta_s-\theta_r) & H_\rho<0\\ \theta S & H_\rho\geq 0\end{cases} \tag{3}$$

$$s_e=\begin{cases}\left(1+\left|aH_p\right|^n\right)^{-m} & h_p<0\\ 1 & h_p\geq 0\end{cases} \tag{4}$$

Specific capacity (C) and relative permeability (k_r) for using in Eq.1 was defined as following:

$$c=\begin{cases}\frac{am}{1-m}(\theta_s-\theta_r)s_e^{1/m}\left(1-s_e^{1/m}\right)^m & H_P<0\\ 0 & H_P\geq 0\end{cases} \tag{5}$$

$$k_r=\begin{cases}s_e\left(1-\left(1-s_e^{1/m}\right)^m\right)^2 & h_p<0\\ 1 & h_p\geq 0\end{cases} \tag{6}$$

In Eq. 3 to 6, constants a, n, m, and 1 specify a particular medium type, and HP = p/(ρ_g) is pressure head, also θr and θs, are residual and saturated liquid volume fractions.

3. Solute Transport

The following equations for the concentrations c_i describe the transport of solutes in variably saturated porous media for the most general case, when the pore space is primarily filled with liquid but also contain pockets or immobile gas:

$$\frac{\partial}{\partial t}(\theta c_i)+\frac{\partial}{\partial t}(\rho_b c_{pi})+\frac{\partial}{\partial t}(a_v c_{Gi})+\nabla(c_i u)=\nabla\cdot[(D_{Di}+D_{ei})\nabla c_i]+R_i+s_i \tag{7}$$

Here c_i denotes the concentration of species i in the liquid (mass per liquid volume), Cp the amount adsorbed to solid particles (mass per dry unit weight of the solid), and cG the concentration in the gas phase (mass per volume gas). The equation for bulk density and gas volume fraction are:

$$P_b=(1-\theta_s)P_p \tag{8}$$

$$a_v=\theta_s-\theta \tag{9}$$

where ρ_p is the solid phase density and 0s is the pore volume fraction (dimensionless).
On the left-hand side of Eq. 7, the first three terms correspond to the accumulation of species mass within the liquid, solid, and gas phases, while the last term describes the convection due to the directional velocity u. On the right-hand side, the first term introduces the spreading off mass from to mechanical mixing (dispersion), as well as from diffusion and volatilization to the gas phase. The tensor is denoted Do and the effective diffusion by De. The last two term describe production or consumption of the species; R, is a reaction rate expression.
In order to solve for the solute concentration of species i, C_i ,the solute mass sobbed to solids C_{pi} and dissolved in the gas-phase $C_{Gi}$are assumed to be functions of C_i. Expanding the time-dependent terms:

$$\frac{\partial}{\partial_t}(\theta c_i) + \frac{\partial}{\partial_t}(P_b C_{Pi}) + \frac{\partial}{\partial_t}(a_v C_{Gi}) = (\theta + P_b K_{Pi} + a_v K_{Gi})\frac{\partial_{Ci}}{\partial_t} + (1 - K_{Gi})C_i\frac{\partial\theta}{\partial_t} + (K_{Gi}C_i - P_P C_{Pi})\frac{\partial\theta_s}{\partial_t} \tag{10}$$

where kpi = ∂cr/∂ci is the sorption isotherm and kGi = ∂cGJ∂ci is the linear volatilization .

$$(\theta + P_b K_{Pi} + a_v K_{Gi})\frac{\partial_{Ci}}{\partial_t} + (1 - \mathrm{K_{Gi}})C_i\frac{\partial\theta}{\partial_t} + (K_{Gi}C_i - P_p C_{Pi}) + \nabla(C_i\vec{u}) = \tag{11}$$

$$\nabla \bullet [(D_{Di} + D_{ei})\nabla C_i] + R_i + S_i$$

3.1. Sorption
The physics interface predefines three relationships to predict the solid concentrations, C_{pi} from the concentration in the liquid phase, ci:

$$C_p = K_F C^N \tag{12}$$

$$C_p = \frac{K_L sc}{1+K_L c} \tag{13}$$

3.2. Dispersion
In the solute Transport interface dispersion is controlled through the dispersion tensor D_D. The tensor formulation by Burnett and Friend can be used:

$$D_{Lxx} = a_1\frac{u^2}{|u|} + a_2\frac{v^2}{|u|} + a_3\frac{w^2}{|u|} \tag{14}$$

$$D_{Lyy} = a_1\frac{v^2}{|u|} + a_2\frac{u^2}{|u|} + a_3\frac{w^2}{|u|} \tag{15}$$

$$D_{Lxy} = D_{Lyx} = (a_1 - a_2)\frac{uv}{|u|} \tag{16}$$

Here the fluid velocities u, v, and w correspond to the components of the velocity vector u in the x, y, and z directions, respectively, and a1 is the longitudinal dispersivity. If z is vertical, a2 and a3, are the dispersivities in the transverse horizontal and transverse vertical directions, respectively.

3.3. Diffusion
In the mass transport in porous media interface, the effective diffusivities are defined as:

$$D_e = \theta\tau_L D_L + a_v \tau_G D_G \tag{17}$$

Here DL and DG are the single phase diffusion coefficients for the species in the liquid and gas phase respectively, and LL and LG are the corresponding tortuosity factors.

$$\tau_L = \theta^{7/3}\theta_s^{-2} \tag{18}$$

$$\tau_G = a_v^{7/3}\theta_s^{-2} \tag{19}$$

3.4. Chemical Reactions

Chemical reactions of all types influence solute transport in the environment. Examples include biodegradation, radioactive decay, and transformation to tracked products, temperature- and pressure - dependent functions, exothermic reactions, and endothermic reactions. The reactions represent change in solute mass per unit volume porous medium per time. For reactions in a liquid phase, multiply the expression by the liquid volume fraction θ.

$$R_i = \theta r_i \tag{20}$$

Here r_i is a reaction rate respectively.
With adding microorganism to the underground water, the hydrocarbon compounds were biodegraded into carbon dioxide and methane [14]. Anaerobic microorganisms produce enzymes in the presence of carbon sources which are responsible for attacking the hydrocarbon molecules.

$$4C_nH_m + (4n - m)H_2O \rightarrow \left(\frac{4n-m}{2}\right) CO_2 + \left(\frac{4n-m}{2}\right) CH_4 \tag{21}$$

The general formula of the chemical reaction kinetic that can be described as the rate of Total Petroleum Hydrocarbon (TPH) reduction is:

$$r_i \left[\frac{mg}{lit.day}\right] = \frac{dc_i}{dt} = -k_i^n \tag{22}$$

Considering the values of reaction orders (n) and reaction rate constants (k), the biodegradati rate is function of pollutant concentration.
Bambang et al. reported reaction rates parameters (k, n) for contaminated soil biodegradation by using of some batteries. As shown in Table 1, Xanthobacter Autotrophicus bacteria could degrade the petroleum oil sludge faster than the others [15].

Table 1. Microorrganisms and reaction rate parameters

Bacteria	Reaction order (n)	Reaction constant (k) $[day^{-1}]$
Xanthobacter Autorophicus	0.8823	0.0324
Pseudomonas pseudoalcaligenes	1.0949	0.00189
Bacillus Me_gaterium	1.3985	0.0204

Suarez and Rifai reported first-order (n=l) biodegradation rates for select petroleum hydrocarbon compounds in groundwater [13]. The reaction constants for biodegradation of different hydrocarbons have been shown in Figure 1.

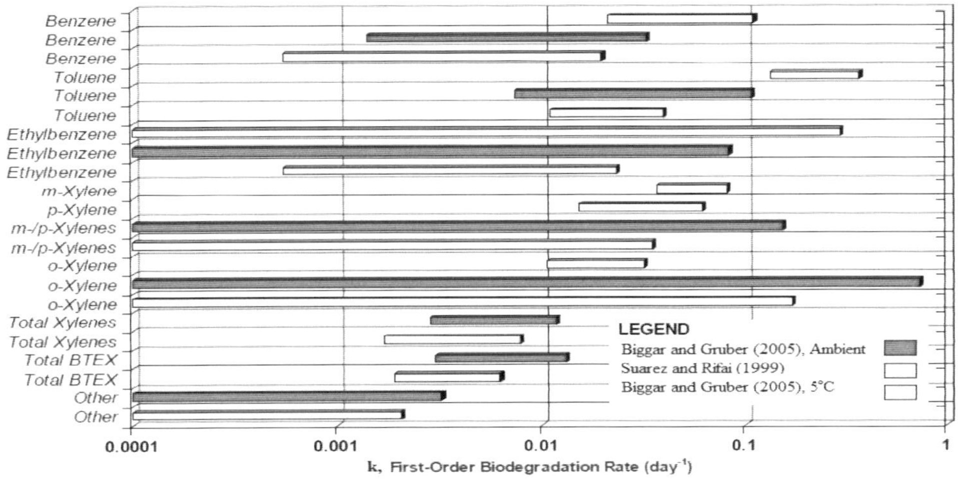

Figure 1: The reaction rate constants of bioremediation reaction for important pollutants

As shown in Figure 1, reaction constant for each pollutant varies with temperature, microorganism type and other parameters. Also Buscheck and Alcantar introduced porous media structure effect on reaction rate constant [16].

4. Experimentation

In this work, an experimental setup prepared for producing valuable data. As shown in Figure 2, a sand pack surrounded with two pools (injection and removal wells) is used for studying pollutant distribution in porous media. Sand pack jointed into pools with couple of sintered filters. The Autotrophicus bacterium was used for biodegradation of oil pollutant. During of experimentation concentration of bacterium is measured and adjusted in specific times. In the injection well, microorganism medium which is needed for their cultivation was controlled by measuring online concentrations of major components in the well every 8 hours. The cultivation medium of Also for reproducing of pollutant zone, pollutant droplet is added for considered zone manually. The used pollutant is synthetically petroleum oil-water solution. Rain fallen (surface water) is simulated in experimental setup with parallel distributers and water supply system. In the removal well contaminant concentration is measured with preparing chemical analysis samples. Physical properties of porous media were shown in Table 2.

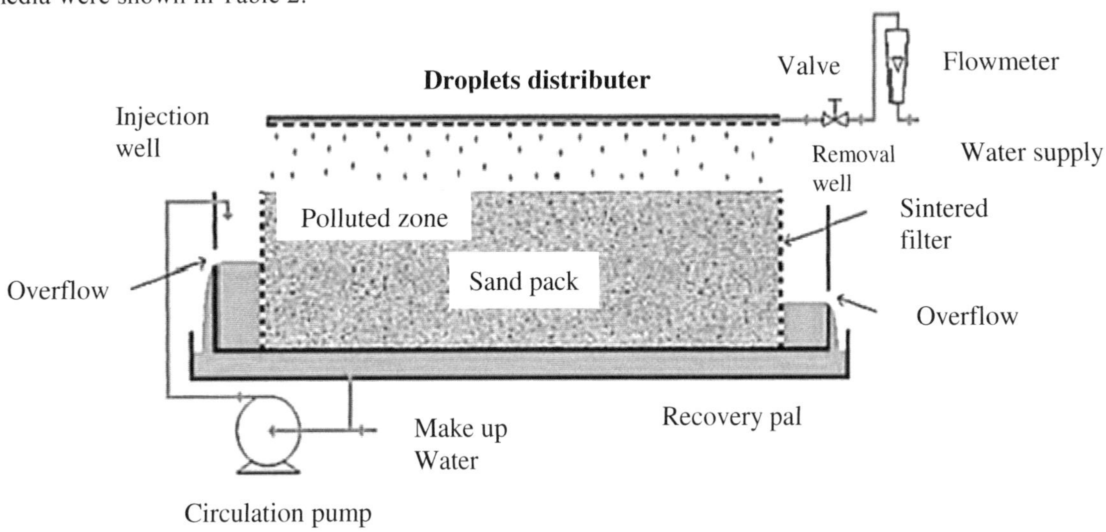

Figure 2. The schematic of experimental setup for bioremediation process

Table 2. The used physical properties in the numerical modeling.

Parameter	unit	value
Conductivity	m/s	2.94x 10-
Voidage	-	0.35
Bulk density	Kg/m3	1400
Height	m	0.2
Width	m	1

5. Results

Numerical modeling of the pollutant dispersion in a porous media is carried out using indoor CFD code based on finite difference method. As shown in Figure 3, the calculation domain is a rectangle porous media that proposed for computation has width and height of L=l.O and H=0.2m, respectively. The domain was divided into uniform 20xlOO grids. Also, the simulation was continued for $3*10^5$ second with time steps equal to 100 second. Here the grid independence is tested by varying the density of computational domain. The grid refinement was continued until the solution parameters such as time averaged pressure drop showed no variations. As discussed previously, physical properties were introduced to the numerical modeling proportional to the experimental work (Table 2). Boundary conditions (B.C.s) for solving of partial differential equations of numerical model (Eq. 1, 11) were listed in Table 3, too.

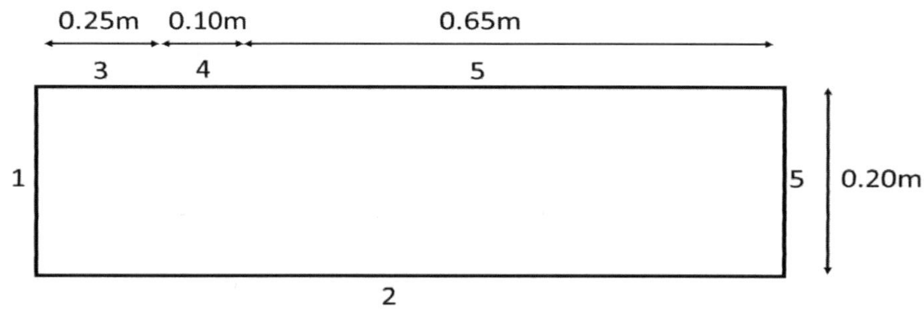

Figure 3. Numerical domain with considered boundary conditions

Table 3. the proposed boundary conditions in the numerical modeling

Surface	B.C.s for hydraulic head	B.C.s for ith pollutant and microorganism transport
1	h=O.lm	Ci=O, Cmic=lO
2	dh/dz=O, Isolated	de/dz, Dc micldz=O, Isolated
3.5	u=l*10^{-6}m/s *	ci, Cmic=O
4	u=l * *I O*$^{-6}$m/s *	c=If), Cmic=O
6	h=0.05m	Dci/dx, Dc mic/dx=O, Outlet

5.1. Water velocity

As shown in Figure 4, superficial velocity of water in the porous media increases with time. At first, water saturation in the aquifer starts to increase and velocity of water reaches to the steady flow condition. At t=l.65xl05s (:=:o2day) an epoch point occurs in velocity trend. Due to a synthetic pressure (or head) gradient within the aquifer that is produced by pump, the water velocity increases with approach of the surface water into aquifer. From Figure 4 it can be find that experimental and model results are in good agreement. Before the epoch point, underestimate of modeling results is because of higher capillary anticipation in the model that prevents entering of water and pushing of trapped air through the porous media or using the lower conductivity value in the modeling. Likewise, after the epoch point upper estimate of model could be due to ignoring of surface evaporation at top of the porous media.

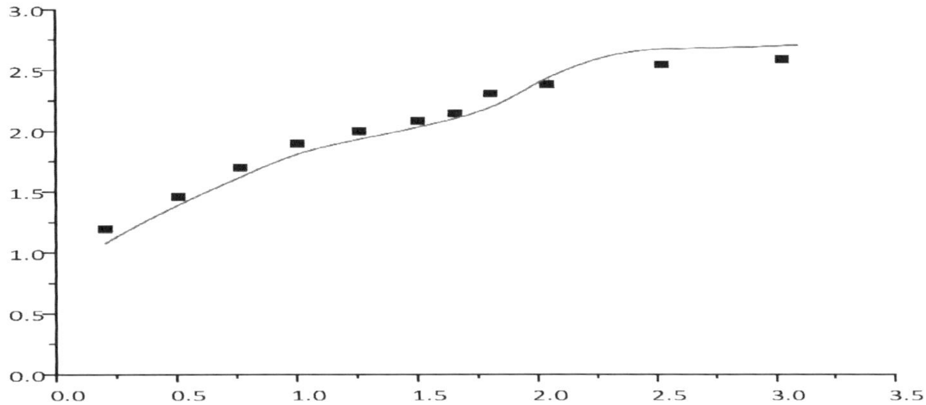

Figure 4. The result of numerical and experimental superficial water velocity along the porous media

The velocity counter of water flow at t=$3*10^5$s was presented in Figure. 5. In addition, the surface water velocity was shown with downer come vectors and aquifer water velocity with declined vectors. It is obvious that the water velocity increases along the aquifer due to addition of the surface waters and decreasing of water table that achieved with hydraulic head gradient.

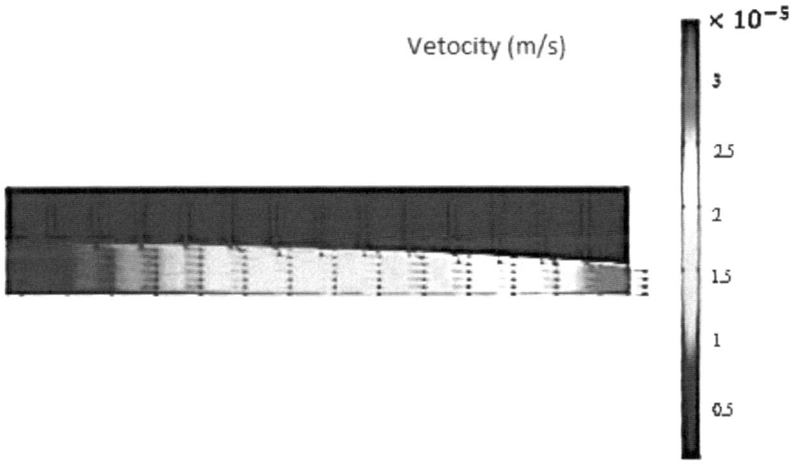

Figure 5. Numerical result of water velocity and saturation distribution

5.2. Contaminant dispersion

Component molar flux (c,u) is an important parameter in mass transfer and finding mass dispersion within a medium. The mass flux multiplying by the crossing area of pollutant gives flow rate of pollutant dispersion in the domain. In Figure 6, contour of pollutant molar flux distribution was shown. Through the unsaturated zone of the porous media, the pollutant introduces to movement with surface water flow. The pollutant penetrates, rises down and reaches to water table (boundary of saturated and unsaturated zones). Fortunately, through the aquifer, the water flow is laminar flow and lower convectional mass transfer occurs in deep direction of aquifer. It is obvious that the larger amount of pollutants is gathered in water table.

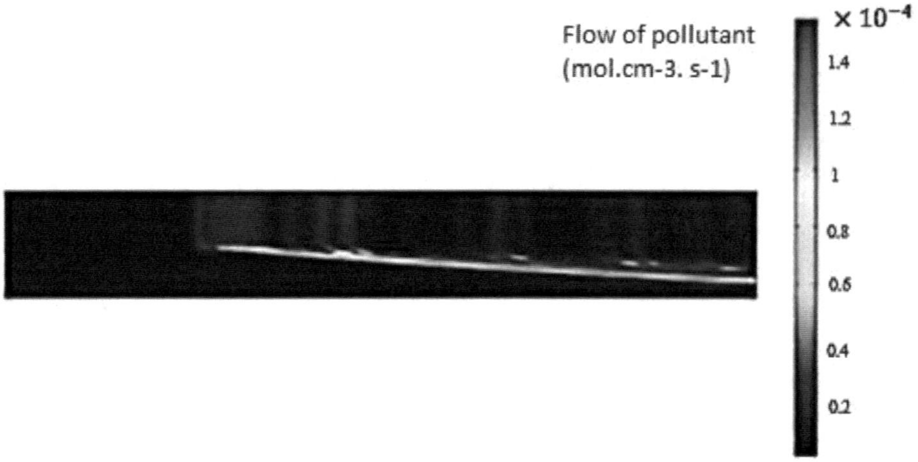

Figure 6. Numerical result of pollutant distribution flux

The water table is contact line of air and water. Then, this is a potential point to application of microorganisms for biodegradation of pollutants within the aquifers, because air could be supplied easier than water saturated zone.

Space/stay time is time required to process one reaction volume of feed that is measured at specified conditions. Here, the reaction volume is aquifer and feed is entered microorganism at injection well. It is obvious that reaction progress increases with increasing of space time and biodegradation yield increases. On the other hand, space time has an inverse relation with entering flow rate of injection well ($\tau = (V/\vartheta)$). This means that higher space time that could be reached with lower injection flow rates that causes the pollutant dispersion in unwanted direction, because pollutant have enough time to diffusion in all directions, and it is not desired. In Figure 7, pollutant flux (measured in removal well) is plotted versus time at various positive heads of injection that produces various entering flow rate in injection well. Also, for showing effect of microorganism, results of a case study without using of microorganism have been shown in Figure 7. As shown, with increasing injection positive head (gives higher flow rate of entering flow) decreases space time and lower biodegradation takes place. In Figure 7, for a specific entering flow rate (head=5m), biodegradation reaction reduced pollutant dispersion flux which presents microorganism performance in the aquifer. Diversion from experimental results were increased with increasing entering positive head, it can be due to non-Darcy flows. For this cases, the head equation (Eq.1) should be modified and with reduction of errors in prediction of velocity profiles better agreement between model and experimental results could be reached.

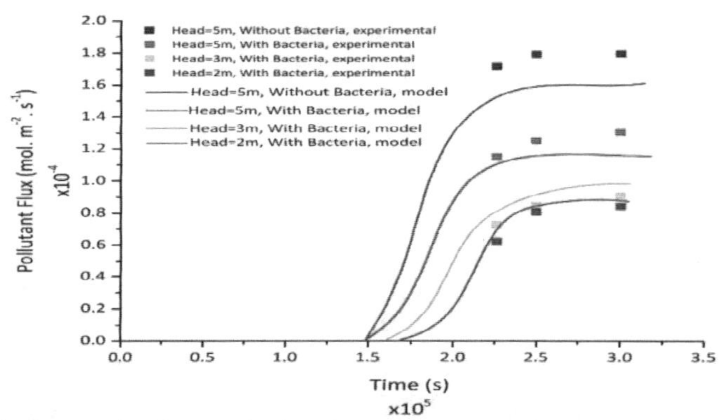

Figure 7. The result of numerical and experimental pollutant dispersion flux along the aquifer

6. Conclusion

In this work, experimental and numerical studies have been done on clarifying of pollutant dispersion in the underground water sources; also the remediation process was introduced for controlling of contaminant growth. The modeling results validated for water velocity within porous media that was proposed for experimentation. In addition, it is found that the pollutant dispersion is concentrated on water table, and accumulation of pollutant is lower in other points of saturated zone. Likewise, space time goes down with flow rate of entering flow and as a result lower biodegradation takes place. Moreover, biodegradation reaction cause to reduction of pollutant dispersion flux that shows better microorganism performance in the aquifer.

7. References

[l] U.S. Environmental Protection Agency Guidelines for delineation of wellhead protection areas, EPA 440/6-87- 010 USPA, Office of Groundwater Protection, Washington, D.C. (1987).
[2] Victor P, Jackson K, The extent of groundwater contamination in the United States. Groundwater contamination, studies in geophysics, National Academy Pres, p. 23-33 (1984).
[3] Liu A, Ming J, A 0. Ramble Science of the Total Environment 346 (2005): 112- 120.
[4] E. Al-A, Ali. M, Ahmed. A, Kamal. H, Advances in Environmental Research 7 (2003): 367-380. [5] Alireza. L, Povel. V, M. Patris. P, Journal of Hazardous Materials 176 (2010): 1006-1017.
[6] Morteza.A, Robin. K, S.O.T. Ogaji, S.D. Probert, Applied Energy 83: (2006) 1258-1264.
[7] A. Hess, Povel. H, Denis. H, J. Zeyer, Journal of Contaminant Hydrology 23: (1996) 329-345.
[8] Maria L, Ewa Z, Ewa K, 'kiewicz M, Ewa s, Muszyn ski Adam, Tabernacka A, Naumczyk Jeremi, eczalik Maciej, Ecological Engineering 37 (2011): 1895- 1900.
[9] C. Ku-Fan, K. C-Ming, C. Chiu-Wen, S.Y. Rao, L. Mu-Sheng, Joliyan. E. Sciences, 22 (2010): 864-871
[10] Willson S. Clinton, Weaver W. James, C. J. Randall, Environmental Modelling & Software 21 (2006): 16-32.
[11] SamanA. M, S. A. Hashemi M, Int. J. Environ. Sci. Tech., 6 (2009): 233-242 .
[12] Chuanbin Du, Dong Liang, Journal of Computational Physics 229 (2010): 4501-4521.
[13] M.P. Suarez, and H.S. R, Bioremediation Journal 3 (1999): 337-362.
[14] Tetr. O, S.M. Hinton, B.J. Silver, H.E. King JR, Coupling hydrocarbon degradation to anaerobic respiration and mineral diagenesis: theoretical constraints, Geobiology 8 (2010): 69-88.
[15] Yudono B, Said Muhammad S, Napoleon A, Fanani Z, J Trop Soils, 16 (2011): 33-38.
[16] Tiam.E. B, and Candy A, Regression techniques and analytical solutions to demonstrate intrinsic bioremediation. In Robin E. H Wilson, D.C. Downey, eds., Intrinsic Bioremediation. Batelle Press, Columbus, Ohio. p. (1995) 109-116.

IOP Conf. Series: Earth and Environmental Science **167** (2018) 012033 doi:10.1088/1755-1315/167/1/012033

A Review of the Effectiveness Study of Botanical Components That Improves Air Filtration

Chang Choo Khean and Tan Yong Yih

Heating, Ventilating, Air-conditioning and Refrigeration Section,
Universiti Kuala Lumpur Malaysia France Institute (UniKL MFI), Bangi 43650, Selangor, Malaysia
E-mail: chang@unikl.edu.my

Abstract. This paper focus on the review of green filtration system that incorporated with evaporative cooling that used to enhance indoor air quality. This system was invented in attempt to thrive a clean environment that becomes a solution for certain places. Indoor air quality (IAQ) and public health risk connected to each other, it is due to the percentage of city population that stays indoors rather than go for outdoors. Indoor air contamination are originated from mixed origin such as volatile organic compounds (VOCs) and indoor airborne particulate matter (PM). There are two (2) methods will be discussed in this review paper. First is the modular green wall and VOCs removal. All trial PM factions are effectively reduced by all plant species in method one (1) which helps in improving indoor air quality (IAQ). The same goes for VOCs removal in method two (2). Our centre of attention is assessing the present condition of study and the prospective research needs.

1. Introduction

Indoor air quality (IAQ) and public health risk connected to each other, it is due to the percentage of city population that stays indoors rather than go for outdoors which is around 85 – 90 % [1]. Data collected from B.F. Yu et al in their research titled "Review of research on air-conditioning and indoor air quality control for human health" in 2009. United States Environmental Protection Agency classify indoor air quality as one of the five national health threat [2] which stated in the Classification by United States Environmental Protection Agency (EPA) in their book "Residential Air Cleaners – 2nd Edition" in 2009. Mediocre air quality seems to be a crucial contributors toward human health which generates unpleasantness, severe and continual diseases such as "Sick Building Syndrome (SBS)". Building occupants encounter ocular, nasal cutaneous irritations, allergies, respiratory dysfunction, headache and even fatigue connected to spending time indoor is used to describe the term sick building syndrome (SBS) [3]. Indoor air contamination are originated from mixed origin such as volatile organic compounds (VOCs) and indoor airborne particulate matter (PM).

Transmit of outside generated particles to indoor raises the indoor particulate matter (PM) concentrations. PM_{10} which incorporated particles smaller than or equivalent to 10 μm in diameter are the most studied done focused on airborne PM range in size. This is because it can infiltrate easily through our breathing and respiratory system [4]. While lung can be penetrated by smaller fine PM fractions, PM2.5 accompanied 2.5 μm aerodynamic diameter. L. C. Ng, et. al in their research in 2015 found that nowadays, modern buildings incorporate with air conditioning system, as well as air tightness assisted in lowering amount of large infiltration particulates outdoor sources [5]. Only some segment of PM from affected air can be filtered due to the incompetent of heating, ventilating and air conditioning (HVAC) systems which already become today's trending. B. Stephens, et. al found that

Minimum Efficiency Reporting Value (MERV) 4, 6, 10 and 11 filters capable in removing a range of particle sizes approximately 0 to 20% efficiencies [6]. Although there are many more high efficiency filters are available in the market today, it created other problems such as exceed usage and energy, costly maintenance and depleted sustainability, forbye inability in capturing gaseous pollutants [7].

According P. Irga et al. in their several studies done, potted plants is proven to phytoremediate various indoor air contaminants comprises of VOCs and carbon dioxide (CO_2) that able to remove along the plants bacterial community and photosynthesis process [8]. One of the main research that had been done in this field showed that 50–70% of VOCs can be reduced using potted plants [9]. Based on Lohr and Pearsons Mims [10], it was found that across foliar confront plus recommended plants accompanied rugged leaf construction such as trichomes efficient in accumulate PM.

This paper presents the effectiveness of botanical components as air filtration media of indoor environment focusing on PM levels and gaseous pollutants alleviation. And overview of the plant species, bio-system arrangement, progress along the years, method of conducting experiment and result is presented. Figure 1 displays the comparison of PM sizes and Figure 2 demonstrates the indoor contaminants resources.

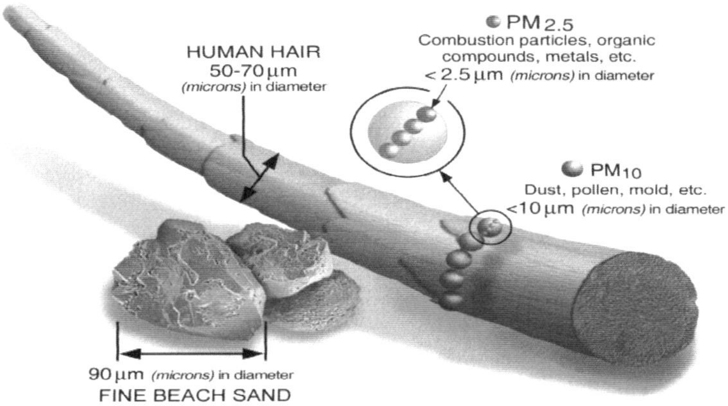

Figure 1. Size comparison for PM particles

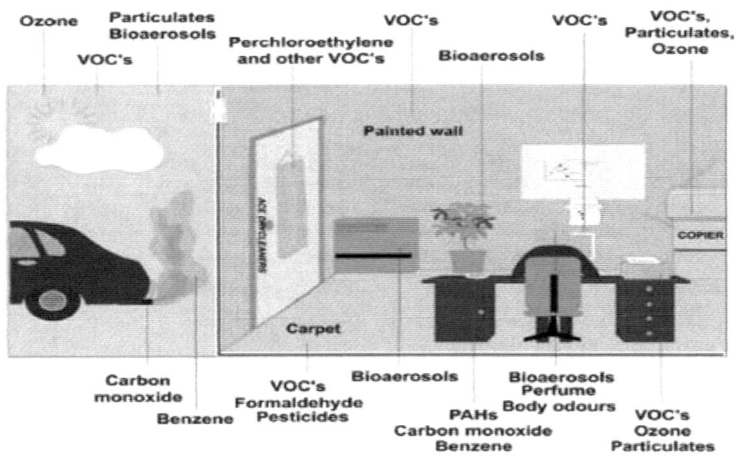

Figure 2. Primary sources of indoor air pollution

2. Botanical Bio-filtration History

For the purpose of this study, bio-filtration and phytoremediation are describes as a hybrid called botanical bio-filtration. Polluted air or water outflow move past a section with intense biological movement where it is being nullify are called bio-filters. Traditional technologies have been replaced to phytoremediation which deem to have non-evasive and worthwhile cost substitute for conditions cleaning up to ten times more economical [11]. T.C. Dollan et al. stated that in the past, United States include phytoremediation within arise list of technical developments [12]. In 1996 in United States, removal of chlorinated VOCs from ground water was done by using poplar trees. In 1980s, the potential of various plants in removing VOCs from indoor were demonstrated by National Aeronautics and Space Administration, US (NASA) scientists [13]. After this, the ability of plants in VOCs removal and reducing PM were tested and demonstrated repeatedly from time to time.

3. Bio-system Arrangement

3.1. Potted Plants (PPs)

In this arrangements, PPs represent the passive type of bio-filter for contaminated air. The water and nutrient must be supplied separately. It also can be placed in exposure chamber which allowing any monitoring to be done as illustrated in Figure 3.

3.2. Plant – assisted (botanical) bio-tricking filters (PBTFs)

PBTFs are called active bio-filtration for a reason. This is because, it equipped with hydroponic plants including filling material root assist and ongoing fed with contaminated air plus nutriment liquid as shown in Figure 4. For indoor air, it is more complement PBTFs rather than PPs as it enable outrageous removal performances.

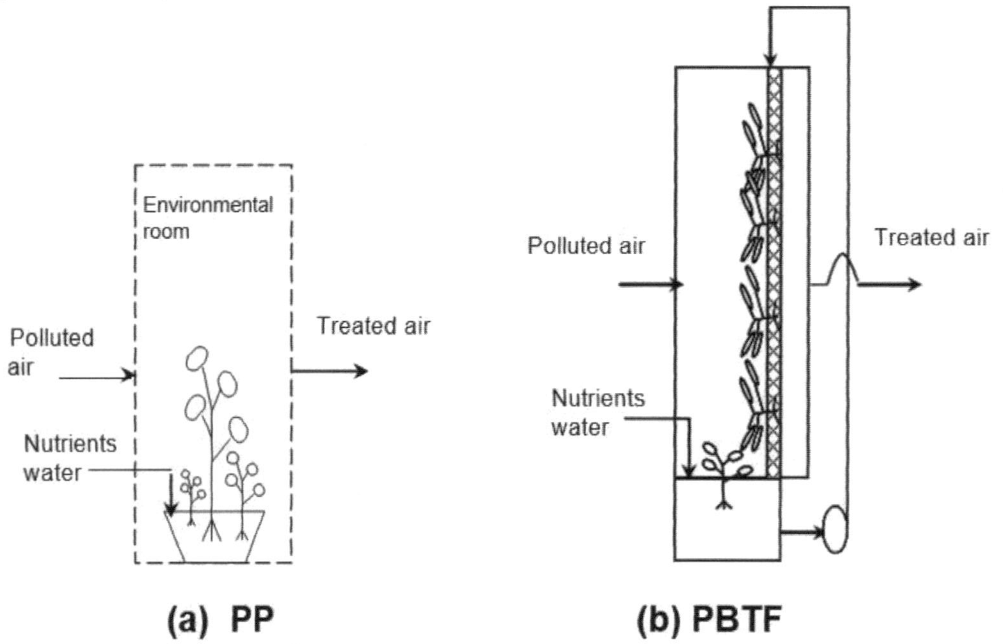

Figure 3. Schematic diagram of botanical bio-filtration systems: (a) PPs and (b) PBTFs.

There are many factors in determining botanical arrangement, such as plant species, growing rates, available funds, location and infrastructure limit.

4. Methods

There are two (2) methods will be discussed in this review paper. First is the modular green wall as suggested by Irga et al [14]. The second is the VOCs removal by W. Scriprapat et al [15].

4.1. Modular Green Wall System

4.1.1. Plant Species

There are seven (7) plant species used in trial for retrieving data of single pass removal efficiency (SPRE). The species are as follow:

i. Fire Flash *(Chlorophytum orchidastrum)*
ii. Fiddleleaf (*Ficus lyrata)*
iii. Goldfish plant (*Nematanthus glabra)*
iv. Lemon button fern (*Nephrolepis cordifolia duffii)*
v. Boston fern (*Nephropelis exaltata bostoniensis)*
vi. Umbrella tree *(Schefflera amate)*
vii. Dwarf umbrella tree *(Schefflera arboricola)*

4.1.2. System Description

It consists of polyethylene module that has 16 holes at front part with an area of 0.25 m^2. Electric powered axial impeller creates suction from rear system and go through plants before released to indoor. Selected plants grow effectively in upright position. More than a year was taken for a system establishment. The setup is shown in Figure 4.

4.1.3. Single Pass Removal Efficiency (SPRE)

Figure 4 shows experiments that were done by using sealed Perspex material chamber with dimension, 0.6m x 0.6m x 0.6m (H) with 216 L capacity. Diffusion of decontaminated airflow past chamber was assisted by the fan within it so that the throw of particles were reduced.

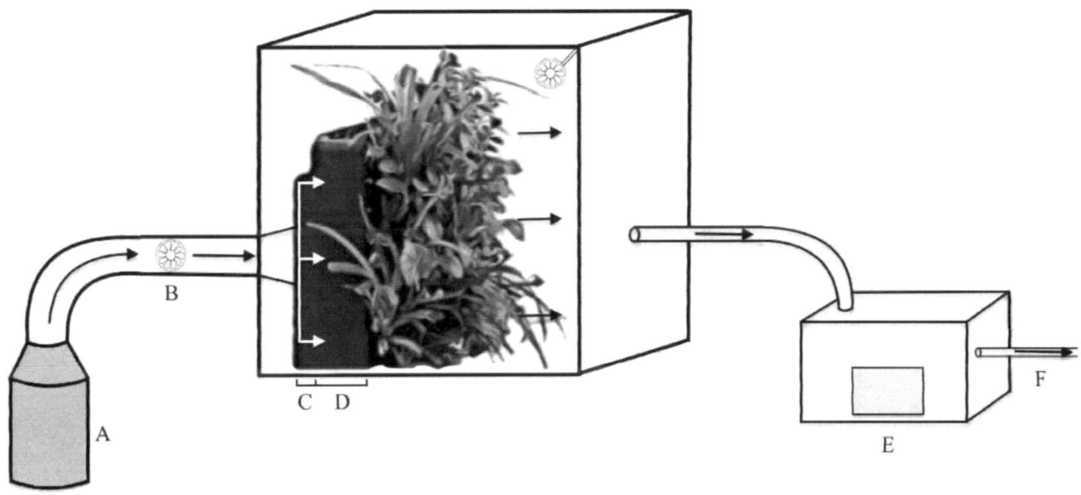

Figure 4. Single pass flow through chamber described in section 4.1.2.

Five (5) PM fractions were recorded for each replicate, $PM_{0.5\text{-}1.0}$, $PM_{1.0\text{-}2.5}$, $PM_{2.5\text{-}5.0}$, $PM_{3.0\text{-}5.0}$and $PM_{5.0\text{-}10.0}$. Equation below is used for SPRE calculation:

Equation 1:

$$\frac{([PM]control - [PM]trial)}{[PM]control} \times 100 = SPRE$$

4.1.4. Other Variables for Experiment

Along with this experiment, various data was also taken, such as plant morphological data (leaf, root and soil), pressure drop (obstruction to airflow against filter module), and data analysis (PERMANOVA used in comparing average SPRE).

4.2. VOCs removal – Toluene and Ethylbenzene

4.2.1. Plant Species

Twelve (12) plant species were chosen for this experiment:

i. Aloe Vera
ii. Sansevieria Masoniana (Snake Plant)
iii. Sansevieria Trifasciata (Viper Bowstring Hemp)
iv. Sansevieria Hyacinthoides (African Bowstring Hemp)
v. Sansevieria Ehrenbergii (Sword Plant)
vi. Kalanchoe Blossfeldiana (Flaming Katy)
vii. Dracaena Deremensis (Lemon Lime)
viii. Dracaena Sandreniana (Lucky Bamboo)
ix. Codiaeum Variegatum (Growing Crotons)
x. Chlorophytum Comosum (Spider Plant)
xi. Cordyline Fraticosa (Firebrand)
xii. Aglaonema Commutatum (Philippine Evergreen)

4.2.2. Screening Plants for VOCs Removal

All plants were tested for similar leaf's area and measured by graph paper. In order to clear soil fragments, plants were cleaned using distilled water. Use of Aluminium foil to cover pot in order to prevent pot absorption.

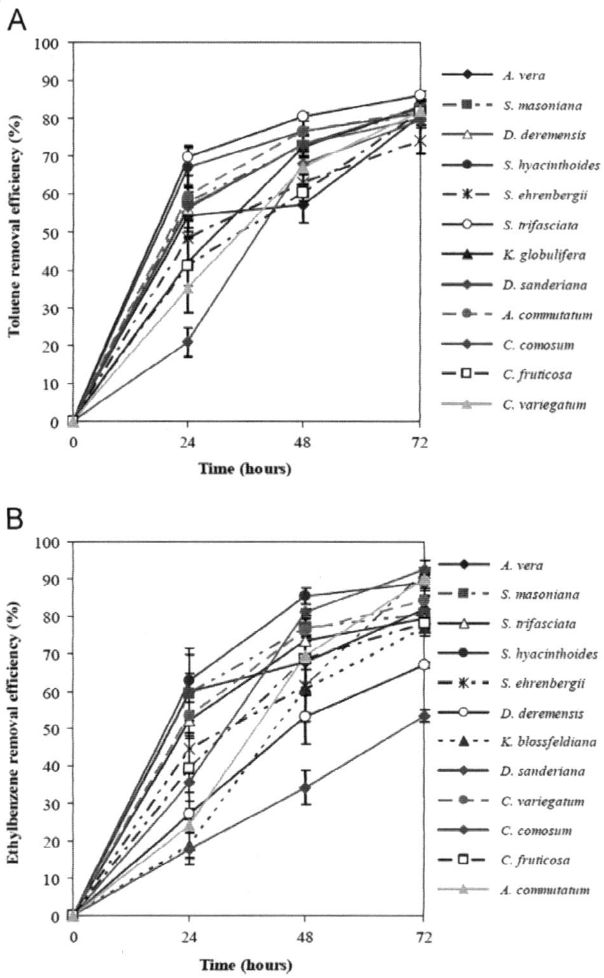

Figure 5. Toluene (A) and Ethylbenzene (B) removal efficiency of plants at 24, 48 and 72 h.

4.2.3. Fumigation

Plant fumigation was experimented by using the glass chamber (Volume 15.6L). Each treatment required three (3) replicate chambers.

4.2.4. Analysis of Gas

Gas chromatography flame ionization detector was used to analyse the concentration of Toluene and Ethylbenzene.

4.2.5. Analysis of Number of Stomata, Cuticle and Phytoxic

Nail varnish was used to duplicate leaf stomata design for stomata figures analyzation. Analytical balance was used to analyse cuticle weight. Plant leaves were measured for photosynthesis and at the same time observe the amount of toxicity.

4.2.6. VOCs Removal Efficiency

From decontaminate chamber, after 72 hours, a gas indication was taken for analyzation purpose as illustrated in Figure 5.

4.2.7. Cuticular Wax Composition Analysis

After 24 hours, in order to remove the waxes, the leaves of designated plants were submerged into hexane.

4.2.8. Analysis in Statistical Form

For complete random design experiment, Analysis of Variance (ANOVA) was used to analyse data statistically.

5. Results and Discussion

According to these two (2) methods, there are many variables to be considered and experimented. To simplify the result gained, both experiments results were separated into different factors:

5.1. Particulate Matters (PM)

All trial PM factions were effectively reduced by all plant species in method one (1) which helped in improving indoor air quality (IAQ). The plant capability as PM filter (SPRE value) also included on their morphological attributes such as roots, structure and leaves.

5.2. Pressure Drop

Escalate filtration capacity also caused by intensifying the pressure drop across root intervention as presented in method one (1). Air that passes through the plant will undergo a resistance to the flow rate and ending up elevated the time taken within plants. It can be said that increased pressure drop relates to SPRE elevated value in mechanical ventilation system.

5.3. VOCs (Toulene and Ethylbenzene) Removal in Plant Screening.

In method two (2), dismissal of Toluene and Ethylbenzene over 12 species had indicated an interesting tendency. Toluene final dismissal was average around 77%, while Ethylbenzene dismissal is more than 70%.

5.4. Cuticular Absorption of VOCs

In absorption between cuticular layer and species, it was found that there was no consequential divergent. As a result, Toluene and Ethylbenzene dismissal and amount of cuticle showed an uncertain relationship.

5.5. Toulene and Ethylbenzene Exposure on Fv/Fm Proportion

Chlorophyll flourescense or Fv/Fm proportion linking difference. Health and purpose of photosynthesis operation and VOCs uptake of plants showed some correlation between them.

6. Conclusion

This review paper emphasizes of botanical bio-filtration in improving indoor air quality (IAQ). Various characteristics such as IAQ issues, bio-system arrangement, methods in conducting experiment and result performance are presented. Based on the methods presented, there are certain variables that can be used in implementing our green filtration system. Technology of botanical has revealed its promising potential in removing VOCs and PM based on the result reviewed. In the next future, the usage of botanical filtration can be focused more on particular places and conditions in improving air quality (indoors or outdoors).

7. References

[1] Yu B F, Hu Z B, Liu M, Yang H L, Kong Q X and Liu Y H Review of research on air-conditioning systems and indoor air quality control for human health 2009 *Int. J. Refrig* **32** 3-20

[2] United States Environmental Protection Agency (EPA) 2009 *Residential Air Cleaners, second ed., A Summary of Available Information* EPA 402-F-09- 002

[3] Wood R, Burchet M, Alquezar R, Orwell R, Tarran J and Torpy F The potted plant microcosm substantially reduces indoor air VOC pollution: I. Office field study 2006 *Water Air Soil Poll.* **175** 163-80

[4] Bari M A and Kindzierski W B Fine particulate matter (PM2.5) in Edmonton, Canada: Source apportionment and potential risk for human health 2016 *Environmental Pollution* **218** 219-29

[5] Ng L C, Persily A K and Emmerich S J IAQ and energy impacts of ventilation strategies and building envelope air tightness in a big box retail building 2015 *Building and Environment* **92** 627-34

[6] Stephens B and Siegel J Ultrafine particle removal by residential heating, ventilating, and air-conditioning filters 2013 *Indoor Air* **23** (6) 488-97

[7] Montgomery J F, Green S I, Rogak S N and Bartlett K Predicting the energy use and operation cost of HVAC air filters 2012 *Energy and Buildings* **47** 643-50

[8] Irga P, Torpy F and Burchett M Can hydro culture be used to enhance the performance of indoor plants for the removal of air pollutants? 2013 *Atmospheric environment* **77** 267-71

[9] Orwell R L, Wood R L, Tarran J, Torpy F and Burchett M D Removal of benzene by the indoor plant 2004 *Water Air Soil Pollut.* **157** 193–207

[10] Lohr V I and Pearson-Mims C H Particulate matter accumulation on horizontal surfaces in interiors: influence of foliage plants 1996 *Atmospheric environment* **30** (14) 2565-68

[11] Pilon-Smits E Phytoremediation 2005 *Annu. Rev. Plant. Biol.* **56** 15–39

[12] Dollan T C and Glynn B A New treatment technologies and some significant regulatory changes 1997 *Environ. Claims J.* **10** (1) 113-22

[13] Wolverton B C, Johnson A and Bounds K *Interior Landscape Plants for Indoor Air Pollution Abatement 1989 Final Report NASA (NASA-TM-101760)* National Aeronautics and Space Administration

[14] Irga P, Paull N, Abdo P and Torpy F An assessment of the atmospheric particle removal efficiency of an in-room botanical bio filter system 2017 *Building and Environment* **115** 281-90

[15] Sriprapat W, Suksabye P and Thiravetyan P Uptake of toluene and ethylbenzene by plant 2014 *Ecotoxicol & Environ. Saf.* **102** 147-51

Production of value-added chemicals from wheat straw lignin by bio-refinery process

Nasim Ramezani[1], Robert Jeng[1], Mohini Sain[1]

1 Center for Biocomposite and Biomaterials Processing, Faculty of Forestry, University of Toronto, ON, M5S 3B3, Canada
E-mails: nasim.ramezani@mail.utoronto.ca

Abstract. Lignocellulosic-based bio-refineries are essential for a secure energy in future. Lignin degradation is required for carbon recycling and production of renewable chemicals. Lignin due to the high level of color and low biodegradability, is categorized as a serious pollutant particularly in the aquatic ecosystem. In this study, biodegradation of organosolv lignin by a white rot fungi, *Bjerkandera adusta,* was verified. The FTIR spectra of lignin, before and after treatment with fungi displayed modification by changing in the structure of lignin. Some of the functional groups have disappeared and some new bands appeared in the spectra. The SDS-PAGE chromatographs also showed that the fungi protein has a molecular weight of 44 kDa. This molecular weight increased to 46 kDa after the treatment of lignin by fungi. The heavier molecular weight occurred due to the bonding of fungi protein with lignin. The results from surface morphology also showed the attachments of lignin molecules to fungi mycelia. As a result, the extracted lignin after treatment with *B. adusta* showed significant changes in its physical and chemical structure which makes it a noble candidate for carrying out value added products and refinery of lignin waste in pulp and paper industries.

1. Introduction

Finding new ways to produce chemicals and fuels from renewable sources, particularly from plant biomass has increased due to the necessity of reduction in greenhouse gas emissions [1]. Lignin is a by-product of pulp and paper industry which is produced in a large scale [2]. Each year 60-100 ktons lignin, 1 Mtonne lignosulfonates, 5-10 ktones sulphur-free soda lignin and 3 ktones Organosolv lignin are produced by Kraft, Sulfite, Soda and Organosolv processes respectively [3]. The lignin produced in Kraft process is generally very condensed and has high amounts of reduced β-O-4 content. Furthermore, the sodium sulphide (Na_2S) consumed in Kraft process causes thiol groups generation. In Organosolv treatment, a greater amount of β-O-4 is preserved due to a milder biomass treatment process [4]. Lignin, the only naturally synthesised polymer, is a cross-linked macromolecule with an aromatic backbone. It is obtained from oxidative coupling of monolignols via radicals made by oxidases. The most common links in lignin structure are β-1, β-5, β-O-4, β-β, 5-5 and 5-O-4 [5]. The existence of different organic and inorganic impurities, distinctive chemical reactivity and non-uniform structure of the lignin macromolecule are the restrictions in the way of converting lignocellulosic biomasses into profitable products. Therefore, the main goal of modern bio-refineries is to overcome these obstacles. For example, by controlling the cleavage of C-O and C-C bonds in a selective depolymerisation method, new categories of aromatic and monomeric species can be obtained [6]. Furthermore, lignin is the most resilient component among lignocellulosic materials. The chemical and physical methods for removing lignin from wastewater have many impediments. One of which is the high price of chemicals applied for this process. The use of absorbents for lignin decoloring, have some other setbacks

such as saturation of absorbent sites due to high Total Suspended Solids (TSS) and Total Dissolves Solids (TDS) [7]. Lignin degradation by microorganisms is a more effective substitute [8].
There are a few anaerobic bacteria of gastrointestinal flora and some species of fungi such as basidiomycetes, brown-rot and white-rot fungi which are capable of producing lignocellulolytic enzymes. Wood rot basidiomycetes decompose all components of plant cell wall to CO_2 and H_2O by two key mechanisms: brown-rot and white-rot decay [8]. Degradation of lignin throughout wood decay process is carried out by basidiomycetes white-rot fungi in nature. White rot fungi, on the other hand, are the only microorganism in nature which are capable of lignin mineralization. Therefore, their ability to degrade lignin can be evaluated in vitro by assessing CO_2 discharged when lignin decomposes thoroughly [9]. White-rot fungi, due to their ability to yield great amounts of extracellular lignocellulolytic enzymes are attractive microorganism for bio-degradation of biomasses. *Bjerkandera* species are intersecting microorganism in biotechnology due to their ability in degrading aromatic xenobiotics in addition to lignin [10]. *B. adusta* is one of the white rot fungi which belongs to basidiomycetes. This fungus produces white rot on trees or wood by penetrating its mycelium in to the cell holes while diffusing ligninolytic enzymes. This fungus breaks down the lignin in the plants cell walls for accessing to the cellulose. Culture medium composition, inoculum concentration, pH, temperature, duration and speed of incubator have major effect on the mycelium growth of *B. adusta*. Ligninolytic enzymes produced by white-rot fungi can oxidise and degrade lignin. Furthermore, they increase the accessibility of the lignin structure. Enzymatic reactions, catalytic oxidation, catalytic reduction and hydrolysis reactions are the key approaches for depolymerisation of lignin [11].
Four enzymes mainly interfere in lignin degradation; manganese peroxidases (MnPs), versatile peroxidases (VPs), lignin peroxidases (LiPs) and laccases (Lac). Due to large size of lignocellulolytic enzymes, their permeability into undecomposed wood cell walls is reduced. Hence, for native lignin decomposition, reactive oxygen species might play an important role. Fungal MnP, LiP and VP are all from the class II peroxidases in the superfamily of heme peroxidases. Lip and MnP are glycoprotein with 38-46 kDa and 40-50 kDa molecular weights respectively [12].
In this study, the biodegradation of the Organosolv lignin to yield value-added chemicals was investigated. For this purpose, white rot fungi, *Bjerkandera adusta* was applied and changes in the structure of lignin were verified.

2. Materials and methods

2.1. *Lignin sample*

In previous study, the lignin was extracted by applying ethanol organosolv treatment from wheat straw. First, the extraction conditions were optimized in a 600-ml Parr reactor. The maximum amount of yield (>90%) and the least amount of carbohydrates impurities and ash was isolated in 200 °C, 400 psi and 120 min. The optimum experiment's conditions in which the highest amount of lignin was yielded, were selected for further extraction. Subsequently, for large-scale production, the lignin extraction was carried out in a 20L reactor [13].

2.2. *Fungi*

Bjerkandera adusta (UAMH 8258) was purchased from University of Alberta Microfungus Collection and Herbarium, Alberta, Canada (UAMH) and was cultivated on a commercial Potato Dextrose Agar (PDA) from HiMedia Laboratories at 26°C. To monitor the alterations in the lignin structure and investigate the enzymes production, the experiment was carried out in a liquid medium. To prepare the inoculum, 10 mm^3 of *B. adusta* agar plugs was suspended in a 5-ml medium. Then the suspension was transferred into a 2000 ml Erlenmeyer flasks containing 1000 ml of Potato Dextrose Broth (PDB) and 2 g of the extracted lignin. The pH was adjusted to 5.5 before autoclaving. All the culture media were sterilized by autoclaving at 121°C for 15 min. The flasks were incubated in a shaker (120 rpm). Two controls were designed for the experiment. One flask was contained PDB and 2g lignin and the other one PDB with *B. adusta* suspension. The cultures were harvested at the 5th and 7th days of incubation to verify the changes in lignin structure and production of enzymes from the fungi.

2.3. Lignin Extraction and Fungi Cultivation

2.3.1. Fungus Modified Lignin Analysis

Each sample was verified for lignin structure, surface morphology and proteins. For this purpose, each culture medium was taken under the biological fume hood and 500 ml of media were collected in the beakers. The contents of culture flasks were filtered with n°1 Whatman filter paper under vacuum. The filtered parts were kept in the freezer. 100 ml NaOH (10N) was added to the solid parts and the mixture were stored overnight in the beaker to dissolve all the lignin inside the mycelia. After 24 hours, the dissolved lignins in NaOH were centrifuged for 20 mins and 10000 rpm. The supernatants were kept in the freezer and the pellets were dried in the oven at 60°C [14]. The samples were named as follow:
Sample #1: *B. adusta* + 2 g lignin in PDB medium after 5 days
Sample #2: *B. adusta* + 2 g lignin in PDB medium after 7 days
Sample #3: *B. adusta* in PDB medium after 5 days (control)
Sample #4: *B. adusta* in PDB medium after 7 days (control)
Sample #5: 2 g lignin in PDB medium (control)

2.3.1.1. Scanning electron microscopy (SEM)
SEM was carried out by a JOEL machine (Hitachi S- 2500, Tokyo, Japan), model 6610LV with a usual SEI Scintillator detector and BSEI diode detectors. Two different magnifications (800X and 6000X) were applied for SEM images. The mycelia of the sample which contained lignin treated by *B. adusta* and the control which contained only fungi were applied for the SEM imaging. The samples were cut carefully to maintain the structure of inner and outer surface of fungi mycelia. The surface was gold coated to bypass electrostatic charge during the study.

2.3.1.2. Fourier transform infrared spectroscopy (FTIR) analysis of lignin sample before and after treatment by fungi
Chemical changes in functional groups and polymer structure in culture media contained sample 1, 2, 3, 4, 5 were analyzed by FTIR machine. The dried samples were mixed with KBr. The samples structures were documented in the range of 4000-400 cm^{-1}. A Bruker Tensor 27, Model Vertex 70 (Germany) with a temperature controller –I0977 detected the FTIR spectra. The spectra were noted with a resolution of 4 cm^{-1} and scan rate of 32 scans per minute.

2.3.1.3. Sodium dodecyl sulfate polyacrylamide gel electrophoresis (SDS-PAGE) of extracted protein
SDS-polyacrylamide gel electrophoresis with marker protein (Bio-Rad) was applied to separate the extracellular proteins. The gel gradient concentration was 4% to 12%. To denature the proteins, 2-mercaptoethanol were applied as reducing agent. A Bis-Tris buffer system at 150 V was used for electrophoresis and Coomassie Brilliant was used for staining.

3. Results and Discussions

3.1. Decolorization of lignin upon the fungi growth in the broth medium

To increase the ability of fungi to degrade lignin, the organosolv lignin with high purity was applied. Figure 1. shows the change in the color of Organosolv lignin culture media after fungal growth. Figure 1. a. shows the culture media before the inoculations. In Figure 1. b. the flask which contained the fungi and lignin (middle) shows the changing in the color of medium. In Figure 1.c. the absorption of the black color particles inside the fungal mycelium, is observed. This could be explained by the decolorizing ability of *B. adusta*. Later by cutting the mycelium and observing them under the SEM the existence of lignin molecules inside the mycelia was confirmed.
Whit-rot fungi, among the potential microbes, due to having a robust lignin degrading enzymatic system are good candidate for lignin degradation and decolorization of pulp and paper waste [15].

a b c

Figure 1. a. the culture medium before inoculation, right: *B. adusta* in PDB (control), middle: *B. adusta* in PDB and lignin, left: lignin and PDB. (control). b. the culture medium after inoculation (5 days). c. *B. adusta* with lignin granule inside their mycelium and decolorize media

3.2. *Surface morphology*

Figure 2. shows the fracture surfaces of the *B. adusta* observed by SEM after 7 days of incubation with lignin. These images clearly show the bonding of extracted lignin particles inside *B. adusta* mycelia. These spheres could not be observed on the control sample (without lignin). The SEM image of *B. adusta* from the media with Organosolv lignin shows the lignin granules with the size of ~4 μm which is compatible with data collected in the previous studies [16].

a b

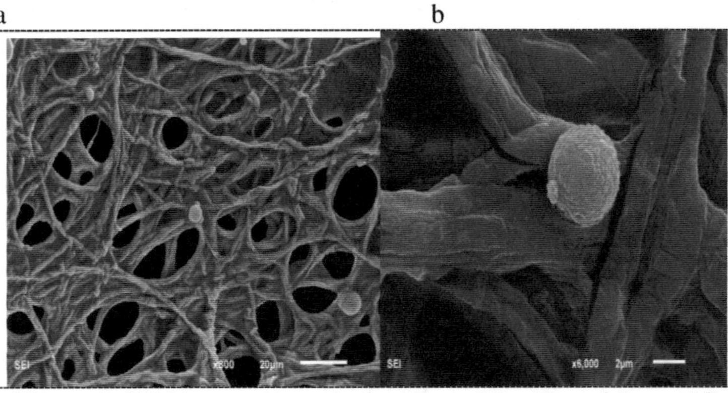

Figure 2. SEM images show the bonding of lignin particles and B. adusta (a) low magnification (b) high magnification.

3.3. *Analysis by fourier transform infrared spectroscopy*

The FTIR analyses were performed on samples obtained from the 5th (#1) and 7th (#2) days of culture media contains *B. adusta,* lignin and PDB (Figure 3). The FTIR analyses also were performed on controls consist of B. *adusta* in PDB culture medium form day 5th (#3), day 7th (#4) and the one with lignin and PDB (#5). FTIR analysis of treated samples from the 5th and 7th days showed certain changes in the spectra as compared with the controls. The FTIR spectra of Organosolv lignin biodegraded by *B. adusta* displayed significant variations from 500 to 4000 cm^{-1}. The control sample which consists of only lignin and PDB shows strong peaks in FTIR spectrum and the results were consistent during the time due to the absence of fungi in medium. When the fungi continued the growth in the flasks consist of fungi and lignin from day 5th and day 7th, some peaks showed only a weak peak in the lignin molecule and some of the peaks disappeared.

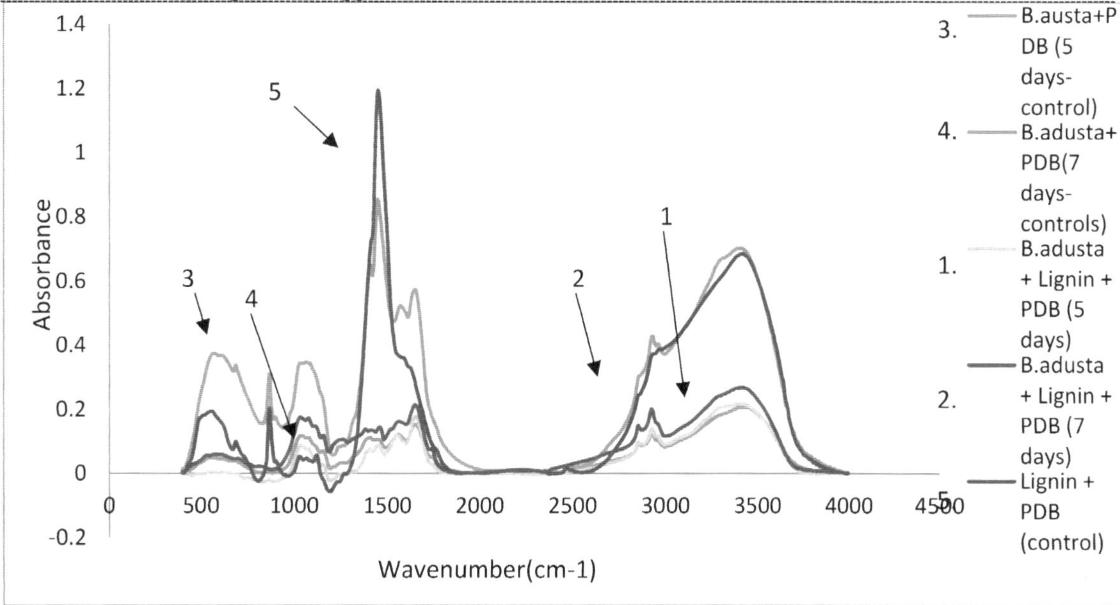

Figure 3. FTIR spectra of treated lignin samples.

The FTIR spectrum of control sample (#5) which contained only lignin and PDB is displayed. The sample showed a peak at 1267 cm^{-1} which is assigned to C-O in guaiacyl ring. This band disappeared in FTIR spectra of sample #1 and #2, which represents the conversion of guaiacyl to other bands. Besides, in sample #5 the band at 1452 cm^{-1} signifies the existence of C-H bend in aromatic ring, this band also disappeared in sample #1 and #2. The band at 3413 cm^{-1} which is associated with oscillation of hydroxyl group is disappeared in the treated samples. Furthermore, the absorption band at 1597 cm^{-1} in sample #5 corresponds to aromatic ring vibration and C=O stretching, this band also could not be detected in the sample #1 and #2. In FTIR spectra of sample #1, the appearance of new C-H bond around 1462cm^{-1} is a direct result of deformation in methyl and methylene groups. Additionally, new aromatic bonds at 1400 cm^{-1} and 2855cm^{-1} of sample #1 spectra are corresponded to C-N stretch of amides and O=C-H:C-H stretch of aldehydes. After treatment of lignin with *B. adusta*, a band at 1654 cm^{-1} originating from conjugated carbonyl in lignin is appeared, which proposed the formation of new carbonyl. By comparing sample #1 with #2, in sample #2, new band at 708 cm^{-1} was observed which was not presented in sample #1. This could be the result of changes in lignin structure during the growth of fungi. By comparing the FTIR spectra in different samples, biodegradation of lignin in sample #1 and #2 could be observed. By comparing the results with sample #3 and sample #4, it can be concluded that fungi growth alone on the PDB has a limited effect on the overall results. Finally, band absorption between 458 and 600 cm^{-1} corresponded to C-Br stretch, which could be detected in all the samples [15], [16].

Studies showed, lignin degradation microorganisms, breakdown guaiacyl unit to vanillic acid as an intermediate [17]. The structure of lignin can be modified in existence of enzymes. For instance, laccases can modify the lignin structure by making dissimilar mesomeric forms that might couple in many potential forming inter-unit bonds. The mesomeric effect can be detected in β-5, β-O-4 and 5-O-4 bonds of lignin molecules [18]. Peroxidases, on the other hand, can oxidase 5-O-4 nonphenolic and phenolic structures in lignin. They also enhance the reactivity of lignin by rising its phenolic content. The biodegradation Organosolv lignin by *B. adusta* can be valuable for a variety of applications such as flocculants, adhesive, sizing agents and corrosion inhibitor [19].

3.4. Gel Electrophoresis

The fungi proteins of lignin treated sample by *B. adusta* (day 7) and the control which contained only fungi (day 7) detached by gel electrophoresis. Figure 4 shows the Gel electrophoresis of lignin treated by fungi and the control which only had fungi.

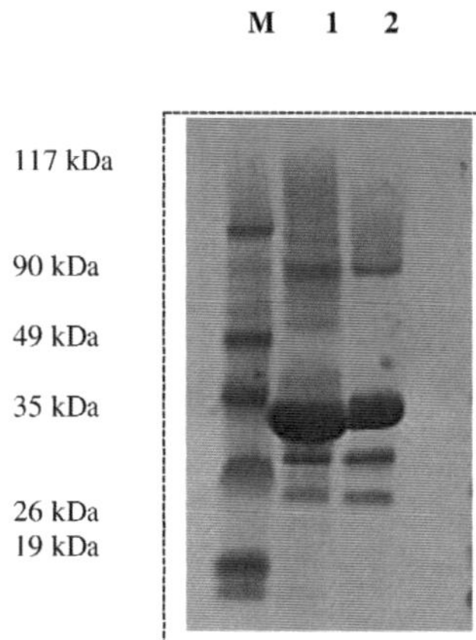

Figure 4. Gel electrophoresis of (M) marker, (1)lignin treated by fungi and (2)control (with only fungi protein).

Although the treated and non-treated samples showed similar protein profiles, but the bands from lignin treated media by fungi expressed more. It also exhibited higher protein concentration. The band positions of treated sample also were slightly different from the control. The molecular mass of proteins produced by treated lignin was ~46kDa and non-treated lignin was ~44kDa which can be defined due to the bonding between protein and polymer. The gel electrophoresis of culture contained lignin without fungi did not show any visible proteins bands. This confirmed that the origin of the protein detected in treated sample and control 1 were from B. adusta activity, not from the lignin itself. Furthermore, the molecular weights of MnP and LiP are 44-45 and 41-43kDa respectively [20], which are closed to what has been shown in our results. The results from gel electrophoresis suggested that the nature of B. adusta proteins, is capable of lignin biodegradation.
The lignin modifying enzymes (LMEs) secrete extracellularly to facilitate decomposing process. These extracellular enzymes degrade lignin effectively but not selectively [9]. *B. adusta* also produces manganese peroxidase- lignin peroxidase hybrid enzyme. These enzyems can oxidize numerous phenolic and nonphenolic precursor such as guaiacol, 2,2 –azinobis (3-ethylbenzthiazoline-6-sulfonic acid) (ABTS) and veratyl alcohol, 2,6-dimethoxyphenol in the lack of Mn(II). MnP and Lip are glycosylated heme-containing enzymes. Mnp, on the other hand, is capable of cleaving and oxidizing lignin by consuming H_2O_2 for its activity [20].

4. Conclusion

Although the total energy demand has grown 24% in worldwide from 2005 to 2015, the bioenergy portion in overall global primary energy utilization has stayed comparatively stable (about 10%). The annual production of lignin in large quantities is more than 70 million tons but only 2% of lignin is applied for industrial purposes and 95% is burnt to produce energy. This is while 80% of world's primary energy supply is still from fossil fuels [21], [22]. The United States and European Union established goals to develop lignin valorization process so it can replace fossil fuels. One of the investment priorities of European Union to achieve a sustainable development within 2020 is lignocellulosic biomass alteration into biofuel which is considered a promising choice to substitute fossil fuels. Lignocellulosic-based bio-refineries are essential for a secure energy in future. Lignin has a critical effect on increasing the amount of liquid transportation fuel accessible from biomass over the carbohydrate particles [23]. Ultimately, lignin is a green biofuel and has a countless market in sustainable energy and bio-polymeric materials [1].

According the results obtained from surface morphology, bonds between lignin particles and mycelia were observed, which appeared essential for lignin degradation. Also, the changes in FTIR spectra of different samples, confirmed the lignin modification during the fungi growth. The results suggest the proposed fungi, *B. adusta*, as a suitable choice for decolourization and removal of lignin wastes. Ultimately, lignin degradation resulted in production of aromatic chemicals which can be applied for further applications.

5. References

[1] Vennestrøm P, Osmundsen C, Christensen C and Taarning E 2011 Beyond petrochemicals: the renewable chemicals industry Angewandte Chemie International Edition **50** 10502-10509.

[2] Ragauskas A, Beckham G, Biddy M, Chandra, Chen F, Davis M, Davison B, Dixon R, Gilna P, Keller M, Langan P, Naskar A, Saddler J, Tschaplinski T, Tuskan G and Wyman C 2014 Lignin valorization: improving lignin processing in the biorefinery. Science **344** 1246843.

[3] Holladay J, Bozell J, White J and Johnson D 2007 Top value-added chemicals from biomass. DOE Report PNNL-16983, 2013-11-20] http://chembioprocess.pnl.gov/staff/staff info.asp.

[4] Bauer S, Sorek H, Mitchell V, Ibáñez A and Wemmer D 2012 Characterization of Miscanthus giganteus lignin isolated by ethanol organosolv process under reflux condition. J. Agric. Food Chem. **60** 8203-8212.

[5] Fitz Patrick M, Champagne P, Cunningham M and Whitney R 2010 A biorefinery processing perspective: treatment of lignocellulosic materials for the production of value-added products, Bioresour. Technol. **101** 8915-8922.

[6] Pollegioni L, Tonin F and Rosini E 2015 Lignin-degrading enzymes, FEBS Journal. **282** 1190-1213.

[7] Chang V and Holtzapple M 2000 Fundamental factors affecting biomass enzymatic reactivity. Appl. Biochem. Biotechnol. **84** 5-37.

[8] Martínez M, Barrasa J, Gutiérrez A, Carlos del Río J and Martínez A 2000 Fungal screening for biological removal of extractives from Eucalyptus globulus wood, Canadian Journal of Botany **77** 1513-1522.

[9] Camarero S, Sarkar S, Ruiz-Duenas F, Martinez M and Martinez A 1999 Description of a versatile peroxidase involved in the natural degradation of lignin that has both manganese peroxidase and lignin peroxidase substrate interaction sites, J. Biol. Chem. **274** 10324-10330.

[10] Heinfling A, Bergbauer M and Szewzyk U 1997 Biodegradation of azo and phthalocyanine dyes by Trametes versicolor and Bjerkandera adusta, Appl. Microbiol. Biotechnol. **48** 261-266.

[11] Pandey M and Kim C 2011 Lignin depolymerization and conversion: a review of thermochemical methods. Chem. Eng. Technol. **34** 29-41.

[12.] Glenn J and Gold M 1985 Purification and characterization of an extracellular Mn (II)-dependent peroxidase from the lignin-degrading basidiomycete, Phanerochaete chrysosporium, Arch. Biochem. Biophys. **242** 329-341.

[13] Ramezani N and Sain M 2018 Optimizing conditions for organosolv lignin extraction from wheat straw using data analysis approach (Submitted).

[14] Negrão D, Sain M, Leão L, Alcides L, Sameni J and Jeng R 2015 Fragmentation of lignin from organosolv black liquor by white rot fungi. BioResources. **10(1):**1553-1573.

[15] Faix O 1991 Classification of lignins from different botanical origins by FT-IR spectroscopy. Holzforschung-International Journal of the Biology, Chemistry, Physics and Technology of Wood **45** 21-28.

[16] Chandra R, Bura R, Mabee W, Berlin A, Pan X and Saddler J 2007 Substrate pretreatment: The key to effective enzymatic hydrolysis of lignocellulosics?, in Biofuels. Springer. pp. 67-93.

[17] Bugg T, Ahmad M, Hardiman E and Rahmanpour R 2011 Pathways for degradation of lignin in bacteria and fungi, Nat. Prod. Rep. **28** 1883-1896.

[18] Gouveia S, Fernández-Costas C, Sanromán M and Moldes D 2012 Enzymatic polymerisation and effect of fractionation of dissolved lignin from Eucalyptus globulus Kraft liquor, Bioresour. Technol. **121** 131-138.

[19] Selig M, Viamajala S, Decker S, Tucker M, Himmel M and Vinzant T 2007 Deposition of lignin droplets produced during dilute acid pretreatment of maize stems retards enzymatic hydrolysis of cellulose, Biotechnol. Prog. **23** 1333-1339.
[20] Hofrichter M 2002 Review: lignin conversion by manganese peroxidase (MnP), Enzyme Microb. Technol. **30** 454-466.
[21] Lora J and Glasser W 2002 Recent industrial applications of lignin: a sustainable alternative to nonrenewable materials. Journal of Polymers and the Environment. **10** 1-2: 39-48. Print.
[22] Sieminski A 2014 International energy outlook, Energy Information Administration (EIA).
[23] Zhao X, Cheng K and Liu D 2009 Organosolv pretreatment of lignocellulosic biomass for enzymatic hydrolysis, Appl. Microbiol. Biotechnol. **82** 815-827.

Acknowledgments
The authors would like to acknowledge the financial support from the Natural Science and Engineering Research Council of Canada (NSERC) and Ontario Research Fund (ORF). We also thank Dr. Krishan Goel, who helped us a lot in this project.

Engine Performance and Emission study of Waste cooking oil and Sewage sludge derived Biodiesel blend

D Priyadarshi[1] and KK Paul[2]

1Research Scholar, Department of Civil Engineering, National Institute of Technology, Rourkela, India, 769008.
2Assistant Professor, Department of Civil Engineering, National Institute of Technology, Rourkela, India, 769008
E-mail: k_karar1@yahoo.co.in

Abstract. Sewage sludge and waste cooking oil based biodiesel (WCOB) recently gaining popularity for a fair amount of lipid content for biodiesel production. In this study, for the first time sewage sludge biodiesel (SSB) based blend was subjected to engine performance analysis. SSB has relatively higher saturated fatty acid content and reduced polyunsaturated fatty acid content. While WCOB has low saturated methyl esters and very high unsaturated methyl ester concentration. Hence, effect of fatty acid structure on performance and emission were also monitored. On the basis of performance WCO30 (30 % v/v WCOB and 70 % v/v diesel blend) found better alternative with low break specific fuel consumption and lesser exhaust gas emission compared to SSB30. However, decent performance was also witnessed for SSB30. It was observed that fatty acid profile has negligible effects on the performance parameters. Whereas, significant variation in NOx, CO and HC emission were noted with change in individual fatty acid methyl ester contents. Utilising SSB30 and WCO30 reduced the CO emission by 12.82% and 10.39% respectively. Saturated fatty acid methyl ester rich SSB30 resulted 9.09% higher NOx emission than petroleum diesel. Hydrocarbon emission were increased by the presence of unsaturated fatty acid methyl esters.

1. Introduction

The growing fuel crisis and limited vegetable oil based biodiesel attracting researchers towards the search of new alternatives. As per World Energy Council (WEC), in 2040 there will be a rise in ratio of the diesel demand to gasoline demand from 1.5 to 3.8 [1]. Among diesel alternative fuels, biodiesel has similar physico-chemical properties to petroleum diesel, high oxygen content, and rich in resources. Therefore, a potential candidate for fossil diesel substitute fuel. Recently, there is a significant increase in biodiesel use in compression ignition engines [2]-[4]. It is reported that 20% biodiesel blend in diesel engine can provide similar performance and emission with that of petroleum diesel [5]. Currently, engine configurations are as such that 100% biodiesel can be used without modification. However, special attention should be paid for transportation and storage of biodiesel. Biodiesel can reduce the emissions of unburnt hydrocarbons (HC), carbon monoxides (CO), sulphates and particulate matter (PM) depending upon methyl esters and engine conditions [6], [7]. Again, biodiesel is renewable. Even if additional energy consumption takes place during production of biodiesel, it is well managed by the very less emission of greenhouse gases during engine operation. It provides clean energy with low particulate emission [8], [9].

For a long time, vegetable oil based biodiesel dominated the biodiesel market worldwide [10]. Whereas, availability of seed for oil extraction is limited. Therefore, less amount of oil production and high processing cost of vegetable oil biodiesel is a major drawback for its commercialization [11]. This is not sufficient to fulfil a large requirement of biodiesel. The search for low cost raw materials has focused towards alternative feedstocks like animal fat, waste cooking oil [12] and sewage sludge [11], [13].

Recently, biodiesel from waste like sewage sludge and waste cooking oil gaining popularity due to its high fatty acid contents [14]. Worldwide more than 20 million tone of dry sewage sludge are produced every year [15]. For some time, treated sludge were used as fertilizer. However, strict regulations on sludge fertilizer introduced due to its heavy metal contents that might degrade the crop quality and damage the soil fertility [16]. Hence, biodiesel production can be a better alternate for reuse of sewage sludge. SSB production follows lipid extraction and transesterification of lipids. Transesterification may be base catalysed or acid catalysed depending upon the free fatty acid (FFA) concentration in lipid. Sludge based lipid found high in FFA as tested in laboratory. Base catalyst doesn't hold good for lipid and oil with high FFA. Acid catalyst is a better option but highly time consuming process [17]. Recently, cosolvent based acid catalysed transesterification getting recognition due to its easy and fast conversion of high percentage of FFA into methyl esters [18], [19]. Cosolvents such as methyl tert butyl ether (MTBE), tetrahydrofuron operates well in low temperature transesterification [19].

In this study an attempt was taken to analyse the performance and emission of biodiesel from sewage sludge and waste cooking oil. Previously no study was performed on engine performance of sewage sludge biodiesel (SSB). SSB has taken along with waste cooking oil biodiesel (WCOB) to investigate the effects of fatty acid profile on performance and emission of DI engine.

2. Sample and methodology

2.1. Test sample preparation

Sewage sludge was collected from the primary settling tank outlet of a municipal sewage treatment plant located in panposh, Rourkela. Lipid extraction was performed following methods adopted by Priyadarshi and Paul (2017). Lipid yield of 25.51 % was measured from sewage sludge. Free fatty acid content more than 3 mg/g KOH was observed in lipid sample. Therefore, lipid was subjected to acid catalysed transesterification in presence of cosolvent (MTBE). Transesterification were performed with 5:1 (v/w) MTBE (methyl tert butyl ether), 15:1 (w/v) methanol, and 1 % (wt%) sulfuric acid in lipid. The mixture was subjected to 500 rpm stirring at a constant temperature of 45 ^{0}C for 4 hours. After phase separation of FAME and glycerol were collected separately. FAME phase was subjected to wet wash with warm distilled water and finally recovered FAME were analysed with GC-MS and GC-FID.

Waste cooking oil biodiesel fulfilling international standard EN14214 were procured. Both SSB and WCO biodiesel were mixed at ratio 3:7 (30% biodiesel and 70% diesel) with petroleum diesel. After blending SSB30 (30% SSB and 70% diesel v/v) and WCO30 (30% WCO biodiesel and 70% diesel v/v) were kept under scrutiny for 10 days. Sample was then subjected to engine performance and emission analysis. Detailed characterisation of diesel, SSB, SSB30, WCOB and WCO30 biodiesel are summarized in Table 1.

Table 1. Properties of diesel, waste cooking oil methyl ester and sewage sludge biodiesel.

Properties	Diesel	SSB	SSB30	WCOB	WCO30
Density, Kg/m^3	820-830	880	846	874	834
Viscosity, cSt	2.5-3.2	4.55	3.7	4.41	3.4
Calorific Value, MJ/kg	42-45	39-42	43.25	38.5-41	43.5
Flash point, ^{0}C	50	178	68	170	63
Iodine value, g I_2 per 100 g	-	85-93	-	95-102	-
Saturated fatty acid, %	-	30-40	-	15-23	-

Mono unsaturated fatty acid (C18:1), %	-	35-44	-	38-45	-
Poly unsaturated fatty acid, %	-	20-25	-	30-37	-

2.2. Engine set up and methodology

Experiments were performed in a single cylinder, four strokes, direct injection air cooled diesel engine with full power of 4.4 kW and maximum speed 1500 rpm. Schematic diagram of the experimental set up was shown in Figure 1. Engine technical specification were given in Table 2. A sensor is fitted to calculate fuel consumption. Air flow intake rate were measured by a U tube manometer. Exhaust gas temperature was measured by a K-type thermocouple. AVL DiGas444 exhaust gas analyser was used to measure exhaust emissions of the engine. Details about the equipment was given in Table 3.

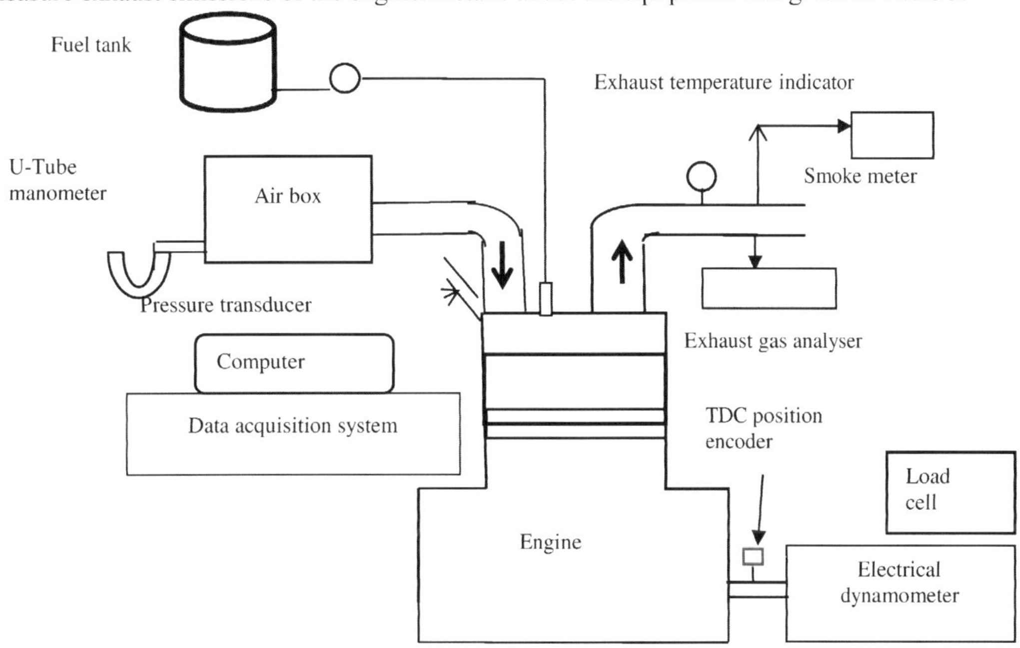

Figure 1. Schematic layout of experimental setup

Table 2. Engine specifications

Make/Model	Kirloskar TAF 1
Rated output	4.4 kW
Rated speed	1500 rpm
Bore	87.5 mm
Stroke	110 mm
Piston type	Bowl-in-piston
Compression ratio	17.5:1
Nozzle Opening pressure	200 bars
Injection timing	23° BTDC
Injection type	Pump-line-nozzle
Nozzle type	Multi hole
No. of holes	3

Table 3. Details of gas analyser

Parameters	Model	Range	Accuracy	Uncertainty
CO	AVL 444 di gas analyzer	0–10%	0.06%	±0.2

HC	0–20000 ppm	±12	±0.2
NOx	0–5000 ppm	±12	±0.2

3. Result and discussion

3.1. Performance parameter

3.1.1. Brake specific fuel consumption (BSFC)

BSFC is a prime parameter that predicts the engine performance. BSFC is the ratio of fuel consumption and brake power of engine. Figure 2 portrays the variation in BSFC observed for SSB30, WCO30 and diesel at different brake power. It is observed that BSFC reduced with increase in brake power. BSFC of 12.99 and 12.57 MJ/kWhr were observed for SSB30 and WCO30 at full load. In the entire process, BSFC recorded were 58.27 MJ/kWhr, 57.58 MJ/kWhr and 66.82 MJ/kWhr for diesel, SSB30 and WCO30 respectively. WCO30 has a more complete combustion than SSB30 due to higher unsaturation in fatty acid content that helps to reduce the BSFC [20]. SSB has high long chain fatty acid concentration along with high density and viscosity sets barrier for complete burn of fuel leads to high BSFC [4], [21]. Significantly higher BSFC in SSB30 biodiesel blend observed. This is because engine supplies fuel on volumetric basis and SSB30, WCO30 density is more than that of petroleum diesel, which delivers additional biodiesel to counteract the lower calorific value [22]. The higher kinematic viscosity of SSB30 causes poor atomization. Hence poor mixing air-fuel. This was the key reason for larger BSFC during engine analysis of SSB30. The higher oxygen in biodiesel blend developed extra lean combustion environment that degraded the engine efficiency, which led to higher BSFC [7], [23].

3.1.2. Brake thermal efficiency

Brake thermal efficiency of engine is inversely proportional to BSFC. Among the tested fuel i.e., diesel, SSB30 and WCO30 brake thermal efficiency of 29.89%, 27.69% and 28.23% respectively were recorded at full load (Figure 3). High viscosity of SSB30 results power loss as it reduces the combustion efficiency due to bad fuel atomization [24], [22].

3.1.3. Exhaust Gas Temperature (EGT)

Variation in exhaust gas temperature in different loading conditions are shown in Figure 4. Mixed effects were observed for biodiesel blends. At 2.2 and 3.3 kW load WCO30 resulted maximum EGT whereas, at 1.1 kW SSB30 has maximum EGT. Relatively equal EGT recorded at full load condition for WCO30 and SSB30. EGT is the amount of heat loss as exhaust gas during engine operation [5]. Increase in EGT were observed with increasing brake power. The recorded EGT values were 303, 339.55 and 340.5 ^{0}C for diesel, SSB30 and WCO30 respectively at full load. An elevated EGT generally observed due to longer combustion duration [25].

3.2. Emission

3.2.1. Nitrogen oxide (NOx) emission

NOx formation takes place by chain reactions of O_2 and N_2 in suitable temperature condition. Generally, NOx emission depends on injection time, injection pressure and fatty acid profile of biodiesel [9]. It can be observed from Figure 5 NOx emission in g/kWhr decreases with increase in brake power). Biodiesel blends recorded significantly high NOx emission than petroleum diesel. The NOx emission at full load for WCO30 (3.59 g/kWhr) and SSB30 (3.51 g/kWhr) are nearly same as shown in Figure 5. Whereas, in all loading condition SSB30 (17.51 g/kWhr) resulted less NOx emission compared to WCO30 (18.21 g/kWhr). SSB30 recorded 9.09 % and WCO30 13.45 % higher NOx emission than petroleum diesel. Higher unsaturation level and hence, high iodine value in WCO causes higher NOx emission [4], [22]. Short chain saturated fatty acid (methyl palmitate) in SSB reduced NOx emission. However more of methyl linoleate (C18:2) in WCO30 increased the NOx emission [9], [4].

3.2.2. Carbon monoxide (CO) emission

CO emission in g/kWhr decreases with increase in brake power are shown in Figure 6. Including all loading condition CO emission of 19.34, 16.86, 17.33 g/kWhr were observed for diesel, SSB30 and WCO30 respectively (Figure 6). Significant reduction in emission were observed in biodiesel blends due to high oxygen content that helps in more complete combustion [26]. CO emission reduced by 12.82% for SSB30 and 10.39% for WCO30. Small difference in reduced emission observed due to the close density values of both biodiesel blends. Presence of large amount of palmitic acid methyl ester and less unsaturated fatty acids can reduce CO emission significantly [4], [7]. Therefore, highly saturated SSB with nearly 30% palmitic acid methyl ester has better CO emission than highly unsaturated WCO biodiesel.

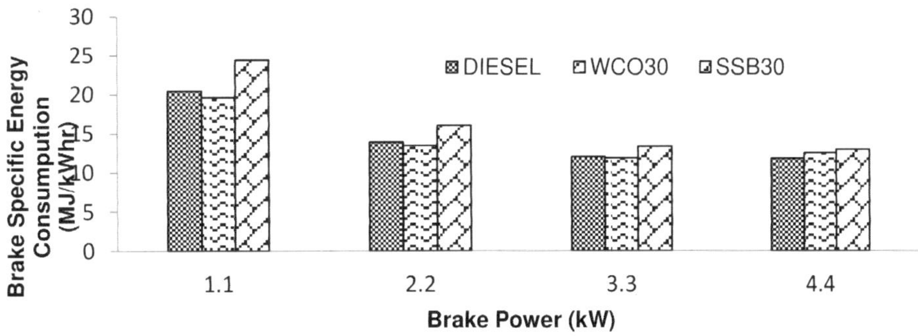

Figure 2. Variation in brake specific fuel consumption with brake power.

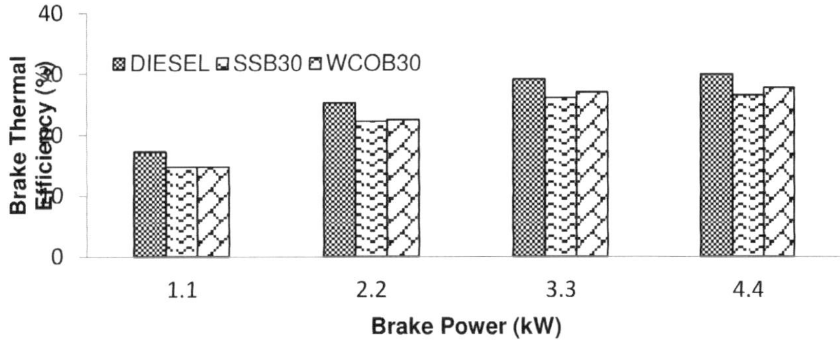

Figure 3. Variation in brake thermal efficiency with brake power.

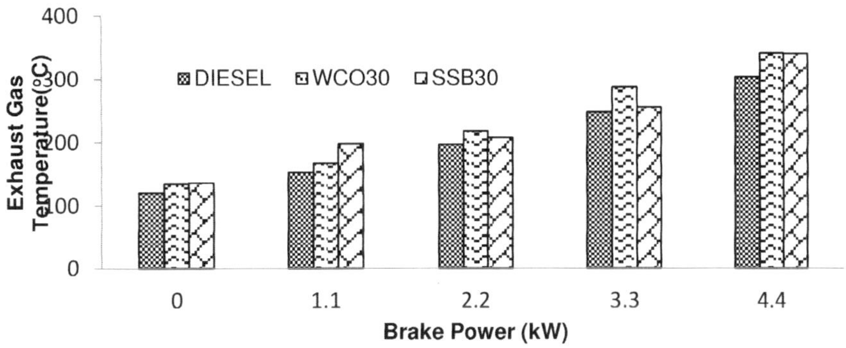

Figure 4. Variation in exhaust gas temperature with brake power.

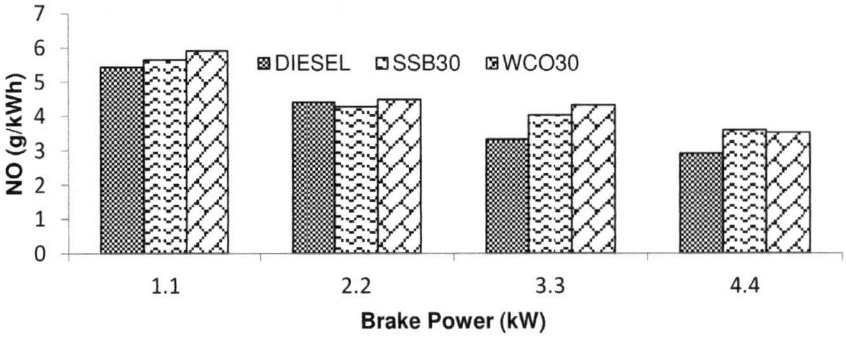

Figure 5. Variation in NOx emission with brake power.

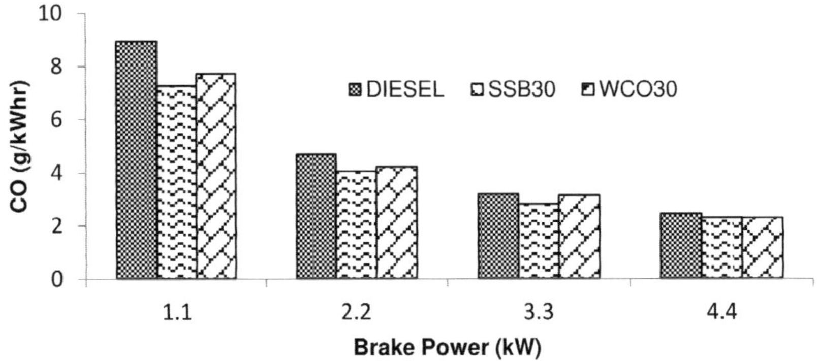

Figure 6. Variation in carbon monoxide (CO) emission with brake power.

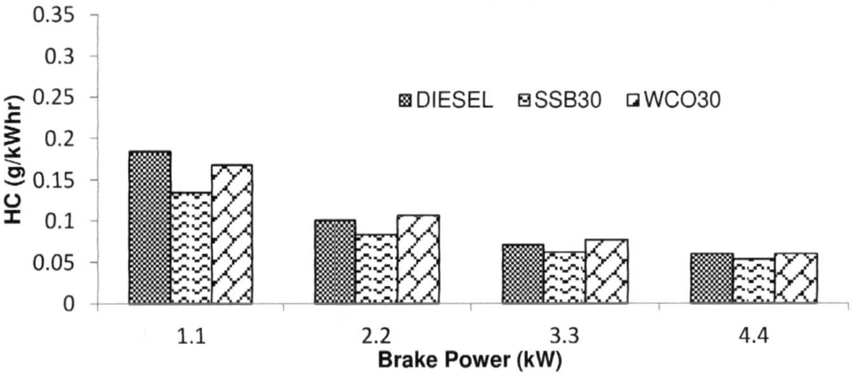

Figure 7. Variation in hydro carbon emission with brake power.

3.2.3. Hydrocarbon (HC) emission

Total HC emission at all loading conditions were shown in Figure 7. Emission of 0.4 g/kWhr, 0.32 g/kWhr and 0.38 g/kWhr were observed for diesel SSB30 and WCO30 respectively. WCO30 and diesel recorded equal amount of emission i.e., 0.059 g/kWhr at full load. Whereas, SSB30 HC emission lowered by 20% and 15.7% from diesel and WCO30 respectively. HC emission in g/kWhr also decreases with increase in brake power. Large concentration of polyunsaturated fatty acid in WCO resulted higher HC emission. Presence of high concentration of methyl palmitate, methyl stearate and methyl oleate can be attributed to the lowest HC emission [3], [4]. Methyl ester of linoleic acid in large amount is responsible for high HC emission by WCO30.

4. Conclusion

While the entire world is in need of energy source and facing fuel crisis alternate sources like waste cooking oil biodiesel and sewage sludge biodiesel can counter the requirements. On the basis of performance parameters WCO30 is a better replacement of diesel, however the emission characteristics of SSB30 is better than WCO30. Prime finding of the study was given below:

- The BSFC for WCO30 found better than petroleum diesel. This is the most influencing characteristic for the commercialisation of WCO30.
- SSB30 has a satisfying BSFC instead of low immiscibility and higher viscosity.
- Fatty acid profile significantly influenced the emission profile. SSB30 with 30 % saturated methyl ester resulted better and cleaner emission than highly unsaturated WCO30 biodiesel blend. Similar monounsaturated methyl ester for both biodiesel blend concluded that the poly unsaturated methyl esters has contributed to higher nitrous oxide, carbon monoxide and hydro carbon emission.
- SSB30 resulted 9.09% hike in NOx emission than diesel and 4.36% lower NOx emission than WCO30.Carbon monoxide emission for SSB30 lowered by 12.86% and 2.48% as compared to diesel and WCO30. SSB30 has a reduced the HC emission of 15.7% and 20% compared to WCO30 and diesel.

5. Notations

SSB	Sewage sludge derived 100% FAME
WCOB	Waste cooking oil derived 100% FAME

6. References

[1] World Energy Council (WEC) 2011 *Global transport scenarios 2050* (London: WEC)

[2] Jayaprasanna Kumar D and Binnal P 2012 Performance evaluation of a single cylinder diesel engine fueled with biodiesel produced from pumpkin oil *J. Sci. and Ind. Res.* **71** 75–78

[3] Liu H, Ma X, Li B, Chen L., Wang Z and Wang J 2017 Combustion and emission characteristics of a direct injection diesel engine fueled with biodiesel and PODE / biodiesel fuel blends *Fuel* **209** 62–68 doi:10.1016/j.fuel.2017.07.066.

[4] Pinzi S, Rounce P, Herreros J M, Tsolakis A and Dorado M P 2015 The effect of biodiesel fatty acid composition on combustion and diesel engine exhaust emissions *Fuel* **104** 170–182 doi:10.1016/j.fuel.2012.08.056.

[5] Yu CW, Bari S and Ameena A 2002 Comparison of combustion characteristics of waste cooking oil with diesel as fuel in a direct injection diesel engine *J Automobile Eng.* **216** 237–43

[6] Gnanasekaran S, Saravanan N and Ilangkumaran M 2016 Influence of injection timing on performance , emission and combustion characteristics of a DI diesel engine running on fish oil biodiesel **116** 1218–1229 doi:10.1016/j.energy.2016.10.039.

[7] Ruhul A, Abedin J, Rahman S M A, Haji B, Masjuki H, Alabdulkarem A, Kalam A and Shancita I 2016 Impact of fatty acid composition and physicochemical properties of Jatropha and Alexandrian laurel biodiesel blends : An analysis of performance and emission characteristics *J. clean. prod.* **133** 1181-1189

[8] Aurélio M, Lagnier B, Ferreira G, Guilherme L, Lamare A, Murta S, Aurelio M and Freitas V De 2017 Comparative study of NOx emissions of biodiesel-diesel blends from soybean, palm and waste frying oils using methyl and ethyl transesterification routes *Fuel* **194** 144–156 doi:10.1016/j.fuel.2016.12.084

[9] Omidvarborna H, Kumar A and Kim D S, 2015 NOx emissions from lowerature combustion of biodiesel made of various feedstocks and blends *Fuel Pro. Tech.* **140** 113–118 doi:10.1016/j.fuproc.2015.08.031

[10] Azeem M W, Hanif M A, Al-Sabahi J N, Khan A A, Naz S and Ijaz A 2016 Production of biodiesel from low priced, renewable and abundant date seed oil *Renew. Energy* **86** 124–132 doi:10.1016/j.renene.2015.08.006.

[11] Olkiewicz M, Torres C M, Jimenez L, Font J and Bengoa C 2016 Scale-up and economic analysis of biodiesel production from municipal primary sewage sludge *Bioresour. Techno.***214** 122–131 doi:10.1016/j.biortech.2016.04.098.

[12] Tan Y H, Abdullah M O, Nolasco-hipolito C, Syuhada N and Zauzi A, 2017 Application of RSM and Taguchi methods for optimizing the transesteri fi cation of waste cooking oil catalyzed by solid ostrich and chicken-eggshell derived CaO *Renew. Energy* **114** 437–447 doi:10.1016/j.renene.2017.07.024.

[13] Urrutia C, Sangaletti-Gerhard N, Cea M, Suazo A, Aliberti A and Navia R 2016 Two step esterification-transesterification process of wet greasy sewage sludge for biodiesel production *Bioresour. Technol.* **200** 1044–1049 doi:10.1016/j.biortech.2015.10.039.

[14] Olkiewicz M, Caporgno M P, Fortuny A, Stuber F, Fabregat A, Font J and Bengoa C 2014 Direct liquid-liquid extraction of lipid from municipal sewage sludge for biodiesel production *Fuel Pro. Tech.* **128** 331–338 doi:10.1016/j.fuproc.2014.07.041.

[15] Melero J A, Sánchez-Vázquez, Vasiliadou I A, Martínez Castillejo F, Bautista L F, Iglesias J, Morales G, Molina R 2015 Municipal sewage sludge to biodiesel by simultaneous extraction and conversion of lipids *Energy Conv. and Manage.* **103** 111–118 doi:10.1016/j.enconman.2015.06.045.

[16] Usman K, Khan S, Ghulam S, Khan M U, Khan N, Khan M A and Khalil S K 2012 Sewage Sludge: An Important Biological Resource for Sustainable Agriculture and Its Environmental Implications *Ame. J. Plant Sci.* **3** 1708-1721 doi:10.4236/ajps.2012.312209.

[17] Priyadarshi, D., Karar, P.K.: 2017 Optimisation of Biodiesel Production Using Taguchi Model *Waste biomass valori.* In press doi: 10.1007/s12649-017-0158-9

[18] Choi OK, Song JS, Cha DK, Lee JW 2014 Biodiesel production from wet municipal sludge: Evaluation of in situ transesterification using xylene as a cosolvent *Bioresour. Technol.* ***166*** 51-56 doi:10.1016/j.biortech.2014.05.001.

[19] Roosta A, Sabzpooshan I 2016 Modeling the effects of cosolvents on biodiesel production *Fuel* **186** 779–786

[20] Heywood J B 1988 *Internal combustion engines fundamentals* (New York: McGraw- Hill)

[21] Hansen K F, Jensen MG 1997 Chemical and biological characteristics of exhaust emissions from a DI diesel engine fuelled with rapeseed oil methyl ester (RME) *SAE paper* 971689

[22] Jinlin X, Grift Tony E, Hansen Alan C 2011 Effect of Biodiesel on Engine Performances and Emissions *Renewable and Sustainable Energy Reviews* **15(2)** 1098–1116 http://dx.doi.org/10.1016/j.rser.2010.11.016.

[23] Ganapathy T, Murugesan K, Gakkhar RP 2009 Performance optimization of Jatropha biodiesel engine model using Taguchi approach *Appl. Energy* **86** 2476–2486 doi:10.1016/j.apenergy.2009.02.008

[24] Aydin H and Bayindir H 2010 Performance and emission analysis of cottonseed oil methyl ester in a diesel engine *Renew Energy* **35** 588–592

[25] Sharma A and Sivalingam M, 2014 Influence of Fuel Injection Timing on the Performance and Emission Characteristics of a Diesel Engine Fueled with Jatropha Methyl Ester- Tyre Pyrolysis Oil Blend *Fuel* **594** 1627–1631 doi:10.4028/www.scientific.net/AMM.592-594.1627

[26] Chauhan B S, Kumar N and Cho H M 2012 A study on the performance and emission of a diesel engine fuelled with Jatropha biodiesel oil and its blends *Energy* **37** 616-622.

IOP Conf. Series: Earth and Environmental Science **167** (2018) 012036 doi:10.1088/1755-1315/167/1/012036

Utilization of kitchen food waste for biodiesel production

S Barik[1], KK Paul[2] and D Priyadarshi[3]

1 M.Tech Research Scholar, Department of Civil Engineering, National Institute of Technology, Rourkela, Odisha, 769008, India
2 Assistant Professor, Department of Civil Engineering, National Institute of Technology, Rourkela 769008, Odisha, India
3 Research Scholar, Department of Civil Engineering, National Institute of Technology, Rourkela, Odisha, 769008, India
E-mail: k_karar1@yahoo.co.in

Abstract. Food waste has led to handling and disposal problems affecting human health and polluting environment. Land disposal is one of the ways of disposing food waste but it produces harmful leachate when rain water falls on it. Reuse of food waste for biodiesel production is one of the key steps to reduce food waste problem and also meet the energy demand. Food waste in the present study was collected from the kitchen of a hostel of an educational institute located in India. This waste was dried by various drying methods to dewater. Lesser is the moisture content, more will be the lipid extraction and hence increased biodiesel production. Lipids were extracted and fatty acids (caproic acid (6:1), lauric acid (12:0), mystric acid (14:0), palmitic acid (16:0), stearic acid (17:0) and oleic acid (18:0) were identified in gas chromatography-mass spectrometry (GC-MS). These fatty acids identified in lipid indicate the potential of food waste for biodiesel production. Transesterification of lipid was performed to produce biodiesel and concentration of fatty acid methyl ester was determined using gas chromatograph flame ionisation detector (GC-FID). Further, biodiesel properties were compared with various standards.

1. Introduction

Worldwide petroleum consumption has increased leading to fossil fuel crisis and energy problem in near future due to rapid urbanization and industrialization. Till now, energy crisis has been met by fossil fuel which is a limited resource. Increased demand of fossil fuel has led to increase in fuel price and serious environmental impacts such as global warming, acidification, deforestation, ozone depletion, eutrophication and photochemical smog. Global fossil fuel energy consumption has doubled in the last few decades. As fossil fuel is a limited source of energy, there is an urgent need to search for an alternative resource that would be economically efficient, socially equitable and environmental friendly [1].

Rapid urbanization and industrialization also generate huge amount of solid waste resulting in handling and disposal problem. Municipal solid waste (MSW) represents a small fraction of total solid waste but attracts more attention because people and environment are adversely affected [2]. In India during 2015, average MSW waste generation was 700 tons per day [3]. Municipal solid waste contains a considerable amount of food waste. Several billion gallons of food wastes are produced from kitchens of residential societies, restaurants, hostels, canteens and food and meat processing industries. Everyday gallons of food wastes generated from different sites are collected and thrown in an open area or dumped in landfill without proper disposal technique [4].

Food wastes adversely affect local water body, land and biodiversity. It causes serious environmental and social problems across the world. The national waste report of Australia in 2008 estimated that food waste make up approximately one-third of MSW and one-fifth of commercial and industrial waste streams. Food waste generated in Australia amount to 7.5 million tons in year 2008 [5]. According to the report released by Hong Kong environment bureau, among 9,000 tons of MSW that is thrown away everyday at landfills, 40% consists of putrescible. Approximately, 90% putrescible was food waste. Other developed countries like Taipei and Seoul produce 1,82,000 tons per year and 7,67,000 tons per year of food waste, respectively. In characterization of municipal solid waste, India comprises maximum of food waste (31.9%) as compared to other wastes such as plastic, textile, paper, glass, cardboard, ash, leather and metal waste [6]. The most common way of food waste disposal is landfill. Landfills by food waste are known to generate carbon dioxide, methane and other toxic gaseous substances [4]. Landfill spreads offensive odour and cause hazardous effects on animals, people and environment. It also generates harmful leachate when rain water falls. This leachate can contaminate soil and ground water [5].
Biodiesel can be produced by pyrolysis, thermal cracking, micro emulsions and transesterification. Transesterification is one of the most common methods for biodiesel production in which oil or fat is reacted with alcohol (methanol or ethanol) in presence of a catalyst to form alkyl ester and glycerol. Primary cost of biodiesel production is affected by cost of feedstock. In this paper an attempt has been made to reuse food waste for biodiesel production by identifying and characterizing the free fatty acids (FFA) present in lipid extracted from kitchen food waste. This innovative study will help to minimize disposal problem and overcome energy crisis. It will also reduce environmental pollution caused by fossil fuel.

2. Materials and Methods

2.1. Sampling site
Food waste is collected from girls hostel of National Institute of Technology, Rourkela, India. This sampling location site is chosen because of huge amount of food waste generation. Food waste after collection is brought to the laboratory for further analysis.

2.2. Food waste analysis
About 2 kg of food waste is collected and brought to the laboratory. Moisture content is removed and optimum temperature was determined. Different drying methods were performed i.e. oven drying (55℃, 75℃, 105℃), freeze drying (-4℃) and sun drying (25-30℃) [7] to determine the optimum temperature. The main challenge to be faced for biodiesel production was efficient lipid extraction as it contains water along food particles. After drying the sample has been grinded to powdered form. The powered sample has been used for lipid extraction.

2.3. Lipid extraction and analysis
Lipid extraction was performed in soxhlet apparatus using methanol as solvent. The extraction of lipid depends on octanol water partition coefficient (K_{OW}). It is defined as the chemical concentration in the octanol phase to its concentration in the aqueous phase of two phase octanol/water system. The organic compound that resides in water phase are hydrophilic and chemical that resides in octanol phase are hydrophobic. After extraction, sample was filtered to separate the liquid and solid particles using whatman 42 filter paper, 125mm (dia). Later, the solvent (methanol) was recovered using rotary evaporator by evaporating at 70℃. The recovered methanol was stored and reused for next extraction. The extracted lipid was stored in a dessicator overnight and weighed to calculate the exact extraction of lipid yield. Lipid was further analysed in gas chromatograph mass spectrophotometer (GC-MS) to determine the presence of free fatty acids which will indicate the potential of food waste for biodiesel production. The residue left after lipid extraction can further be analysed and reuse in pharmaceutical to prepare medicine, drugs and plant nutrients.

2.4. Transesterification and FAME analysis

Transesterification of food waste was performed using acid catalyst. A known amount of lipid was taken in a round bottom flask. Solution of methanol (solvent) and sulphuric acid (catalyst) was added to lipid and kept closed. The solution kept in a heating plate maintaining a temperature of 65℃ and mixed properly for 2-2.5 h. Usually excess methanol is used for complete conversion of fat or oil to fatty acid methyl ester (FAME). After FAME production the sample was cooled and transferred to separatory funnel. The product was left undisturbed for 24 h for separation of biodiesel and glycerol. Methanol present in ester layer was recovered using rotary evaporator at 70℃.

One of the important steps in biodiesel production is washing of biodiesel to remove catalyst and other impurities. Removal of glycerol and glycerides is a key step because quality of biodiesel is dependent on glycerol content. High concentration of glycerol in biodiesel can cause problem during storage and increase in aldehyde emissions [8]. Washing of biodiesel include wet washing and dry washing. Wet washing means adding distilled water to biodiesel and dry washing means adding adsorbent like magnesol, anhydrous sodium sulphate, resin etc. In this study, wet washing of biodiesel has been performed after transesterification also called water wash method. Wet wash method has been selected because this is the oldest and most common method of washing. Moreover, the waste water recovered after washing can be reused for irrigation purpose after minimal treatment. Other treatment methods generate hazardous wastes that are difficult to disposed off. The purpose of water wash method is to remove unreacted alcohol, catalyst or glycerine present in biodiesel. It will also remove if any soap is present. After glycerine settles in the bottom of the separatory funnel, biodiesel is separated and pH is measured. The pH of the unwashed biodiesel was found close to 9. Slowly warm distilled water was added in equal proportion to biodiesel. The solution was allowed to settle by transferring to separatory funnel to separate water and biodiesel. It will take around 24hrs to settle down. Once the mixture has settled down, biodiesel will float on top of water. After washing, the pH of biodiesel was found to be 7. Washing can be performed multiple times to purify the biodiesel.

The purified biodiesel was analysed in gas chromatograph flame ionization detector (GC-FID) to determine the concentration of fatty acid methyl ester. Further, biodiesel properties were compared with various standards.

3. Results and Discussion

3.1. Moisture content and lipid analysis

Removal of moisture content is a vital step in lipid yield and biodiesel production because water present in sample may inhibit the solvent to penetrate inside the sample by surrounding the food particles. The moisture of the sample is dependent on drying method and temperature.

Different drying methods have been selected to determine the optimized temperature for less moisture content and more lipid yield. In general, higher the drying temperature, lesser is the moisture content, more amount of lipid yield and hence biodiesel production [7]. Sun drying method gave moisture content of 4.6% in 240h at drying temperature of 25-30℃ with lipid yield of 16.5%. This method is dependent on weather condition and time consuming. Freeze drying method gave moisture content of 7.5% in 48 h. In oven drying 105℃ gave less moisture content of 0.1% compared to 55℃ (2.4%) and 75℃ (1.5%). When different drying methods are compared, it is observed that oven drying method at 105℃ gave less moisture content (0.1%) with lipid yield of 37.3%. Hence, the optimized temperature for drying is 105℃. very high temperature in oven drying will burn the sample and break the fatty acid chain because for biodiesel production long chain fatty acids are required.

To study the effect of lipid yield due to variation in food waste composition, the food waste sample was collected weekly (7 days). The lipid yield due to food waste composition has been discussed in Table 1.

Food waste mainly consist of rice, pulses, vegetables along with paneer in vegetarian food whereas non vegetarian food consist of meat, fish, chicken along with oil and fats. Fat content in 100 g of chicken is 14 g, 100 g of mutton contains 21g fats, 100 g of fish contains 12 g fats, and 100 g of paneer contains 20.8 g fats. It is observed that lipid yield is more in case of vegetarian and non-vegetarian compared to vegetarian only because of more amount of oil and fats. According to literature, more the amount of oil/fats, more is the lipid yield extraction and hence, more amount of biodiesel yield.

Table 1. Effect of lipid yield due to food waste composition

S.N.	Days	Food waste composition	Lipid yield (%)
1.	Monday	Vegetarian	32.5
2.	Tuesday	Vegetarian and non-vegetarian	37.3
3.	Wednesday	Vegetarian and non-vegetarian	36.8
4.	Thursday	Vegetarian	30.2
5.	Friday	Vegetarian and non-vegetarian	36.4
6.	Saturday	Vegetarian and non-vegetarian	37.3
7.	Sunday	Vegetarian and non-vegetarian	37.2

3.2. Lipid analysis

Lipid extraction was performed in soxhlet apparatus using methanol as solvent. Lipids are of two types i.e., neutral lipid and polar lipid. Neutral lipids are insoluble in water but soluble in organic solvents such as methanol, chloroform, hexane, etc. Solvent extraction of lipid is performed to separate the components dissolved in mixture. The separation of specific compound is possible because solvent dissolve the targeted compounds. Targeted compounds such as triglyceride can be separated by mixing the solution to a temperature near to boiling of solvent which will evaporate solvent but targeted compounds will not [9]. The role of methanol is to extract triglyceride, phospholipids, cholesterol, etc [10]. The extracted lipid was analysed in gas chromatograph mass spectrophotometer (GC-MS) to determine the presence of free fatty acids. Table 2 discuss about the free fatty acids identified in lipid analysis. These identified free fatty acids indicate the potential of food waste for biodiesel production.

Table 2. Free fatty acids identified in lipid analysis

S.N.	Retention time	Organic compounds
1.	4.070	Caproic acid (C6:0)
2.	9.191	Lauric acid (C12:0)
3.	11.181	Mystric acid (C14:0)
4.	12.595	Palmitic acid (C16:0)
5.	13.741	Stearic acid (C17:0)
6.	13.853	Oleic acid (C18:0)

3.3. Transesterification and FAME analysis

Transesterification of lipid was performed using methanol as solvent and sulphuric acid as catalyst. The reaction was performed in the ratio 11:1 (methanol: lipid molar ratio) at 60℃ for 2-3 h and catalyst of 2.5 wt % was added. It is always advisable to perform the reaction close to the boiling point of methanol because of proper misciblibility of methanol with lipid. Reaction rate increases with increasing temperature resulting in increase in solubility of methanol in oil-rich phase. Higher reaction temperature results in shorter reaction time in mass transfer controlled reaction because transesterification reaction is more preferred at higher temperature [11]. Usually the transesterification should be performed below the boiling point in order to prevent evaporation of methanol. It is recommended to perform reaction below boiling point by various literatures. Kusdiana and Saka (2004) also reported that in supercritical state, methanol acts as an acid catalyst in transesterification reaction [12]. Temperature increases the energy of the reacting molecule and also improves the miscibility of alcoholic polar media into a non-polar oily phase resulting in much faster reaction [13]. Due to proper mixing of solvent and lipid in presence of catalyst for time i.e. 2-3 h result in faster reaction rate and complete conversion of biodiesel. The biodiesel yield of 32.3% was obtained in the ratio 11:1 at 60℃ for 2-3h and catalyst of 2.5 wt%. The produced biodiesel was washed properly using distilled water to remove impurities and obtained pure biodiesel. The pure biodiesel was further analysed in gas chromatograph flame ionization detector (GC-FID) to determine the concentration of fatty acid methyl ester (FAME). The organic compound was identified by comparing the peaks of obtained FAME of food waste with that of standard FAME. Figure 1 shows the peak of the identified fatty acid methyl ester. Table 3 discuss about the wt% of fatty acid methyl ester identified in GC analysis.

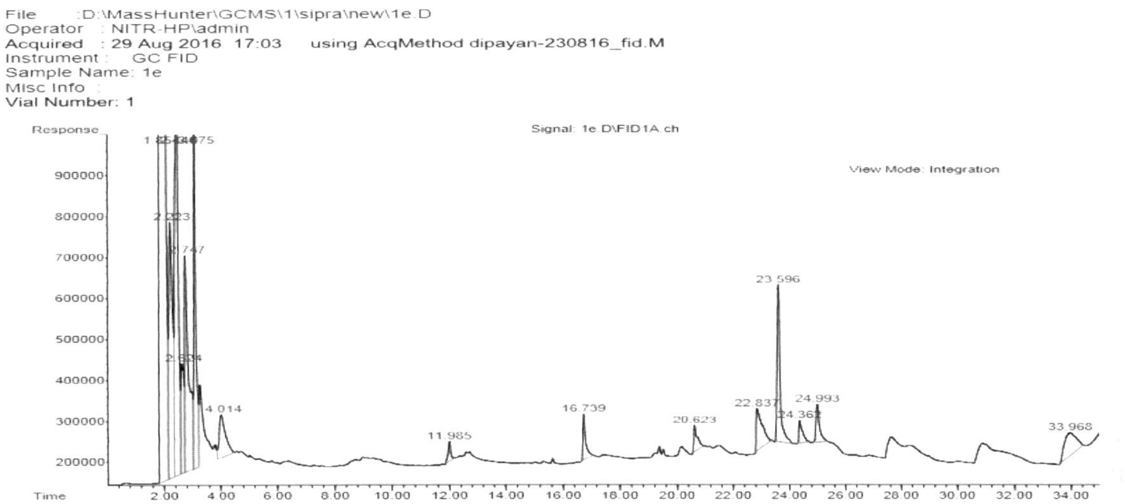

Figure 1. GC-FID FAME analysis of food waste

Table 3. Wt% of fatty acid methyl ester identified in GC analysis.

Fatty acid profil	**Retention time**	**FAME**	**Wt %**
1.	4.014	Caproic acid methyl ester (C6:0)	17.64
2.	11.985	Lauric acid methyl ester (C12:0)	4.04
3.	16.739	Mystric acid methyl ester (C14:0)	9.23
4.	22.837	Palmitic acid methyl ester (C16:0)	21.23
5.	23.596	Palmitoleic acid methyl ester	11.06
6.	24.362	Stearic acid methyl ester (C18:0)	19.19
7.	24.993	Oleic acid methyl ester (C18:1)	15.67

Fatty acids identified may be both saturated and unsaturated. Fatty acids mostly have carbon atoms between C12- C22. Saturated fatty acids identified in biodiesel include caproic acid (6:0), lauric acid (12:0), mystric acid (14:0), palmitic acid (16:0) and stearic acid (17:0). Unsaturated fatty acids have one carbon-carbon double bond which can be present in different positions. The most common unsaturated fatty acids contain chain length of 16-22 and a double bond with cis configuration. Unsaturated fatty acid identified in biodiesel is oleic acid (18:0). It was found that saturated fatty acids dominate in FAME analysis of lipid. Due to saturated components, biodiesel produced from food waste will undergo cold flow problem. Several studies were conducted by various researchers to improve cold flow properties of biodiesel such as use of additives to reduce intermolecular bond, decrease the crystalline temperature and combining biodiesel with petro diesel, as well as the use of thermal cracking process, ozonation technique and winterization technique to reduce the concentration of saturated fatty acid esters. However, specific method or additive that can improve cold flow behaviour of all types of biodiesel is not available. Cold flow enhancers are used to improve the cold flow properties of biodiesel [14]. According to Refaat (2009), length of fatty acids chains as well as its composition plays a substantial role in cold flow properties of biodiesel [15]. National biodiesel board (2013) also stated that properties of biodiesel depends on the raw material (type of grease, fat or oil) from which they are produced [16].

3.4. Biodiesel properties

The obtained biodiesel was further analysed to determine its physical and chemical properties and then compared with various standards as summarized in Table 4.

Table 4. Comparison of biodiesel properties with ASTM, EN and IS standard

S.N.	Properties	ASTM D6751	EN14214	IS 15607 (2005)	Obtained value
1.	Density at 15°C (kg/m^3)	875-900	860-900	860-900	872
2.	Kinematic viscosity (mm^2/s)	1.9-6.0	3.5-5.0	2.5-6.0	2.2
3.	Acid value (mg of KOH/g), Max	0.5	0.5	0.5	0.63
4.	Pour point (°C)	-3 to 12	-	-	7
5.	Calorific value (MJ/kg)	-	35	-	31.38
6.	Metals (Na+K) (mg/kg)	5	5	-	1.5
7.	Metals (Ca+Mg) (mg/kg)	5	5	-	1.25
8.	Ash content, percent by mass, Max	0.02	0.02	0.02	0.0072
9.	Cloud point (°C)	-	-	-	12
10.	Flash point (°C), Min	130	101	120	164

Density of biodiesel produced from food waste is found to be 872 kg/m^3 which is within the limit according to ASTM, EN and IS standard. The values depend on their fatty acid composition as well as on their purity. One of the most important properties of biodiesel is kinematic viscosity. The kinematic viscosity of biodiesel is found to be 2.2 mm^2/s which in the range of 1.9-6.0 according to ASTM D 6751 according to test method D445. This implies that biodiesel enhanced fluidity of fuel for diesel engine and good spray pattern that would generate across the combustion chamber, allowing proper mixing with air [17]. The acid value for biodiesel is 0.63mg KOH/g and it exceeds the EN 14214 and ASTM D6751 standard according to which acid value should not exceed the value 0.5mg KOH/g. as triglycerides present in oil and fats converted to fatty acids causing an increase in acid number. Pour point is the lowest temperature at which oil starts to melt or pour. It is the measure of ability of a fuel to operate under cold weather conditions. The pour point of biodiesel is found to be 7°C which is within the range of ASTM D6751 (-3 to 12 °C). During cold weather biodiesel causes fuel starvation and operational problem which cease the fuel flow in engine. Calorific value is the measurement of heat or energy produced and is measured either as gross calorific value or net calorific value. The gross calorific value of biodiesel measured by bomb calorimeter is found to be 31.38MJ/kg which is less than the standard i.e., 3 5 M J /kg according to EN 14214 standard. Calorific value of diesel fuel is 45.5MJ/kg. Less calorific value means more fuel consumption.

If metal contamination can occur, metal deactivators can be used to chelate transition and inhibit catalytic oxidation and polymerization effect. Alkali metals could also form soap and cause insoluble of diesel blends. Alkali metals of (Na+K) were found to 1.5 mg/kg and (Ca+Mg) were found to be 1.25mg/kg which is within the range of 5mg/kg according to standard EN 14538. Flash point is the minimum temperature at which vapour given off by a fuel when heated will flash with a test flame held above the surface without catching fire and plays a vital role in fire hazard of fuel. It gives the idea about the boiling point of liquid fuel, volatility and explosive hazards. Flash point (164 °C) of biodiesel obtained from food waste is within the limit according to ASTM, EN and IS standard. As the biodiesel starts to freeze, it will form crystals that start clumping together. Cloud point of obtained biodiesel is 12 °C i.e., below this temperature biodiesel will form a cloudy appearance. Ash point describes the amount of inorganic contaminants such as abrasive solids, catalyst residue, soluble metals contained in fuel. Ash content of biodiesel is found to be 0.0072 which is within the standard value (0.02).

4. Conclusion

To overcome fossil fuel crisis problem, environmental pollution and meet the energy demand one of the best way is to replace non-renewable resource with renewable resource. Various researches have been performed to identify the best alternative way of reusing renewable resources, most importantly waste product for biodiesel production. Waste products like sewage sludge, industrial waste, waste cooking oil etc., are zero cost raw materials that can be used for biodiesel production which will help to minimize waste generation, handling and disposal problem. In this study an innovative approach to utilize kitchen food waste for biodiesel production has been initiated. Food waste was dewatered and lipid was extracted using methanol as solvent with yield of 37.3%. Lipid analysis was performed in GC-MS to identify the fatty acids. Lipids were further transesterified using methanol as solvent and sulphuric acid as catalyst producing FAME yield of 33.2%. The extracted biodiesel was analysed in GC-FID to determine the concentration of free fatty acid methyl ester. Physical and chemical properties of biodiesel were performed and found satisfied by comparing with different standards.

5. References

[1] Elsolh N E M, 2011 The manufacture of biodiesel from used vegetable oil *A thesis submitted to the faculty of engineering at Kassel and Cairo Universities for the degree of Master of Science* **12.**

[2] Goel S, 2008 Municipal solid waste management in India, A Critical review *Journal of Environmental Science and Engineering* **50** 319-328.

[3] Pandey S K, Tyagi P and Gupta A K, 2007 Municipal solid waste management in Ghazipur city- a case study *Journal of Agricultural and Biological Science* **2**, 41-43.

[4] Karmee S K, 2016 Liquid biofuels from food waste: Current trends, prospect and limitation. *Renewable and Sustainable Energy Reviews* **53** 945–953.

[5] Karmee S K and Lin C S K, 2014 Valorisation of food waste to biofuel: Current trends and technological challenges. *Sustainable Chemical Processes* **2** (22).

[6] Srivastava R, Krishna V and Sonkar I, 2014 Characterization and management of municipal solid waste: a case study of Vanaras city, India *International Journal of Current Research and Academic Review* **2** 10-16.

[7] Olkiewicz M, Caporgno M P, Fortuny AStuber, F Fabregeat A, Font J and Bengoa C 2014 Direct liquid – liquid extraction of lipid from municipal sewage sludge for biodiesel production *Fuel Proc. Technol.* **128** 331-338.

[8] Faccini C S, Cunha M E, Moraes M S A, Krause L C, Manique M C, Rodrigues M R A, Benvenutti E V and Caramao E B, 2011 Dry washing of biodiesel purification: a comparative study of adsorbents *Journal of the Brazilian Chemical Society* **22**(3) 558-563.

[9] Hutton MW 2009 Extraction and characterization of lipids from microalgae grown on municipal waste water. *A Master's Thesis Presented to the Faculty of California Polytechnic State University San Luis Obispo.*

[10] Ferraza T P L, Fiu´zaa M C, Santosb M L A, Pontes de Carvalhoa, Soares N M 2004 Comparison of six methods for the extraction of lipids from serum in terms of effectiveness and protein preservation. *J. Biochem. Biophys. Methods* **58** 187–193.

[11] Noureddini H and Zhu D 1997 Kinetics of transesterification of soybean oil. *J. Am. Oil Chem.* **74** 1457-1463.

[12] Kusdiana D and Saka S 2004 Effects of water on biodiesel fuel production by supercritical methanol treatment. *Bioresour. Technol.* **91** 289–295.

[13] Ogbu I M and Ajiw V I E 2013 Biodiesel production via esterification of free fatty acids from cucurbita pepo l. seed oil: kinetic studies. *Int. J. Sci. Technol.* **2** 616-621.

[14] Monirul I M, Masjuki H H, Kalam M A, Zulkifli N W M, Rashedul H K, Rashed M M and Mosar M H 2015 A comprehensive review on biodiesel cold flow properties and oxidation stability along with their improvement processes *Royal Soc. Chem. Adv.* **5** 86631-86655.
[15] Refaat A A, 2009 Correlation between the chemical structure of biodiesel and its physical properties *Int. J. Environ. Sci. Technol.* **6** 677-694.
[16] National Biodiesel Education Program Diesel Fuel Cold Flow Properties. Biodiesel Short course –Cold Flow, Biodiesel Education 2013 Retrieved October 25, 2016, from *http://web.cals.uidaho.edu/biodiesel/biodiesel-shortcourse-cold-flow.*
[17] Gandure J, Ketlogetswe C and Temu A 2013 Fuel properties of jatropha methyl ester and its blends with petroleum diesel *J. Eng. Applied Sci.* **8** 900-908.

IOP Conf. Series: Earth and Environmental Science **167** (2018) 012037 doi:10.1088/1755-1315/167/1/012037

Comparative Energy Analysis of Photovoltaic Module

S Madi[1,2] and A Kheldoun[1]

1 Signals and Systems Laboratory, Institute of Electrical and Electronic Engineering, University M'Hamed Bougara of Boumerdes, Algeria.
2 Faculty of Sciences and Applied Sciences, University Akli Mohand Oulhadj–Bouira, Algeria.
E-mail: madi.saida@yahoo.fr ; aissa.kheldoun@univ-boumerdes.dz

Abstract. This work introduces a comparative energy analysis of photovoltaic module using its different single diode equivalent electrical circuits (five-parameter model, four-parameter model, and ideal model). These equivalent circuits are used for different purposes according to the objective. The ideal model is still used in many applications particularly power system analysis and water pumping. Three commercially available (SUNTECH) photovoltaic modules (50W, 70W and 130 W) are used in this analysis at standard test conditions. Using those three PV modules; the relative errors of the maximum output power mismatch are highlighted in the three cases. Obtained results show that the five-parameter model gives the most accurate maximum power prediction in comparison to that of manufacturer's data sheet, while the ideal model shows a remarkable relative error at the maximum output power which is up to 8.93013%. This evaluation allows users deciding on the selection of appropriate equivalent circuit to be used in a given application.

1. Introduction

Solar energy has been recognized as one of the alternative solutions to replace fossil based power plants. Its abundance, noiselessness, clean source, free cost, advantages among many others of this source of electric energy. Photovoltaic is one way to convert sunlight energy to electricity. The number of PV modules depends on the power demand, voltage requirement and site's solar radiation availability. The PV generator can be standalone or grid tied system. For the two architectures, analysis, design, optimization, or control, the use of the equivalent circuit of the PV module is required. Different equivalent circuits of PV module are available which differ in the number parameters and the number of diodes used in the modeling. This paper investigates equivalent circuits of single-diode model. Chouder A. *et al.* [1] presented a simulation study and experimental validation of a PV grid connected system with a rated power of 3.2 KW. A PV module, forming part of the whole PV array, is modeled by a single diode five-parameter model and the model has been validated experimentally by carrying out outdoor I–V characteristic measurements. Similarly Elyes*et al.* [2] used the five-parameter model for PV generator when they reported the development of a statistical fault detection approach for monitoring the performances in a grid connected PV system by detecting faults on the DC side and diagnosing the type of detected fault. To simplify analysis and computation, four parameter equivalent circuit is preferred. For instance, Biji*et al.* [3] modeled and simulated the PV pumping system for maximum efficiency using for-parameter model of PV generator. Djeriou *et al.* [4] applied a maximum power point tracker (MPPT) technique for stand-alone photovoltaic water pumping system to improve the overall operating efficiency system. The latter used a four-parameter single diode model to simulate the whole system. The PV generator of the system is modeled using the four-parameter model. Neglecting PV cell physical phenomena such as contact resistances, the current lost by photocell sides as well as the ageing of cells [5], [6] while introducing the capacitor (C) allows to overcome the problem of causal [7]. The PV generator equivalent to a current source shunted by a junction diode which is known to be the ideal model. This model is employed to analyze photovoltaic system being associated to storage batteries where modeling using bond graph methodology is investigated in [5], [8]. In addition, Sieckeret *et al.* [9] presented extensive reviews of various cooling techniques used to enhance the performance of a PV system where the PV generator is modeled using the ideal model. Similarly Mezghani *et al.* [6] used the ideal model of PV generator to model the hybrid pumping system using the bond graph approach. Anbarasi Jebaselvi *et al.* [10] provided overviews of the performance analysis of hybrid power systems, control methodologies and modeling techniques so as to get an optimum output power. The latter study is based on the modeling of the PV generator using the ideal equivalent circuit.

Many researchers have focused on the analysis of neglecting model parameters such as series and parallel resistance. Madansure *et al.* [7] presented the PV system behavior modeling using bond graph approach. They introduced different bond graph based model representations of the PV cell using different equivalent electrical circuits namely five-parameter model, diode capacitance representation, Norton representation, Thevenin's representation and Gyrator representations. Besheeret *al.* [11] presented thorough comparative analysis a study for various kinds of single-diode model based photovoltaic power source. The analysis was done for different values of temperature as well as solar irradiation. Ahmad H. Besheer *et al.* [12] presented a comparative analysis study for the PV module based on single diode model representation with various levels of complexity. The analysis is based on four single diode model and the experiment validated data for two different kinds of PV module available in the market. The relative errors between each PV single diode model output and the experimentally validated data are computed at low, medium and high temperature and irradiance level as well as at standard test conditions. The variation of the temperature and the solar irradiation are not done simultaneously which may not reflect the correct corresponding relative error between the dynamic performance of such photovoltaic models and the experimental data from the manufacturer's data sheet under varying atmospheric conditions. Silvano [13] proposed two mathematical models of PV cell, namely: the complete model (five-parameter model) and the simplified model (four-parameter model) where the values of the relative error between the dynamic performance of such photovoltaic models and the experimental data from the manufacturer's data sheet under varying atmospheric conditions are not highlighted.

This work presents a comparative energy analysis of commercial (SUNTECH) photovoltaic module of 50, 70 and 130W using three different representations of equivalent electrical circuits (five-parameter model, four-parameter model, and ideal model).Analyses are done at standard test conditions in order to provide comparisons between the obtained results using the three representations and the values given in manufacturing data sheet. The remaining of the paper is organized as follows: Section 2 provides the modeling of PV module based on bond graph methodology and three different equivalent electrical circuits of PV module. Section 3 shows the obtained results using the three representations at STC. Finally, a conclusion is drawn to end up the paper.

2. Modeling of photovoltaic module

Bond graphs actually indicate the structure through which energy is exchanged [14]. It represents the power transfer within a system by using a series of connections called power bonds [15], [16]. It utilizes nine basic elements or building blocks which may represent physical subsystems, components or phenomena in every energy domain. The elements are classified in three categories, the passive elements and the active elements and the elements of the junction. These elements in addition to the detectors are used to represent the phenomenon that connects the generalized variables. Since this technique is based on the electrical circuit of the system. In this paper the modeling of PV modules is done based on its equivalent electrical circuit and using bond graph methodology.

The equivalent electrical circuit of the photovoltaic (PV) module consists of a photovoltaic current source I_{phM} in parallel with a diode (D_M) and shunt resistance (R_{ShM}) expressing a leakage current and a series resistance (R_{SM}) describing internal resistance to the current flow, as shown in Figure 1. (a). The bond graph representation of this later is shown in Figure 1. (b). Where; the bond graph representation of the photocurrent source is a source flow (S_f: I_{phM}), while the shunt and series resistances both are represented by a resistance (R: R_{ShM} , R: R_{SM}). The diode (D_M) is represented by a resistance (R: R_{DM}) with considering the Shochley's diode equation.

Based on Figure 1, and using the characteristic equations of the junctions in consideration of the causality, the resulting expression of the PV module's output voltage, $V_1(e_6)$, is given by the following equation [14]

$$V_1 = V_{ThM}\, ln\left(\frac{I_{phM}-\frac{V_M}{R_{ShM}}-\frac{R_{SM}}{R_{ShM}}I_1-I_1}{I_{satM}}+1\right)-R_{SM}I_1 \quad (1)$$

From the previous equation one can deduce that the PV module current equation, $I_1(f_6)$, can be deduced as follows

$$I_1 = I_{PhM} - I_{satM}\left(exp^{\frac{V_1+R_{SM}I_1}{V_{ThM}}} - 1\right) - \frac{V_1+R_{SM}I_1}{R_{ShM}} \quad (2)$$

The model represented by equations **(1)–(2)** is usually carried out using Kirchhoff's laws and Shochley'sdiode equation. Multiplying the PV module current, given by the above equation by its output voltage yields the output power of the used PV module, as given in the following equation

$$P_1 = I_1 * V_1 \quad (3)$$

Since the shunt resistance is very large it can be removed from the equivalent electrical circuit of the PV module shown in Figure 1. Thus, a simplified equivalent electrical circuit of PV module is obtained as shown in Figure 2.

Based on Figure 2, this latest figure and using the characteristic equations of the junctions in consideration of the causality, the PV module current, I_2, equation is obtained as a function of the PV module output voltage, V_2, as given in the following equation

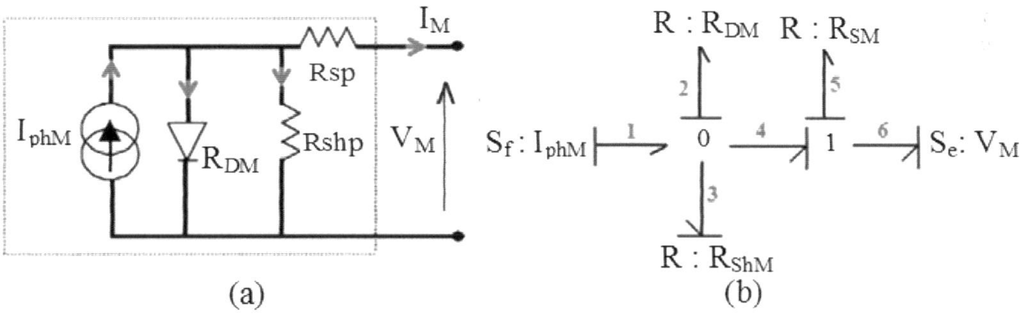

Figure 1. (a) Equivalent electrical circuit of the PV module and (b) its bond graph model.

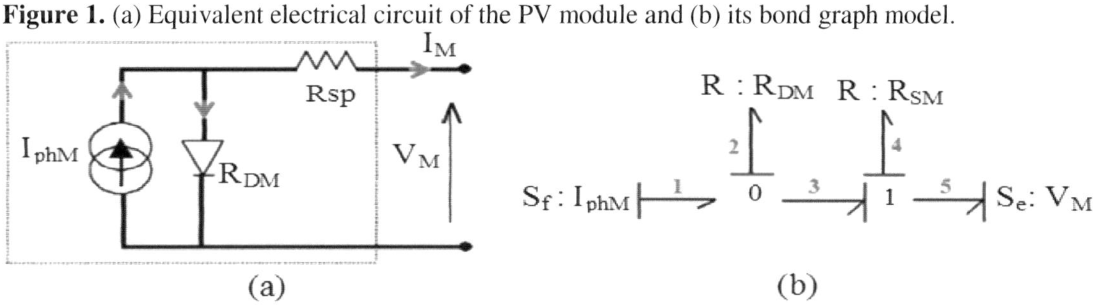

Figure 2. (a) Equivalent electrical circuit of the PV module and (b) its bond graph model with neglecting the shunt resistance.

$$I_2 = I_{PhM} - I_{satM}\left(exp^{\frac{V_2+R_{SM}I_2}{V_{ThM}}} - 1\right) \tag{4}$$

Multiplying the photovoltaic module current, given by the above equation by its output voltage yields the output power of the used photovoltaic module, as given in the following equation

$$P_2 = I_2 * V_2 \tag{5}$$

Further simplification can be done in the equivalent electrical circuit of the photovoltaic cell by neglecting the serial resistance Rs as it is very small. And therefore, in the equivalent electrical circuit of the photovoltaic module the series resistance can be neglected. This model is shown in Figure 3. (a). Based on this latter and using the characteristic equations of the junctions in consideration of the causality, the PV module current, I_3, equation is obtained as a function of the PV module output voltage, V_3, as given in the following equation.

$$I_3 = I_{PhM} - I_{satM}\left(exp^{\frac{V_3}{V_{ThM}}} - 1\right) \tag{6}$$

Multiplying the photovoltaic module current, given by the above equation by its output voltage yields the output power of the used photovoltaic module, as given in the following equation

$$P_3 = I_3 * V_3 \tag{7}$$

3. Results and discussion

In this section, a commercially available (SUNTECH) photovoltaic module of 50, 70 and 130 W is analyzed at standard test conditions for the three cases. For each PV module the I-V curves and the P-V curves are plotted for the three cases where the differences in maximum power points are highlighted (see Figure 4-Figure 6). The curves plotted in Figure 4-Figure 6 show that the five-parameter and four-parameter models are very close to each other while the ideal model is far from them with remarkable higher maximum power prediction.

The absolute relative error of the PV module parameters ($\Delta E\%$) was calculated to reflect the accuracy of the estimated parameters using the following equation.

$$E\ \% = \frac{|E_{simulated} - E_{measured}|}{E_{measured}} * 100 \tag{8}$$

Where, $\Delta E_1\%$, $\Delta E_2\%$, and $\Delta E_3\%$ represent the absolute relative errors between the manufacturer datasheet and the calculated data for case 1, case 2 and case3, respectively.

Obtained results are reported in Table 1-Table 3. From those tables, it is clear that the obtained maximum power using the five-parameter model is very close to that given in the manufacturing data sheet of the PV module. In other words, the absolute relative error of the maximum power is 0.01176% for PV module of 50W, 0.01109% for PV module of 70W and 0.001607% for PV module of 130W.

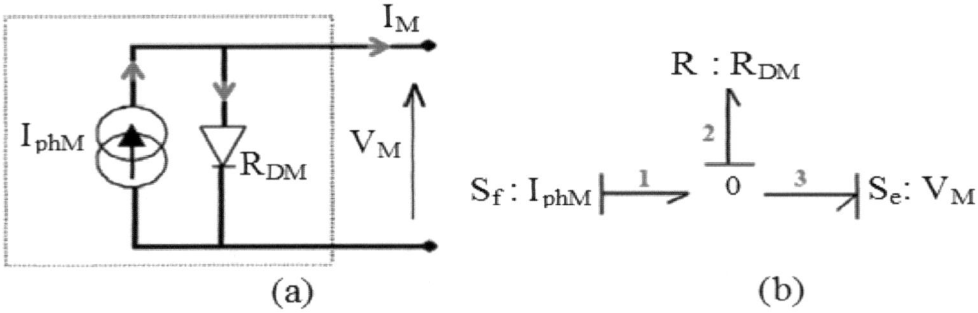

Figure 3. (a) Equivalent electrical circuit of the PV module and (b) its bond graph model with neglecting both the shunt and the series resistances.

Four-parameter model introduces an additional maximum power with respect to the five-parameter model appears in the values of the absolute relative error of the maximum power which is 0.00684% for PV module of 50W, 0.00501% for PV module of 70W and 0.001331% for PV module of 130W.

Neglecting series and shunt resistances in the the PV module's equivalent circuit results in a considerable value of the absolute relative error. In other words, the absolute error of the maximum power is computed as 6.27636% for PV module of 50W, 6.81198% for PV module of 70W and 8.927192% for PV module of 130W with respect to the four-parameter. In addition, the absolute relative error of the maximum power is 6.2832% for PV module of 50W, 6.81699% for PV module of 70W and 8.928523% for PV module of 130W with respect to the five-parameter. Therefore; it is important to mention that neglecting one the equivalent circuit resistances results in mistaken estimation of the available maximum power. In other words, the absolute relative error of the maximum output power of PV generator will have an additional value more than 6% of that given in its data sheet. And therefore; the corresponding absolute relative error of the fill factor is higher by more than 6% with respect to the corresponding value of datasheet.

Table 1.SUNTECH photovoltaic module of 50 W at STC

SUNTECH PV module 50W	V_{Pmax} (V)	I_{Pmax} (A)	P_{Mmax} (W)	FF	R_{SM} (Ω)	R_{ShM} (KΩ)
Case1	17.5	2.9136	50.988	0.7472	0.3733	79.2
Case 2	17.5	2.9138	50.9915	0.7473	0.3733	
Case 3	18.45	2.9372	54.1913	0.7941		
Data sheet	17.4	2.93	50.982	0.7471		
ΔE_1 %	0.5747	0.5597	0.01176	0.0133		
ΔE_2 %	0.5747	0.5529	0.0186	0.0267		
ΔE_3 %	6.03448	0.24573	6.29496	6.2909		

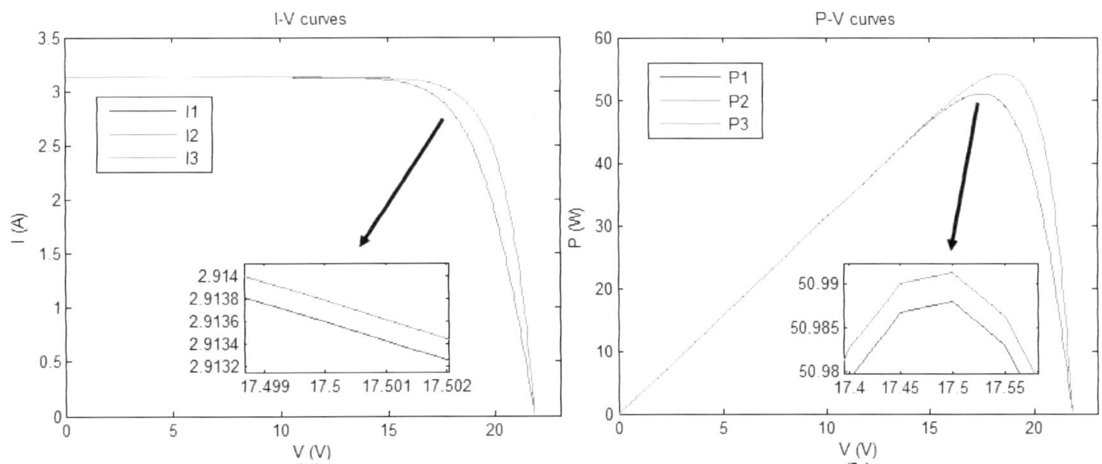

(a) (b)

Figure 4. SUNTECH photovoltaic module of 50 W at STC: (a) I-V curves and (b) P-V curves.

Table 2.SUNTECH photovoltaic module of 70 W at STC

SUNTECH PV module 70 W	V_{Pmax} (V)	I_{Pmax} (A)	P_{Mmax} (W)	FF	R_{SM} (Ω)	R_{ShM} (KΩ)
Case1	17.6	3.9751	69.96176	0.74474	0.2986	96.77
Case 2	17.6	3.9753	69.96528	0.74478	0.2986	
Case 3	18.63	4.0113	74.730519	0.7955		
Data sheet	17.8	3.93	69.954	0.7446		
ΔE_1 %	1.1235	1.1475	0.01109	0.0188		
ΔE_2 %	1.1235	1.1526	0.0161	0.02417		
ΔE_3 %	4.6629	2.0687	6.82808	6.83588		

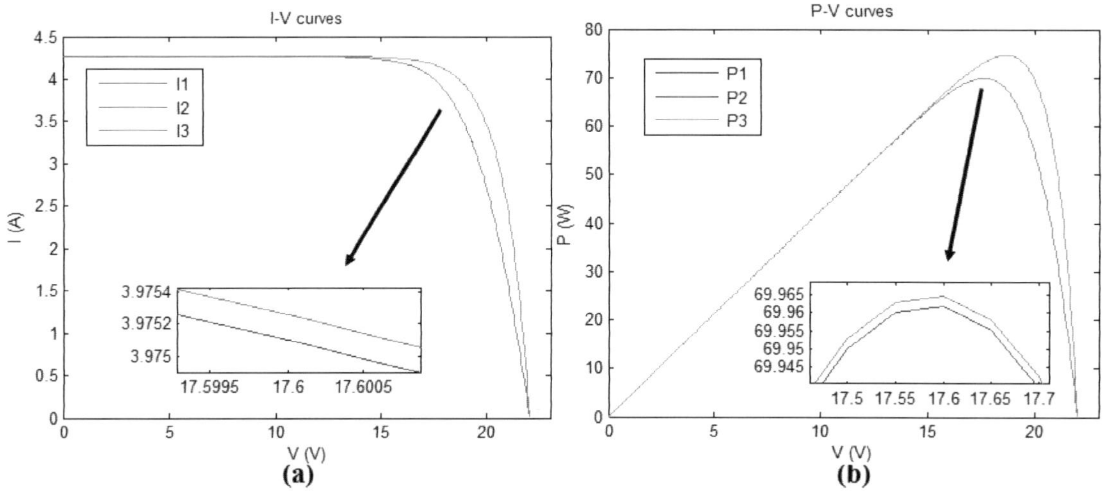

Figure 5.SUNTECH photovoltaic module of 70 W at STC: (a) I-V curves and (b) P-V curves.

Table 3.SUNTECH photovoltaic module of 130 W at STC

SUNTECH PV module 130W	V_{Pmax} (V)	I_{Pmax} (A)	P_{Mmax} (W)	FF	R_{SM} (Ω)	R_{ShM} (KΩ)
Case1	17.3	7.5133	129.98009	0.730307	0.203	189.1
Case 2	17.3	7.5134	129.98182	0.730317	0.203	
Case 3	18.65	7.5917	141.585205	0.795511		
Data sheet	17.4	7.47	129.978	0.730295		
ΔE_1 %	0.5747	0.57965	0.001607	0.001643		
ΔE_2 %	0.5747	0.58809	0.002938	0.003024		
ΔE_3 %	7.1839	1.62918	8.93013	8.930089		

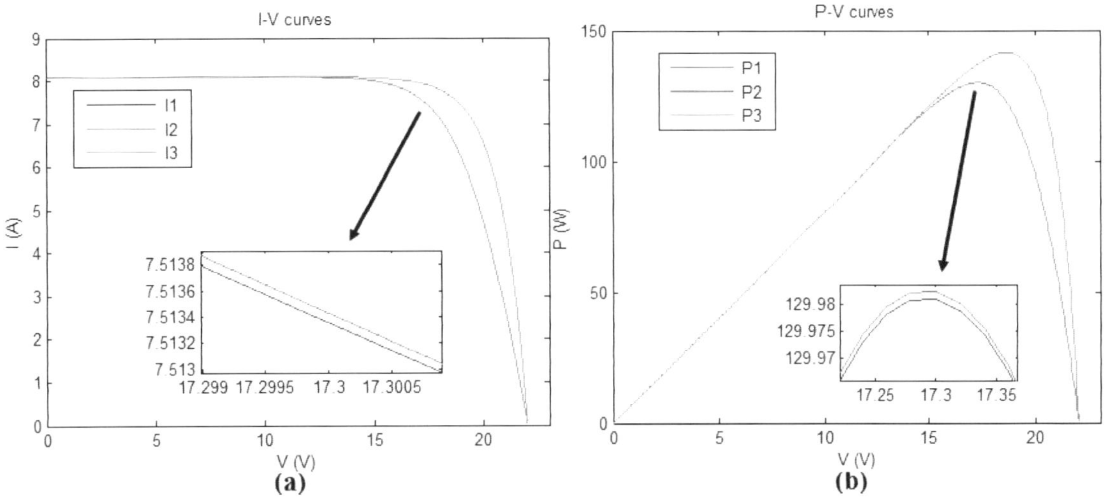

Figure 6.SUNTECH photovoltaic module of 130 W at STC: (a) I-V curves and (b) P-V curves.

4. Conclusion

In the present paper, a comparative energy analysis of three PV modules is done at standard test conditions based on different equivalent electrical circuits which are five-parameter model, four-parameter model, and ideal model. These equivalent circuits are the most encountered in published studies on the analysis, optimization, design and control of photovoltaic systems.

It has been found that decreasing the number of equivalent circuit parameters decreases the accuracy maximum power prediction. Even though it simplifies the computation but neglecting this effect is not without consequences on the analysis. The selection of equivalent circuit for some application where computation is burden becomes a compromise. As prediction of the PV generator's maximum power is a part of the computation, therefore using simplified PV model becomes an advantage. However, power predictions mismatch may affect the accuracy of analysis or simulation. This study quantifies the amount of error that can be introduced by using simplified models with respect to the accurate one.

5. References

[1] Chouder A, Silvestre S, Sadaoui N and Rahmani L 2012 Modeling and simulation of a grid connected PV system based on the evaluation of main PV module parameters *Simulation Modelling Practice and Theory* 20 pp 46–58

[2] Garoudja E, Harrou F, Sun Y, Kara K, Chouder A and Silvestre S 2017 Statistical fault detection in photovoltaic systems *Solar Energy 150* pp 485–499

[3] Biji G 2012 Modelling and simulation of PV based pumping system for maximum efficiency *EEE*

[4] Djerou S, Kheldoun A and Mellit A 2017 Efficiency improvement in induction motor-driven solar water pumping system using golden section search algorithm *Arabian Journal for Science and Engineering (Arab J Sci Eng)*

[5] Mezghanni D, Andoulsi R, Mami A and Dauphin-Tanguy G 2007 Bond graph modelling of a photovoltaic system feeding an induction motor-pump *Simulation Modelling Practice and Theory 15* pp 1224–1238

[6] Mezghani D, Othmani H and Sassi F 2017 A New Optimum Frequency Controller of Hybrid Pumping System: Bond Graph Modeling-Simulation and Practice with ARDUINO Board, *(IJACSA) International Journal of Advanced Computer Science and Applications, Vol. 8, N°1* pp 78–87

[7] Madansure V. N, Banerjee S, Mukherjee A and Chattopadhyay P. K 1997 Bond graph modelling and simulation of spice-pounding machines fed from a photovoltaic source *International Journal of Energy Research 21* pp 683-694
[8] Zitouni N, Khiari B, Andoulsi R, Sellami A, Mami A and Hssen A 2011 Modelling and non linear control of a photovoltaic system with storage batteries: A bond graph approach *.IJCSNS International Journal of Computer Science and Network Security, Vol.11, N°6* pp 105-114
[9] Siecker J, Kusakana K and Numbi B. P 2017 A review of solar photovoltaic systems cooling technologies *Renewable and Sustainable Energy Reviews 79* pp 192–203
[10] Anbarasi Jebaselvi G. D and Paramasivam S 2013 Analysis on renewable energy systems *Renewable and Sustainable Energy Reviews 28* pp 625–634
[11] Besheer A. H, Kassem A. M and Abdelaziz A. Y 2014 Single-diode Model Based Photovoltaic Module: Analysis and Comparison Approach *Electric Power Components and Systems 42* pp 1289–1300
[12] Besheer A. H and Abdelaziz A. Y 2014 A Comparative Analysis for Different Kinds of Single Diode Model Photovoltaic Module, *(ISGT ASIA) IEEE innovation smart grid technologies – Asia* pp 41–46
[13] Vergura S 2016 A Complete and Simplified Datasheet-Based Model of PV Cells in Variable Environmental Conditions for Circuit Simulation *Energies 9, 326*
[14] Madi S and Kheldoun A 2017 Bond graph based modeling for parameter identification of photovoltaic module *Energy 141* pp 1456–1465
[15] Kypuros J. A 2008 System Dynamics: Bond Graphs, Analaysis, and Control Design for Undergraduate Engineering Students *Ph.D. Department of Mechanical Engineering, University of Texas-Pan American, Edinburg, TX*
[16] Dauphin-Tanguy G 2000 The bond graph *HERMES Science Europe Ltd, Paris-France*

Nomenclatures:

Symbol	**Definition**
D_M	Sensitive diode to the light in PV module;
FF	PV module fill factor;
I	PV module current, (A);
I_{Pmax}	PV module current at the MPP, (A);
I_{phM}	Photocurrent produced by PV module, (A);
I_{satM}	Reverse saturation current of diode in PV module l, (A);
K_b	Boltzmann's constant ($K_b = 1.38065 * 10^{-23}\ J/K$) ;
P_{Mmax}	Maximum experimental peak output power of PV module, MPP, (W);
R	Bond graph representation of resistance, (Ω);
R_{DM}	Bond graph representation of the diode DM in PV module, (Ω);
R_{ShM}	Parallel (shunt) resistance of PV module, (Ω);
R_{SM}	Series resistance of PV module, (Ω);
S_f	Bond graph representation of current source, source flow; (A);
STC	Standard test conditions ($G_{STC} = \frac{1000\text{W}}{\text{m}^2}$; $T_{STC} = 25°C$);
T	PV module's working temperature, (K);
V	Output voltage of the PV module, (V);
V_{Pmax}	Voltage at the MPP of PV module, (V);
V_{ThM}	Thermal voltage and it is equal to $\frac{n_M K_b T}{q}$, (V);
e	Bond graph representation of voltage; (V);
f	Bond graph representation of current; (A);
n_M	Quality factor of the diode (D_M);
q	Magnitude of electronic charge, (q=1.602*10−19 C);
$\Delta E\%$	Absolute relative error of the PV module parameters;

A soft application for surface solar radiation estimation in Algeria

Rahli Hamida Samiha[1] and Benharrats Farah[2]

1, 2 Centre of Space Techniques (CTS) Algerian Space Agency (ASAL)
E-mail: hrahli@cts.asal.dz

Abstract. In this article, we have developed an application of four theoretical models that allows to estimate the solar radiation received on the ground for any region of the Algerian territory. The first model makes it possible to estimate the global solar radiation received on the ground. This model is a spectral model for the determination of direct and diffuse irradiance on a horizontal and sloping surface and for a clear sky. The second model is a Bird & Hulstrom time model that estimates solar radiation at ground level for a certain period (Hours, days, months, ... etc.). The third model is the Davies & Hay model, which makes it possible to trace the rates of absorption and diffusion of solar radiation by the various atmospheric constituents as a function of time and make comparisons. The last model is Lacis & Hansen which allows to trace the global radiation for a certain period. These models are based on the determination of the transmission coefficients of the various atmospheric constituents. These coefficients require the availability of current meteorological parameters (relative humidity, ambient temperature, atmospheric pressure, etc.) and geographical parameters of the site (latitude, longitude and altitude). The application developed made it possible to study several models for estimating global solar radiation at a few Algerian sites and to draw a conclusion on the most appropriate models of estimations.

1. Introduction

Solar energy occupies one of the most important places among the various possible sources of energy. An accurate knowledge of the distribution of solar radiation at a given geographic location is of major importance for the development of solar energy devices and for evaluating their performance. Unfortunately, for Algeria measurements of solar radiation are not readily available. For this reason, other solutions are often used to estimate the components of solar radiation on the ground.

Total solar irradiance is divided into two components: [1] the beam component from direct irradiation and the diffuse component. The ensemble forms the global radiation, which is determined by three groups of factors: the geometrical relations between the sun and the surface of the earth, the attenuation due to the crossing of the atmosphere, and the topographical factors. These parameters include day number, average temperature, relative humidity, altitude, longitude, zenith angle, and albedo of the soil.

These components can be measured directly by different measuring instruments or they can be estimated by analytical models that are established after several experimental measurements [2].

The objective of this work is to develop an application that allows to estimate the solar radiation received on the ground on any region of Algeria. Four models have been implemented by the Java language, a spectral model (Bird & Riordan) and three time models (Bird & Hulstrom, Davies & Hay and Lacis & Hansen), the curves of the different components of solar radiation for each model are presented by a graph.

In the following, an evaluation of the chosen models will be carried out in order to know the main difference between them. A graphic study was developed to judge the behaviour of the models and to know their strengths and their shortcomings.

2. Bird & Riordan model

This model is a spectral model for the determination of direct and diffuse irradiance on a horizontal and sloping surface and for a clear sky.

Modifications have been made to the method used for the calculation of direct irradiance. These modifications include the addition of an Earth-Sun distance correction factor, the use of Leckner's expression of the water vapour transmittance with some modification of the Leckner absorption coefficients, and the use of the expression of Robinson's ozone mass given by Iqbal [3].

For this model, global solar irradiation for a wavelength λ is the sum of direct irradiance and diffuse irradiance:

$$G_\lambda = I_{d\lambda} + D_\lambda \tag{1}$$

Determination of direct irradiance

The direct irradiance for a wavelength λ is given by the following relation:

$$I_{d\lambda} = H_{o\lambda}\, D\, T_{r\lambda}\, T_{a\lambda}\, T_{w\lambda}\, T_{o\lambda}\, T_{u\lambda} \tag{2}$$

Where $H_{o\lambda}$ is the extra-terrestrial irradiance at the mean earth-sun distance for a wavelength λ; D is the correction factor of the earth-sun distance; and $T_{r\lambda}$, $T_{a\lambda}$, $T_{w\lambda}$, $T_{o\lambda}$, $T_{u\lambda}$ are the transmission functions of the atmosphere at the wavelength λ, respectively for the molecular diffusion (Rayleigh), the attenuation by the aerosols, the absorption by the vapour of water, absorption by ozone, and absorption by uniform gas mixtures [4].

Determination of diffuse irradiance

Diffuse irradiance is induced by the diffusion effects of air molecules and aerosols. It should also be noted that aerosols are elemental atmospheric particles that cause a significant problem in the evaluation of solar irradiance, particularly in the study of diffuse irradiance, and this difficulty lies in the evaluation of their number, size, distribution and their optical properties.

In the case of this Bird model, the diffuse irradiance D_λ on a horizontal surface is the sum of three components:

1. The Rayleigh $D_{r\lambda}$ diffusion component.
2. The component of diffusion by $D_{a\lambda}$ aerosols.
3. The component that takes into account multiple reflections of the radiation between the ground and the atmosphere $D_{g\lambda}$.

In this case, the total diffuse radiation D_λ will be given by the sum of the different components, namely:

$$D_\lambda = D_{r\lambda} + D_{a\lambda} + D_{g\lambda} \tag{17}$$

According to the modified Bird spectral model, the expressions of the diffuse radiation components are given by:

$$D_{r\lambda} = H_{o\lambda}\, D\, cos\theta_Z\, T_{o\lambda}\, T_{u\lambda}\, T_{w\lambda}\, T_{aa\lambda}\left(1 - T_{r\lambda}^{0.95}\right) 0.5\, C_S \tag{18}$$

$$D_{a\lambda} = H_{o\lambda}\, D\, cos\theta_Z\, T_{o\lambda}\, T_{u\lambda}\, T_{w\lambda}\, T_{r\lambda}^{1.5}\,(1 - T_{as\lambda})\, 0.5\, F_S\, C_S \tag{19}$$

3. Davies and Hay model

The Hay-Davies model is a model for radiation insolation (direct and diffuse). It is composed of the isotropic component and the circumsolar component. It is given by the relationship [5] and horizon brightening is not taken into account. The equations used in this model were partially the result of comparing several existing models [6].

3.1. Calculation of direct radiation

The relationship proposed by Davies & Hay for the calculation of direct radiation is expressed by equation (20):

$$I = I_{sc} \times [(1-\alpha_0) \times \tau_r - \alpha_w]\tau_a \times \cos\theta_z \quad (20)$$

Where, , $\alpha_0$is the coefficient corresponding to the absorption of direct solar radiation by the ozone layer; and τ_r the coefficient of transmission after molecular diffusion or Rayleigh scattering is expressed by equation (21)

$$\tau_r = 0.972 - 0.08262m_a^2 + 0.00095m_a^3 + 0.000437m_a^4 \quad (21)$$

m_a is the corrected air mass expressed by the following relation:

$$m_a = \frac{p}{p_0} m_r = \frac{\exp(-0.0001184 * z)}{\cos(\theta_z) + 0.15(93.885 - \theta_z)^{-1.253}} \quad (22)$$

$$\tau_a = (0.12445\,\alpha - 0.0162) + (1.003 - 0.125\alpha)\exp[-\beta m_a\,(1.089\,\alpha + 0.5123) \quad (23)$$

Such that: α and β are respectively the coefficient characterizing sky visibility and the Angstrom cloud coefficient. Their calculation is based on the experimental determination of two attenuation coefficients $ka_{\lambda 1}$ and $ka_{\lambda 2}$corresponding to the wavelengths in which the absorption by water vapour does not occur (λ_1 = 380 nm and λ_2 = 500 nm respectively).

3.2.Calculation of diffuse radiation

The diffuse radiation on a horizontal plane is the sum of the three diffuse components: D_r, D_a and D_m.

$$D_r = I_{sc}\tau_0\tau_a(1-\tau_r)0.5 \times \cos\theta_z \quad (25)$$

It represents the diffuse radiation after the Rayleigh scattering, of which τ_0 is the transmission coefficient after absorption by ozone.

$$\tau_0 = 1 - \alpha_0 \quad (26)$$

$$D_a = I_{sc}\,(\tau_0\tau_r - \alpha_\omega) \times [F_c \times \alpha_\omega(1-\tau_a)]\cos\theta_z \quad (27)$$

3.3. Calculation of global radiation

The global radiation on a horizontal plane is calculated by:

$$G = I + D \quad (28)$$

4. Bird & Hulstrom Model

In 1981, Bird and Hulstrom [6] developed a mathematical model that calculates the incident solar radiation on the ground. This model takes into account the effects of diffusion and absorption that the solar radiation undergoes during its crossing of the atmosphere [7]. It is based on the determination of the transmission coefficients of the different atmospheric constituents. These coefficients require the availability of current meteorological parameters (relative humidity, ambient temperature, atmospheric pressure, etc.) and geographical parameters (latitude, longitude and altitude).

5. Lacis &Hansen Model

The Lacis & Hansen model is a semi-empirical model, it allows to calculate global solar radiation on a horizontal plane [8]-[10]. The input parameters of this model are: Solar constant, day number of the year, zenith angle, ambient temperature, relative humidity, atmospheric pressure, latitude, longitude, and altitude of the place and the albedo of the soil.

6. Presentation of the software

The software was developed with the Java language under the NetBeans environment. It allows the user to choose between four models of solar radiation estimation. Each model has a graphical interface where the user must choose a region and enter the input parameters. The Figure 1 gives an overview on the developed software interface.

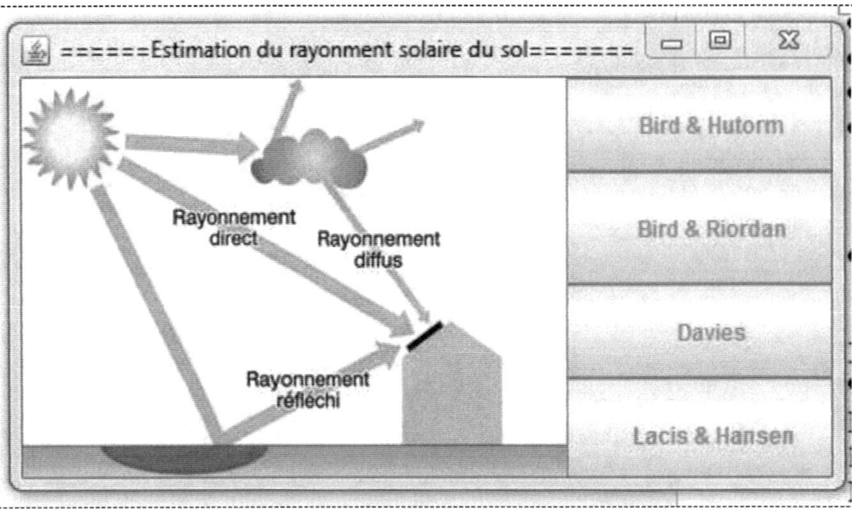

Figure 1. main interface of the software

6.1. Bird & Riordan Example

Seven parameters are to be introduced in this model, meteorological parameters (average temperature and relative humidity of the air) and geographical parameters (latitude, longitude and altitude). The absorption coefficients as a function of the wavelength and describing the different elements of the atmosphere. Figure 2 shows an overview of the parameters to be introduced for the Bird & Riordan model.

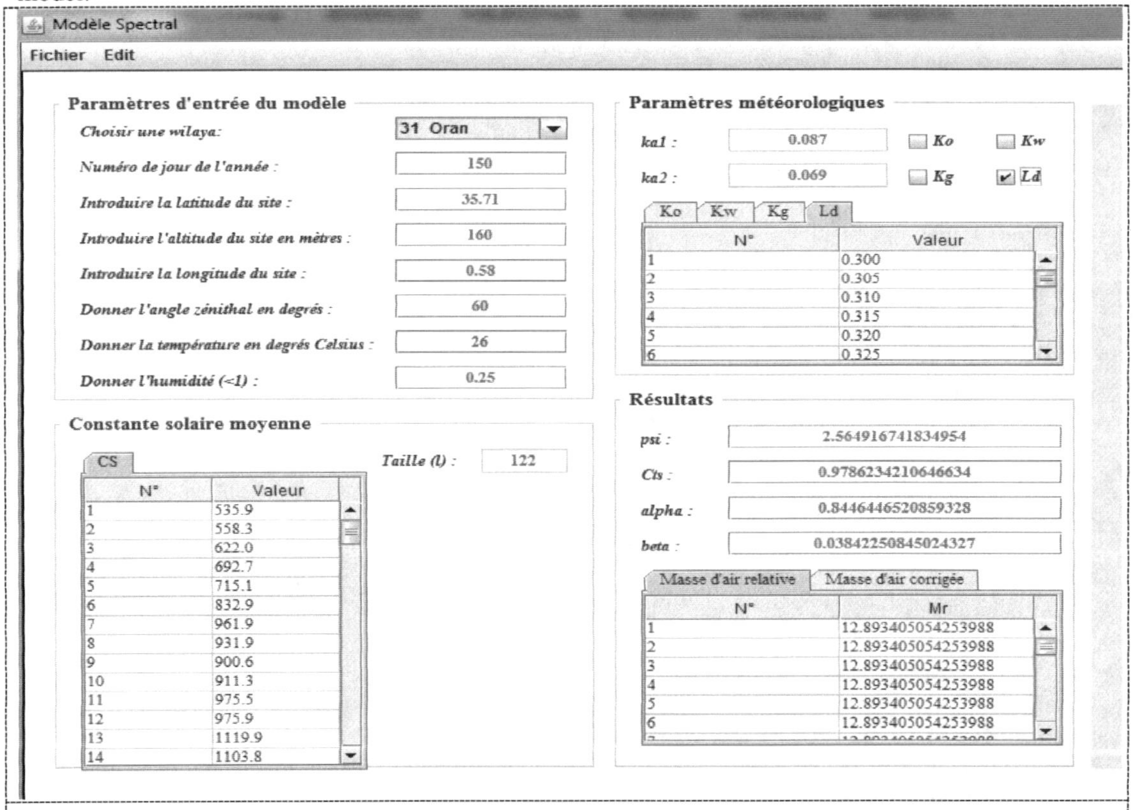

Figure 2. Module overview of Bird & Riordan model

The Figure 3 shows the results of the Bird spectral model for the Oran region.

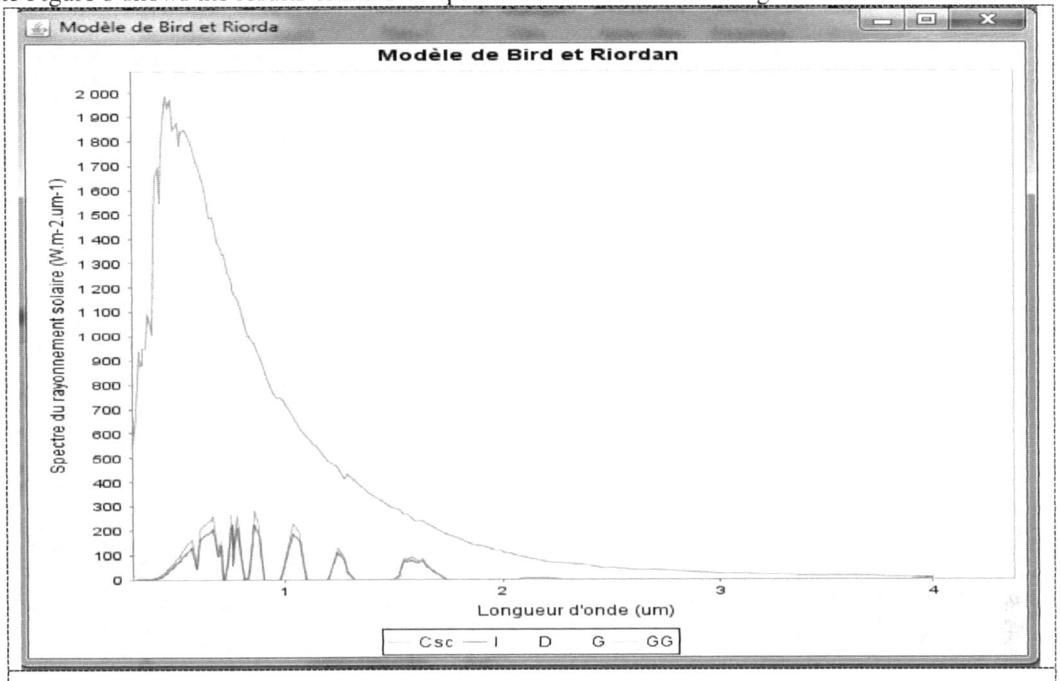

Figure 3. Radiation estimated by the spectral model of Bird & Riordan

6.2. Bird &Hulstorm Example:

The Figure 4 shows an overview of the time model for the region of Tamanrasset.

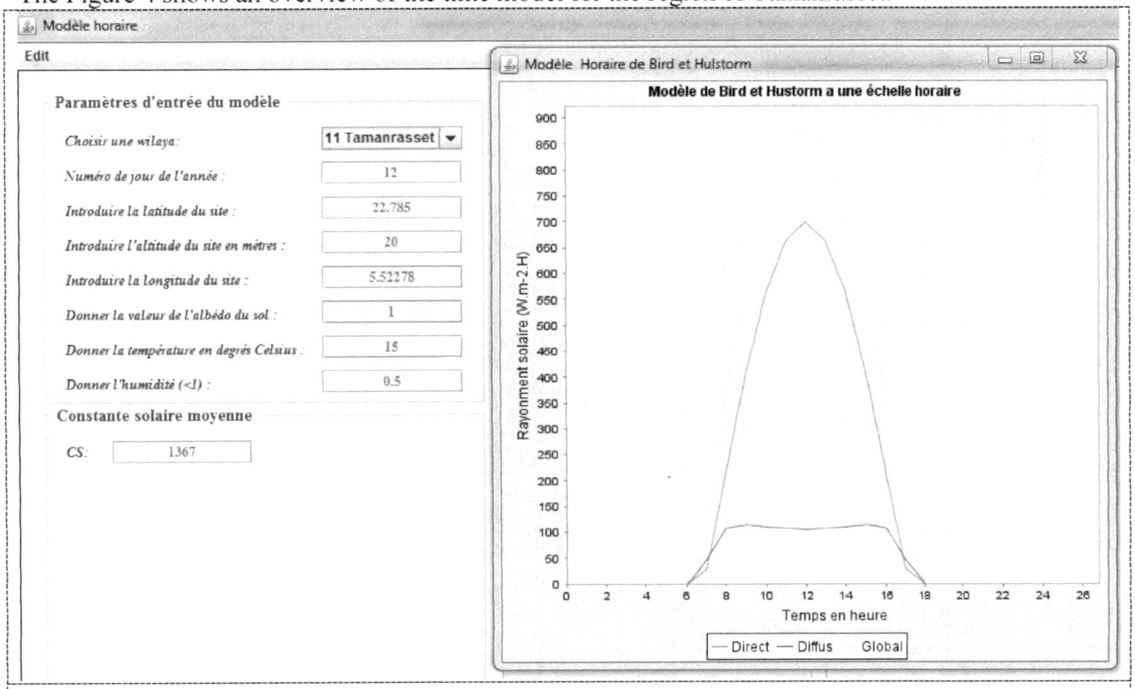

Figure 4. Module overview of Bird & Hustorm model and estimated radiation from Tamanrasset region

6.3. Davies & Hay Example:

The Figure 5 shows the graph of the Davies & Hay model for the Ghardaia region.

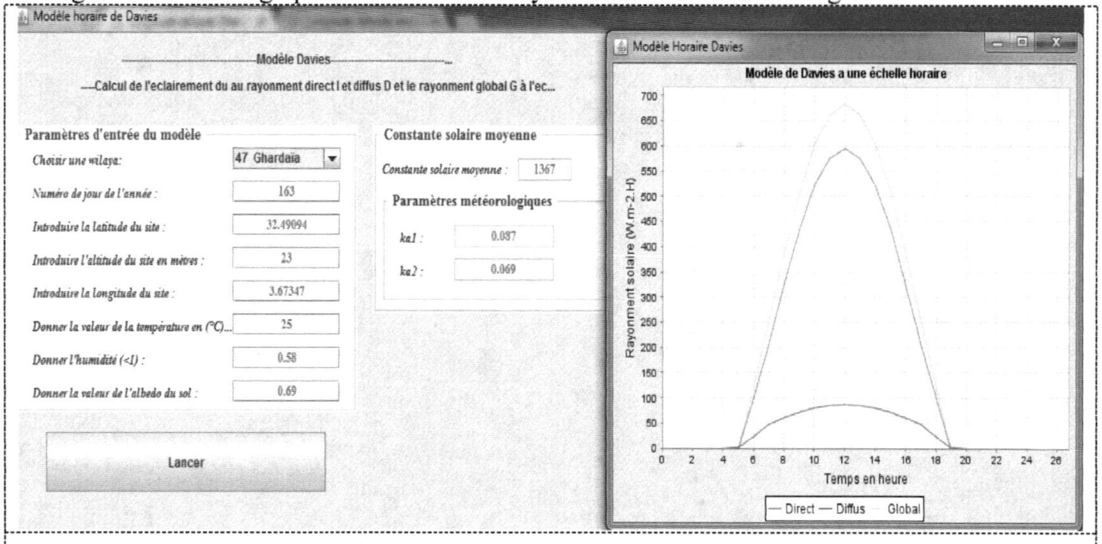

Figure 5. Module overview of Davies & Hay model and estimated radiation from Ghardaia region

6.4. Hansen & Lancen Example:

The Figure 6 shows the graph of the Hansen & Lancen model for the Bechar region.

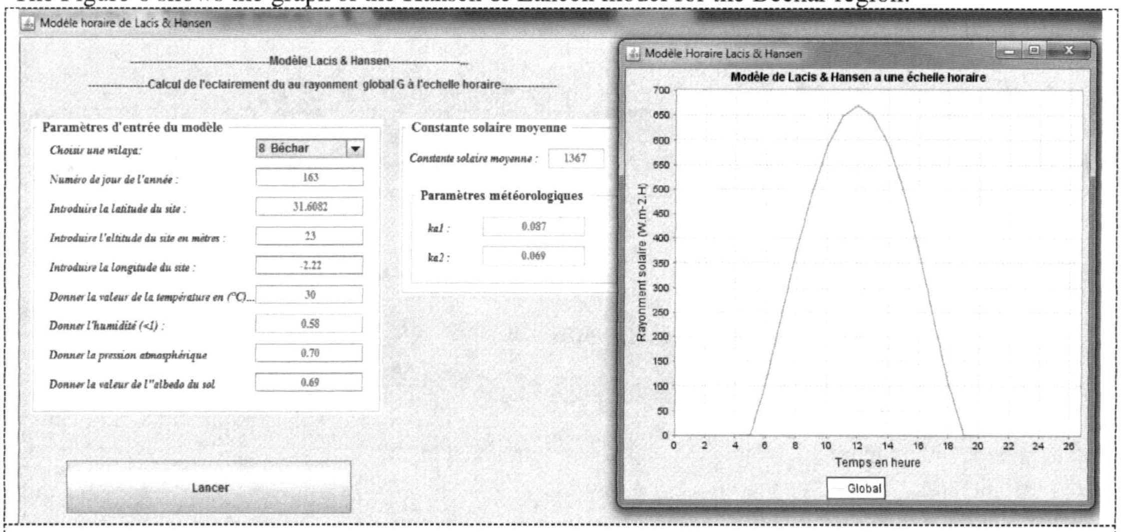

Figure6. Module overview of Hansen &Lancen model and estimated radiation from Bechar region

7. Conclusion

In this work, a model of solar radiation by four models a spectral model Bird and Riordan, and three time model (Bird & Hulstorm, Davies, Lacis & Hansen) was presented and simulated. From the models considered it was possible to calculate the three components of solar radiation; direct, diffuse and global in the conditions of the clear sky.

From the results that we have reached and based on the comments we have made, we find that generally each of the models considered in this study can calculate each component of solar radiation on an inclined plane and on a horizontal plane.

These models have the advantage of being simple, they use mathematical equations not too heavy for the study the estimate of the solar radiation. The designed software can be applied for any region and any day of the year.

As a result, it should be noted that given the results obtained and presented above, the Bird and Hulstrom model has relatively better results. So given its universality and view that it holds in a more detailed manner the phenomena of attenuation of the solar radiation, we recommend its estimate of the various components of the solar radiation.

8. References

[1] Loutzenhiser P G, Manz H C. Felsmann C, Strachan P A, Frank T and Maxwell G M 2007, Empirical validation of models to compute solar, irradiance on inclined surfaces for building energy simulation, (Solar Energy 81 254–267)

[2] Benbouza C 2014, Production and performance of integrated hybrid water (PV / T) sensors in the building-case of Algeria (Memory of Magister on climate engineering).

[3] Daryl R M 2013, solar radiation: Practical modelling for renewable Energy applications. Energy and the Environment, (Series editor Abbas Ghassemi. New Mexico State University. CRC Press, Taylor and Francis group).

[4] Mouhous C S 2012, Statistical study of solar radiation on an inclined plane, (Memory of Magister. University AbouBeker Belkaid, Tlemcen, Algeria).

[5] Davies J A and Hay J E 1979 Calculation of the solar radiation incident on a horizontal surface. In: Proceedings, (first Canadian solar radiation data workshop Canadian Atmospheric Environment Service).

[6] Bird R E and Hulstrom R L 1981, A Simplified Clear Sky Model for Direct and Diffuse Insolation on Horizontal Surfaces, (Solar Energy Research Institute, Golden, CO TR-642-761, 33 pp).

[7] Safaripour M H and Mehrabian M A 2011, predicting the direct, diffuse, and global solar radiation on a horizontal surface and comparing with real data, (Heat Mass Transfer. DOI 10.1007/s00231-011-0814-8).

[8] Robert L A and Scott A Wells 2007, A Comparison of Five Models for Estimating Clear-Sky Solar Radiation, (Civil and Environmental Engineering Faculty Publications).

[9] Mesri M M, Rougab I, Cheknane A and Bachari N I 2012, Estimation of solar radiation on the ground by semi-empirical models, (Revue of Renewable Energies Vol. 15– 463).

[10] Lacis A L and Hansen J E 1974 A parameterization for absorption of solar radiation in the earth's atmosphere (J Atmospheric Sci 31:118–133).

IOP Conf. Series: Earth and Environmental Science **167** (2018) 012039 doi:10.1088/1755-1315/167/1/012039

Increasing hybrid pv/wind/diesel generator power output with increased pv module efficiency.

L O Idoko[1], O Anaya-Lara[1] and A Attya[2]
1 Department of EEE, University of Strathclyde. Glasgow, UK.
2 School of Computing & Engineering, University of Huddersfield, UK.
E-mail: oneplus8@yahoo.com

Abstract. This research work focuses on increasing the power output of a hybrid energy system by giving a boost to the energy supplied from the PV module. It embraces the cooling of the PV module using a multi-concept cooling technique. An experiment was set up to reduce the surface temperature of the PV module to 20°C in order to increase its efficiency and hence power output. The experiment was performed using two 250 watts PV modules, water spraying was carried out on one of the modules which, was also attached with an Aluminium heat sink at the rear, while the other module was mounted without any form of cooling. The modelling of the hybrid energy system was done using MATLAB Simulink. An equation for PV module power output was used with a derating factor of 95% and the result of the experiment shows that power output of 262.4 watts is achievable, and the simulation result shows an increase in the power contributed by the module.

1. Introduction

The growth of any developing nation depends largely on the accessibility of adequate Power, as the availability of adequate power is a necessity, for an appreciable economic and technological development to be achieved.

Sources of energy supply from natural resources such as wind and sunlight provides a clean alternative source of power in locations which experiences high sunshine and high wind speed and as a result, energy from renewable sources can compete favourably with energy generation from fossil fuel, which usually introduce pollution to the environment [1].

The PV module is a good source of energy in Nigeria and other parts of the world, which experiences long periods of sunshine. In locations where long period of sunshine hours is experienced during the day, the excess sunlight is not fully utilized due to the effect of temperature on the PV module.

As the intensity of the sun increases, the temperature of the module surface also increases. Temperatures about 25°C causes the power output from the solar panel to reduce. Imagine a situation where the temperature of the surface of the module increases above 45°C, a lot of the incident radiation is converted to electricity and heat energy.

The energy converted to heat energy is wasted in the process, hence, the power output generated by the PV module reduces as the temperature of the module surface increases. In order to make the module more efficient, cooling of the surface of a PV module is a necessity [2]. A recent research, [3] shows that when you place a PV module under sunlight, only about 31% of the energy from the sun is converted to useful energy, the remaining percentage is converted to heat energy which usually causes the temperature of the module to increase and this gives rise to the following challenges:

- There is a reduction in the electricity produced by the module
- The increase in the module temperature due to the energy wasted as heat can damage the material used to fabricate the module.

- The conversion efficiency of the PV module also reduces.

In order for us to maximize the use of the available energy from the sun to generate electricity, cooling of the PV module is a necessity.

1.1. Multi-concept cooling technique

This research work embraces a multi-concept cooling technique to help reduce the energy wasted as heat energy when a PV module is placed under sunlight. The aim is to extract the excess energy, which causes the temperature of the PV module surface to rise to a level, where the power output reduces due to a reduction in the conversion efficiency of the module. The concept involves:

- Cooling the surface of the PV module to a temperature of 20°C using water
- Attaching aluminium heat sink at the rear of the PV module.
- Mounting of the module at a considerable height above the ground to enhance air-cooling.

Cooling of the module surface with water helps to reduce the temperature of the module surface considerably; this helps to increase the conversion efficiency of the PV module. Conversely, the aluminium heat sink attached to the rear of the module serves two functions: it helps to extract heat from the rear of the module and, it also helps the module to maintain a uniform temperature.

1.2. New features and advantages of proposed method.

The new features of the proposed method include the followings: cooling of the module is achieved using multi-concept cooling technique, ice blocks is added to the cooling water to reduce its temperature and cooling is applied at the surface as well as the rear of the module.

Conversely, the main advantages of this method are as follows: maintaining the temperature of the module surface at 20°C ensures the module yields steady increase in power output, at an irradiance of 1000w/m^2, more power than the module rated capacity is achieved and in the absence of any of the three concepts of cooling involved, a certain amount of cooling can be achieved.

1.3. The efficiency of PV module

The efficiency, η of a PV module is generally defined as in [2], as the ratio of the energy produced by the module to the energy obtained from the sun. It can be represented mathematically as equation (1)

$$\eta = E_{pd} / E_{ob} = P_{\max} / E.A \quad (1)$$

Where P_{max} is the maximum power obtained from the module, the solar irradiance at STC in W/m2 is represented by E, while A represents the module surface area in m^2.

E_{pd} and E_{ob} represent the energy produced by the module and the energy obtained from the sun respectively. The efficiency of a PV module can also be represented as equation (2).

$$\eta_m = \eta_{rT}[1 - \beta(T_m - T_{rT})] \quad (2)$$

Where η_m is the module efficiency, η_{rT} represents the module efficiency at 25°C, T_m is the module temperature, β is the temperature coefficient of power and T_{rT} is module reference temperature, which is usually 25°C.

1.4. The effect of Temperature on PV module

In order to account for the effect of temperature on a PV module, the PV module power output as in [4] is expressed using equation (3).

$$P_{PV} = Y_{PV}.f_{PV}(G_T / G_{T,STC})[1 + \alpha_P(T_C - T_{C,STC})] \quad (3)$$

Where

f_{PV} represents the derating factor (%) of the PV module,

Y_{PV} is the rated capacity of the PV array, which implies the module's output power under standard test conditions (KW)

G_T represents the solar radiation incident on the PV array in the current time step (KW/m^2)

$G_{T,STC}$ represents the incident radiation at Standard Test Conditions (1Km/m^2)

α_P represents the temperature coefficient of power (%/°C)

T_C represents the PV cell temperature in the current time step (°C)

$T_{C,STC}$ represents the PV cell temperature under standard test conditions (25°C)

1.5. Research location

The experiment carried out in order to increase the efficiency of the PV module was performed in Sokoto state, in northern Nigeria located within Latitude 13.1274 and Longitude 5.2046, an area which, experiences a long period of intense sunlight. The sunshine hours data for Sokoto, Sokoto State, Nigeria is as shown in Figure 1.

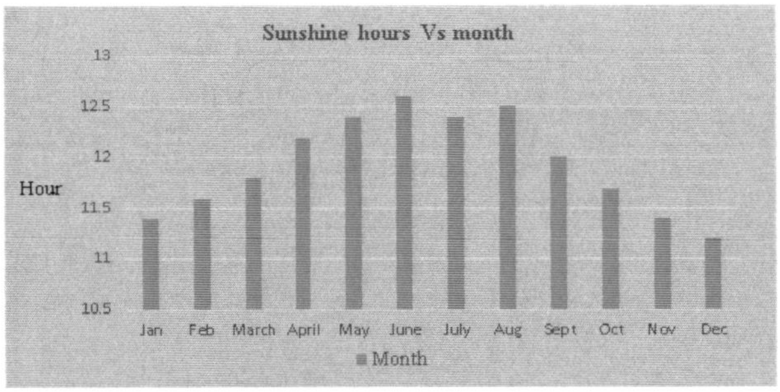

Figure 1. Sunshine hours in Sokoto in the 2016.

The aim of the experiment is to improve the efficiency of the PV module and in the process, increase the amount of energy contributed to the hybrid energy system by the PV.

2. Materials

The experiment was carried out successfully using the following devices as shown in Table 1

Table 1. Devices used for the experiment.

Device	Quantity
250 watts PV module	2
Aluminium heat sink	1
10 litres water spraying can	1
AL 300 data logger (6-channels)	1
Digital Multimeter	4
Infrared thermometer	1
Meteon Pyranometer and display unit	1
Long nose plier	1
Multipurpose plier	1
A set of spanners	1
Measuring tape	1

The experiment requires the use of both 250 watts, as one of the modules was mounted without an attached aluminium heat while the other has an attached aluminium heat sink, the dimension of the aluminium heat sink used in this project is as shown as in Table 2

Table 2. Properties of Aluminium heat sink.

Aluminium section	Size
Length of Al sheet	154cm
Height/Length of Al fin	84cm
Thickness of Al fine	1mm
Depth of Al fin	10cm
Spacing between Al fins	3.12mm
Width of Al fin base	2.4cm

3. Method

The experiment was setup as shown in Figure 2. One of the module is mounted without any form of attachment while the other module is mounted with the heat sink attached to the rear of the module. In order to create room for air-cooling, both modules were mounted at a height of 137cm above the ground level, the Al 300 six-channel data logger, Pyranometer, variable resistors and digital Multimeters were all mounted properly as shown in Figure 2 and readings were taken.

Figure 2. Project setup

The function of the Al heat sink attached to the rear of one the module, is to extract heat from the module and keep the temperature uniform while, cooling of the surface of the PV module was carried out by spraying the surface of the PV module with water.
Ice blocks were introduced into the cooling water to reduce the temperature of the cooling water, spraying of the module was done with the cold water and a module surface temperature of about 20°C was achieved. Spraying of the module was carried out as shown in Figure 3.

Figure 3. Spraying of PV module with cold water.

The results of the experiment are as shown in Table 3, the power output of the PV module at a derating factor of 95% was calculated using equation (3). The calculation was carried out for the module without aluminium heat sink and cooling, the module with Aluminium heat sink + water-cooling, and finally it was used to compute the power output of the module when the module surface temperature was reduced to 20°C
Using a derating factor of 95% to comply with the PV manufacturer's data sheet which, has a power tolerance of ± 5% gives Table 3

Table 3. Module power output at a derating factor of 95%.

Time	Ambient temp (°C)	Panel without Al heat sink, surface temp (°C)	Module + Al heat sink & water-cooling surface temp (°C)	Solar radiation (Wm-2)	Module without Al heat sink power output (w) at f_{pv} = 95%	Module + Al heat sink & water-cooling, power output (w) at f_{pv} = 95%	Module + Al heat sink & water-cooling of module surface to 20°C power output (w) at f_{pv} = 95%
09:00	36	40.3	40.3	407	90.16	90.16	
09:15	36.9	46.7	46.3	558	119.9	120.1	
09:30	36.8	45	44	736	159.4	160.2	
09:45	38	50.1	49.5	779	164.6	165.1	
10:00	40	55	53.4	789	162.7	164	
10:15	40.9	53.3	51	826	171.7	173.7	
10:30	43.3	54.1	52	897	185.8	187.7	
10:45	43.9	56.8	54.5	932	190.4	192.6	
11:00	44	58.6	53	955	193.3	198.9	
11:15	46	59.4	54.2	960	193.5	198.7	
11:30	47.6	61	54	1017	203.3	210.7	
11:45	47.9	61.2	54.3	1019	203.5	210.8	
12:00	49	63	55.9	1021	201.9	209.5	
12:15	47.7	60	57.2	1072	215.4	218.5	
12:30	43.6	56.6	55	1097	224.3	226.1	
12:45	49.5	64.1	39	1081	212.6	240.9	262.4
13:00	45.3	54	50	1126	233.3	238	
13:15	46.7	55.2	51	1085	223.4	228.2	
13:30	41.1	51.4	48	1073	225.2	229	
13:45	41.1	51.5	40	982	206	217.8	238.4
14:00	43.5	58	49.5	1087	220.7	230.3	
14:15	46	59.7	50	1065	214.3	225.1	
14:30	48	62	53	1054	209.6	219.5	
14:45	52.5	66	44	1056	205.6	229.8	256.3
15:00	41.2	54.7	52.8	658	135.9	137.2	
15:15	42.7	54	53	879	182.1	183	
15:30	47.6	58	56	831	168.7	170.4	
15:45	45	58	43	810	164.4	177.1	196.6
16:00	44	56	51	705	144.6	148.3	

A section through Table 3 which, indicates the difference in the power output produced by the module when subjected to different conditions is as shown in Table 4

Table 4. Power outputs obtained from the experiment

Time	Module without Al heat sink power output (w) at f_{pv} = 95%	Module + Al heat sink & water-cooling, power output (w) at f_{pv} = 95%	Module + Al heat sink & water-cooling of module surface to 20°C power output (w) at f_{pv} = 95%
12:45	212.6	240.9	262.4
13:45	206	217.8	238.4
14:45	205.6	229.8	256.3
15:45	164.4	177.1	196.6

From Table 4 above, more power is obtained with the module + Al heat sink & water cooling, but maximum power output of 262.4 w was achieved with cooling the PV module + Al heat sink & water-cooling of module surface to 20°C.

In order to implement the effect of the increase in power output using MATLAB, the module operating temperature set to 20°C, is used as an input to the PV array MATLAB Simulink block.

4. Model composition of hybrid pv/wind/diesel generator

The research project comprises of 60kw PV array, 50kw diesel generator, 50kw PMSG driven wind turbine, 480 Vision 6FM200D batteries made up of a string size of 30 with 16 strings in parallel, each battery have a nominal voltage of 12volts and a nominal capacity of 200AH (2.4KWh). The configuration of the hybrid energy system is as shown in Figure 4

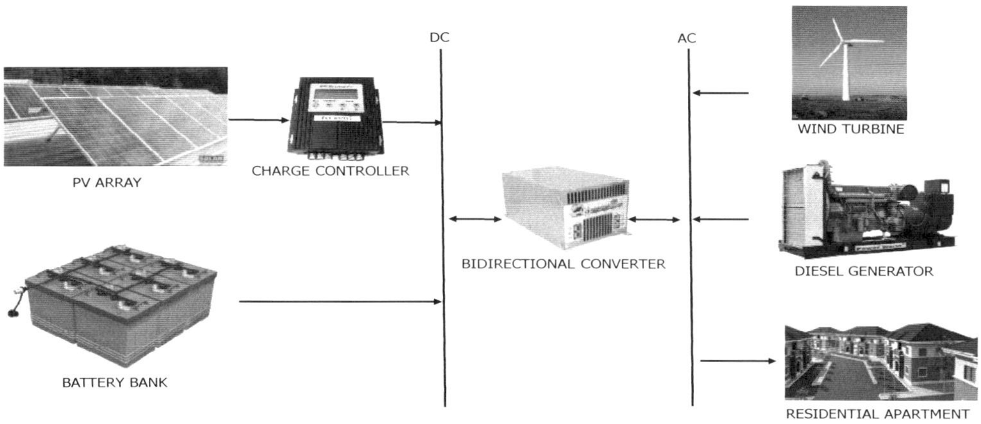

Figure 4. Hybrid energy system configuration

4.1. Design of 60kw PV array

The maximum power delivered by a photovoltaic array as in [5] is given by the expression

$$P_{mp} = (N_p \times I_{mp}).(N_s \times V_{mp}). \tag{4}$$

Where the N_p and N_s stand for the number PV modules connected in parallel and in series respectively. I_{mp} and V_{mp} stand for the maximum current and voltage at MPPT respectively. This is because adding PV modules in series increases the voltage while adding PV modules in parallel increases the current. In order to achieve a PV array with a capacity of 60KW, a PV array voltage of 614V was considered by connecting 20 PV modules in series, conversely 12 PV modules are connected in parallel and each module has a maximum voltage, V_{mp} of 30.7v and a maximum current, I_{mp} of 8.15A at maximum power.

4.2. Sizing of battery bank

The size of the battery bank can be calculated as in [6] using equation (5).

$$B_{cp} = \frac{D_d . D_A}{\eta . V_{sys} . DoD} \tag{5}$$

Where DoD, η, V_{sys}, B_{cp}, D_d and D_A stand for depth of discharge, efficiency, Battery capacity, daily load demand and battery days of autonomy respectively. Battery round trip efficiency of 85% was selected for this project.

Sizing of battery bank for this project does not require any further design as the sizing was achieved with the help of HOMER software simulator. A total of 480 batteries, each with nominal voltage of 12V and a nominal capacity of 200Ah (2,4Kwh) is required for this project and it comprises of a string size of 30 with 16 battery strings in parallel.

Batteries add up in series to increase the voltage while the amperage remains the same, conversely, they add up in parallel to increase current while the voltage remains the same. A string size of 30 implies connecting 30 batteries in series.

This generates 30 (12V), at an amperage of 200Ah and hence, the bus voltage is 360V, hence a battery string size of 30 generates 360V, 200Ah. Sixteen, 16 parallel strings of battery connected in parallel yields 360V, 16 (200Ah). Therefore, the 480 batteries required for this research work generate a bus voltage of 360V and an amperage of 3200Ah.

4.3. Boost converter design

In this case, the boost converter is used to give the PV voltage, Vpv a boost from 614V to 700V, which is the selected V_{dc} of the VSC DC bus.

The boost converter duty cycle (D) is expressed as in [7] as

$$D = \left(\frac{V_{dc} - V_{pv}}{V_{dc}} \right) = \left(\frac{700 - 614}{614} \right) = 0.12 \tag{6}$$

The capacity of the boost converter inductor is given as

$$L_{bc} = \left(\frac{V_{in} D}{f_{sw} \Delta I_{pv}} \right) = \left(\frac{614 \times 0.12}{10 \times 10^3 \times \frac{5}{100} \left(\frac{60 \times 10^3}{614} \right)} \right) = 1.508 mH \tag{7}$$

Where ΔI_{pv} is considered as the input current ripples (5%) of I_{pv} and f_{sw} represents the switching frequency in hertz, the value of the inductance, L_{bc} is selected as 1.6mH.

4.4. DC link capacitor design

The size of the DC link capacitor can be calculated as in [7] using the expression

$$C_1 = C_2 = \left(\frac{I_d}{2(2 \times \omega \times \Delta V_{dcrip}} \right) = \left(\frac{P / V_{dc}}{2(2 \times \omega \times \Delta V_{dcrip})} \right) = 4875 \mu F \tag{8}$$

Where V_{dcrip} is the ripple voltage, its value is 2% of the V_{dc}, I_d represents the VSC DC link current, the angular frequency is represented by ω , the value of the selected DC link capacitors are C1=C2=500μF

Interfacing Inductor design and selection

The value of the interfacing inductor is achieved using the expression in [7] as

$$L_{int} = \left(\frac{\sqrt{3} \times m \times V_{dc}}{12 \times O_f \times f_{sw} \times I_{Rr}} \right) = \left(\frac{\sqrt{3} \times 0.90 \times 700}{12 \times 1.2 \times 900 \times 0.25 \times 86.60} \right) = 3.89 mH \tag{9}$$

Where O_f represents the overloading factor, m represents the modulation index, fsw represents the switching frequency, Vdc represents the DC link voltage, I_{Rr} represents the ripple in the grid current, its value is 25% of the grid current, IR.

$$I_R = \left(\frac{60 \times 10^3}{\sqrt{3} \times 400} \right) = 86.60 A \tag{10}$$

4.5. Modelling of diesel generator

The diesel generator is made up a diesel engine and an alternator for electricity generation; it consist of a diesel engine, a synchronous generator, an excitation system and speed controller. The diesel engine and governor system exert control on the speed of the governor to supply mechanical power. Its diesel engine comprises of the governor and an internal combustion (IC) engine.

The diesel generator ensures the conversion of energy from fuel such as diesel or bio-diesel into mechanical energy with the help of its internal combustion engine, after which, this mechanical energy is converted into electrical energy with the help of its electric machine operating as a generator. The governor is made up of actuator and speed controller, the maintenance of a constant speed during the period of operation of a diesel generator is maintained by the governor [8] [9]

The transfer function of the actuator as well as that of the regulator as in [8] is as shown in equation 11 and 12 respectively

$$H_a = \frac{(1 + T_{a1} s)}{s(1 + T_{a2} s) + (1 + T_{a3} s)} \tag{11}$$

$$H_r = \frac{\Upsilon_r (1 + T_{ra} s)}{(1 + T_{r1} s + T_{r2} s^2)} \tag{12}$$

Where, T_{a1}, T_{a2} and T_{a3} represent the time constants of the actuator, Υ_r represents the gain of the regulator while, the time constants of the regulator is represented by T_{r1}, T_{r2} and T_{r3}.

Conversely, the speed regulation and diesel engine of the diesel generator can be described using differential equations as in [8] as equation (13) and equation (14).

$$\frac{dm_B}{dt} = \frac{1}{\tau_2} (K_2 P_c - \frac{K_2}{\omega_{ref} R} \Delta\omega - m_B) \tag{13}$$

$$\frac{dPc}{dt} = \frac{-K1}{\omega_{ref}}\Delta\omega \tag{14}$$

Where the rate of fuel consumption of the diesel engine is denoted m_B, ω_{ref} is the engine reference speed in rad/sec, R is the permanent speed drop of the diesel engine, K_1 is the summing loop amplification factor of the governor, the gain and dead time of the engine is represented by K_2 and τ_2 respectively.

The dead time, τ_2 can be represented as in [8] as equation (15).

$$\tau_2 = \frac{60S_t}{2Nn} + \frac{60}{4N} \tag{15}$$

N refers to the speed in rev/min, n is the quantity of cylinders, S_t represents the number of stroke engine, in this case, S_t=4 and due to combustion, the generated mechanical power is expressed as equation (16)

$$P = E_1 m_B \eta \tag{16}$$

η represents the efficiency and E_1 represents the proportionality constant.

4.6. Modelling of wind energy conversion system

With the help of the wind turbine, wind energy is converted into mechanical power and hence. The mechanical power is transformed into electrical power. The value of the mechanical power can be computed using the equation (17) [10]

$$P_m = 0.5\rho A C_p(\lambda,\beta) V_{wind} \tag{17}$$

Where ρ represents the density of air, its value is within the range of 1.22-1.3Kg/ms, V_{wind} refers to the speed of the wind in m/s, A represents the swept area of the turbine blades (m^2), the coefficient of power is denoted by C_p (λ,β), it's influenced by two factors, the pitch angle of the blade, β and the speed slip ratio, mathematically, the tip speed ratio is defined as

$$\lambda = \Omega R / V_{wind} \tag{18}$$

R is the radius of the blade (m) and Ω refers to the angular speed (m/s)

The coefficient of power is defined as in [10] as

$$C_p(\lambda,\beta) = C_1\left(\frac{C_2}{\lambda_1} - C_3\beta - C_4\right)\exp\left(\frac{-C_5}{\lambda_1}\right) + C_6\lambda \tag{19}$$

Where

$$\frac{1}{\lambda 1} = \left(\frac{1}{\lambda + 0.08\beta} - \frac{0.035}{\beta^3 + 1}\right) \tag{20}$$

The rotational torque can be expressed as

$$T_m = P_m / \Omega \tag{21}$$

And the optimal angular speed is represented by the relation

$$\Omega opt = \lambda_{opt} V_{wind} / R \tag{22}$$

In order to achieve the maximum mechanical power, the following relation is used

$$P_{m\_max} = 0.5\rho A C_{p\max} V^3{}_{wind} \tag{23}$$

4.6.1. PMSG driven wind turbine

The PMSG driven wind turbine was chosen for this project as a result of certain advantages in [11] as follows:

- There are absence of brush/slip ring.
- The mechanical stress is low
- The rotor does not experience any form of copper loss
- Reactive/active power controllability is higher

The sign of the torque input of the PMSG is what determines its mode of operation, if its sign is negative, the PMSG operates as a generator and if its sign is positive, the PMSG operates as a motor [12].

The implementation of PMSG model is done in the dq-coordinates, which implies AC states does not exist in the model. Its mathematical equation is developed from the synchronous rotating reference frame by ensuring that the d-axis aligns with the rotor flux direction [10] [13].

$$V_q = R_s i_q + L_q \frac{di_q}{dt} - \omega_e L_d i_d + \omega_e \varphi_m \tag{24}$$

$$V_d = R_s i_a + L_d \frac{di_a}{dt} - \omega_e L_q i_q \tag{25}$$

Where V_q and V_d represent the q and d components of the voltages in the stator, i_q and i_d represent the q and d components of the currents in the stator, L_q and L_d represent the inductances (H) of the machine, φ_m represents the magnetic flux (wb), ω_e represents the electrical speed (rad/s).
In order to obtain the electrical torque, the following expression is used.

$$Te = \frac{3}{2} P \{\varphi_m i_q + (L_d - L_q) L_d i_q\} \tag{26}$$

P in the expression above represents the poles pair and the machine rotor dynamics can be represented by the following expression

$$T_m - T_e = B\omega_r + J \frac{d\omega_r}{dt} \tag{27}$$

Where J represents the inertia of the rotor (kgm^2), ω_r is the speed of the rotor (rad/s), B represents the friction in the rotor Kgm^2/s, the mechanical torque that was produced by the wind is represented by T_m
Assuming L_d=L_q=0, and the d-reference current, id^*=0, then the (L_d-L_q)$i_d i_q$ becomes negligible.

4.7. Control of the hybrid energy system

The battery bank performs two roles in this project as it serves a sink, and also as a source, hence it charges or discharges when the network experiences surplus or lack of power from the sources of energy due to the situation of the weather. When the amount of electricity supplied by the sources of power is greater the load on the network, the battery controller charges the battery and when the load on the network is more than the electricity supply from the sources, the controller causes the battery bank to discharge [14].
The control of the battery storage is done in such a way as to ensure the DC voltage is stabilized and this is achieved through the control of the DC/DC bidirectional converter in the network.
When there is a shortage of power from the sources, the bidirectional converter serves as a boost converter to provide power by discharging the battery and when the power supplied from the sources is in excess, it operates as a buck converter to charge the battery [15].
The hybrid system is also, incorporated with a dump load, the dump load is made up of a bank of resistors and a power converter, the capacity of the dump load power is such that it exceeds the amount of power produced by the wind, PV and diesel generator by 30%. It is to ensure the control of the network should, a condition arises when there is no load and the battery storage fails or is fully charged.
The flow of power in the network can be represented as

$$P_{network} = P_{pv} + P_{wt} + P_{diesel} - P_{load} \tag{28}$$

Where, the net power generated is represented by $P_{network}$, P_{pv} is the amount of power supplied by the PV, P_{wt} represents the power supplied by the wind turbine and P_{load} is the power of the load
When $P_{network} < 0$, it means the power supplied by the energy sources is not enough and the battery provides power to make up for the shortage of power, but when $P_{network} > 0$, it implies the amount of power produced by the energy sources is in excess, the excess power is stored in the battery, but the moment it becomes more than the rated power of the batteries in the battery bank, a fraction of the excess power is transferred to the system dump load.

5. Results

The results obtained from the simulations carried out from the project implementation using MATLAB simulink are as shown in Figure 5.

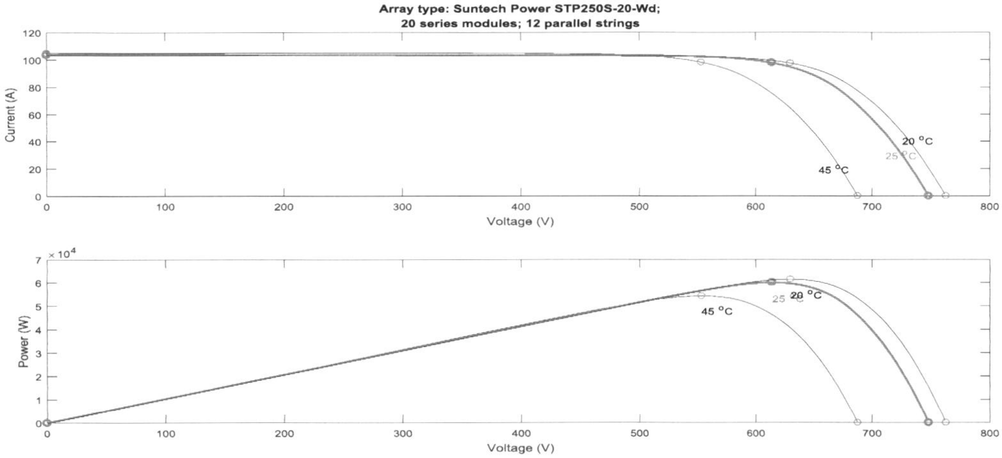

Figure 5. I-V and P-V curve of 60KW PV array.

The I-V and P-V curve below shows the usual characteristic of a PV module, which generate less power as the temperature of the module increase.

The research location for this project usually experiences low wind speed unlike some part of Europe where the wind speed is high and as a result the simulation was carried out at a wind speed of 6.1m/s and at a constant irradiance of 1000w/m^2. The result is as shown in Figure 6 and Figure 7.

Figure 6. 60KW PV array power output at 45°C.

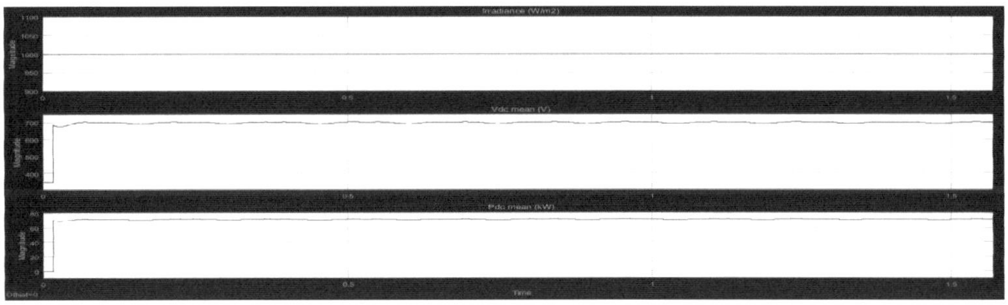

Figure 7. 60KW PV array power output at 20°C.

From Figure 6 and Figure 7, it can be observed that the power output produced by the PV array at a temperature of 20°C is more, this is because, the PV module works better when the module surface temperature is low

The measured Vdc and the reference Vdc of the network were compared and the values obtained are very close as shown in Figure 8

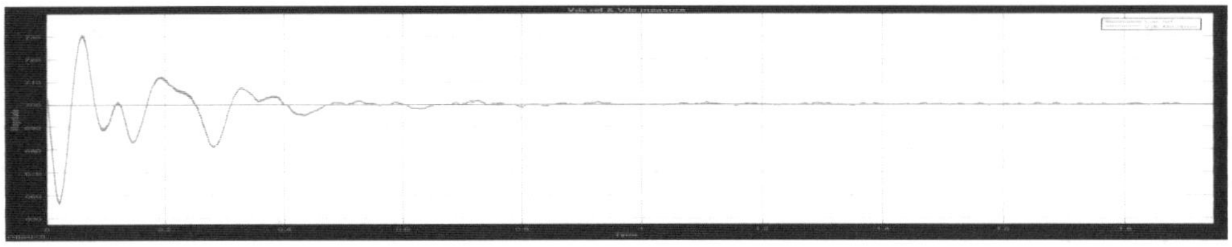

Figure 8. Vdc reference and Vdc measured

The diesel generator rotor speed in pu and the stator currents are as shown in Figure 9 and Figure 10 respectively

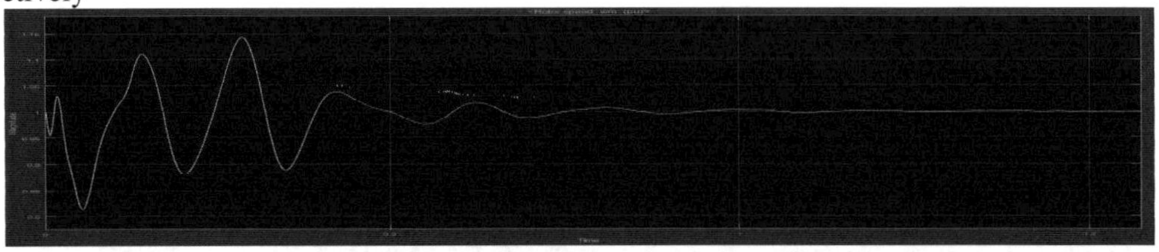

Figure 9. Diesel generator rotor speed.

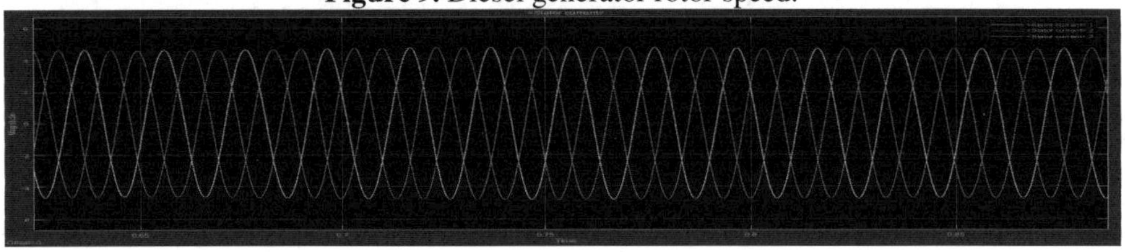

Figure 10. Diesel generator stator currents

The currents, voltages, the active and reactive power of the wind turbine are as shown in Figure 11.

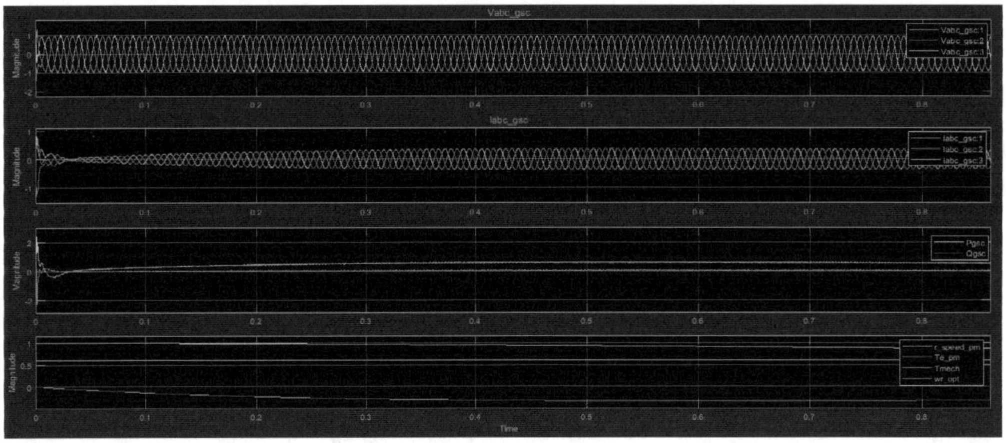

Figure 11. Wind turbine currents, voltages, active and reactive power output

6. Conclusion

The reduction in the surface temperature of the PV module increases the efficiency of the PV module, this results in an increase in the power output obtained from the experiment. The experiment shows that an increase in power output from 250watts to 262.4 watts is achievable. The result of the simulation also shows that at a temperature of 20°C, the PV module contributes more power to the hybrid energy system than at a temperature of 45°C. Hence, there is an increase in the hybrid PV/Wind/Diesel generator power output with increased module efficiency. Future works in this project involve the provision of voltage control of the hybrid energy system.

7. References

[1] Engin M. Sizing and simulation of PV-wind hybrid power system. Int J Photoenergy. 2013;2013

[2] Chikate B V, Sadawarte Y A. The Factors Affecting the Performance of Solar Cell. Int J Comput Appl Sci Technol [Internet]. 2015 [cited 2017 Sep 16];975–8887. Available from: http://research.ijcaonline.org/icquest2015/number1/icquest2776.pdf

[3] Koteswararao B, Radha K, Vijay P and Raja N. Experimental Analysis of solar panel efficiency with different modes of cooling. 2016;8(3):1451–6.

[4] Homer Energy. How homer calculates the PV array power output [Internet]. Homer Energy Support. 2014 [cited 2017 Oct 31]. p. 1. Available from: https://www.homerenergy.com/support/docs/3.10/how_homer_calculates_the_pv_array_power_output.html

[5] Kumar S, Verma A K, Hussain I and Singh B. Performance of Grid Interfaced Solar PV System under Variable Solar Intensity. [cited 2017 Nov 22]; Available from: https://www.researchgate.net/profile/Ikhlaq_Hussain/publication/280562137_Performance_of_grid_interfaced_solar_PV_system_under_variable_solar_intensity/links/55b9c8c608aed621de0879dc/Performance-of-grid-interfaced-solar-PV-system-under-variable-solar-inte

[6] Akinyele D O and Rayudu R K. Community-based hybrid electricity supply system: A practical and comparative approach. Appl Energy [Internet]. Elsevier Ltd; 2016;171:608–28. Available from: http://dx.doi.org/10.1016/j.apenergy.2016.03.031

[7] Mohan N, Undeland T M and Robbins W P. Power electronics : converters, applications, and design. 3rd ed. John Wiley & Sons; 2003. 824 p.

[8] Faten Hosney Fahmy. PV/Diesel Hybrid System for Fuel Production from Waste Plastics Recycling. Ijmer [Internet]. 2014;4(12):68–79. Available from: http://www.ijmer.com/papers/Vol4_Issue11/Version-1/H04011_01-6879.pdf

[9] Pachori A, Suhane P and Scholar P. Design and modelling of standalone hybrid power system with MATLAB/simulink. Int J Sci Res Manag Stud [Internet]. [cited 2017 Dec 3];1(2):2349–3771. Available from: http://www.ijsrms.com/media/3n2-IJSRMS0102116_v1_is2_65-71.pdf

[10] Babu N R and Arulmozhivarman P. Wind energy conversion systems - A technical review. J Eng Sci Technol. 2013;8(4):493–507.

[11] Kim H S and Lu D D. Wind Energy Conversion System from Electrical Perspective — A Survey. 2010;2010(November):119–31. Available from: http://www.scirp.org/journal/PaperInformation.aspx?PaperID=3324

[12] Mishra S, Gupta M, Garg A, Goel R and Mishra V K. Modeling and Simulation of Solar Photo-Voltaic and PMSG Based Wind Hybrid System. IEEE Students' Conference on Electrical, Electronics and Computer Science [Internet]. 2014 [cited 2017 Nov 19]. Available from: http://ieeexplore.ieee.org/stamp/stamp.jsp?arnumber=6804430

[13] Gajewski P and Pieńkowski K. Analysis of a wind energy converter system with PMSG generator. 2015;

[14] Samrat N H, Ahmad N Bin, Choudhury I A and Taha Z B. Modeling , Control , and Simulation of Battery Storage Photovoltaic-Wave Energy Hybrid Renewable Power Generation Systems for Island Electrification in Malaysia Modeling , Control , and Simulation of Battery Storage Photovoltaic-Wave Energy Hybrid Renewab. Sci World J. 2014;2014(February 2016):22.

[15] Yasin A, Napoli G, Ferraro M and Antonucci V. Modelling and control of a residential wind/PV/battery hybrid power system with performance analysis. J Appl Sci [Internet]. 2011 [cited 2017 Sep 6];11(22):3663–76. Available from: http://www.scialert.net/qredirect.php?doi=jas.2011.3663.3676&linkid=pdf

Risk Assessment of Bulgarian Electricity Network under Natural Disasters

P Zlateva[1] and D Velev[2]

1 IR, Bulgarian Academy of Sciences, Sofia, Bulgaria
2 University of National and World Economy, Sofia, Bulgaria
E-mail: plamzlateva@abv.bg, dgvelev@unwe.bg

Abstract. The electricity transmission and distribution networks are heavily affected from natural disasters. Electricity networks can also incur multiple or multi-risks, including their conjoint and cascading effects, or systemic risks. The aim this paper is to propose an approach for risk assessment of Bulgarian electricity transmission and distribution network under natural disasters. The proposed approach for risk assessment is based on guidelines and techniques of risk management standards ISO 31000:2009 and EN 31010:2010. Here, the risk assessment is the overall process of risk identification, risk analysis and risk evaluation. The risk assessment results can support the all key stakeholders to take more informed decision about effective protection of the electricity networks from natural disasters.

1. Introduction

In recent years, natural disasters have caused severe damage to the energy systems across the world [1], [2]. The electricity transmission and distribution networks are heavily affected [3], [4]. Thus, the number and duration of temporary power breakouts are increased [5], [6]. There is also a rise in the intensity and severity of these accidents [7]. Considering the importance of power supply for many systems that use energy, even minor breakouts can cause cascading consequences and intensify the initial negative impact [8], [9].

Although the electricity breakouts from natural disasters occur rarely, they are capable to cause significant disruptions in the Bulgarian economy and society due to the high degree of systemic interconnection. The energy system is strongly connected with other systems such as transport, water supply, communications, health care, and more. Therefore, power breakout for any duration can cause ubiquitous direct and indirect negative consequences. Therefore, modern economies and societies can properly function only with continuous power supply.

Natural disasters are difficult to avoid. For these reasons United Nations (UN), European Union (EU) and any governments develop strategies to disaster risk reduction in regard to the increased natural disasters [10], [11]. European Commission in the end of 2010 proposed official working paper "Risk Assessment and Mapping Guidelines for Disaster Management" [12].

The established trend of increasing natural disasters in the future is likely to lead to more damages of the electricity transmission and distribution networks, such as outcrop of cables because of erosion or damage to the transport infrastructure; increase of the losses on the power lines; increased of the damages to the overhead electrical infrastructure from extreme storms and winds; direct mechanical damages to the electricity network caused by a short circuit of power lines; indirect mechanical damage and short-circuits caused by the disruption of overhead lines due to fallen trees and debris; material damage (including falling) of overhead lines and pillars caused by ice, etc. [13], [14]

Nowadays, because of the possible losses mentioned above discussions and activities related to the protection of the electricity transmission and distribution networks from the impacts of natural disasters are especially topical. This requires strengthened cooperation between all key stakeholders: national governments, regional and non-governmental organizations, the business community, academia, development agencies, and financial institutions.
Protection of the electricity networks from natural disasters included risk assessment (to identify threats /natural disasters/, assess vulnerabilities, identify and quantify potential losses); risk preparedness, prevention, and mitigation (including technical and physical protection measures and planning, as well as organizational measures, capacity building, early warning, and internal controls); risk management (disaster management); reconditioning (backup supply and provisional repair); and risk recovery (reconstruction, financing, repairing, and restoring) [13], [14].
Protection of the electricity transmission and distribution networks is a challenging task due to the system complexity and the many elements included such as generators, transformers, and high- and low-voltage transmission and distribution lines. All these components are interdependent and include a large number of elements, such as interconnectors, edges, and nodes [15]. Each element can become vulnerable to existing and newly natural hazards. Electricity networks can also incur multiple or multi-risks, including their conjoint and cascading effects, or systemic risks.
The aim this paper is to propose an approach for risk assessment of Bulgarian electricity transmission and distribution network under natural disasters. The proposed approach for risk assessment is based on guidelines and techniques of risk management standards ISO 31000:2009 and EN 31010:2010.

2. Essence of the risk assessment process

Risk management standards ISO 31000:2009 "Risk management — Principles and guidelines" and IEC/ISO 31010:2010 "Risk management - Risk assessment techniques" are a useful tool in representing and logically organizing risk management process in a way that makes decision-making open to inputs from different stakeholders [16], [17].
The contribution of risk assessment to the risk management process according IEC/ISO "Risk management - Risk assessment techniques" is shown on Figure 1. Here, the risk assessment is the overall process of risk identification, risk analysis and risk evaluation.

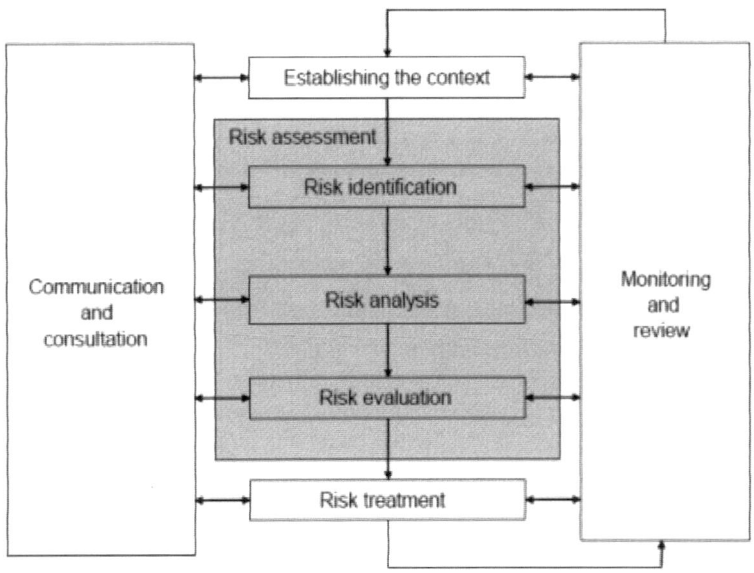

Figure 1. Contribution of risk assessment to the risk management process.

It is necessary to point that the risk assessment may require a multidisciplinary approach since risks from natural hazards ordinary cover a wide range of causes and consequences.

Usually for the purpose of risk assessment from natural hazards the following term definitions are used [17], [18]:

- *Hazard* is a dangerous phenomenon, substance, human activity or condition that may cause loss of life, injury or other health impacts, property damage, loss of livelihoods and services, social and economic disruption, or environmental damage.
- *Natural hazard*: Natural process or phenomenon that may cause loss of life, injury or other health impacts, property damage, loss of livelihoods and services, social and economic disruption, or environmental damage. Natural hazard events can be characterized by their magnitude or intensity, speed of onset, duration, and area of extent.
- *Exposure*: People, property, systems, or other elements present in hazard zones that are thereby subject to potential losses.
- *Vulnerability*: The characteristics and circumstances of a community, system or asset that make it susceptible to the damaging effects of a hazard. In probabilistic/quantitative risk assessments, the term vulnerability expresses the part or percentage of *Exposure* that is likely to be lost due to a certain hazard.
- *Risk* is a combination of the consequences of an event (*Hazard*) and the associated likelihood/probability of its occurrence.
- *Risk assessment* is the overall process of risk identification, risk analysis, and risk evaluation.
- *Risk identification* is the process of finding, recognizing and describing risks.
- *Risk analysis* is the process to comprehend the nature of risk and to determine the level of risk.
- *Risk evaluation* is the process of comparing the results of risk analysis with risk criteria to determine whether the risk and/or its magnitude is acceptable or tolerable.
- *Risk criteria* are the terms of reference against which the significance of a risk is evaluated.
- *Consequences* are the negative effects of a disaster expressed in terms of human impacts, economic and environmental impacts, and political/social impacts.
- *Hazard assessments* determine the probability of occurrence of a certain hazard of certain intensity.

According EN 31010:2009 [17] in situations where the likelihood of occurrence of a hazard of certain intensity can be quantified investigators refer to the term probability of occurrence. When the extent of the impacts is independent of the probability of occurrence of the hazard, which is often the case for purely natural hazards, such as earthquakes or storms, risk can be expressed as [16]:

$$Risk = probability\ of\ occurrence * hazard\ impact$$

or

$$R = f(P * C),$$

where R is the risk; P – the probability of occurrence of the natural hazard; C – the consequences (natural hazard impact).

Risk matrix is very helpful in the risk assessment process. In particular the risk matrix or so-called consequence/ probability matrix is a means of combining qualitative or semi-quantitative ratings of consequence and probability to produce a level of risk.

The format of the risk matrix depends on the context in which it is used. The scale used may have 5 or more points. The matrix may be set up to give extra weight to the impact or to the likelihood, or it may be symmetrical.

Usually in risk assessment process the risk matrix 5x5 is used (Figure 2).

Here the *Probability levels* (Relative likelihood) are graded as "*Very low*", "*Low*", "Medium", "*High*" and "*Very high*".

Consequences (Relative impact) are also graded as "*Very low*", "*Low*", "*Medium*", "*High*" and "*Very high*".

The *Risk levels, R* are defined as "*Low*", "*Medium*", "*High*" and "*Very high*".

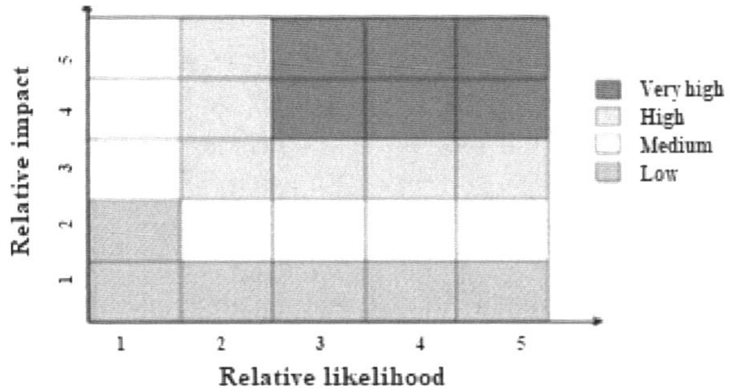

Figure 2. Risk matrix 5x5: Consequences (Relative impact) / Probability (Relative likelihood).

3. Risk identification of the Bulgarian electricity network from natural disasters

According to the above definition, risk identification is the process of finding, recognizing and describing risks. In cases where the risk is induced by natural disasters, then the process of risk identification requires descriptions of all potential natural hazards, which cause these disasters.

In this paper, the risk identification of Bulgarian electricity transmission and distribution network requires descriptions of the natural hazards occurring on the territory of the country.

The predominant natural hazards on Bulgarian territory are identified as follows:

- Geological processes and phenomena: Earthquakes; Slope failures (landslides, landslips, creep, falls, flows, subsidence); Mud-rock flows (seli); Erosion and abrasion; Storm surge.
- Hydrological processes and phenomena: Floods; Dry periods; Snow flows and glaciations; Icings.
- Meteorological processes and phenomena: Strong wind; Extreme temperatures; Freezings, Drought, Tornado phenomena; Dust storms; Hailstorms; Wet snow; Fog (coastal, evaporation, radiation, valley, upslope); Thunderstorm; Silver thaw; Wild land fire.

It can be pointed some representative examples of identified risks from natural disasters for the Bulgarian electricity network as follows

- Losses from transmission of electricity due to extremely high temperatures (including hot days and heat waves).
- Damage to the electricity transmission and distribution network from icing due to extremely low temperatures (including cold days and cold airwaves).
- Flooding of the electricity transmission and distribution network (including electricity substations) leading to power breakouts and/or higher capital costs for equipment repairs due to more rainfall and high humidity.
- Droughts that cause soil downgrades, damaging underground electricity transmission and distribution network
- Strong winds and storms damage the electricity transmission and distribution network (for example, from fallen trees), leading to power breakouts and/or higher capital costs for repairs of the equipment.
- Landslides also damage the electricity transmission and distribution network, resulting in power breakouts and/or higher capital costs for equipment repairs.
- The power lines could exceed the established maximum temperatures in the technical documentation and violate the requirement for a minimum distance from the earth's surface due to thermal expansion, which could lead to a power breakout.

4. Risk analysis of the Bulgarian electricity network from natural disasters

The essence of the proposed approach is consists in the determination of the levels of risk for the electricity transmission and distribution network as a whole and/or its individual elements from the natural disasters in a given geographic region for a certain time interval. Here, according to the definition the risk analysis is the process to determine the levels of risk.

IOP Conf. Series: Earth and Environmental Science **167** (2018) 012040 doi:10.1088/1755-1315/167/1/012040

In this study, it is considered that the natural hazard that causes the corresponding natural disaster has four levels of intensity.
Furthermore it is necessary to note that the adequate risk analysis requires producing distinct risk matrices for each intensity levels of the natural hazard. Usually it is considered the following four levels of the natural hazard intensity: (1) Low hazard intensity, (2) Medium hazard intensity, (3) High hazard intensity, (4) Very high hazard intensity (Table 1).

Table 1. Hazard intensity levels

Relative intensity	Hazard intensity levels, *H*
(4)	Very high hazard intensity
(3)	High hazard intensity
(2)	Medium hazard intensity
(1)	Low hazard intensity

In this case, the risk is defined as follow:

$$R = \sum_{k=1}^{4} R_k \tag{1}$$

where R_k is the determined risk corresponding to k intensity level of natural hazard, k=1,...,4.
The risk R_k is obtained by following product

$$R_k = P_k . C_k \qquad k=1,..., 4 \tag{2}$$

where P_k is the occurrence probability of the natural hazard with k intensity level; C_k is the consequences caused by action of the of the natural hazard with k intensity level.
The values of the consequences are expressed in monetary terms (USD, Euro, etc.).
The consequences can be defined as: the value of non-earned revenue corresponding to the amount of unsold electricity or the value of the alternative costs for repair or recovery of the damaged network elements.
The calculated value of consequences C_k and the given value of the probability P_k for occurrence of the natural hazard with k intensity level in the considered time interval are substituted in (2) to calculated corresponding risk level R_k.
Than using (1) it is calculated the total risk assessment R of a monitored object (whole electricity transmission and distribution network or its individual elements) from the natural disasters in a given geographic region) from natural hazard with four intensity levels for a certain time interval.
Each of the resulting risk levels R and R_k, k=1,...,4 can be presented as a distinct risk matrix as the proposed on Figure 2.
First from Table 2 it can be determined the particular level of the probability P_k by using predefined range of each of the five levels. The constants IP_i , i=1,…5 are previously defined on the base of the real data or expert knowledge.
Second by analogy with the probability from Table 3 it can be determined the particular level of the consequences C_k by using predefined range of each of the five levels. The constants IC_i, i=1,…5 are previously given.
Third from Table 4 for each of the resulting risk assessments R and R_k, k=1,...,4, can be determined the particular risk levels by using predefined range of each of the four levels. The constants IR_i , i=1,…4 are previously given.
The proposed risk assessment with these determinations of the particular risk levels R and R_k, k=1,...,4, leads to more effectiveness of the risk management about natural hazards.

Table 2. Relative likelihood / Probability levels

Relative likelihood	Probability levels, P_k	Probability value intervals
(1)	Very low probability	$IP_4 < P_k$
(2)	Low probability	$IP_3 < P_k \leq IP_4$
(3)	Medium probability	$IP_2 < P_k \leq IP_3$
(4)	High probability	$IP_1 < P_k \leq IP_2$
(5)	Very high probability	$P_k \leq IP_1$

Table 3. The Consequences levels

Relative impact	Aggregated loss / Consequences levels, C_k	Consequence value intervals
(1)	Very low probability	$IC_4 < C_k$
(2)	Low probability	$IC_3 < C_k \leq IC_4$
(3)	Medium probability	$IC_2 < C_k \leq IC_3$
(4)	High probability	$IC_1 < C_k \leq IC_2$
(5)	Very high probability	$C_k \leq IC_1$

Table 4. The Risk levels

Risk levels, *R or* R_k	Risk value intervals
Very high risk	$IR_3 < R_k$
High risk	$IR_2 < R_k \leq IR_3$
Medium risk	$IR_1 < R_k \leq IR_2$
Low risk	$R_k \leq IR_1$

5. Risk evaluation of the Bulgarian electricity network from natural disasters

The risk evaluation is carried out by comparing the determined risk levels according Table 4 with specific risk criteria to decide whether the risk is acceptable or inadmissible

It is necessary to ensure the sustainability of Bulgarian electricity network to natural disasters because it is part of the national critical infrastructure.

For this reason, the network needs to be maintained and exploited, following building standards, regulations and good practices. The risk assessment shall be carried out taking into account: the parameters of the electricity network, the characteristics of the potential natural disasters, the peculiarities of the geographical region, the historical and forecast data available. The results are compared with defined criteria for safety and adaptation of the electricity network and its elements.

6. Conclusion

The risk assessment of electricity network as an essential part of risk management from natural disaster is a significant problem of the present day. This paper propose an approach for risk assessment of Bulgarian electricity transmission and distribution network under natural disasters. It is considered that the natural hazard that causes the corresponding natural disaster has four levels of intensity. The proposed approach for risk assessment is based on guidelines and techniques of risk management standards ISO 31000:2009 and EN 31010:2010. Here, the risk assessment is the overall process of risk identification, risk analysis and risk evaluation. The risk assessment results can support the all key stakeholders to take more informed decision about effective protection of the electricity networks from natural disasters.

7. References

[1] URS 2010 Adapting Energy, Transport and Water Infrastructure to the long-term impacts of climate change. Ref. No RMP/5456. (http://archive.defra.gov.uk/environment/climate/documents/infrastructurefull- report.pdf)
[2] Ebinger J and Vergara W 2011 Climate Impacts on Energy Systems. Key issues for energy sector adaptation. *World Bank study*. Washington. http://www.esmap.org/esmap/sites/esmap.org/files/EBook Climate%20Impacts%20on%20Energy%20Systems..pdf)
[3] Padli J, Habibullah M and Baharom A 2010 Economic impact of natural disasters' fatalities, *International Journal of Social Economics* **37**(6) pp 429 – 441
[4] You H and Zhao F 2013 Analysis of the causes of damages to power facilities in the M7.0 Lushan Earthquake, *Electric Power Construction* **34**(8) pp 100-104
[5] Power Outage, *Disaster Survival Resources* 2017, http://www.disaster-survival-resources.com/
[6] Schmidthaler M, Reichl J and Cohen J 2014 Assessing the Socio-Economic Effects of Power outages in the European Union Ad Hoc Using www.blackoutsimulator.com, *CIRED Workshop* - Rome, 11-12 June 2014, Paper 0486.
[7] Poljanšek K, Bono F and Gutiérrez E 2012 Seismic risk assessment of interdependent critical infrastructure systems: The case of European gas and electricity networks. *Earthquake Engineering and Structural Dynamics* 41 (1) pp 61–79
[8] Anderson S 2016 *School Power Outages Mean More than Canceled Classes*, Last updated on September 23, 2016, http://www.powerprotectionresource.com/articles/425439-school-power-outages-mean-more-than-canceled-classes.htm
[9] Kunz M. et al. 2012 CEDIM FDA-Report on Hurricane Sandy 22-30 October 2012, 2ndReport
[10] United Nations Office for Disaster Risk Reduction (UNISDR) 2015 Proceedings Third UN *Word Conference on Disaster Risk Reduction*, 14-18 March 2015, Sendai, Japan.
[11] Prevention Web - Strengthening climate change adaptation through effective disaster risk reduction, http://www.preventionweb.net
[12] European Commission 2010 *Risk Assessment and Mapping Guidelines for Disaster Management*, working paper.
[13] Organization for Security and Co-operation in Europe (OSCE) 2016 *Protecting Electricity Networks from Natural Hazards*, Vienna, Austria.
[14] Farrokh Nadi 2014 Experience of Nordic Region &Recent Developments in Risk Assessment for Electricity Networks, *OSCE Expert Workshop*: Sharing Best Practices to Protect Electricity Networks from Natural Disasters, 2nd July 2014, Vienna, Austria
[15] Berkes F and Ross H 2013 Community resilience: toward an integrated approach, *Society and Natural Resources* **26** pp 5-20
[16] ISO 31000:2009 *Risk management — Principles and guidelines.*
[17] IEC/ISO 31010:2009 *Risk management - Risk assessment techniques.*
[18] UNISDR 2009 *UN International Strategy for Disaster Reduction Sec*, 15 January 2009, www.unisdr.org/eng/library/lib-terminology-eng.htm.

Acknowledgments
The authors express their gratitude to the Bulgarian National Science Fund for the partial financial support under Grant № DNTS/ China 01/6 from 2014, Bulgarian–Chinese scientific cooperation

Research on Construction Mechanical Behaviour of Connecting Aisle in Shield Tunnel

J C Sun[1], H Zhang[1,*], T Peng[2], C L Xiao[1]

1 MOE Key Laboratory of Transportation Tunnel Engineering, Ministry of Education, Southwest Jiaotong University, Chengdu, Sichuan, China
2 Department of Road and Bridge Engineering, Vocational and Technical College of Communications, Chengdu City, Sichuan Province, China
E-mail: tunnelzh@home.swjtu.edu.cn

Abstract. Based on connecting aisle of shield tunnel projects in Shenzhen Metro Line 5, in accordance with the complex space structure formed by shield tunnel and connecting aisle ,the study chose a three-dimensional numerical model linked by structure and surrounding rock for the calculation and analysis of the stress effect. It showed that due to the connecting aisle construction, the symmetrical deformation of shield tunnel was transformed into unsymmetrical deformation and the crossing segment produced a phenomenon of stress concentration, which caused the production of the upper and lower section of the overall tension of adverse stress form in side opening segment and contact channel structure side themselves. The tension zone of the whole structure was to connect through the lower half of the channel and the surrounding part of segment, so highly attention should be paid to the force structure and waterproof both in design and construction period.

1. Introduction

The connecting aisle of urban subway is generally in the middle part of running tunnel and the lowest position of line, and the connecting aisle construction complicated with water-collecting pump and drainage pump in order to link two tunnel, water collection, water drainage, fire proof and so on. The construction of connecting aisle is bound to change the forced state in shield tunnel, form the very adverse force way at cross connection structure, produce stress concentration and make the cross segment connected with connecting aisle become the weak link in the force[1], [2]. Many studies of mechanical behaviour in segment construction were carried out by domestic and foreign scholars, HE et al.[3] researched the segment stress change by shield tunnel connecting aisle construction. HE et al.[4] monitored and numerical simulation the segment mechanical characteristics in No.1 line of Nanjing subway construction, the segment stress characteristics of shield tunnel crossing sandy soil was studied. YE et al. [5] analyzed the influencing factors of shield segment during construction period and pointed out that construction load of lining segment has typical three-dimensional characteristics and uncertainty. Based on fluid-solid coupling, XIA et al.[6] used numerical analysis and field measurement to research structural internal force of shield tunnel construction under high hydraulic pressure. ZHOU et al.[7] conducted in-situ test on shield segment of dot shield tunnel in Shanghai metro. Through site monitoring of deep shield tunnel lining segments, Jin SAITO and Atsushi KOIZUMI [8] proposed segment erection process was a three-dimensional mechanical behaviour and researched on segment stress change during construction. Jin SAITO and Shigreu KUROSAKI [9] monitored segment stress of deep tunnel and simulated segment construction process under large depth by three-dimensional finite element. Because of the connecting aisle construction ineffective, No.4 line of Shanghai metro accident in July 1, 2003 , resulting in direct economic losses of up to 1.5 billion yuan, caused a national shock. But in the domestic design, the main tunnel shield variation of displacement and stress is not very clear in the connecting aisle construction by mining method[10], [11]. In order to ensure the safety of connecting aisle construction and reduce the impact of the main tunnel with connecting aisle construction, the process of displacement and mechanical changes during segment removal, excavation, lining application is studied by using FLAC 3D, and the feasibility of the connecting aisle construction is verified to lay a foundation for further study.

2. Engineering background

2.1. Project description

The right line of Shenzhen Subway Line 5 Yi Jing station to Huang Beiling station is 1051m, and the left line is 1078m. The shield zone includes two portal parts and one connecting aisle.

According to human evacuation, the clearance of connecting aisle is determined as arches-straight wall section, size of 2.1m (wide) × 2.3m (high). The connecting aisle and pumping station are constructed by mining method with composite lining, leading conduit as an auxiliary reinforcement measure and grouting with cement mortar for the external formation. The connecting aisle and pump house are in rocks of grade Ⅴ, and covering soil is 13m. The upper covering soil of shallow tunnel is about 2m, then as following: soft soil layer, gravelly cohesive soil layer, completely weathered granite, intermediary weathered granite. The cross section and the longitudinal section of the connecting aisle is shown in Figure 1 and Figure 2.

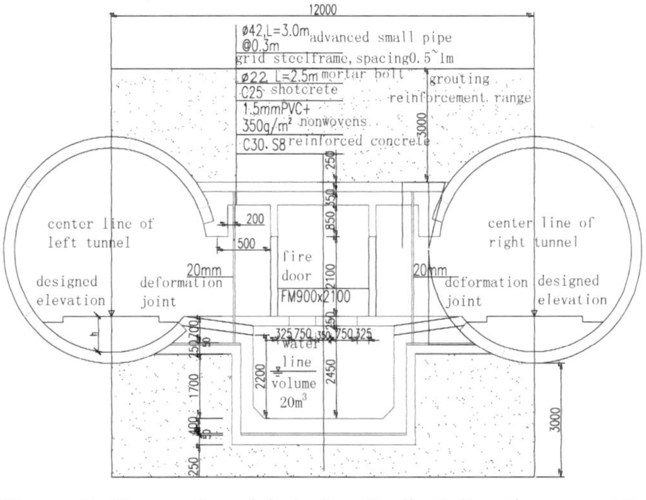

Figure 1. Connecting aisle in longitudinal direction

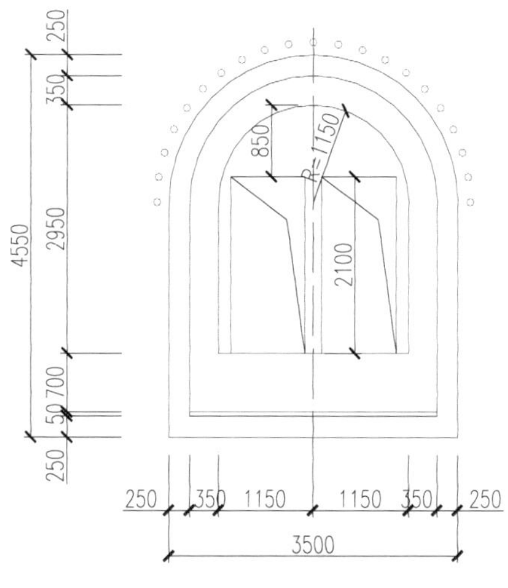

Figure 2. Connecting aisle in cross direction

2.2. Construction Procedures of Connecting Aisle

According to the construction conditions, combined with the subway connecting aisle construction experience, through technical and economic comparison, the construction scheme was using the soil grouting reinforcement and tunnel excavation. Grouting reinforcement and excavation were carried out in the tunnel, main construction procedures were as follows: ①Ground consolidation in the tunnel(Figure 3-a); ②Erection of support in the shield tunnel(Figure 3-b); ③Check whether the effect of soil reinforcement to meet the strength, impermeability requirements; ④Removal of the opening segment in right line(Figure 3-c); ⑤Construction of the protective door; ⑥Construction of connecting aisle lining, secondary lining(Figure 4-d), tunnel portal part(Second lining at the bottom of the pumping station is not needed); ⑦ Construction of pump house initial lining; ⑧ Wallboard construction; ⑨Removal of the tunnel support. Figure1 shows the main construction process.

(a) Ground consolidation in tunnel

(b) Support erection in tunnel

(c) Removal opening segment

(d) Lining of connected aisle

Figure 3. Construction procedure of connecting aisle

3. Numerical analysis of shield tunnel and connecting aisle construction

3.1. Computational mechanics model and related parameters

According to design working drawing of connecting aisle reinforcement section, numerical calculation model of connecting aisle is shown in Figure 4 and the cross connection structure is shown in Figure 5. The influence scope of stress and displacement is in the range of 3 to 5 times tunnel diameter after excavation, in order to eliminate the calculation error induced by size effect, along the tunnelling direction, the size is 60m(Equivalent to 40 rings, each ring is 1.5m) and the size is 80m along the traverse of tunnel section. However, along the vertical direction, the thickness of the casing soil can be determined according to the actual depth of connecting aisle to determine. And downward should consider the calculating influence scope, the choice of strata thickness is 22m. The study adopted ideal elastic-plastic constitutive relation and Drucker-Prager yield criterion for numerical simulation. Due to the effect of segment joint, the reduction factor of segment stiffness is taking value 0.75 s. Table 1 shows the calculation parameters of related materials.

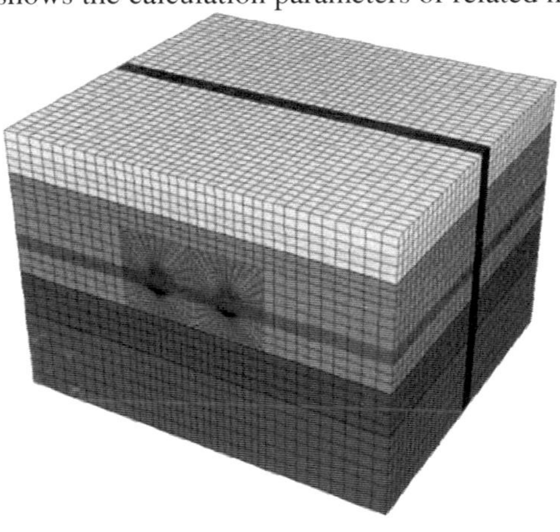

Figure 4. Three-dimension model

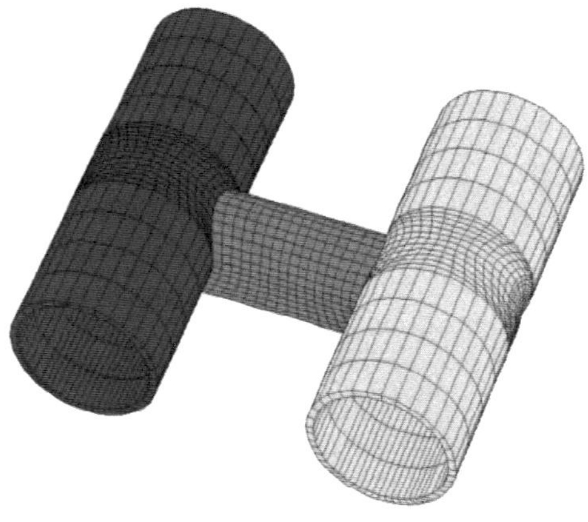

Figure 5. Mesh of cross connection

Table 1. Computation parameters of various materials

Soil layer	Density (kg/m^3)	C (KPa)	Φ (°)	Poisson ratio μ	Elastic modulus (MPa)
Plain fill Q^{ml}	2300	18	14	0.42	8
Silty clay Q_4^{al+pl}	2000	30	5	0.38	18
Gravelly clay Q^{el}	1810	31	22	0.33	33
Completely decomposed granite γ_5^3	1770	45	26	0.31	80
Intermediary weathered migmatite γ_5^3	2570	1500	35	0.28	8000
Segment	2500	/	/	0.17	31500
Initial support of connecting aisle	2500	/	/	0.2	28 000
Secondary lining of connecting aisle	2500	/	/	0.2	31 500

3.2. Boundary conditions and initial stress field

Adopted displacement boundary conditions, computational model applies vertical constraints in the bottom of the model and horizontal constraints in both sides of the model, while the upper surface is free surface. Because of additional load on ground surface, the load P0=20kPa is applied at the top of the model. The initial stress field is generated by the dead weight and tectonic stress is not considered.

3.3. Numerical simulation process

In order to master the complex mechanic characteristic of spatial cross connection structure composed of shield tunnel and connecting aisle, the numerical simulation process of connecting aisle construction is as follows:

To determine the release ratio of stress at excavation face, the displacement release ratio is needed to obtain. Base on the numerical calculation of node displacement results, the release ratio of horizontal and vertical displacement is obtained from the average displacement release factor of each node. It is suggested that the load release ratio and its displacement release ratio is approximately same.

①Shield tunnel excavation under the rock wool-hole state, the release rate of surrounding rocks load is 30%; after the tunnel segment application, the release rate is 70%.

②The mining method is used in connecting aisle construction, it is assumed that the whole construction process is divided into three steps, each step 2m.

③Connecting aisle excavation under wool-hole state, the release ratio of stress is 30%,after connecting aisle initial support application, the release ratio of stress is 70%.

4. Analysis of calculation results

4.1. Analysis of Structural Stress

The principal stress distribution of segment is shown in Table 2, after the finish of shield tunnel construction but before the connecting aisle construction, there is a greater tensile stress in the vault and arch bottom of the main tunnel, the maximum tensile stress is 0.27MPa and the maximum compressive stress is 3.69MPa in both hance sides. The tensile stress of cross connection segment structure is increasing in different degree both in the open hole side and away from the side due to the connecting aisle construction, but the compressive stress of the segment in open-hole side is decreasing gradually with the construction progress, while the compressive stress of the segment away from the open-hole side is increasing gradually, shown in Figure 6 and Figure 7.

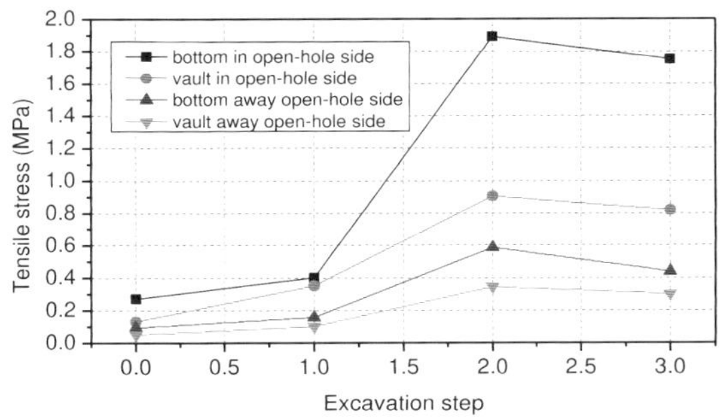

Figure 6. Tensile stress in segment

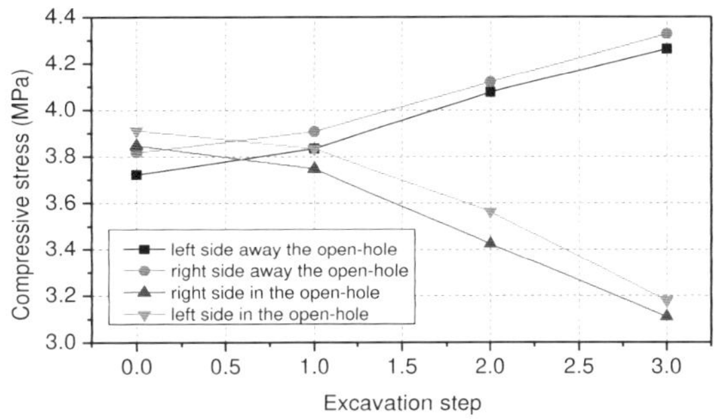

Figure 7. Compressive stress in segment

Table 2. Stress of the segment

Construction steps	Tensile stress/MPa		Compressive stress /MPa	
	Maximum	Position	Maximum	Position
Complete double line of	0.27	tunnel floor	3.69	hance
Excavation step 1 of	0.4	bottom in open-	3.83	horizontal position in open-
Excavation step 2 of	1.89	bottom in open-	4.08	horizontal position in open-
Excavation step 3 of	1.75	bottom in open-	4.26	horizontal position away

Because of the connecting aisle construction, the original force way of main tunnel is under stress redistribution, which will cause the production of stress concentration in cross segment, and will also cause the occurrence of adverse stress form which the whole tension is in upper and lower sections in the open-hole side segment. Therefore, it is very necessary to apply effective strengthening and enough supporting in the segment vicinity of the connecting aisle during the construction. The principal stress distribution of the connecting aisle in every construction stage is shown in Table 3, with the construction progress, the variation range of the maximum tensile stress in the structure decreased gradually and the maximum value is in arch bottom vicinity of the open-hole side; while maximum compressive stress increased gradually and the maximum value is in hance away from open-hole side. There is a very large tension stress zone in structure itself caused by the construction of connecting aisle, which caused overall tension situation of structure of the upper and lower section, and the situation is much unfavorable.

Table 3. Stress of connecting aisle

Construction steps	Tensile stress/KPa		Compressive stress /MPa	
	Maximum	Position	Maximum	Position
Excavation step 1 of connecting aisle	2.04	tunnel floor	2.84	hance
Excavation step 2 of connecting aisle	2.79	tunnel floor	2.96	hance
Excavation step 3 of connecting aisle	2.65	tunnel floor	3.43	hance

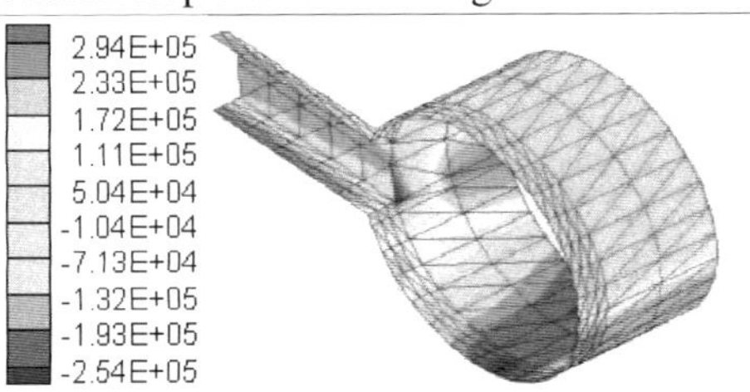

Figure 8. Moment at connecting aisle

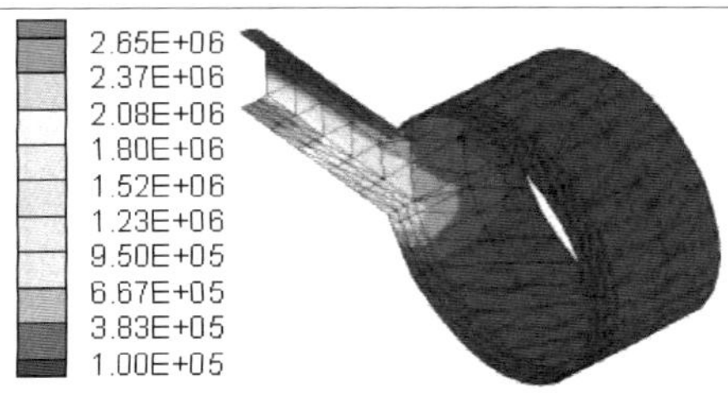

Figure 9. Tensile stress at connecting aisle

It can be seen from Figure 8 and Figure 9, the arch bottom and vault in the connecting aisle is the max region of positive moment, but the moment when mutation is in the lower part of the interface, there is appearance of a larger negative moment. The tensile area of the whole structure was lower half part of connecting aisle and segment in tunnel intersection, highly attention should be paid to both structure stress and waterproof functions of these regions in design and construction period.

4.2. Analysis of Structural deformation

It is shown in Table 4 and Table 5, during the connecting aisle excavation, the symmetric deformation of shield tunnel is converted into asymmetric deformation and is developing toward the side of connecting aisle. The structural stiffness is decreasing due to the segment demolition and connecting aisle construction, as a result, the deformation of the segment close to connecting aisle is significantly bigger than the segment in the other side. During excavation, the increasing amount of deformation of the segment in open-hole side is almost three times bigger than the segment in the other side. Because of connecting aisle construction, there is a remarkable additional deformation along the vertical direction and horizontal direction in the cross segment, especially in the open-hole side. The maximum additional deformation in vertical direction is 2.19mm and the maximum additional deformation in horizontal direction is 9.06mm. It can be seen from Table 5, in the process of excavation, the primary deformation of the structure is vertical deformation in vault, the maximum settlement is 24.39mm and the deformation in other part of the structure is relatively small. With the increasing of construction progress, the deformation in every part of the structure is increasing at different degrees except for uplift in the arch bottom.

Table 4. Deformation of the segment

Construction steps	**Vertical deformation(mm)**		**Horizontal deformation(mm)**		**Position**
	Tunnel roof	Tunnel floor	Tunnel roof	Tunnel floor	
Complete double line of shield tunnel	-4.83	-2.16	5.44	0.33	open-hole side
	-4.81	-2.19	-5.52	-0.35	away open-hole side
Excavation step 1 of connecting aisle	-6.89	-1.19	-3.46	-0.24	open-hole side
	-5.25	-2.21	-5.42	-0.29	away open-hole side
Excavation step 2 of connecting aisle	-6.93	-1.16	-3.54	-0.25	open-hole side
	-5.37	-2.18	-5.61	-0.31	away open-hole side
Excavation step 3 of connecting aisle	-7.02	-1.15	-3.62	-0.28	open-hole side
	-5.48	-2.07	-5.43	-0.22	away open-hole side

Table 5. Deformation of the connecting aisle

Construction steps	**Vertical deformation**		**Horizontal deformation**	
	Tunnel roof	Tunnel floor	Arch foot at the top	Arch foot at the bottom
Excavation step 1 of connecting aisle	5.78	0.52	1.15	0.33
Excavation step 2 of connecting aisle	16.77	0.53	4.67	0.35
Excavation step 3 of connecting aisle	24.39	0.55	6.32	0.39

4.3. Comparison of calculated value and monitoring value on ground settlement.

Figure 10 is the monitoring point arrangement on ground settlement of connecting aisle, to analyze the difference of calculation results and measured results based on numerical simulation. The maximum settlement value of monitoring section in connecting aisle by numerical simulation is 6.71mm, the maximum settlement is above the axis center of connecting aisle, shown in Figure10 and Figure 11.

Field monitoring data shows that the maximum settlement is 5.93mm, from the position and value of the maximum settlement, the position of maximum settlement is consistent with calculation results. The maximum settlement by numerical simulation is close to the monitoring value and calculated value is 10% larger than the monitoring value. From the influence scope, calculated results are similar with monitoring results.

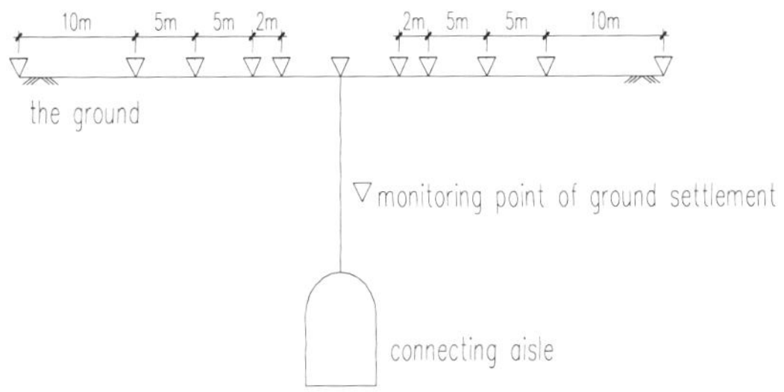

Figure 10. Monitoring points layout

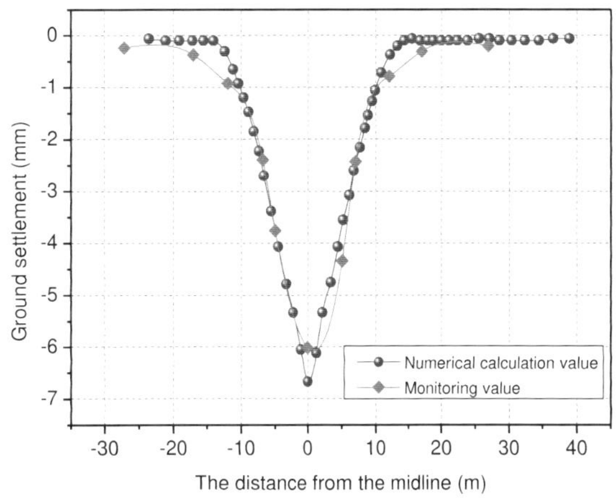

Figure 11. Calculated and measured value of settlement

5. Conclusions

By applying numerical simulation to further investigate mechanical behavior of complicated spatial structure composed of shield tunnel and connecting aisle, four conclusions are listed as follows:

①With the construction process, the deformation of cross structural of shield tunnel and connected aisle change from symmetric deformation to unsymmetrical deformation of connected aisle, and the local deformation is significantly in cross connection segment.

②Because of the connecting aisle construction, the original force way of main tunnel is under stress redistribution, which will cause the production of stress concentration in cross segment, and will cause the occurrence of the adverse force way in open-hole side and connecting aisle segment.

③The symmetrical deformation of shield tunnel was transformed into unsymmetrical deformation and stress concentration of connecting aisle due to the connect aisle construction. The adverse force way was occurred in segment with openings and connecting aisle.

④The tensile area of the whole structure was lower half part of connecting aisle and segment in tunnel intersection, highly attention should be paid to both structure stress and waterproof functions of these regions in design and construction period.

6. References

[1] Zhang Z Q and He C 2005 Study on construction of cross connection of shield tunnel and connecting aisle by freezing method[J]. *Chinese Journal of Rock Mechanics and Engineering*. 24, 18(Sep.2005), 3211-3217.

[2] Zhang Z Q and He C. 2002 Research on mechanical behaviour of construction of link line and shield tunnel on No.2 line of Guangzhou metro[J]. *Journal of the China Railway Society*. 24, 6(Dec.2002), 89-92.

[3] He M D, Le G P and Liu J 2007 Channel of shield tunnel construction segment longitudinal stress change[J]. *Urban Rapid Rail Transit*.20, 3(Jun.2007), 56-59.

[4] He C and Zeng D Y. 2007. Research on mechanical characteristics of metro shield tunnel segment in sandy strata[J]. *Rock and Soil Mechanics*. 28, 5(May.2007), 909-914.

[5] Ye F, He C and Wang S M. 2011 Analysis of mechanical characteristic of shield tunnel segments lining and its influence during construction[J]. *Rock and Soil Mechanics*. 32, 6(Jun.2011), 1801-1807+1812.

[6] Xia W Y, He C, Yan Q X, et al. 2007. Structural internal force analysis of rock shield tunnel under high water pressure during construction[J]. *Chinese Journal of Rock Mechanics and Engineering*. S2 (Dec.2007), 3727-3731.

[7] Zhou W B, Zheng Y F and T Li 2005 Mechanics characteristics study of segment in double-o-tube tunnel under In-situ test method during its construction process [J]. *CHINESE QUARTERLY OF MECHANICS*. 3 (Sep.2005), 459-463.

[8] Jin S, Shigeru K, Akira T et al 2008. The damage control of the segment during tunnelling in a large depth shield tunnel[J]. *Doboku Gakkai Ronbunshuu F*. 64, 2(Jun.2008), 173-184.

[9] Jin S, Shigeru K, Akira T, et al. 2007. Damage factors of the segment during tunnelling in a large depth shield tunnel[J]. *Doboku Gakkai Ronbunshuu F*. 63, 2(Jun.2007), 200-211.

[10] Wang H and Zhu W B 2004 Study on the control technology of frozen method for connecting passage in shield-driven metro tunnel in soft soil[J]. *Modern Tunnelling Technology*.41, 3 (Jun.2004), 17-21.

[11] Li X L, Peng L M, Wu T, et al 2009 Numerical simulation on mechanical behaviour of connecting passage in shield tunnel [J]. *RAILWAY STANDARD DESIGN*. 4(Aug.2009), 102-105.

Analysis of Deformation Law of Tunnel Surrounding Rock based on Visualization of Rock Mass Structure

H Zhang[1], J C Sun[1,*], X C LIU[1], M Z Gao[2], S G Chen[1]

1 MOE Key Laboratory of Transportation Tunnel Engineering, Ministry of Education, Southwest Jiaotong University, Chengdu, Sichuan, China
2 College of Water Resource & Hydropower, Sichuan University, Chengdu, Sichuan, China
E-mail: jianchunsun@my.swjtu.edu.cn

Abstract. In order to accurately obtain the characteristic parameters of the rock mass structural plane of the tunnel and provide effective basis for numerical calculation, based on tunnel of Guiguang railway, this paper measured the structural characteristics of typical tunnel side wall and excavating face by using Sirovision non-contact measurement technology of 3d photogrammetry and accomplished digital geological logging, information statistics of structural plane attitude, rock stability analysis. By using the discrete element software UDEC, the whole process of tunnel excavation, primary support and secondary lining are simulated. The study show that the surrounding rock deformation, internal force of primary support, axial force of anchor bolt will improve with the increase of fragmentation degrees of surrounding rock. The primary support is the determinant of constraint of surrounding rock deformation and bearing the load after tunnel excavation. Secondary support only as a permanent support structure, as a safety margin, the internal force is small compared with the primary support, but taking into account the decreasing of physical and mechanical parameters of the surrounding rock excavation, rheology of soil and reducing of mechanical properties of primary support material. The secondary lining should be calculated and designed as force structure.

1. Introduction

The methods used to collect the structure plane information of rock mass including drilling core jointing collection method [1], scan line method, precise scan line method [2], and the sampling window method [3]. With the advent of the information age, the use of photogrammetry technology [4] solution is a popular method to calculate the rock mass scale information including the orientation of structure plane, spacing and roughness of the rock mass. Wang Fengyan et al. [5] used digital close-shot pho-togrammetry to obtain the geometrical information of rock mass structural plane. Tian Shengli et al. [6] applied digital close-up photogrammetry technology to tunnel and underground engineering deformation monitoring. Zhou Chunlin et al. [7] pro-posed the improved camera pose acquisition method and implemented the non-contact measurement of the rock structure surface. Wang Guohui et al. [8] have in-troduced digital camera to the field of deformation monitoring. The marketization of Australian CAE Sirovision system and Austrian ShapeMetrix3D system has further promoted the research on the engineering application of close-up photogrammetry. Xu Shuai et al. [9] introduced the composition and post-processing software of CAE Sirovision, and applied it in the field of mining structure surface collection. On the basis of the ShapeMetrix3D system, Wang Shuhong et al. [10] proposed the cutting rock mass structural plane method based on virtual grid and constructed

IOP Conf. Series: Earth and Environmental Science **167** (2018) 012042 doi:10.1088/1755-1315/167/1/012042

the GeoSMA-3D system and applied close-shot photogrammetry to the field of critical mass analysis. However, there are few reports on the analysis of tunnel structure surface.

2. Introduction of 3D Photogrammetry System SIROVISION

SIROVISION is a non-contact measurement based on the digital close-up photo-grammetry technology. It is the three-dimensional digital photogrammetry and rock mass structure analysis system, which is applied in the field of geotechnical engineer-ing. Its main components are shown in Figure 1 and Figure 2.The image of ground mass is obtained by high precision digital camera and after software processing, 3D image is obtained, and structural parameter analysis can be carried out for rock mass. Image of the object, as shown in Figure 3. In essence, the close-up photogrammetry is similar to the forward intersection method in engineering surveying which is based on the different pictures of the same scene taken on different positions, and according to the geometric relationship implied by the corresponding image points of stereopair, the three-dimensional coordinates of the object point are determined by the method of the forward intersection, as shown in Figure 4.

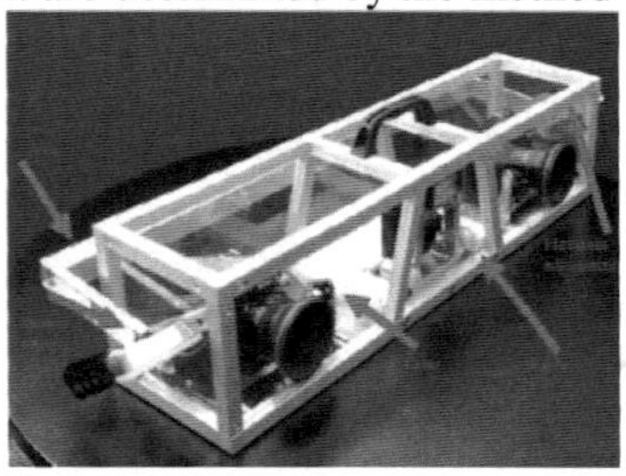

Figure 1. The main components of main engine

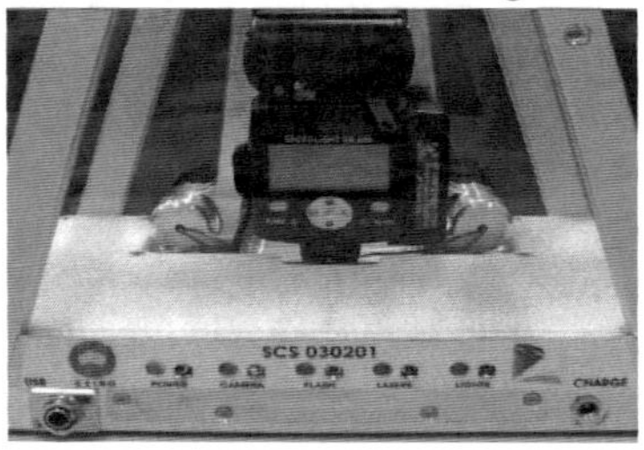

Figure 2. The control panel

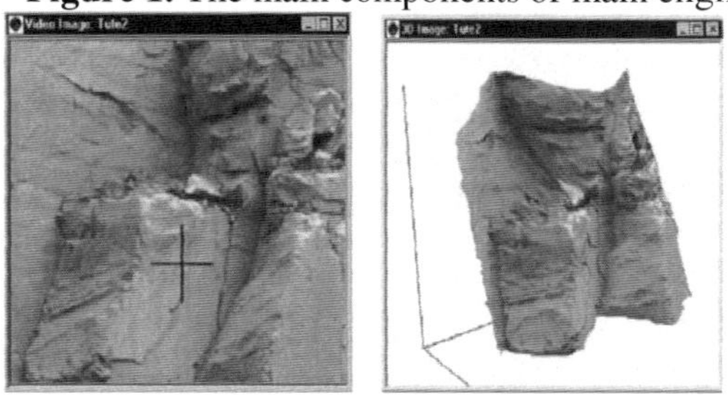

Figure 3. Camera photos and 3D digitized results

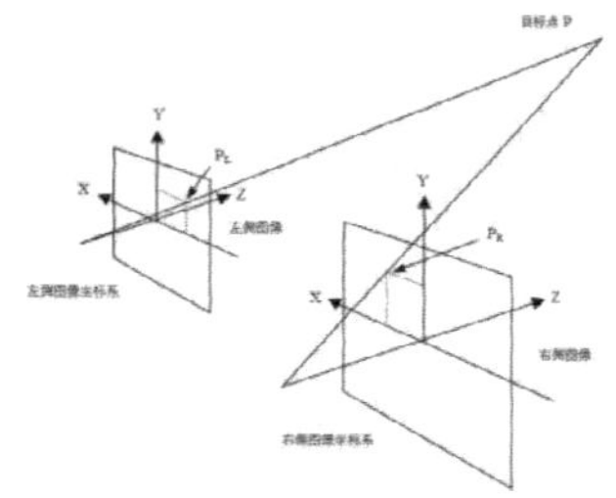

Figure 4. Space point location map

3. Application of SIROVISION system in tunnel construction

The SIROVISION system is used for the measurement of the structural characteristics of excavating face and side wall in Tuanzhai tunnel of Guiguang railway. The main experimental instruments and ancillary tools for the process include, Stereo Camera System0303(SCS0303), Siro3D software, SIROJOINT software, compass, total station (with a range), flexible rule, tape rule, camera, lacquer, etc. The horizontal distance of the rock body and the camera directly affects the data effect, and the shortest distance (S) and the baseline length (d) are better in the range of 6 ~ 7.

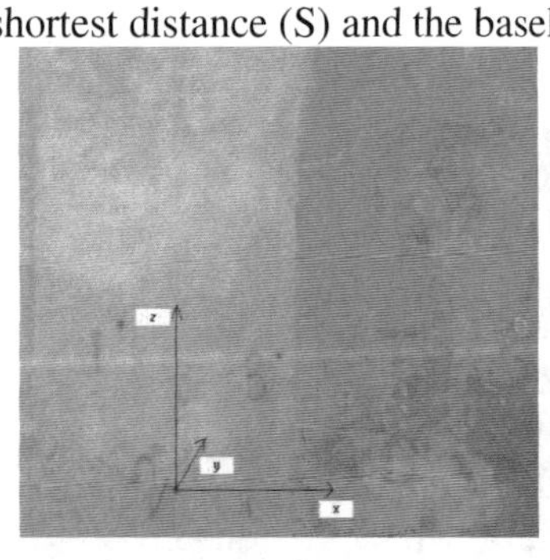

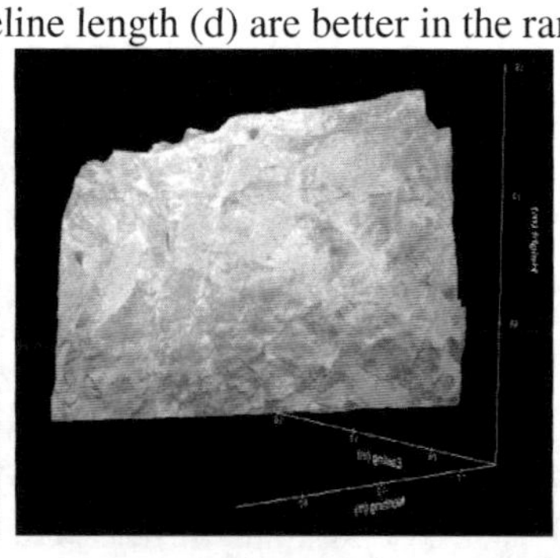

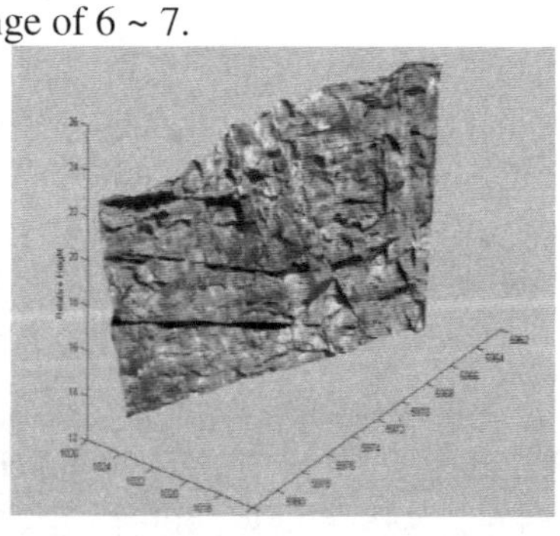

Figure 5. Coordinate reference points **Figure 6.** Image processing results **Figure 7.** Siro3D 3d image

Three coordinate reference points are sprayed with red paint in measuring the side wall and 2# point is selected as the origin of coordinates. The penetration direction of excavating face is in the direction of X; The vertical side wall is in the direction of Y, and the vertical direction is the direction of Z, as shown in Figure 5 and Figure 6. Using Siro3D software to process the photos of SCS0303, generate 3D photos, as shown in Figure 7; Using Sirojoint software to process 3D photos, the structure surface is measured and the results needed are exported, as shown in Figure 8 to 12.

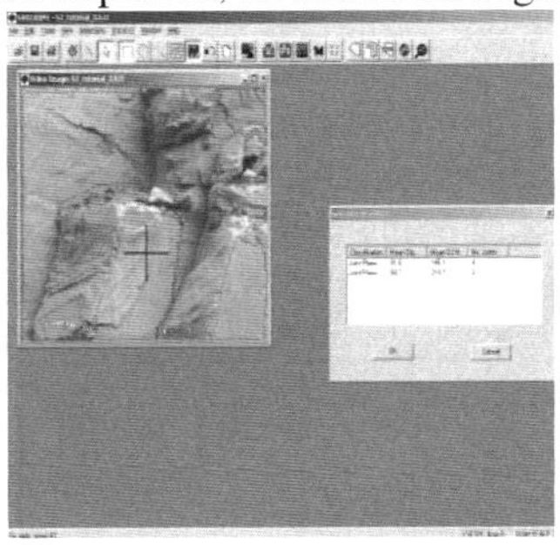

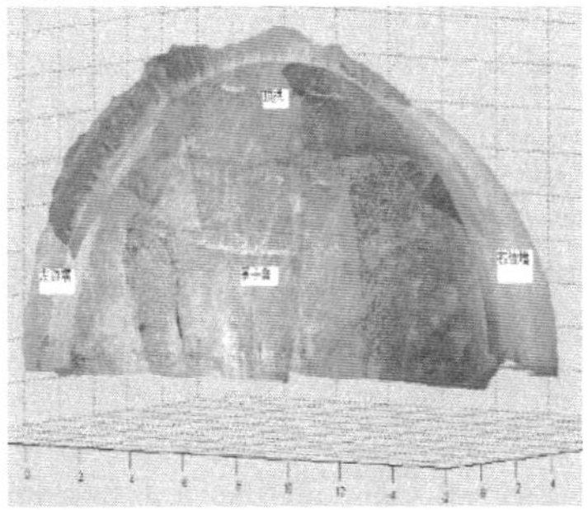

Figure 8. Trace calibration **Figure 9.** Joint surface statistics **Figure 10.** 3D effect of Tuanzhai tunnel

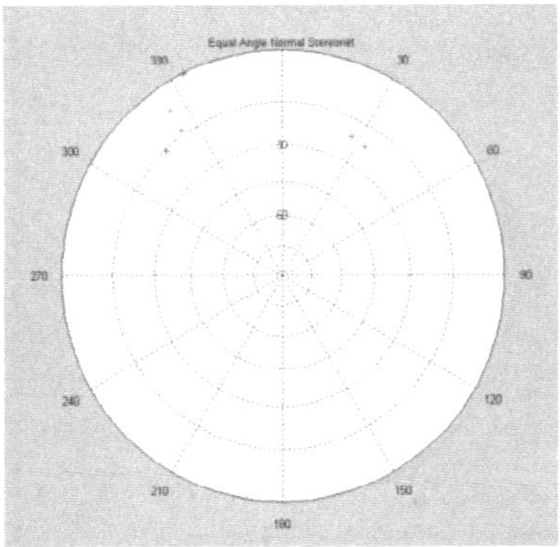

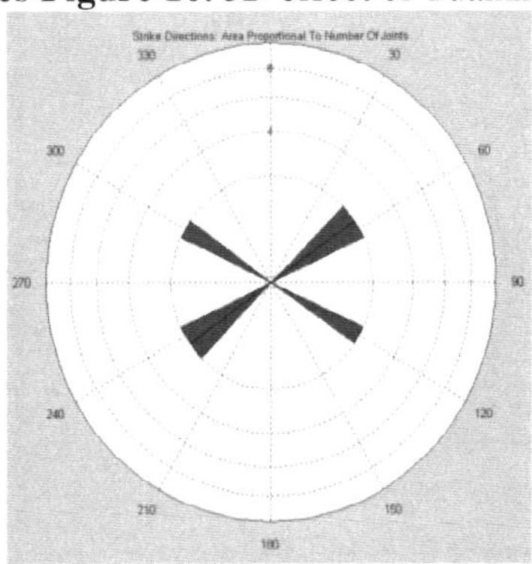

Figure 11. Polar diagram of joint surface **Figure 12.** Axial rose diagram of joint surface

Based on the above measurements, it is not difficult to draw the main structural plane of the excavating face of which direction is about 150~165 ° and orientation is about 245 ~ 245 °,and inclination is about 60~75 °.The structural plane orientation is parallel to the tunnel axis and the working face is relatively stable, and the possibility of a large landslide in the excavating face is unlikely. Excavation sidewall and constraints of structural plane trace generally less than 20 cm, and the longest is 29.6 cm which belongs to Ⅳ, Ⅴ structural plane and it's not going to go through the tunnel and only the local rock is cut into a block. The measuring result derived from Sirojiont as shown in Table 1

Table 1. Statistical results of structural surface features on heading face

Dip (°)	Dip_Direction (°)	Name	Length (m)
78.4	212.4	9	0.095
63.7	238.3	1	0.273
50.7	61.6	3	0.06
73.8	252.4	4	0.244
75.5	92	5	0.156
44.4	265.3	6	0.012
49.6	304.5	7	0.019
64.1	236.7	10	0.146
70.3	230.1	11	0.103
71.3	45.3	12	0.296
52	217.4	13	0.154
71.9	264.2	14	0.117
74.9	226.8	15	0.161

4. Study on tunnel deformation in discontinuous media

4.1. Model establishment and cases setting

The span of the tunnel is 13.46 m, with a height of 12.4 m. The model of numerical simulation is 90m wide and 110m high, which satisfies the requirements of at least 5 times diameter of tunnel in numerical simulation of continuous media. In order to fully grasp the applicability of the design of the support structure to the whole section, three working conditions are considered for the variation of the inclination and spacing of the discontinuous surface. The UDEC model is shown in Figure 13. The joint setting of different working conditions is shown in Table 2.

Table 2. The dip and spacing values of discontinuity in three cases

Case	Dip of joint1	Dip of joint2	fault	Space of joint1	Space of joint2
case1	0°	-65°	none	1m	4m
case2	0°	-65°	none	1m	2m
case3	0°	-65°	-75°	1m	2m

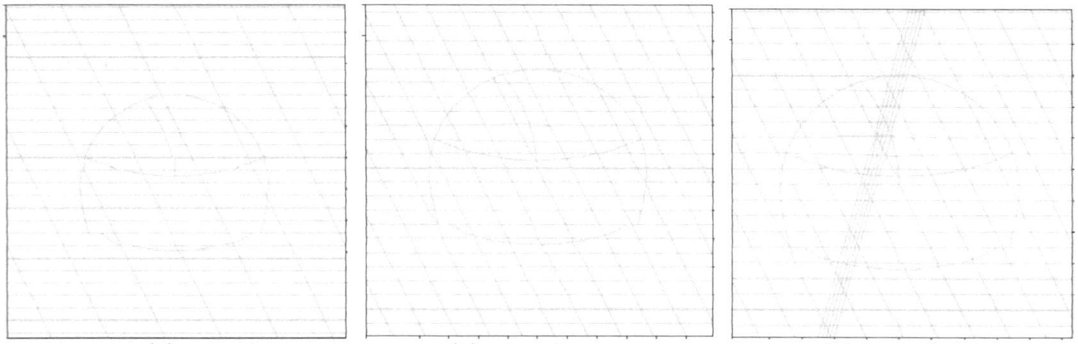

(a) case1 (b) case2 (c) case3

Figure 13. Model of different cases

4.2. Numerical calculation parameters

This inspection is combined with the specific situation of the engineering construction, and refer to other engineering experience, so there is 30% release of excavation load after the tunnel excavation and before primary support by rockbolt and shotcrete; There is another 60% release of excavation load after the bolt and steel support construction and before Second lining, and after that the remaining 10% of excavation load is released. The calculated parameters are shown in Table 3 and Table 4 respectively. The tunnel is buried at 540m and the pressure coefficient is 0.8.

Table 3. Calculation parameters of formation

Stratum	Density(kg/m^3)	Shear modulus (GPa)	Bulk modulus (GPa)
limestone	2600	7.53	2.72

Table4. Calculation parameters of Joints and faults

Parameters	Radial stiffness	Shear stiffness	Cohesion	Internal friction angle	Tensile strength
joint	10 GPa/m	10 GPa/m	0 MPa	20	0 MPa
fault	10 GPa/m	10 GPa/m	0 MPa	0	0 MPa

4.3. Deformation law of surrounding rock in tunnel excavation

Through the calculation of three cases, the expression of tunnel after excavation and before primary support is vault and arch bottom displacement is larger, and when the tunnel is after excavation and before primary support, the displacement of tunnel vault is the largest, 4.4 mm, 4.8 mm, 5.6 mm, respectively as shown in Figure 14. Tunnel primary support after the primary support and before second lining applied, most of the displacement of surrounding rock is occurred in this stage, working condition of the maximum displacement is 14.8 mm, 16.3 mm, 18.1 mm, respectively as shown in Figure 15. After second lining applied, with the release of the final 10% initial stress, the working condition of the maximum displacement increases to 16.9 mm, 18.6mm, 19.3 mm respectively, and after the second lining applied, the surrounding rock final deformation as shown in Figure 16.

Figure 17 shows the displacement around the tunnel entrance changes rapidly before support. Most of the displacement is mainly occurred after the primary support and before second lining applied. After the second lining applied, the displacement change is very small. By comparing with the vault subsidence monitoring value of Figure 18 shows that calculated values and measured values are within 15% and the calculated value is slightly larger than the measured values, which explain the procedure of discrete element method can effectively simulate the deformation of tunnel surrounding rock after excavation and dynamic guidance can be made to the construction.

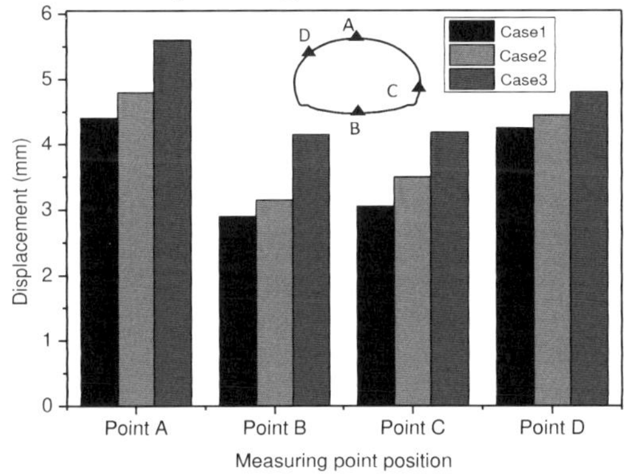

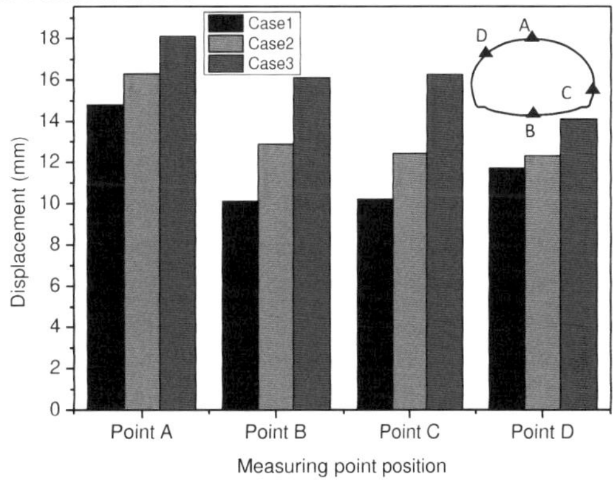

Figure 14. Monitoring value before primary support **Figure 15.** Monitoring value after second lining

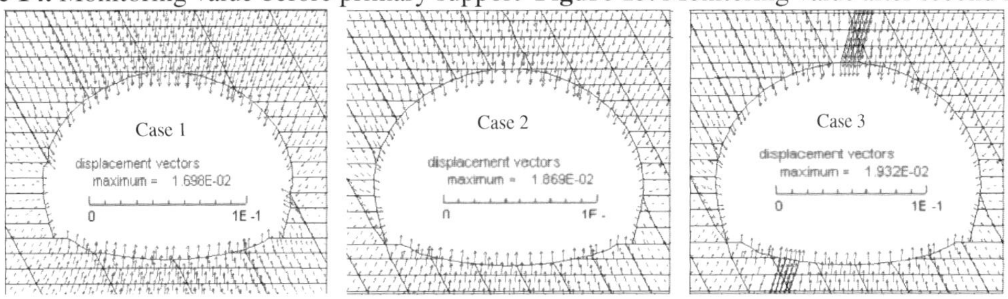

Figure 16. Surrounding rock deformation under different conditions after secondary lining

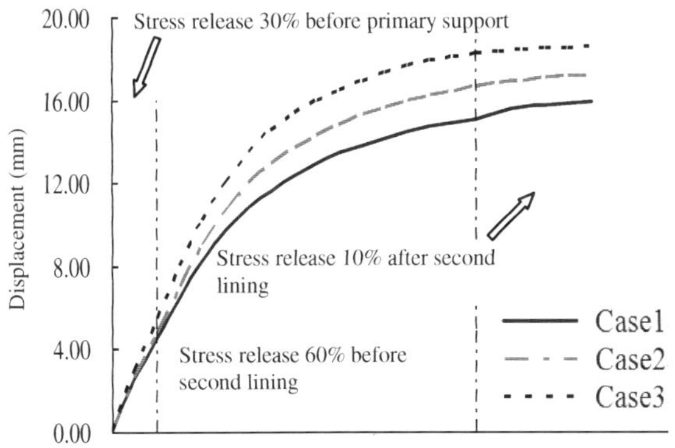

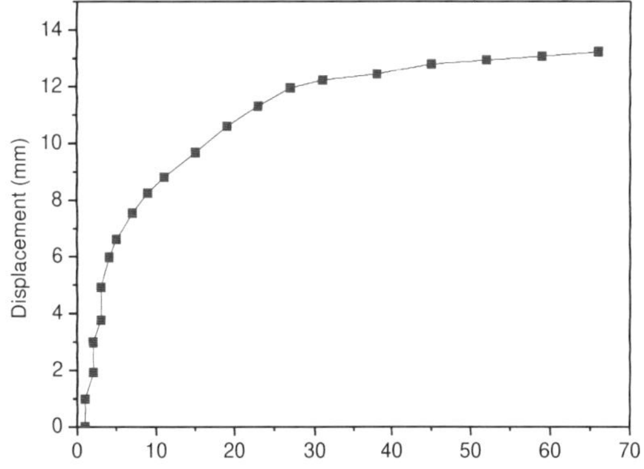

Figure 17. Displacement and time curve of tunnel vault **Figure 18.** Vault crown settlement

4.4. The principal stress distribution in tunnel excavation

Figure 19 to 21 is the principal stress distribution of the tunnel under different stages in three cases. As three conditions show the vault and arch bottom displacement are larger, so the stress of the vault and arch bottom also release more stress and lead to minor stress value. The displacement of the arch is smaller and the main stress is greater. The stress at the arch of the tunnel arch is the most stress at the primary support of the tunnel after excavation. As the surrounding rock continues to deform, the stress of the tunnel section increases further. After the second lining applied, the stress increases slightly with the release of the final 10% initial stress.

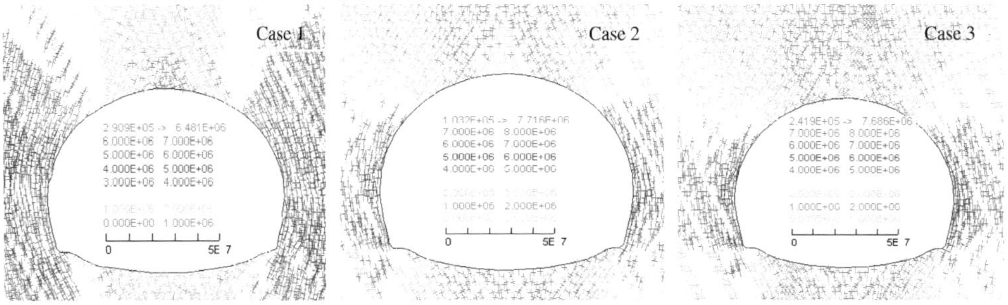

Figure 19. Principal stress after tunnel excavation before primary support

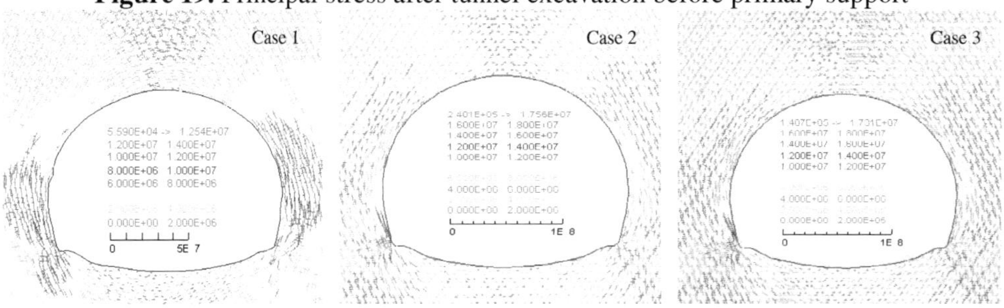

Figure 20. Principal stress after primary support before secondary lining

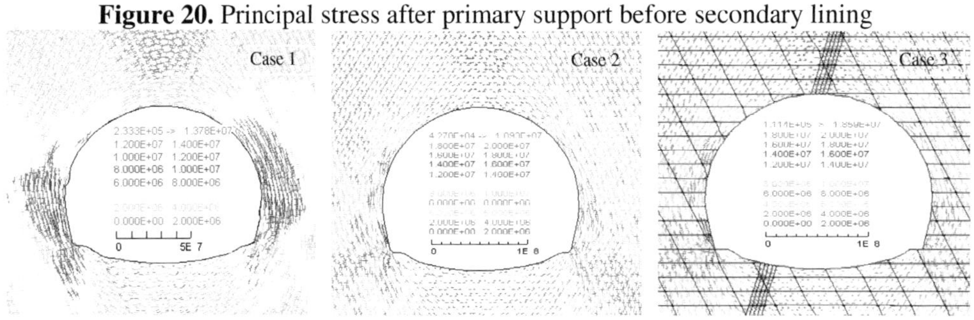

Figure 21. Principal stress after secondary lining

5. Conclusion

(1) The displacement of arch bottom increases obviously with the increase of rock fracture. After the initial support, the release displacement is about 90%, and the newly added displacement is small after the second lining. The deformation of surrounding rock, initial internal force and bolt axial force increase with the increase of the fracture degree of surrounding rock.

(2) Primary support after tunnel excavation plays a determinant of the deformation constraint and load of surrounding rock. Secondary support is only as a permanent support structure and safe reserve, and its internal force is relatively small compared with the primary support. As considering the physical mechanical parameters of surrounding rock are reduced after excavation, and the rheological properties of the rock mass and the mechanical properties of early supporting materials are weakened, secondary support should not be ignored and should be calculated according to the mechanical structure design.

(3) Three-dimensional photogrammetry can not only realize geological backup, but also can make the maximum program reduction of excavated tunnel. It can be combined with field monitoring data and explain the deformation mechanism of surrounding rock. The next step is to develop a third-party software interface and mechanical analysis can be directly imported to guide the construction.

6. References

[1] Wang C Y, Hu P L and Sun W C 2010 Method for evaluating rock mass integrity based on borehole camera technology [J]. *Rock and Soil Mechanics*. 31 (4) : 1326-1330.

[2] Li L, Wang H J and Wang Y M 2011 Discussion on description methods of rock fracture structure [J]. *Metal Mine*. (6): 1-5.

[3] Shen Y J, Xu G L, Dong J X, et al 2011 Relationship between mean trace length of joints and location of sampling window [J]. *Chinese Journal of Rock Mechanics and Engineering*. 30(3): 596-602.

[4] Zhao X D, Liu J, Zhang H X, et al 2014 Rock mass structural plane digital recognition and stope stability classification based on the photographic surveying method [J]. *Journal of Mining & Safety Engineering*. 31(1): 127-133.

[5] Wang F Y, Chen J P, Fu X H, et al 2008 Study of geometrical information of obtaining rock mass discontinuities based on VirtuoZo [J]. *Chinese Journal of Rock Mechanics and Engineering*. 27(1): 169-175.

[6] Tian S L, Ge X R and Tu Z J 2006 Testing study of digital close-range photogrammetry for measuring deformations of tunnel and underground spaces [J]. *Chinese Journal of Rock Mechanics and Engineering*. 25 (7): 1309.

[7] Zhou C L, Zhu H H, Zhao W 2010 Non-contact measurement of rock mass discontinuity occurrence with binocular system [J]. *Chinese Journal of Rock Mechanics and Engineering*. 29 (1): 111-117.

[8] Wang G H, Ma L, Yang T F, et al 2012 Study and application of deformation monitoring to tunnel with amateur camera [J]. *Chinese Journal of Rock Mechanics and Engineering* 24(S2):5885-5889.

[9] Xu S, Zhang C, An Long, et al 2015 Error analysis and correction method of sirovision telemetry system of jointed rock mass [J]. *Metal Mine*. (8): 110-115.

[10] Wang S H, Zhang H and Zhang Y Q, et al 2011 Random structural plane cutting of a rock slope spatial block model and its key blocks analysis [J]. *Journal of Northeastern University* (Natural Science), 32 (3) : 431-434.

An Analysis of the Carbon Fixation Change according to the Greenbelt Deregulation Using InVEST Model –In Case of Anyang and Gwacheon City-

JiYoung Choi[1] and SangDon Lee[2]

1 Ph, D Student in Department of Environmental Science and Engineering, Ewha Womans University, Seoul, South Korea.
2 Professor in Department of Environmental Science and Engineering, Ewha Womans University, Seoul, South Korea.
E-mail: jiyoungchoi1310@gmail.com

Abstract. The greenbelt regions in South Korea are now under the pressure of development as the greenbelt around urban areas is deregulated. Due to the greenbelt deregulation, the environment formerly conserved in the greenbelt is being turned into bare ground, which is reserved for development, or has already been developed. Government deregulated 10% of the greenbelt in Gyeonggi Province in 2015 to solve the housing problems and secure industrial areas. In view of the environmental conservation and development policies, the impact of greenbelt deregulation was analyzed in the present study by comparing the amount of carbon fixation before and after the greenbelt deregulation by using the InVEST model. In Anyang city, the ratio of urbanization was significantly increased for ten years within the temporal scope of the present study and the carbon fixation was decreased by 48485.05 Mg of C. In Gwacheon city, the carbon fixation was decreased by 8492.17 Mg of C in the same period. The result of the present study may provide fundamental data for analyzing the economic value of the greenbelt that may be deregulated in the future and estimating the impact that land development policies may have on the environment and the society.

1. Introduction

South Korea's greenbelt was firstly designated in 1976 around Seoul in the regions between the radii of 15 and 25 km from the center of the city in the width of 10 km to secure green spaces and prevent the extension of urban areas. Later, the greenbelt was extended to the radius of 40 km from the center of Seoul to protect green spaces and drinking water sources[1]. However, because of the conflicts that continuously occurred in the greenbelt regions about the land use, 28.5% of the greenbelt throughout South Korea was deregulated in 2001 to resolve the civil complaints and increase housing reliability[2] In 2010, 107.42 km^2 of the greenbelt was deregulated(table 1), and the largest portion of the deregulated greenbelt was in Gyeonggi Province (34.01%).

Deregulation of greenbelt causes not only expansion of urban areas but also reckless development due to excessive land use[3], [4], which may cause continuous reckless development in surrounding regions because of the conversion of the land use aimed at urbanization.

The purpose of the present study is to analyze the benefits of natural assets obtained from the presence of greenbelt in terms of carbon fixation with reference to the time of greenbelt deregulation. The national land use policy was analyzed in the present study with regard to the land use planning and environmental policy making in the temporal scope between 1990 when the greenbelt was conserved due to the need for sustainable land conservation and late 2000s when the greenbelt was started to be deregulated.

Table 1. The Status of Deregulation of Development Restriction Areas (Standard in 2010).

Section	**Seoul**	**Daejeon**	**Gwangju**	**Busan**	**Ulsan**	**Daegu**	**Incheon**
Area (km^2)	0.65	4.53	13.03	9.04	4.53	7.53	2.10
Percentage (%)	0.61	4.22	12.13	8.42	4.22	7.01	1.95
Section	**Gyeonggi**	**Chungnam**	**Chugbuk**	**Gyeongnam**	**Jeonnam**	**Gyeongbuk**	**Total**
Area (km^2)	36.53	1.56	1.53	9.04	14.53	2.61	107.42
Percentage (%)	34.01	1.45	1.42	8.42	13.53	2.43	100

Date: J J Yoon (2011) reconstruction[3].

2. Materials and Methods

2.1. Study Area

The spatial range is Gwacheon and Anyang city in Gyeonggi Province(figure 1). Research study area was selected in Gyeonggi Province with the highest rate of regional changes due to the deregulation the development-restricted zones. Among them, Gwacheon city is a major area where the environment is still preserved within greenbelt regulation. And Anyang city is selected due to the most variable land use area, when most of its greenbelts were released in 2000.

As about 92% of Gwacheon city is restricted to development with greenbelt, only about 8% of the areas are released, therefore they are preserved with greenbelt. Especially, as Anyang city geographically close to nearby industrial areas, a large-scale industrial space will be created to meet demand for related business partners. So, InVEST model was applied to these two areas where land use is significantly different.

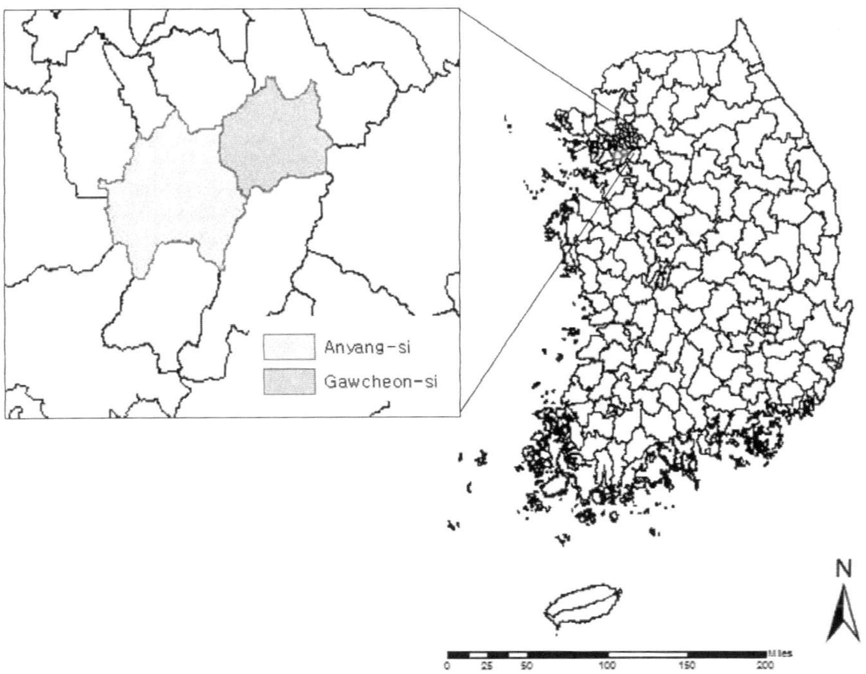

Figure 1. Study area – Anyang and Gawcheon city.

2.2. Research Methods

2.2.1 InVEST Model

InVEST (Integrated Valuation of Ecosystem Service and Tradeoff) model[5] is developed by Stanford University jointed National Capital Project and WWF (World Wide Fund For Nature) to identify between the natural assets of ecosystem and the economic relations. Based on result of ecosystem services assessment, policy-makers and stakeholders should set environmental management goals and create scenario in order to achieve the goals[6], [7].

The InVEST Version 3.3.3 Carbon model can present future changes as well as current values and analyze the changes and trends of ecosystem services based on a variety of spatio-time data.

Therefore based on the 1990 and late 2000s of Land-use map and 4 coefficient of carbon pool, it was calculated estimation of Carbon Fixation.

2.2.2 InVEST Carbon Model

Raster data file set drawn from Land-use Map and Carbon pool table to represent each of the land-use map classes are required for modelling by using Arc GIS Map 10.5.

Also another input data is the table of carbon pools (aboveground mass, belowground mass, soil, dead mass) (figure 2). This value was compiled from evaluation of published scientific literature on carbon storage to land-use classes and was obtained through the IPCC (Intergovernmental Panel on Climate Change)[8], [9], [10], [11].

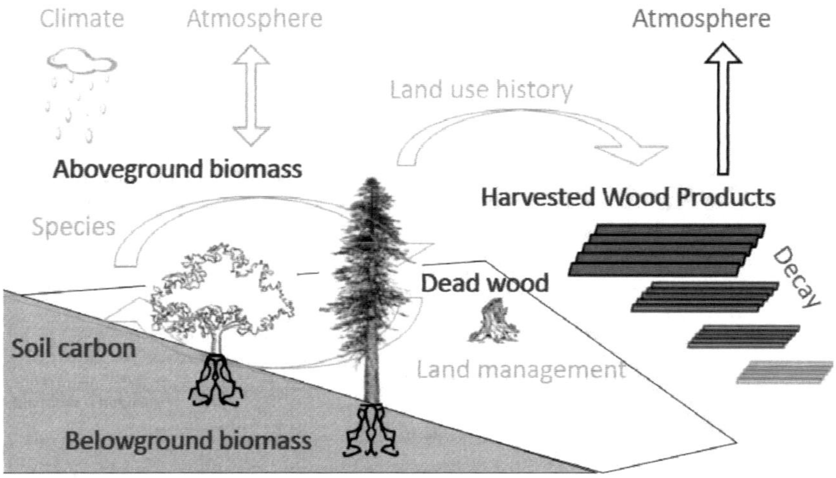

Figure 2. InVEST Carbon model Process[7].

3. Results and Discussion

figure 3 shows the amount of carbon fixation in Gwacheon and Anyang city since 1990-2000s. Carbon fixation of water, wet-land, grass and forest regions was having high fixation amount and as the ratio of urban and barren regions was high, the carbon fixation tends to decrease. The model result of Gwacheon city was estimated to be 649266.44 Mg of C in 1990, and the carbon fixation in late 2000s was estimated to be 640774.25 Mg of C by decreasing 8492.17 Mg of C from the previous decade. Also, Anyang's carbon fixation in 1990 was derived from 881866.88 Ma of C, and carbon fixation in late 2000s was estimated to 833381.81 Mg of C by decreasing 48485.05 Mg of C between 1990 and late 2000s.

The result is an analysis based on land-use cover map (LULC), causes of the changes can be interpreted based on them. Anyang city shows that the proportion of urban areas has changed significantly around city' center. And Gwacheon city was analysed to have maintained the ratio of the

IOP Conf. Series: Earth and Environmental Science **167** (2018) 012043 doi:10.1088/1755-1315/167/1/012043

forest area, but the barren region around the center was replaced by the urban area according to the time flows from 1990 to late 2000s.

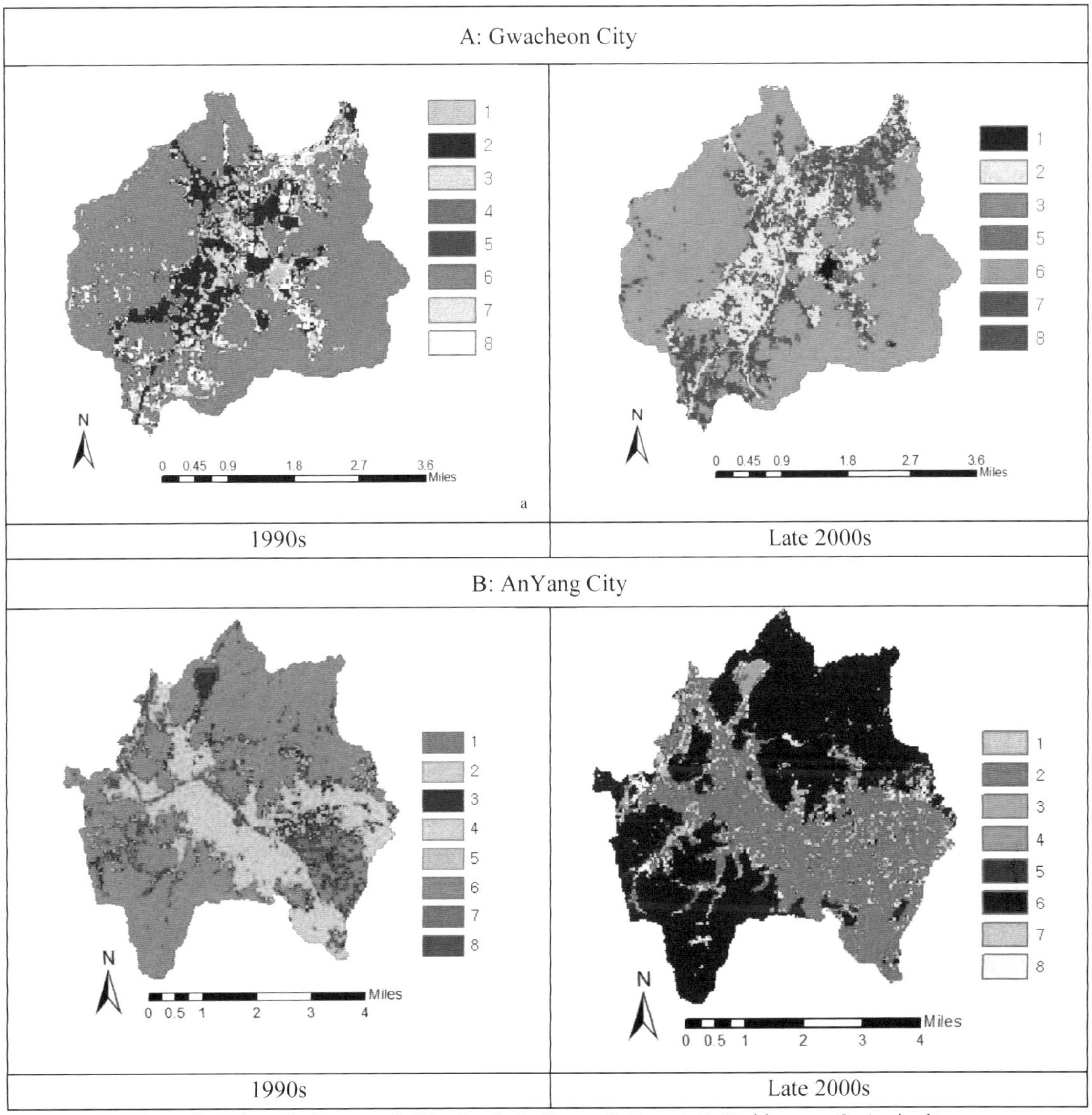

[a]1: Water, 2: Urban, 3: Barren, 4: Wet-land, 5: Grass, 6: Forest, 7: Paddy area, 8: Agriculture area.

Figure 3. Map of Carbon Fixation Result by InVEST Carbon model.

4. Conclusion

The study conducted a change in the research area by analysing the carbon fixation due to the deregulation of the Korean greenbelt. Due to the greenbelt deregulation, the environment protected by greenbelt releases has been developed into a barren and urban area and is under threat. Later, The South Korean government additionally deregulated 10% of the greenbelt in Gyeonggi Province in 2015 to solve the housing problems and secure industrial areas.

Accordingly, the impact of greenbelt deregulation was analyzed by comparing the amount of carbon fixation before and after the greenbelt deregulation by using the InVEST model for environmental conservation and development policies. In Anyang city, the carbon fixation was decreased by

48485.05 Mg of C in 1990-2000s, and In case of Gwacheon city, it was decreased by 8492.17 Mg of C in the same period.

The results of this study can be applied to fundamental data for analyzing the economic value of the greenbelt that may be deregulated in the future and estimating the impact that land development policies may have on the environment and the society. And For more perfect assessment, it needs to combine soil analyses in terms of carbon fixation analysis in future studies.

5. References

[1] Ryu D H and Lee D K 2013 Evaluation on Economic Value of the Greenbelt's Ecosystem Services in the Seoul Metropolitan Region *J. Korea. National Association for Urban Planning and Design.* **48** 3 pp 279-292
[2] Moon M S 2016 A Study on the Industrial Space Improvement and the Establishment of the Development Restriction Area. Gyeonggi Development Institute. **251**
[3] Yoon J J 2011 Development Restriction Area 1971-2011. Korea Land and Housing Institute.
[4] Development of Decision Supporting Framework to Enhance Natural Capital Sustainability: Focusing on Ecosystem Service Analysis. 2015 Korea Environment Institute
[5] Sharp R, Tallis H T, Ricketts T, Guerry A D, Wppd S A, Chaplin-Kramer R, Nelson E, Wolny S, P;wero N, Vigerstol K, Pennington D, Mendoza G, Aukema J, Foster J, Forrest J, Cameron D, Arkema K, Lonsdorf E, Kennedy C, Verutes G, Kim CK, Guannel G, Papenfus M, Tofr J, Marsik M, Bernhardt J, Griffin R, Glowinski K, Chaumount N, Perelaman A, Lacayo M, Mandle L, Hamel P, Vogl A L, Rogers L, Bierbower W 2015 InVEST User's Guide. Stanford University, University of Minnesota, The Nature Conservancy, World Wildlife Fund
[6] IPCC 2001 Climate Change 2001: The Scientific Basis. Third Assessment Report of the Intergovernmental Panel on Climate Change. Cambridge University Press, New York. pp 365.
[7] Natural Capital Project 2012 Informing Land-Use Plans in Central Sumatra. INVEST User Guide Release +VERSION
[8] Kim C K 2014 Use the Framework of Ecosystem Service to Inform Sustainable Development of Marine and Coastal Environment. J. Society for Marine Environment and Energy. **11** 4-4
[9] European Commission, Organisation for Economic Co-operation and Development, United Nations, World Bank 2013 System of Environmental-Economic Accounting 2012 Experimental Ecosystem Accounting 1-204
[10] Korea Environment Institute 2015 Development of Decision Supporting Framework to Enhance Natural Capital Sustainability: Focusing on Ecosystem Service Analysis. Korea Environment Institute
[11] Lee H W, Kim C K, Hong H J, Roh Y H, Kang S I, Kim J H, Shin S C, Lee S J 2015 Development of Decision Supporting Framework to Enhance Natural Capital Sustainability: Focusing on Ecosystem Service Analysis. Korea Environment Institute. 3479-3651

Acknowledgments
This research was supported by Ministry of Science, ICT, and Future Planning in Korea (2015M3C8A6A06012735).

IOP Conf. Series: Earth and Environmental Science **167** (2018) 012044 doi:10.1088/1755-1315/167/1/012044

A Methodology for Designing an Information System for Road Infrastructure Monitoring and Traffic Management in Disasters

D Dimitrov[1], P Zlateva[2] and D Velev[3]

1 Todor Kableshkov University of Transport, Sofia, Bulgaria
2 IR, Bulgarian Academy of Sciences, Sofia, Bulgaria
3 University of National and World Economy, Sofia, Bulgaria
E-mail: ddimitrov@vtu.bg, plamzlateva@abv.bg, dgvelev@unwe.bg

Abstract. The impact of natural disasters and other adverse events on road infrastructure leads to a partial reduction or complete renunciation of its working capacity. Currently, there is no online-based information system to disclose and manage traffic in the event of a problem or disruption of a transport segment from the road infrastructure. The main purpose of the paper is to present a common methodology for designing a cloud-based online information system for road infrastructure monitoring and traffic management in disasters. The methodology of the information system is based on traditional geographic information system (GIS) solutions, Web-based technologies and GPS navigation systems.

1. Introduction

Successful implementation of passenger and freight transport is a complex task that is directly dependent on the working capacity and quality of road infrastructure [1], [2]. In a number of cases, when a disaster occurs, road infrastructure is compromised and it has a reduced or totally discontinued ability to perform its functions [3]. In such cases, it is important for the disaster services to properly register and map them and to take the necessary steps to overcome the disaster situation, and to inform the public of the dangers and actions they have taken [4], [5]. Specifically, for transport infrastructure, this means that the affected transport sections must be marked on the relevant map and more specific information concerning the time of occurrence, estimated time to solve the situation and other important information that has to be entered into the database in which these events are recorded [6], [7].

In the event of a disaster in the area where it has occurred and developed, it is necessary from the operational point of view to carry out a number of measures for reallocation of the traffic on alternative arcs of the transport network with a view to minimizing the complexity of the operation of the transport. Recently, the so-called "intelligent transport systems" have been developed to achieve dynamic driving and navigation [8], [9]. By their nature, they are still in the area of traffic safety and they have information-advising role on the navigation of individual means of transport. In these information-advising systems, the so-called transport and flow-through tasks that provide alternative pathways in the event of a problem with a given point on the network, based mostly on the criterion of "shortest" or "fastest" traffic workaround.

Regarding the magnitude of traffic in the road network, there are a number of developments in the field of real-world information systems and applications, while in terms of the impact of disasters on traffic, things have not yet developed enough and there is a front for systematization and development.

The fact is that there is still no developed unified and commonly accepted disaster mapping system, and there are separate developments mainly in the developed countries of Europe and the world.
The purpose of the paper is to present a common methodology for designing a cloud-based online information system for road infrastructure monitoring and traffic management in disasters. The main idea is to analyse and classify the adverse events resulting from disasters, taking into account their impact on the road infrastructure's working ability in transport. Common methodological guidelines are also formulated to create a cloud-based online information system that will improve traffic management in transport. The chosen model for operating the system is based on GPS information and navigation systems, GIS systems for mapping events at individual transport points and sections of the road network.

2. Classification, registration and mapping of disasters

The essence and main elements of classification, registration and mapping of disasters can be summarized as follows:

- Disaster is an event where there is a significant disturbance in the normal functioning of society that is caused by natural phenomena and / or human activity and it results in negative consequences for people's life or health, infrastructure, property, the economy and the environment. This creates a crisis situation that needs to be dealt with and managed, as well as it should be mapped and registered.
- Classification of disasters is done on certain features that characterize the origin of the event (natural and anthropogenic), the territorial scope (local, regional, global), the mode of occurrence (fast and slow) and duration (sudden, rapid and prolonged). The natural events are geophysical, meteorological, hydrological, climatic and biological, while the anthropogenic (as a result of human activities) are radiation accidents, industrial accidents, transport accidents and catastrophes, terrorist acts, humanitarian and social crises, fires, smuggling, etc.
- Mapping for disaster protection and crisis management is performed: before a disaster, during a disaster, after a disaster. It is also necessary to take into account the timing of the event and the time of eventual elimination of the consequences of the event. In addition, the individual stages of recovery and actions and information messages, instructions of a methodical and other nature can be registered.
- The mapping signage system, the meaning of which is a set of graphic signs and their meanings, including the rules for making the signs and applying them to the particular map, must be designed in such a way as to ensure readability, visibility, distinctiveness and consistency between the signs and their corresponding objects of reality [10]. Although there is no such a unified and universal standardized sign system, a simplified model of such a color rendering system can be used in this case on the elements of the transport infrastructure. Offered on a given point or section of the road, it is has been marked with the following colors: green, orange, red, and blue with "X" indicating the ability of the point or section to have maximum, disturbed, minimal working capacity or completely out of action.

The acquired accurate information serves the relevant specialized authorities to effectively manage and make quick and adequate solutions to the situation. This information serves as a means of risk assessment, planning of preventive measures, planning of protection and rescue activities, assessment of the affected areas, such as planning recovery activities.
The transport infrastructure is central to the overall state infrastructure and it is also under management, control and restoration of its working capacity. It is particularly important that the events of registering the problems in its working capacity or the faults caused by the disasters are known to the respective services and persons involved in the monitoring and restoration, as well as to the general public in order to prevent subsequent accidents and problems with the people, who are in or passing through a given region or transport segment.

3. A model for the operation of the proposed information system

The general model for the operation of the proposed information system is presented on Figure 1. This model represents possible options for both predicting potential hazards and disaster-related and post-disaster-related actions.

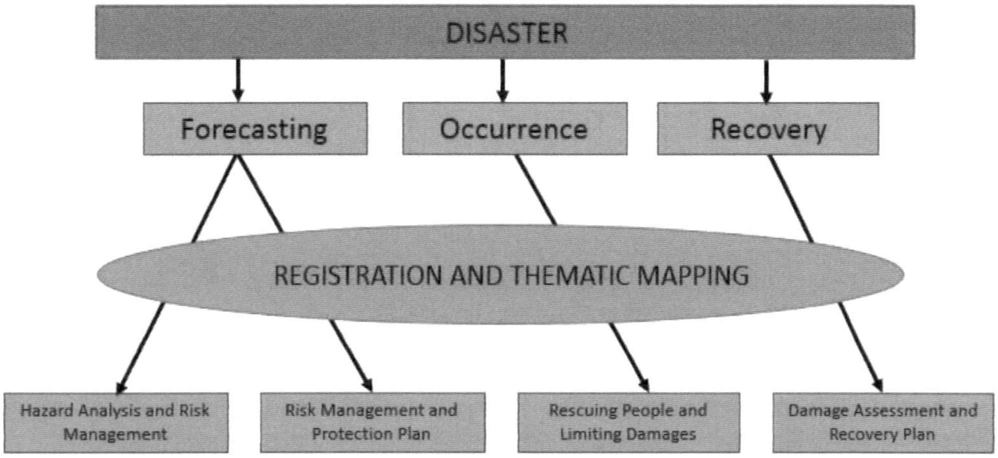

Figure 1. Model for registration and thematic mapping of disasters and actions for their forecasting, time actions and disaster recovery.

The following expression presents the main components of the recorded disaster as part of the set of all recorded disasters:

$$Disaster \in DISASTERS \left\{ \begin{array}{c} disasterID(uniqueID), location(area), \\ time(begin, end), damages(people, equipment), \\ cash(direct, indirect), moreInfo(freeText) \end{array} \right\} \quad (1)$$

Generally, for each disaster registration, the area is mapped where damages are expected or occurred, the start and the end of the disaster period, breakdowns and casualties, direct and indirect cost of prevention, and an explanatory text describing another additional information related to the disaster itself.

A classification that should cover the proposed information system from the point of view of the mapping of the events for prevention and disaster recovery is presented on Figure 2.

An exemplary graphic model for representing the road network with the marked traffic zones, which are directly and indirectly affected by a disaster is given on Figure 3. The following areas of the road infrastructure are formed:

$$DisasterZone_i \begin{cases} All\ DangerZone_j, & where\ j = 0,1,2,\dots,J \\ All\ DifficultZone_k, & where\ k = 0,1,2,\dots,K \end{cases} \quad (2)$$

When using a ready-made GIS base, it is necessary for each element of the graphic road infrastructure to add information about the type of events caused by the disasters, the characteristics of thece disasters, the affected road sections, the corresponding time intervals and other useful information.

Figure 3 shows a model for the so-called "Quick" and reference maps that reflect the immediate impact of disaster events.

This graphical model is based on the Google Maps APIs toolkit [11]. An alternative GIS toolkit may also be used to develop the software, but this should be the result of a more thorough study of alternative technologies and the capabilities of their application development environments.

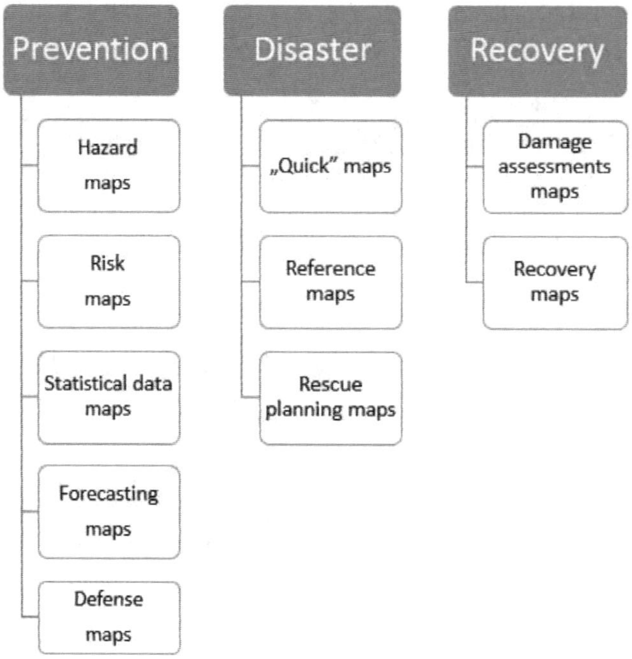

Figure 2. Classification of event mapping and disaster recovery events.

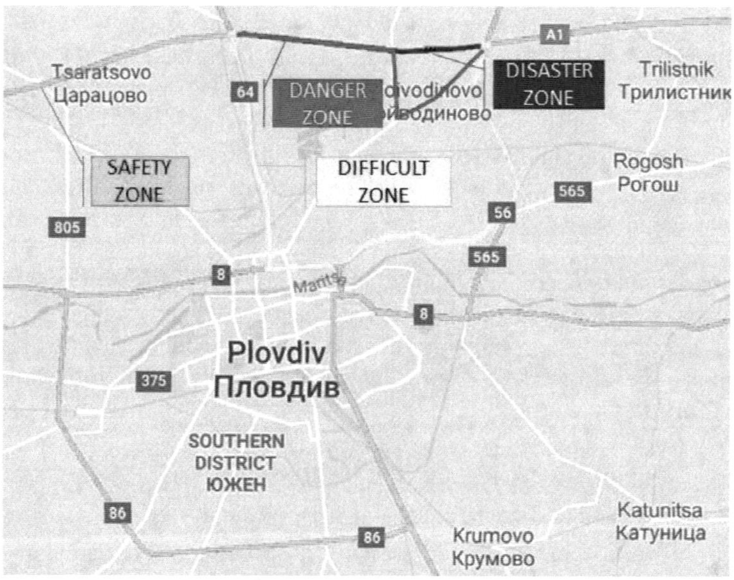

Figure 3. Graphical presentation of the road infrastructure by marking the areas affected by disasters.

4. Main methodological guidelines for designing of the information system

The presented issue for information service for crisis situations is quite complex for technical implementation and in any case it will be related to the building of a number of regional information systems, which in synchrony exchange and transmit data to each other, i.e. to work similarly to the model of the weather forecasting system.

This means they should be connected to sensors that record the disaster events and to have the appropriate governmental and other services to enter information into the database, including after receiving civilian signals and information. The information service of such systems is a responsibility of the respective specialized institutions for Civil Protection [12], [13] and Road Infrastructure system [14], [15].

Main methodological guidelines with regard to software technology:

- The environment should be built on modern cloud technologies including server, database and geo-data virtualization;
- The client part is a Web based interface, fully compatible with WWW Consortium (W3C) standards, supported by all popular Web browsers.
- A reference interface APIs should be setup that implement connection to GPS-based navigation systems.

5. Conclusion

The paper discusses the methodological issues for the registration of adverse events caused by natural disasters and crises and which affect the road infrastructure's efficiency. The issue of disaster thematic mapping, which is at the heart of the planning of prevention, action and disaster recovery, is considered.

The main focus of the paper is on the functioning of road infrastructure, which is related to the natural disaster events. They lead to a partial reduction or complete destruction of the infrastructure capacity, resulting in short and long interruptions of transport performance as well as to the heavier decommissioning of large-scale road infrastructure.

It is pointed out that all stakeholders should take into account the impact on road infrastructure efficiency of adverse events caused by disasters. Common methodological guidelines and requirements for their information provision and a cloud-based online information system have been formulated, which will improve traffic management in transport. The model of this information system is based on traditional GIS solutions, and a visualization is planned to be achieved through Web-based technology tools. This includes the use of information resources from modern GPS navigation systems.

6. References

[1] Tzvetkova S 2017 Increasing the social effectiveness of public transport, CBU, *Int. conference on Innovations in Science and Education*, March 22-24, Prague, Czech Republic, pp 486-489

[2] Karagyozov K, Razmov T, Todorova M, Varadinova J and Dzhaleva-Chonkova A 2012 Impact of Natural Disasters on Transport Systems – Case Studies from Bulgaria, *Report from the International panel of the WEATHER project funded* by the European Commission's 7th framework programme, Todor Kableshkov University of Transport, Sofia

[3] Peng C, Regmi A D, Qiang Z, Yu L, Xiaoqing C and Deqiang C 2017 Natural hazards and disaster risk in one belt one road corridors. In: Mikos M, Tiwari B, Yin Y and Sassa K (eds) *Advancing Culture of Living with Landslides*. WLF 2017. Springer, Cham

[4] Shibayama T 2017 Japan's transport planning at national level, natural disasters, and their interplays, *European Transport Research Review* **9**: 44 p.18

[5] Demirel H, Kompil M and Nemry F 2015 A framework to analyze the vulnerability of European road networks due to Sea-Level Rise (SLR) and sea storm surges, *Transportation Research Part* A: *Policy and Practice* **81** (11) pp 62-76

[6] Bíl M, R. Kubeček V, Bílová M and Sedoník J 2015 Evaluating road network damage caused by natural disasters in the Czech Republic between 1997 and 2010, *Transportation Research* Part A: *Policy and Practice* **80** (10) pp 90-103

[7] Gasimova E 2014 The damages caused to automobile roads by natural disasters and the measures of protection in Azerbaijan Republic, *International Journal of Business, Humanities and Technology* **4** (1) pp 136-139

[8] Dimitrov D and Kirchev T 2013 Application of Modern Information Tools in an Operational Transport Management, *Mechanics, Transport, Communications Academic journal* **11** (3) Article No 0783

[9] Tzvetkova S 2017 Improving the quality of urban passenger transport in the city of Sofia by incorporating intelligent transport systems, *Economic and Social Alternatives* **4** pp 29-42

[10] Marinova S 2014 *Thematic mapping and visualization for early warning and crisis management*, PhD. Thesis, UACG, Sofia
[11] Google Maps APIs 2017 *Build the next generation of location experiences* https://developers.google.com/maps/
[12] Parliament, Republic of Bulgaria 2017 *Law on Disaster Protection*, State newspaper (DV) no. 102 from 19 December 2006; Last amendment DV no. 97 from 5 December 2017
[13] Council of Ministers, Republic of Bulgaria 2013 *National Program for Disaster Protection 2014-2018*
[14] Parliament, Republic of Bulgaria 2017 *Law on the Roads*, State newspaper (DV) no. 26 from 29 March 2000; Last amendment DV no.96 from 1 December 2017
[15] Road Infrastructure Executive Agency, Republic of Bulgaria 2009 *Technical rules and requirements for the maintenance of roads.*

Acknowledgments
The authors express their gratitude to the Bulgarian National Science Fund for the partial financial support under Grant № DNTS/ China 01/6 from 2014, Bulgarian–Chinese scientific cooperation